Y0-FBG-563
SHENANDOAH COLLEGE
LIBRARY
WINCHESTER, VA
WITHDRAWN

PROBABILITY AND STATISTICS

AN UNDERGRADUATE COURSE

SHENANDOAH COLLEGE
LIBRARY
WINCHESTER, VA.

SHENANDOAH UNIVERSITY LIBRARY
WINCHESTER, VA 22601

PROBABILITY AND STATISTICS

AN UNDERGRADUATE COURSE

MEYER DWASS NORTHWESTERN UNIVERSITY

W. A. BENJAMIN, INC., Menlo Park, California

QA 273 .D887 1970
519 D960p
Dwass, Meyer.
Probability and statistics: an undergraduate course.

PROBABILITY AND STATISTICS: An Undergraduate Course

Copyright © 1970 by W. A. Benjamin, Inc.
All rights reserved
Standard Book Number 8053-2376-7 (Clothbound Edition)
Library of Congress Catalog Card Number 77-102179
Manufactured in the United States of America
1234-K3120

W. A. BENJAMIN, INC., Menlo Park, California

for
Shirley
Golda
Emily
Michael
and
Claudia

16.65 Goldberg 3-29-78

PREFACE

The theory of probability is a branch of mathematics that describes "chance phenomena." Its origins mainly lie in 16th and 17th century speculations about gambling. Gambling still provides some stimulus for the subject, but the present century has seen the probabilistic description of erratic phenomena permeate physics, biology, economics, and virtually every branch of science and technology. From the view of probability theory there may be important regularities within a mass of wildly moving and seemingly unpredictable gas molecules as well as within a multitude of conflicting human attitudes and opinions.

In practical scientific situations it may not always be possible to set down a probability model that fully describes a chance phenomenon, in the sense that there may be competing models that serve equally well until the experimenter can obtain more detailed information. The theory of statistics attempts to develop means of analyzing data from chance experiments in order to infer an appropriate underlying probability model.

This book evolved out of a set of notes used in teaching a three quarter course in probability and statistics. This is an undergraduate course taught in a mathematics department, although typically the audience consists not only of undergraduate mathematics majors but also of undergraduates and graduate students in fields as diverse as the natural sciences, the social sciences, engineering, and business.

Chapters 1 through 11 develop some of the main ideas of probability theory and Chapters 12 through 18 are concerned with the application of these ideas to statistics. Before using this book, I often used a text in discrete probability for the first quarter and then turned to one or another statistics text for continuous probability and for statistics. However, I usually found the treatment of the continuous case inadequate and deplored the lack of a unified text suitable for the full year.

The prerequisite for this book is a year or a year and a half of calculus. This should include some power series and some multiple integration. At Northwestern the first year of calculus includes a quarter of linear algebra. This seems to be the trend in many universities and is exploited in Chapter 17, though a review of the parts of vector space theory needed there is given in Chapter 16.

There is purposely more material in the book than can be covered in a single year, thus allowing the instructor to favor some of his own tastes and providing some variety if the course is taught several times. Following is a list of sections that can be omitted without seriously disturbing the continuity of the remaining material. Also given are several sample programs for using the book.

Usually I spend the first two quarters on probability and the third quarter on statistics, which are roughly the proportions for probability and statistics in the book, though sometimes I increase the statistics part of the course. Of the two sample programs that follow, one gives greater emphasis to probability, the other to statistics.

I personally enjoy combinatorial problems and tend to spend a fair amount of time on them, but I do not think that combinatorics should overwhelm a student's first exposure to probability. Too often students leave a first course in probability with the notion that combinatorics and probability are one and the same. For this reason random variables and expected value are introduced very early, thus allowing combinatorial problems to be done with a few basic probability tools in hand. If the instructor wants to deemphasize combinatorics, then a good part of Chapter 3 can be omitted.

The exercises at the ends of the chapters form an essential part of the text. It is recommended that students work as many of these as they can and that instructors integrate the exercises into their teaching. Many of the exercises contain hints and a clue to difficulty often lies in the elaborateness of the hint. There may be some benefit in first trying an exercise without consulting the hint. There is a supplementary solution book in which the exercises have been worked out. If the reader reaches an impasse, he can consult the solution book. In any case he may want to consult the solution book for an approach that may differ from his own.

The following sections or groups of sections can be omitted without disturbing the main line of the text:

(3.1, 3.2), (3.3, 3.4, 3.5), (3.6), (3.7), (3.8, 3.9, 3.10), (3.10),
(4.4, 4.5)
(5.2), (that part of 5.3 that follows Example 3), (5.5)
(6.5, 6.6), (that part of 6.7 that follows Example 4)
(7.10)
(8.2), (8.3), (8.5, 8.6, 8.7)
(9.2, 9.3), (9.6), (9.9)
(All of Chapter 11), (11.5, 11.6, 11.7), (11.8), (11.9, 11.10, 11.11),
(11.12, 11.13, 11.14)
(12.5, 12.6)
(13.5, 13.6, 13.7), (13.7)
(All of Chapter 14)
(15.5), (15.6, 15.7), (15.7), (15.8)
(All of Chapters 16 and 17), (16.7, 17.5, 17.6, 17.7, 17.8), (17.8)
(All of Chapter 18), (18.5), (18.6), (18.7, 18.8, 18.9, 18.10, 18.11)

Specifically, the following omissions will provide a year course that emphasizes statistics:

Omit 3.3, 3.4, 3.5, 3.6, 3.7; 4.4, 4.5; Chapter 11.

The following omissions will provide a year course which emphasizes probability:

Omit Chapter 14, Chapter 16, Chapter 17.

Of course, there are many other programs that can be constructed using the list of sections recommended for possible omission.

Finally, I would like to acknowledge my indebtedness to Wassily Hoeffding, Lloyd Fisher, Jean Gibbons, and John Dinkel for many helpful technical suggestions, and to Sandy Clark for typing the manuscript.

Meyer Dwass

Evanston, Illinois
December 1969

NOTE TO THE INSTRUCTOR

This book is for a year's course in probability and statistics for undergraduate mathematics majors, students in behavioral science, physical science, or engineering. Since students of such a course are usually very diverse in their mathematical preparation and areas of interest, there is purposely more material in this book than can be covered in a single year. This should give you the opportunity to find a path through the book that satisfies your own needs and tastes. Some of the possibilities are the following:

A. Chapters 1–11 can be the basis for a first half-year of probability and Chapters 12–18 for a second half-year of statistics. This will require that some topics be omitted throughout. The preface has a list of various topics that can be omitted without disturbing the continuity of the text.

B. It is also possible to design a year's course which gives greater emphasis to probability than to statistics by covering more topics in Chapters 1–11 and omitting more topics from Chapters 12–18. Again the list of topics for omission should be consulted in the preface.

CONTENTS

Chapter One

THE PROBABILITY MODEL

In this first chapter we set down axioms for a discrete probability space, which gives a model for chance experiments having a finite or countably infinite number of possible outcomes. At first we concentrate on examples such as throwing dice and coins and drawing numbers from a box. These have the advantages of being easy to visualize and of lying within our common experience. By way of motivating the axioms, we first describe the frequency theory of probability, which is an intuitive, empirical way of interpreting probabilities (Note 1).

1.1 CHANCE EXPERIMENTS

We encounter chance phenomena daily. We are at ease with expressions such as "the chances are," "the odds are against it," "that is most improbable," and "this event is very likely." Our experience is rich with what we will formally call "chance experiments," some examples of which follow.

Two dice are thrown.
A coin is thrown.
A ticket is drawn from a box of lottery tickets.

These experiments have three points in common:

1. In none of them is it possible to predict beforehand how the experiment will turn out.
2. It is possible, however, to make a list of all the possible outcomes of each experiment.
3. One has "feelings" or "beliefs" based on experience and in-

tuition about the "likelihoods" or "probabilities" of the various outcomes.

These common features of chance experiments provide a basis for a mathematical model. The mathematical development will begin in Section 1.3. Meanwhile, here are several additional examples of chance experiments discussed from a purely intuitive point of view. In each of these examples we list the possible outcomes and the probabilities conventionally assigned to these outcomes.

Example 1 A coin is thrown twice. Denoting head by H and tail by T, we give the list of possible outcomes and their probabilities.

Possible outcomes	*Probabilities*
H H	1/4
H T	1/4
T H	1/4
T T	1/4

Example 2 A coin is thrown until a head appears. The outcome is taken to be the number of throws required to obtain a head. The possible outcomes can be identified with the set of all positive integers. Outcome 1 means a head appeared on the first trial; outcome 2 means the first two trials were tail, head; outcome 3 means the first three trials were tail, tail, head, and so forth.

Possible outcomes	*Probabilities*
1	1/2
2	1/4
3	1/8
.	.
.	.
.	.
k	$1/2^k$
.	.
.	.
.	.

Notice that the sum of these probabilities is 1, since

$$r + r^2 + r^3 + \cdots = \frac{r}{(1-r)} \quad \text{when} \quad |r| < 1 \quad \text{(Note 2)}$$

Example 3 A 10-faced die with faces marked 0, 1, 2, 3, 4, 5, 6, 7, 8, 9 is thrown repeatedly without a stop. (Of course, this is a fiction, since no man is immortal.) The outcome of the experiment is the resulting infinite sequence of integers. Suppose that a decimal point is placed in front of the sequence and the result is read as a real number between 0 and 1.

Possible outcomes	*Probabilities*
Every real number between 0 and 1 is a possible outcome.	The probability assigned to any single outcome is 0. More interesting are the probabilities assigned to sets of outcomes. If $0 < a < b < 1$, then the probability assigned to the interval $[a, b)$ is $b - a$, the length of the interval. For example, the probability assigned to the interval $[0.374, 0.549)$ is $0.549 - 0.374 = 0.175$. This number is reasonable because an outcome is in this set only if the first three throws are any of the 175 triples (3, 7, 4), (3, 7, 5), . . . , (5, 4, 8) out of the 1000 possible triples. The probabilities of more complicated sets are inferred from the probabilities of intervals.

Example 1 has a finite number of possible outcomes, and Example 2 has a countably infinite number of possible outcomes. This means that the outcomes can be placed in a one-to-one correspondence with the set of all positive integers 1, 2, 3, ... We will say that there are a countable number of outcomes if their number is either finite or countably infinite. The theory for the countable case is developed in Chapters 1 through 6. The number of outcomes in Example 3 is infinite but not countable, because there are more than a countable number of real numbers in the interval (0, 1) (Note 3). The mathematical theory for such experiments begins in Chapter 7.

Apparently different chance experiments may be described by the same model of outcomes and probabilities. For instance, consider the following. A psychologist runs a rat through a maze. On the basis of past experience, he postulates that it is as likely as not that the rat will choose the correct path through the maze. Suppose the result of the experiment is the number of successive attempts required until the correct path is finally run. If the rat has no memory of his past experiences then this chance experiment is really the same as the one in Example 2 (Note 4).

1.2 THE FREQUENCY THEORY OF PROBABILITY

There is a way of interpreting probabilities empirically called the *frequency theory of probability.* Consider a chance experiment with a countable number of distinct possible outcomes, labeled w_1, $w_2, \ldots$. If the chance experiment is repeated N times, then some of the N results of the experiment may be of type w_1, some of type w_2, and so forth. Let N_i denote the *number* of results of type w_i, so that

$$N_1 + N_2 + \cdots = N$$

The *proportion* of results of type w_i is N_i/N, so that

$$\frac{N_1}{N} + \frac{N_2}{N} + \cdots = 1$$

The frequency theory of probability asserts that the following limits exist as N goes to infinity:

$$\lim_{N\to\infty} \frac{N_i}{N} = p_i, \qquad i = 1, 2, \ldots$$

The limiting values $p_1, p_2, \ldots$ are called the *theoretical probabilities* of the outcomes $w_1, w_2, \ldots$, respectively. These assertions are really not about mathematical limits but rather about empirical phenomena. To verify these limits one would have to be able to repeat the chance experiment an infinite number of times, which is impossible. Attempts have been made to place the frequency theory of probability on a sound logical basis; these efforts have been beset by great difficulties, and the logical basis of the theory will not be pursued here. For present purposes, it is useful to interpret (1) in a purely intuitive way: Associated with the outcomes $w_1, w_2, \ldots$ are *theoretical probabilities* $p_1, p_2, \ldots$. *When the experiment is repeated a "large" number of times, the observed proportions* $(N_1/N), (N_2/N), \ldots$, *are "close" to these theoretical probabilities.*

It is useful to consider *sets* of outcomes as well as individual outcomes. Sets of outcomes are called *events.* By the probability of an event will be meant the sum of the probabilities of the individual outcomes that make up the event. From the point of view of the frequency theory, when the same experiment is repeated N times,

TABLE 1.1

w_i	*Possible outcomes*	N_i = *number of observed outcomes of type* 1	N_i/N	p_i = *theoretical probability*
w_1	$= (H, H)$	23	0.23	0.25
w_2	$= (H, T)$	28	0.28	0.25
w_3	$= (T, H)$	25	0.25	0.25
w_4	$= (T, T)$	24	0.24	0.25

the proportion of the N results that are of one of the types in the event will be "close" to the theoretical probability of the event, if N is "large." To be specific, suppose the event consists of the two outcomes w_1 and w_2. Then the proportion of the N results that are of type w_1 or w_2 is $(N_1/N) + (N_2/N)$, which is "close" to $p_1 + p_2$. Thus, $p_1 + p_2$ is called the probability of the event $\{w_1, w_2\}$. Most of the properties of the mathematical model that will be developed can be motivated by the frequency theory. For instance, if A and B are *disjoint* events, then the probability of $A \cup B$ is required to be the same as the sum of the probabilities assigned to A and B separately. This is reasonable, since the observed proportion of the N outcomes that are either in A or in B is the sum of the proportions that are in A or B separately.

It will be useful for the reader to try to motivate theoretical facts about probability theory by means of the frequency theory. Such motivations will be explicitly brought out from time to time.

Quite often in scientific problems where probabilities are difficult to determine mathematically, the frequency theory is used to obtain an empirical approximation. The actual experiment or a simulation thereof is repeated a large number of times. The observed proportion of the outcomes in the event of interest is determined and is used as an approximation to the theoretical probability. Such techniques are widely used and are called *Monte Carlo methods* (Note 5).

Example 1 Each member of a class of 100 students threw a coin twice. The observed proportions of times that the various outcomes were observed are given in Table 1.1.

1.3 DISCRETE SAMPLE SPACES

The purpose of this section is to introduce some standard terminology used in the mathematical model for chance experiments—namely, sample point, sample space, and event.

sample point, sample space

The basic object in the mathematical model is the *sample point,* which is just a formal name for *possible outcome.* The collection of all sample points is denoted by $\mathfrak{X}$ and is called the *sample space.* A sample space with a countable number of sample points is called a *discrete sample space. Through Chapter 6, the only sample spaces to be considered are those with a countable number of sample points.*

Example 1 The experiment of Example 1, Section 1.2, in which a coin is thrown twice, is described by a finite sample space. The sample points are

$$w_1 = (H, H), \qquad w_2 = (H, T), \qquad w_3 = (T, H), \qquad w_4 = (T, T)$$

Example 2 The experiment of Example 2, Section 1.1, in which a coin is tossed until a head appears, is described by a countably infinite sample space. The sample points can be labeled by the positive integers as

$$w_1 = 1, \qquad w_2 = 2, \qquad \ldots$$

The words used in probability theory often reflect its involvement with the real world. Thus, in a probability setting, "point" becomes "sample point" and "space" becomes "sample space." A set of sample points is also given a name having a special empirical flavor—*event.* An event is simply a set of sample points—any set of sample points. Later, in Chapter 7, when $\mathfrak{X}$ may consist of more than a countable number of sample points, some discretion will be necessary; not every set of sample points will be suitable for probability theory purposes. For the time being, however, any set may be considered an event. In particular, the whole sample space $\mathfrak{X}$ is an event, and $\mathfrak{X}$ is sometimes called the *sure* or *certain event.* In probability theory, the empty set $\varnothing$ is called the *impossible event.* Because events are sets, the notions and notations of elementary set theory are appropriate to their discussion. Thus, if A, B are events, then

A', the complement of A
$A \cup B$, the union of A and B
$A \cap B$, the intersection of A and B

are all events. Often the simplest way to describe an event is to list its component sample points, as in the following examples (Note 3).

TABLE 1.2

	w_1	w_2	w_3
$\varnothing$	0	0	0
A	1	0	0
B	0	1	0
C	0	0	1
D	1	1	0
E	1	0	1
F	0	1	1
$\mathfrak{X}$	1	1	1

Example 3 Suppose $\mathfrak{X}$ consists of the three sample points w_1, w_2, w_3. Counting $\varnothing$, there are eight distinct events:

$\varnothing$ = impossible event
$A = \{w_1\}$
$B = \{w_2\}$
$C = \{w_3\}$
$D = \{w_1, w_2\}$
$E = \{w_1, w_3\}$
$F = \{w_2, w_3\}$
$\mathfrak{X} = \{w_1, w_2, w_3\}$ = certain event

These events are also described by Table 1.2. The symbol 1 in the w_i column means that w_i is one of the points making up the event. The symbol 0 in that column means that w_i is not in the event. There are $2^3 = 8$ distinct triples made up of 1's and 0's (see Exercise 9b).

1.4 PROBABILITY MEASURE

To every event A there will be assigned a number $P(A)$ called "the probability of A." The function P, which is a *set function* because it is defined over all the subsets of $\mathfrak{X}$, is called a probability measure. The requirements imposed on the probability measure P follow.

properties of probability measure

(a) For any event A, $0 \leq P(A) \leq 1$.
(b) $P(\mathfrak{X}) = 1$, $P(\varnothing) = 0$.
(c′) If A, B are disjoint events, then $P(A \cup B) = P(A) + P(B)$.
(c) More generally, if $A_1, A_2, \ldots$ is a sequence of mutually disjoint events, then

$$P(A_1 \cup A_2 \cup \cdots) = P(A_1) + P(A_2) + \cdots$$

Some elementary consequences of (a), (b), and (c) follow, the proofs of which are left to the reader.

(d) $P(A') = 1 - P(A)$.

(e) $P(A \cup B) = P(A) + P(B) - P(A \cap B)$ (see Figure 1.1).

We should emphasize that (a), (b), and (c) are axioms, whereas (d) and (e) are consequences of these axioms. These properties are motivated by the frequency theory of probability, presented in Section 1.2. For instance, from the frequency theory point of view, if the experiment is repeated many times, then the *proportion of times* that the outcomes in A are actually observed will be close to the theoretical probability $P(A)$. Inasmuch as a proportion is of necessity between 0 and 1, the same requirement is imposed on $P(A)$. The reasonableness of every one of the requirements (a) through (c) can be similarly motivated by the frequency theory.

Because $\mathfrak{X}$ is a discrete sample space, the probabilities of the one-point sets, the sample points $w_1, w_2, \ldots$ play a special role. Denote by $p_1, p_2, \ldots$ the probabilities of the individual sample points $w_1, w_2, \ldots$, respectively. That is,

$$p_1 = P(\{w_1\}), \qquad p_2 = P(\{w_2\}), \qquad \ldots$$

It follows from (c) that

$$P(A) = \sum_{w_i \text{ in } A} p_i$$

That is, $P(A)$ is the sum of the probabilities of all the sample points that make up A. In particular,

$$P(\mathfrak{X}) = 1 = p_1 + p_2 + \cdots$$

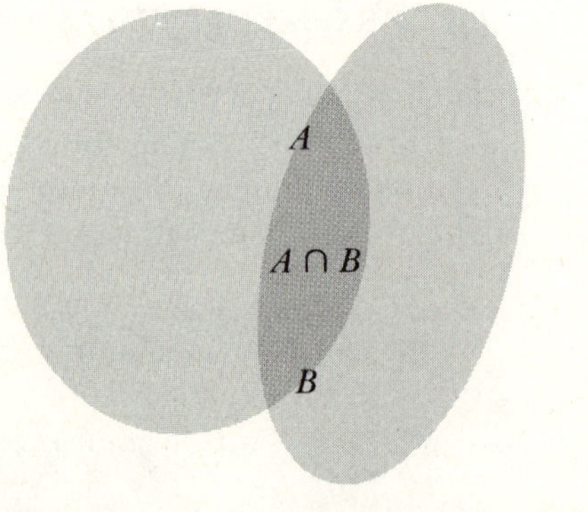

FIGURE 1.1 **$P(A) + P(B)$ adds the probability of $A \cap B$ twice when it should only be added once. $P(A) + P(B) - P(A \cap B)$ gives the probability of $A \cup B$.**

That is, the sum of the p_i values over all the sample points is 1.

The preceding motivates the typical way in which probabilities are assigned to a discrete sample space $\mathfrak{X} = \{w_1, w_2, \ldots\}$. Probabilities are first assigned to the one-point events by $P(\{w_i\}) = p_i$, where $p_1, p_2, \ldots$ are nonnegative numbers that sum to 1. The probability of any event A is then defined by (1). It is left to the reader to verify that this definition of P satisfies (a), (b), and (c) (Note 6).

Some remarks are in order about events of probability zero. We have pointed out that the impossible event is required to have probability zero. Thus, if a die has its faces marked 1, 2, 3, 4, 5, and 6, then the probability of a total of 13 points in two throws is zero. On the other hand, even though an event is not impossible, an appropriate probability measure may still assign to it the probability zero. For instance, suppose the die is thrown by a dishonest player who has the skill of never allowing a 6 to appear. The probability of a total of 12 points in two throws would have to be zero.

The sample space $\mathfrak{X}$ equipped with its probability measure P is called a *probability space.* A formal way of saying the same thing is that the pair of objects $(\mathfrak{X}, P)$ is called a probability space. Because $\mathfrak{X}$ is countable, $(\mathfrak{X}, P)$ is called a *discrete probability space.*

Example 1 Suppose $\mathfrak{X}$ is a sample space consisting of four sample points (a, b, c, d), which are assigned the probabilities 0.2, 0.5, 0.1, and 0.2, respectively. There are 16 events altogether. A list of these events and their probabilities appears in Table 1.3.

In applying probability theory to concrete chance experiments, one starts by assigning a "reasonable" probability measure to the

TABLE 1.3

Event	*Probability*	*Event*	*Probability*
$\varnothing$	0	(b, c)	0.6
(a)	0.2	(b, d)	0.7
(b)	0.5	(c, d)	0.3
(c)	0.1	(a, b, c)	0.8
(d)	0.2	(a, b, d)	0.9
(a, b)	0.7	(a, c, d)	0.5
(a, c)	0.3	(b, c, d)	0.8
(a, d)	0.4	$(a, b, c, d) = \mathfrak{X}$	1.0

sample space. Originally, probability theory only considered experiments with a *finite* number of outcomes with the probability value assigned to each outcome being the same. In other words, to each of the n sample points $w_1, \ldots, w_n$ is assigned the probability $1/n$. Then $P(A)$, the probability of a set A, is evaluated by *counting* the number of sample points that make up A and dividing this count by n. This so-called case of "equal likelihood" still provides many of the examples for discrete probability theory (Note 7).

Example 2 An "honest" six-faced die is thrown. The word "honest" means that each of the six outcomes 1, 2, 3, 4, 5, and 6 is assigned the same probability 1/6. If A is the event that the result of the throw is an even number, then $P(A) = 3/6$, because $A = \{2, 4, 6\}$.

Example 3 Four cards marked 1, 2, 3, and 4 are laid out "at random" in a row. The outcomes of the experiment are the 24 permutations of the four integers (1, 2, 3, 4). The phrase "at random" means that each of the 24 sample points is assigned the same probability 1/24. If A is the event that 1 appears in the first position, then $P(A) = 1/4$. This statement is true because there are six permutations of 1, 2, 3, 4 in which 1 is in the first position. Hence, $P(A) = 6/24 = 1/4$ (see Exercise 14).

Example 4 Three balls marked 1, 2, and 3 are randomly assigned to two cells marked a and b. The possible dispersions of the balls are given in Table 1.4.

The usual reasonable probability model assigns probability 1/8 to each of the eight triples (a, a, a), (a, a, b), . . . , (b, b, b). Suppose now that the balls are not marked 1, 2, and 3 but are indistinguish-

TABLE 1.4

Ball 1	*Ball 2*	*Ball 3*
a	a	a
a	a	b
a	b	a
b	a	a
b	b	a
b	a	b
a	b	b
b	b	b

TABLE 1.5

Number of balls in cell a	*Number of balls in cell b*	*Probability*
3	0	1/8
2	1	3/8
1	2	3/8
0	3	1/8

able one from the other. Table 1.4 no longer makes sense in this context, but the outcomes of the experiment can be described by giving the numbers of balls that fall in cells a and b, as in Table 1.5. To justify the probabilities assigned to the four outcomes (3, 0), (2, 1), (1, 2), and (0, 3), make believe that the balls have been temporarily marked 1, 2, and 3. The probabilities of the originally described eight outcomes should be 1/8 each; after all, the behavior of the balls should not be influenced by whether or not they are marked. If the markings are now erased, the original outcomes "collapse" into new outcomes, as summarized in Table 1.6.

"Misguided" students will often assign the unreasonable probabilities 1/4, 1/4, 1/4, 1/4 to each of the outcomes (3, 0), (2, 1), (1, 2), (0, 3). It turns out that in statistical mechanics certain types of "misguided" particles behave in exactly this way (see the discussion about Bose–Einstein probabilities in Section 3.7).

1.5 A REVIEW OF SOME COMBINATORICS

Many examples in elementary probability theory involve finite sample spaces where each of n sample points has the same prob-

TABLE 1.6

Outcomes if the balls are identified				*Number of balls*		
Ball 1	*Ball 2*	*Ball 3*	*Probabilities*	*In cell a*	*In cell b*	*Probabilities*
a	a	a	1/8	3	0	1/8
a	a	b	1/8			
a	b	a	1/8	2	1	3/8
b	a	a	1/8			
b	b	a	1/8			
b	a	b	1/8	1	2	3/8
a	b	b	1/8			
b	b	b	1/8	0	3	1/8

ability $1/n$. In such cases, evaluating probabilities of events often requires facts about permutations and combinations. For this reason, we present a short review of such facts (Note 8).

I. Let $(a_1, \ldots, a_n)$ and $(b_1, \ldots, b_m)$ be two finite sets of objects. The number of pairs of the form (a_i, b_j) is nm.

Example 1 The six possible pairs made up from (a_1, a_2, a_3) and (b_1, b_2) are (a_1, b_1), (a_1, b_2), (a_2, b_1), (a_2, b_2), (a_3, b_1), and (a_3, b_2).

II. Let $(a_{11}, \ldots, a_{1n_1}), \ldots, (a_{k1}, \ldots, a_{kn_k})$ be k finite sets of objects. The number of k-tuples of the form $(a_{1i_1}, a_{2i_2}, \ldots, a_{ki_k})$ is $n_1 n_2 \cdots n_k$, and I is a special case of II for $k = 2$.

Example 2 There are 10^3 triples of the form (i, j, k) in which each entry is one of the 10 symbols 1, 2, 3, 4, 5, 6, 7, 8, 9, 0.

III. Let $(a_1, \ldots, a_n)$ be n distinct symbols. Let k be no greater than n. The number of distinct k-tuples of these symbols, *repetitions not being allowed*, is $n(n-1) \cdots (n-k+1)$. This number is written for short as $n^{(k)}$ and is read as the "*kth factorial power of n.*" Do not confuse $n^{(k)}$ with the conventional power n^k.

Example 3 The $3^{(2)} = 6$ distinct pairs made up from the symbols (a, b, c) without repeating the use of a symbol are (a, b), (a, c), (b, a), (b, c), (c, a), and (c, b).

IV. Let $(a_1, \ldots, a_n)$ be n distinct symbols. The number of distinct arrangements of these symbols is $n(n-1) \cdots 3 \cdot 2 \cdot 1$, which is written as $n!$ and is read as "*n factorial.*" Note IV is a special case of III for $k = n$. Table 1.7 shows some values of $n!$ ($0! = 1$ by convention).

Example 4 The $3! = 6$ distinct arrangements of the symbols (a, b, c) are (a, b, c), (a, c, b), (b, a, c), (b, c, a), (c, a, b), (c, b, a).

TABLE 1.7

n	$n!$
0	1
1	1
2	2
3	6
4	24
5	120
6	720
7	5,020
8	40,160
9	361,440
10	3,614,400

V. Let $(a_1, \ldots, a_n)$ be n distinct symbols. Let k be no greater than n. The number of distinct k-tuples of these symbols, repetitions not being allowed and different orderings of the same k symbols not being counted separately, is

$$\frac{n^{(k)}}{k!} = \frac{n(n-1)\cdots(n-k+1)}{k(k-1)\cdots 3\cdot 2\cdot 1}$$

This is written as $\binom{n}{k}$ and is read "binomial coefficient n, k."

Table 1.8 shows some values of $\binom{n}{k}$.

TABLE 1.8 The binomial coefficients $\binom{n}{k}$

n	k = 0	1	2	3	4	5	6	7	8	9
1	1	1								
2	1	2	1							
3	1	3	3	1						
4	1	4	6	4	1					
5	1	5	10	10	5	1				
6	1	6	15	20	15	6	1			
7	1	7	21	35	35	21	7	1		
8	1	8	28	56	70	56	28	8	1	
9	1	9	36	84	126	126	84	36	9	1

In Table 1.8, the entries for n, k, and $n, k + 1$ add up to the entry for $n + 1, k + 1$. For instance, in the boxed-in entries, $28 + 56 = 84$ (see Exercise 8). The entries across for any n sum to 2^n. For instance, for $n = 4$, $1 + 4 + 6 + 4 + 1 = 16 = 2^4$ (see Exercise 5).

There is an alternative way of interpreting the count in V which is useful.

V(a). The number of distinct k-tuples of the form $(a_{i_1}, \ldots, a_{i_k})$, where $i_1 < i_2 < \cdots < i_k$ are k indices from among 1, 2, . . . , n, *arranged in order*, is $\binom{n}{k}$.

Example 5 The number of committees of size 3 that it is possible to list using an available 10 persons is $\binom{10}{3} = 120$. If a committee also implies a chain of command and lists members as first member, second member, and third member, there are $10^{(3)} = 720$ possible committees.

Example 6 In a certain telephone system, telephone numbers must be of the form (i, j, k), where the symbols are from the set 0, 1, 2, 3, 4, 5, 6, 7, 8, 9; they must also be arranged in increasing order, $i < j < k$. There are $\binom{10}{3} = 120$ numbers possible. If the restriction of increasing order is removed, there are $10^{(3)} = 720$ telephone numbers possible.

VI. Suppose there are n symbols, k of which are indistinguishable a's and $(n - k)$ of which are indistinguishable b's. There are $\binom{n}{k}$ distinct arrangements of these symbols. (By V, there are $\binom{n}{k}$ choices of positions for the a's among the n places.)

Example 7 The $\binom{4}{2} = 6$ arrangements of the symbols (a, a, b, b) are (a, a, b, b), (a, b, a, b), (a, b, b, a), (b, a, a, b), (b, a, b, a), (b, b, a, a).

VII. We have $\binom{n}{k} = \binom{n}{n-k}$. Hence, $n^{(k)}/k! = n^{(n-k)}/(n-k)! = \binom{n}{k}$. Any reasonable interpretation of $\binom{n}{n}$ in V or V(a) or VI is that it should equal 1—hence, the convention that $0! = 1$.

Example 8 The easy way of computing $\binom{100}{98}$ is from $\binom{100}{98} = \binom{100}{2} = 100^{(2)}/2! = 100 \cdot 99/2 = 495$, rather than from $\binom{100}{98} = 100^{(98)}/98!$

VIII. Suppose there are k_1 indistinguishable a's, k_2 indistinguishable b's, . . . , k_r indistinguishable z's. The number of distinct arrangements of these $k_1 + k_2 + \cdots + k_r = n$ symbols is $n!/k_1!k_2! \cdots k_r!$ This number is sometimes denoted $\binom{n}{k_1 k_2, \ldots, k_r}$ and is read as the "multinomial coefficient n; $k_1, k_2, \ldots, k_r$." Notice that if $n = 2$, then $\binom{n}{k, n-k} = \binom{n}{k}$.

Example 9 The $\binom{4}{2, 1, 1} = 12$ arrangements of the symbols (a, a, b, c) are (a, a, b, c), (a, a, c, b), (a, b, a, c), (a, c, a, b), (a, b, c, a), (a, c, b, a), (b, a, a, c), (c, a, a, b), (b, a, c, a), (c, a, b, a), (b, c, a, a), (c, b, a, a).

IX. If n is a positive integer, then

$$(a+b)^n = \sum_{k=0}^{n} \binom{n}{k} a^k b^{n-k}$$

This is called the *binomial expansion*.

Example 10 $(a+b)^4 = b^4 + 4ab^3 + 6a^2b^2 + 4a^3b + a^4$

X. If n is a positive integer, then

$$(a_1 + a_2 + \cdots + a_r)^n = \Sigma \binom{n}{k_1, k_2, \ldots, k_r} a_1^{k_1} a_2^{k_2} \cdots a_r^{k_r}$$

where the sum is over all r-tuples of nonnegative integers, $k_1, \ldots, k_r$ that sum to n. This is called the *multinomial expansion.*

Example 11 $(a+b+c)^3 = a^3+b^3+c^3+3a^2b+3ab^2+3a^2c+3ac^2+3b^2c+3bc^2+6abc$

SUMMARY

The first six chapters of this book concern chance experiments with a finite or countably infinite number of outcomes. The mathematical model for a chance experiment is a *probability space* $(\mathfrak{X}, P)$. The basic set of outcomes is represented by $\mathfrak{X} = (w_1, w_2, \ldots)$, called the *sample space.* The w_i's represent the individual outcomes and are called *sample points.* The *probability measure* P assigns to any set of sample points A (also called *event* A) a number $P(A)$, which is the probability of A. Thus, P is a set function satisfying the following axioms:

$$0 \leqslant P(A) \leqslant 1, \qquad P(\mathfrak{X}) = 1, \qquad P(\emptyset) = 0$$

and

$$P(A_1 \cup A_2 \cup \cdots) = P(A_1) + P(A_2) + \cdots$$

when $A_1, A_2, \ldots$ are disjoint. One consequence of the axioms is $P(A') = 1 - P(A)$. An easy way of assigning a probability measure that satisfies the axioms is to first assign probabilities $p_1, p_2, \ldots$ to the one-point events $w_1, w_2, \ldots$. The p_i's must be nonnegative numbers that sum to one. The probability of an arbitrary event A is then defined to be the sum of all p_i's corresponding to the w_i's that make up A.

The axioms for a probability space are motivated by the *frequency theory of probability.* The frequency theory is an empirical interpretation that considers what happens when a chance experiment is repeated a large number of times. Among these repetitions we consider the *proportion* of times we have actually observed outcomes in the set A. According to the frequency theory, this proportion will approach the limit $P(A)$ as the experiment is repeated indefinitely. Thus, the theoretical probability $P(A)$ is an idealization of the proportion of times an experiment results in outcomes in A over a large number of repetitions.

EXERCISES

Give mathematical models for the chance experiments in Exercises 1 through 4. In each case, describe a listing of the sample points and reasonable probabilities to assign to them. Unless otherwise stated, the word *die* will always refer to a conventional six-faced die with faces marked 1, 2, 3, 4, 5, and 6.

1 (a) A box contains five balls numbered 1, 2, 3, 4, 5. A ball is drawn at random.
(b) A box contains five balls numbered 1, 2, 3, 4, 5. A ball is drawn at random, replaced, and a second drawing is made at random.
(c) A box contains five balls numbered 1, 2, 3, 4, 5. A ball is drawn at random, is not replaced, and a second drawing is made at random.

2 (a) Five cards marked 1, 2, 3, 4, 5 are shuffled and laid out in a row.
(b) Five cards marked 1, 2, 3, 4, 5 are randomly assigned to the positions marked in Figure 1.2.
(c) Five cards marked 1, 2, 3, 4, 5 are randomly assigned to the positions marked in Figure 1.3.

3 (a) A coin is tossed five times.
(b) A die is tossed five times.
(c) A die having faces marked 1, 1, 2, 2, 3, 4 is tossed five times.

4 (a) A coin is tossed until a head appears.
(b) A die is tossed until a 1 appears.
(c) A die is tossed until either a 1 or a 2 appears.

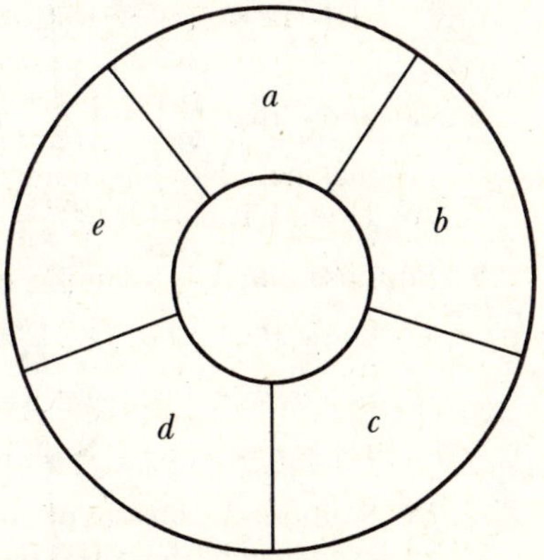

FIGURE 1.2 When the five cards marked 1, 2, 3, 4, 5 are assigned, rotations of the same circular arrangement are not distinguishable from one another. Thus, (1, 3, 2, 4, 5) in (*a, b, c, d, e*) is not recognizably different from (5, 1, 3, 2, 4) in (*a, b, c, d, e*).

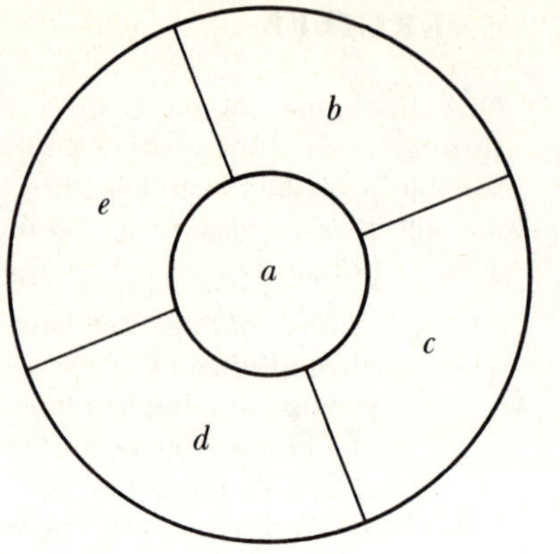

FIGURE 1.3 Rotations of the same circular arrangement are not distinguishable from one another. Thus (1, 3, 2, 4, 5) in (*a*, *b*, *c*, *d*, *e*) is not recognizably different from (1, 5, 3, 2, 4) in (*a*, *b*, *c*, *d*, *e*).

5 Show that $\Sigma_{i=0}^{n} \binom{n}{i} = 2^n$. (Hint: In IX, Section 1.5, let $a = b = 1$.)

6 Show that $\Sigma_{i=0}^{n} (-1)^i \binom{n}{i} = 0$. (Hint: In IX, Section 1.5, let $a = 1$, $b = -1$.)

7 Show that $\Sigma_{i=0}^{n} i\binom{n}{i} = n2^{n-1}$. [Hint: Notice that $i\binom{n}{i} = n\binom{n-1}{i-1}$ and use Exercise 5. Alternatively, differentiate both sides of $\binom{n}{0} + \binom{n}{1}t + \cdots + \binom{n}{n}t^n = (1+t)^n$ with respect to t, and set $t = 1$.]

8 (a) Show that $\binom{n}{i} + \binom{n}{i+1} = \binom{n+1}{i+1}$, which gives a fast way of making tables of binomial coefficients (see Table 1.8).
(b) Extend Table 1.8 to $n = 10$ and 11 using (a).

9 Suppose that $\mathfrak{X}$ is a sample space consisting of n sample points.
(a) Show that there are $\binom{n}{k}$ distinct events consisting of k sample points, $k = 1, 2, \ldots$. Conclude from Exercise 5 that there are $1 + \binom{n}{1} + \cdots + \binom{n}{n} = (1+1)^n = 2^n$ distinct events, counting $\varnothing$.
(b) Suppose $\mathfrak{X}$ consists of the n sample points $w_1, w_2, \ldots, w_n$. An event can be described by a sequence of n 0's and 1's. For instance, the event consisting of w_1 and w_2 is represented by the sequence 1, 1, 0, 0, . . . , 0. The two 1's mean that the first two sample points are in the event, and the string of 0's means that none of the remaining points are in the event. In general, if $(e_1, \ldots, e_n)$ represents the event, then $e_i = 1$ means that w_i is in the event

and $e_i = 0$ means that w_i is not in the event. Using this scheme for representing events, show directly that there are 2^n distinct events. [What about the sequence $(0, 0, \ldots, 0)$? What event does it represent?] Compare with Example 3, Section 1.3.

(c) Suppose that A is an event consisting of k sample points ($0 < k < n$). Show that there are 2^{n-k} distinct events that contain A.

10 Five balls are randomly distributed among three boxes designated a, b, and c. Evaluate the probabilities of the following events.

(a) Box a is empty.
(b) Box a and only box a is empty.
(c) Exactly one box is empty.
(d) At least one box is empty.
(e) No box is empty.
(f) Two boxes are empty.
(g) Three boxes are empty.
(h) Either Box a or box b are empty.

(Hint: Let a sample point be the 5-tuple of box assignments to the five balls, keeping track of the order of the balls. Thus (a, a, a, b, c) means ball 1 in box a, ball 2 in box a, ball 3 in box a, ball 4 in box b, ball 5 in box c. There are 3^5 sample points, and we interpret "randomly distributed" to mean that each sample point has probability $1/3^5$. The answers are in Exercise 11.)

11 Repeat Exercise 10 with n balls and three boxes. Check your general expressions against your answers to Exercise 10. [Answers: (a) $2^n/3^n$, (b) $(2^n - 2)/3^n$, (c) $(2^n - 2)/3^{n-1}$, (d) $(2^n - 1)/3^{n-1}$, (e) $1 - (2^n - 1)/3^{n-1}$, (f) $1/3^{n-1}$, (g) 0, (h) $2(2/3)^n - (1/3)^n$.]

12 Five cards marked 1, 2, 3, 4, 5 are randomly laid out in a row. Evaluate the probabilities of the following events.

(a) Card 1 appears in the first position.
(b) Card 1 is immediately followed by card 2.
(c) Card 1 is not in the first position and card 2 is not in the second position.
(d) There are exactly three matches. (A match means that card i is in position i.)

[Answers: (a) $4!/5! = 1/5$, (b) $1/5$, (c) $13/20$, (d) $1/12$.]

13 A conventional deck of 52 playing cards is shuffled and 13 cards are dealt to each of four players. Designate the players as A, B, C, and D. Evaluate the probabilities of the following events.

(a) Player A gets all the hearts, B gets all the diamonds, C gets all the clubs, and D gets all the spades.
(b) Each player gets cards of only one suit.
(c) Player A gets no aces.
(d) Players A and B get a total of 15 red cards between them.

[Answers: (a) $(13!)^4/52!$, (b) $24(13!)^4/52!$, (c) $\binom{48}{13}\Big/\binom{52}{13}$, (d) $\binom{26}{15}\binom{26}{11}\Big/\binom{52}{26}$.]

14 Suppose n cards marked $1, 2, \ldots, n$ are laid out in a row at random. Let A be the event that card 1 appears in the first position and let B be the event that card 2 appears in the second position.

(a) Show that $P(A) = P(B) = 1/n$ (compare with Example 3, Section 1.4).
(b) Show that $P(A \cap B) = 1/n(n-1)$.
(c) Show that $P(A \cup B) = (2n-3)/n(n-1)$.

15 (a) Show that $n^{(k)} = 0$ when $k > n$ (n and k are positive integers). [Hint: One of the terms in $n(n-1) \cdots (n-k+1)$ is zero.]
(b) If $\binom{n}{k}$ is defined as $n^{(k)}/k!$ when $k > n$, then this binomial coefficient equals zero.

NOTES

1 Gerolamo Cardano, 1501–1576, an Italian physician, is thought to be the pioneer of probability theory. His book called *Liber de Ludo Aleae,* was written as a handbook for gamblers. For an interesting biography by a modern mathematician, see [1], which contains a translation from the Latin of *Liber de Ludo Aleae.* Cardano's manuscript remained unpublished until 87 years after his death. The initially felt stimulus for probability theory arose in a correspondence between Blaise Pascal (1623–1662) and Pierre de Fermat (1608–1665).

2 Geometric series are in most calculus books. In particular, see Section 18.1 of [2] and Section 10.8 of [3].

3 For a discussion of set theory and elementary facts about countability, see Chapter 1, Part I of [4]. Example 6, Section 1.14 of [4] proves the noncountability of the real numbers.

4 For some general reviews of the scope of probability theory, see the *Scientific American* articles [5], [6], and [7].

5 An important attempt to place the frequency theory on a sound mathematical basis is the work of Richard von Mises (1883–1953), summarized in Chapter I of [8].

6 Additional discussions of the basic axiomatics of discrete probability spaces can be found in Chapter 1 of [9], Chapter 1 of [10], and Chapter 1, Part II of [4].

7 "Honest" coins, "honest" dice, and "random" drawings from decks of cards or collections of balls are convenient fictions in that they provide probability spaces the sample points of which are equally probable. We shall use such models extensively throughout the book, although in practice their appropriateness may often be dubious. Historically, such examples form the heart of elementary probability theory. For an extensive collection of exercises using these models, see the old but useful and still available books [11] and [12].

8 For an additional review of combinatoric manipulations, see Chapter II of [9] and Appendix A of [10].

Chapter Two

RANDOM VARIABLES

Outcomes of a chance experiment need not be numbers, but it is very useful to assign numbers to these outcomes. Formally, we are dealing with a numerically valued function over the sample space; such a function is called a *random variable*. By *expected value* is meant a weighted average of a numerical function over the sample space. This chapter develops basic facts about random variables and their expected values.

2.1 DEFINITION OF A RANDOM VARIABLE

Often an experimenter represents experimental outcomes by summarizing numbers. The original outcomes may be quite complicated and the numbers that replace them may help clarify the result of the experiment. For example, a medical researcher selects a person at random from a specified population. Associated with each person drawn might be his cholesterol count, or his weight, or his white blood cell count, and so forth. If the experiment is throwing a coin three times, a summarizing number of interest might be the number of heads obtained. Mathematically, a random variable is simply a numerically valued function defined over the sample space. It is customary to denote random variables by uppercase letters from the "latter half" of the alphabet, such as S, T, U, V, W, X, Y, and Z, possibly with subscripts as $S_1, S_2, \ldots$. It is useful to maintain the distinction between a random variable and the numerical value it assigns to a sample point. For example, $X(w)$ is a real number, namely the value assigned to the sample point w by the function X (Note 1).

definition of a random variable

A random variable is a real-valued function defined on a sample space.

Only because we deal with discrete sample spaces can we allow random variables to be *arbitrary* real-valued functions. For non-discrete sample spaces, additional requirements must be imposed on the functions. This topic is discussed in Chapter 7.

Example 1 A six-faced die is thrown. The numbers 1, 2, 3, 4, 5, 6 conventionally assigned to the faces represent the values of one random variable of interest in this experiment. If the sample points representing the outcomes are w_1, w_2, w_3, w_4, w_5, w_6, then a random variable X is defined by $X(w_1) = 1$, $X(w_2) = 2$, $X(w_3) = 3$, $X(w_4) = 4$, $X(w_5) = 5$, and $X(w_6) = 6$ (see Figure 2.1). Intrinsically, the outcomes of the experiment are by nature not numerical, but could be described by colors or in other ways. The conventional markings of 1, 2, 3, 4, 5, 6 describe only one particular random variable. Mark-

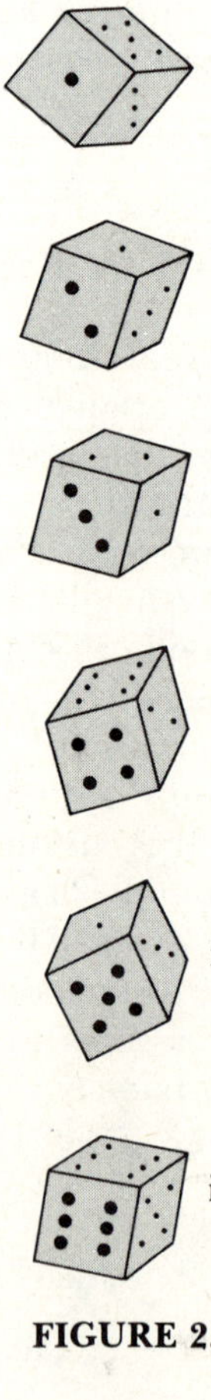

is represented by w_1 $X(w_1) = 1$

is represented by w_2 $X(w_2) = 2$

is represented by w_3 $X(w_3) = 3$

is represented by w_4 $X(w_4) = 4$

is represented by w_5 $X(w_5) = 5$

is represented by w_6 $X(w_6) = 6$

FIGURE 2.1

TABLE 2.1

Sample points	X_1	X_2	X_3	X	Y	Z
$H\ H\ H$	1	1	1	3	2	2
$H\ H\ T$	1	1	0	2	2	1
$H\ T\ H$	1	0	1	2	1	1
$H\ T\ T$	1	0	0	1	1	0
$T\ H\ H$	0	1	1	2	1	2
$T\ H\ T$	0	1	0	1	1	1
$T\ T\ H$	0	0	1	1	0	1
$T\ T\ T$	0	0	0	0	0	0

ing the faces 1, 1, −9, π, 345, $-\cos(32°)$ defines still another random variable.

Often it is of interest to define *several* random variables over the same sample space, as in the following example.

Example 2 A coin is thrown three times. Define random variables X_1, X_2, X_3, X, Y, Z as follows:

$$X_i = \begin{cases} 1 & \text{if } i\text{th toss is a head} \\ 0 & \text{otherwise} \end{cases} \qquad i = 1, 2, 3$$

X = total number of heads

Y = number of heads in the first two tosses

Z = number of heads in the last two tosses

A listing of the eight sample points and values of the defined random variables is given in Table 2.1.

2.2 THE NOTATION (X IN E)

In this section we introduce some notation relating to random variables.

The symbol $(X = a)$ denotes an event, namely the set of all sample points w, such that $X(w) = a$. Obvious variations on this notation will be used. For instance, $(X < a)$ denotes the set of all w such that $X(w) < a$, and so forth. In general, if E is a set of real numbers, then (X in E) denotes the set of w such that $X(w)$ is in the set E.

If there are no sample points w such that $X(w) = a$, then $(X = a)$ is the impossible event—that is, the empty set. For a discrete sample space *there are at most a countable number* of distinct real values $a_1, a_2, \ldots$, such that none of the sets $(X = a_1)$, $(X = a_2)$, . . . are empty. These numbers $a_1, a_2, \ldots$ are called the *possible values* of the random variable X (see Figure 2.2).

A random variable with a countable number of possible values is called a *discrete random variable* to distinguish it from other types that will arise later.

Since $(X \text{ in } E)$ is an event, the notation $P(X \text{ in } E)$ is well defined, which is the probability of the set of all sample points w such that $X(w)$ is in E.

Example 1 An honest coin is thrown three times. Suppose that each of the eight sample points has probability 1/8. Denote these sample points as follows:

$$w_1 = (H, H, H), \quad w_2 = (H, H, T), \quad w_3 = (H, T, H),$$
$$w_4 = (H, T, T), \quad w_5 = (T, H, H), \quad w_6 = (T, H, T),$$
$$w_7 = (T, T, H), \quad w_8 = (T, T, T).$$

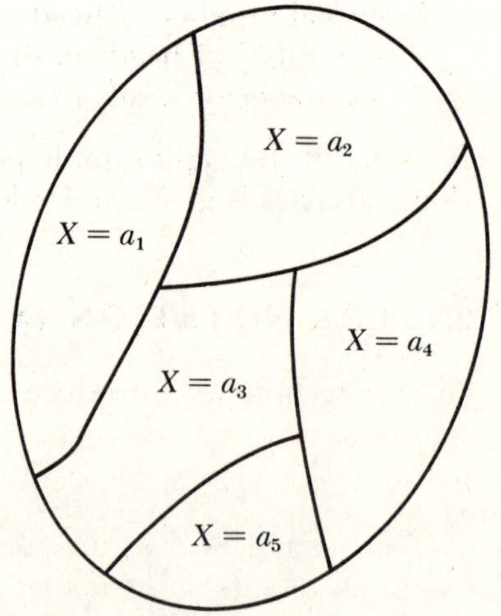

FIGURE 2.2 The set of all sample points w such that $X(w) = a_i$ is $(X = a_i)$. If $a_1, a_2, \ldots$ are the distinct possible values of X, then the events $(X = a_1)$, $(X = a_2)$, . . . are disjoint. Their union is the whole sample space $\mathfrak{X}$.

Let X be the total number of heads. The following illustrate the notation (X in E): $(X = 2) = (w_2, w_3, w_5)$; $(x < 2) = (w_4, w_6, w_7, w_8)$; ($X$ is an odd integer) $= (w_1, w_4, w_6, w_7)$. The probabilities of these events are as follows: $P(X = 2) = 3/8$; $P(X < 2) = 4/8$; $P(X$ is an odd integer$) = 4/8$. The possible values of X are 0, 1, 2, and 3.

Example 2 A biologist breeds rabbits for six months and wants to make a probability model to describe various aspects of the breeding process. One measurement of interest will be X, the proportion of males born among the total births. The possible values of X are of the form m/n where m and n are nonnegative integers and $m \leqslant n$.

2.3 FUNCTIONS OF RANDOM VARIABLES

There are standard ways of combining numerically valued functions into new functions. We will discuss some of these operations for random variables. Suppose that X and Y are two random variables defined on the same sample space. By $X + Y$ is meant a new random variable that assigns to the sample point w the value $X(w) + Y(w)$. Of course, this is what is always meant by the sum of two real-valued functions—the function that assigns to each point the sum of the numbers assigned by each of the two functions separately. The same definition is appropriate for a sum of n random variables. The random variable $X_1 + \cdots + X_n$ assigns to the sample point w the sum of the values that $X_1, \ldots, X_n$ separately assign to that point.

Various other operations of forming new random variables from previously defined ones are possible. We trust that the notation will always leave clear just which random variable is being defined. For instance, $3X$ is the random variable assigning to w the value $3X(w)$; X^2 assigns to w the value $[X(w)]^2$; XY assigns to w the value $X(w)Y(w)$; and so forth. Of course, the new random variable may not always be well defined. For instance, the random variable X/Y is well defined only if $Y(w)$ is not equal to 0 for any sample point w.

Example 1 A coin is thrown three times. Let X_i equal 1 if the ith throw is a head and 0 otherwise, and let $i = 1, 2, 3$. Table 2.2 shows the values of several random variables related to X_1, X_2, and X_3.

TABLE 2.2

Sample points	X_1	X_2	X_3	$X_1 + X_2 + X_3$	$X_1/(X_2 + 1)$	$X_1X_2X_3$
H H H	1	1	1	3	1/2	1
H H T	1	1	0	2	1/2	0
H T H	1	0	1	2	1	0
H T T	1	0	0	1	1	0
T H H	0	1	1	2	0	0
T H T	0	1	0	1	0	0
T T H	0	0	1	1	0	0
T T T	0	0	0	0	0	0

Example 2 An economist makes a model to describe steel production in 10 major steel centers. If X_i represents the production in the ith center for a given year, then $X_1 + \cdots + X_{10}$ is the total production for that year over all 10 centers.

2.4 EXPECTED VALUE OF A RANDOM VARIABLE

The intuitive idea of expected value has its origins in games of chance. Suppose a gambler repeatedly plays a game the outcomes of which are w_1, w_2, and w_3, with probabilities 0.1, 0.2, and 0.7, respectively. With outcome w_1 he wins twenty dollars, with outcome w_2 he wins ten dollars, and with outcome w_3 he loses ten dollars. What is his long-range average gain (or loss) over many games? Suppose that he plays a total of N games with N_1 games resulting in w_1, N_2 in w_2, and N_3 in w_3. Of course, $N_1 + N_2 + N_3 = N$. The average gain in dollars over the N plays is $(20N_1 + 10N_2 - 10N_3)/N$. According to the frequency theory of probability (Section 1.2), the observed proportions N_1/N, N_2/N, N_3/N are for large N, close to the theoretical probabilities 0.1, 0.2, 0.7, respectively. Hence, the average gain in dollars is close to $20(0.1) + 10(0.2) - 10(0.7) = -3$ when N is large. In other words, on the average the player loses three dollars per game. (Why does he keep playing?) If the random variable X is the gain to the player, the expected value of X is defined to be $X(w_1)(0.1) + X(w_2)(0.2) + X(w_3)(0.7) = 20(0.1) + 10(0.2) - 10(0.7) = -3$ (Note 2). The general definition of expected value follows.

definition of expected value

Suppose $\mathfrak{X} = (w_1, w_2, \ldots)$ is a discrete sample space with the probability measure P assigning the value p_i to w_i, $i = 1, 2, \ldots$. If X is a random variable, then the *expected value* (or expectation) of X, denoted EX, is defined to be

$$EX = X(w_1)p_1 + X(w_2)p_2 + \cdots \tag{1}$$

assuming that

$$|X(w_1)|p_1 + |X(w_2)|p_2 + \cdots < \infty \tag{2}$$

If (2) does not hold, the expected value is said not to exist.

Equation (2) requires that the sum in Eq. (1) be *absolutely convergent.* A succinct way of expressing this statement is that EX is said to exist if and only if $E|X| < \infty$. This requirement is imposed for the following reason—usually there is no "natural" way of ordering the sample points $w_1, w_2, \ldots$. It would be very unpleasant if the value of EX depended on the order in which the sample points were labeled. The requirement of absolute convergence is, in effect, equivalent to requiring that EX have a value that is not influenced by the particular arrangement of the sample points. It is a basic fact that an infinite series is absolutely convergent if and only if any rearrangement of the series converges. Moreover, the sum is always the same (Note 3).

Let us take a closer look at what happens when $E|X| = \infty$. Let A be the set of those sample points w such that $X(w) > 0$, and let B be those sample points such that $X(w) < 0$. Define

$$E(X;\, A) = \sum_{w_i \text{ in } A} X(w_i)p_i, \qquad E(X;\, B) = \sum_{w_i \text{ in } B} X(w_i)p_i$$

It follows that

$$EX = E(X;\, A) + E(X;\, B)$$

and

$$E|X| = E(X;\, A) - E(X;\, B)$$

The equation $E|X| = \infty$ means that either $E(X; A) = \infty$ or $E(X; B) = -\infty$. However, if $E(X; A) = \infty$ but $E(X; B)$ is finite, it makes sense to say that $EX = \infty$. (Similarly, if $E(X; A)$ is finite but $E(X; B) = -\infty$, it makes sense to say that $EX = -\infty$.) In particular, if X is a nonnegative random variable and EX fails to exist, we say that $EX = \infty$ (see Example 5).

If $\mathfrak{X}$ consists of a finite number of sample points $w_1, \ldots, w_n$, then $|X(w_1)|p_1 + \cdots + |X(w_n)|p_n$ is finite; therefore, for finite sample spaces EX always exists. (See Exercises 22 and 23 for some variations on this fact.)

The frequency theory interpretation of expected value given at the beginning of this section for a particular game of chance holds for any chance experiment. Consider an experiment with outcomes $w_1, w_2, \ldots$ and probabilities $p_1, p_2, \ldots$. If this experiment is repeated N times, producing N_1 outcomes of type w_1, N_2 outcomes of type w_2, and so forth, the frequency theory of probability says that N_i/N is close to p_i for N large. The actual average of the X values of the N observations is then $[X(w_1)N_1 + X(w_2)N_2 + \cdots]/N$. By the frequency theory, this value is close to $X(w_1)p_1 + X(w_2)p_2 + \cdots$, which by definition is EX. Thus, the frequency theory says that over a large number of repetitions of a chance experiment the observed average of X values is close to some idealized value. This value is EX. These ideas are made precise in the discussion of the law of large numbers in Chapter 10.

Example 1 Each member of a class of 100 students threw a coin twice. The number of heads obtained in the two throws is X. Table 1.1, Section 1.2 showed the numbers of observations of different types that were actually observed. The average of the 100 observed X values is

$$\frac{2(23) + 1(28) + 1(25) + 0(24)}{100} = 0.99$$

The "theoretical average," EX, is

$$[2(0.25) + 1(0.25) + 1(0.25) + 0(0.25)] = 1.00$$

Example 2 A box contains six balls numbered 1, 2, 2, 3, 4, 5. A ball is drawn at random. The expected number to appear is $(1 + 2 + 2 + 3 + 4 + 5)/6 = 17/6$.

Example 3 In a population of n persons, the ages of the individuals are $a_1, a_2, \ldots, a_n$. A person is drawn at random. Let X denote the age of the person drawn. Then $EX = (a_1 + a_2 + \cdots + a_n)/n$. Notice that EX is the ordinary *average* of ages in the population.

Example 4 A coin is thrown until a head appears. The expected number of tosses required to obtain a head is

$$1(\tfrac{1}{2}) + 2(\tfrac{1}{2})^2 + 3(\tfrac{1}{2})^3 + \cdots \tag{3}$$

To sum (3), first use the fact that

$$r + r^2 + r^3 + \cdots = \frac{r}{1 - r} = \frac{1}{1 - r} - 1 \tag{4}$$

whenever $|r| < 1$. Equation (4) is just one form of the familiar sum of a geometric progression (Note 2, Chapter 1). Equation (4), evaluated for $r = 1/2$, shows that the probabilities $1/2$, $(1/2)^2$, $(1/2)^3$, . . . sum to 1. Differentiate both sides of (4) with respect to r and multiply by r to obtain

$$r + 2r^2 + 3r^3 + \cdots = \frac{r}{(1-r)^2}$$

To evaluate (3), set $r = 1/2$ and the expected value turns out to be 2. On the average, a coin must be thrown two times to obtain a head. (See the hint for method 2, Exercise 9, for another way of evaluating this expected value.) (See Note 4.)

Example 5 A box contains one white ball and one black ball. A ball is drawn at random. If the ball drawn is white, the procedure stops; if it is black, the drawn ball is replaced and an additional black ball is placed in the box. Again, a ball is drawn at random. If the ball drawn is white, the procedure stops; if it is black, the ball is replaced, an additional black ball is placed in the box, and another drawing is made; and this continues until the white ball is finally drawn. The possible outcomes of the experiment are taken to be the number of drawings required to obtain a white ball. The possible outcomes are, then, the integers 1, 2, 3, 4, ... Reasonable probabilities to assign to these outcomes are $1/2$, $(1/2)(1/3)$, $(1/2)(2/3)(1/4)$, $(1/2)(2/3)(3/4)(1/5)$, . . . , which are the same as $1/2 = 1 - 1/2$, $(1/2)(1/3) = 1/2 - 1/3$, $(1/3)(1/4) = 1/3 - 1/4$, and so forth. (If you are not convinced that these probability assignments are reasonable, take them for granted until you have studied Chapter 4. In Exercise 7(b), Chapter 4, we ask you to return to this problem.) Notice that the sum of probabilities of the sample points is

$$\left(1 - \frac{1}{2}\right) + \left(\frac{1}{2} - \frac{1}{3}\right) + \left(\frac{1}{3} - \frac{1}{4}\right) + \cdots = 1$$

The expected number of drawings to obtain the white ball is

$$1\left(\frac{1}{2}\right) + 2\left(\frac{1}{2}\right)\left(\frac{1}{3}\right) + 3\left(\frac{1}{3}\right)\left(\frac{1}{4}\right) + \cdots = \frac{1}{2} + \frac{1}{3} + \frac{1}{4} + \cdots = \infty$$

(Note 5.) The expected value does not exist, but it makes sense to say that on the average it requires an infinite number of trials to draw the white ball.

2.5 LINEARITY OF EXPECTED VALUE

According to the frequency theory of probability, EX is an idealization of the average of X values that are actually observed over a large number of repetitions of the chance experiment. Suppose now that there are two numbers associated with the outcomes—an X number and a Y number. Consider now the random variable $X + Y$, which assigns to the outcome w the value $X(w) + Y(w)$. The average of the values of $X + Y$ over all observations is just the sum of the average of the X values and the average of the Y values. The comparable fact for the theoretical averages, the expected values, is that the expected value of a sum is the sum of the expected values.

Theorem

Suppose that X and Y are random variables the expected values of which exist, and that a and b are constants. Then the expected value of $aX + bY$ exists and

$$E(aX + bY) = aEX + bEY \tag{1}$$

Proof. That EX and EY exist means that $\Sigma_i\,|X(w_i)|p_i$ and $\Sigma_i\,|Y(w_i)|p_i$ are each finite. By the triangle inequality, $|aX(w_i) + bY(w_i)| \leq |a||X(w_i)| + |b||Y(w_i)|$. Hence, $\Sigma_i\,|aX(w_i) + bY(w_i)|p_i$ is finite, and thus $E(aX + bY)$ exists. Inasmuch as $\Sigma_i\,[aX(w_i) + bY(w_i)]p_i = a\,\Sigma_i\,X(w_i)p_i + b\,\Sigma_i\,Y(w_i)p_i$ for absolutely convergent series, the proof is complete (Note 6). ▮

If there are only a finite number of sample points, then the expected values always exist and the preceding proof is particularly simple.

By a similar argument, if $X_1, \ldots, X_n$ are random variables the expected values of which exist, and if $a_1, \ldots, a_n$ are constants, then the expected value of $a_1X_1 + \cdots + a_nX_n$ exists and

$$E(a_1X_1 + \cdots + a_nX_n) = a_1EX_1 + \cdots + a_nEX_n \tag{2}$$

Example 1

A die is thrown twice. Suppose that each of the 36 sample points has probability 1/36. Let X be the number on the first throw and Y the number on the second throw. Since $EX = EY = 7/2$, $E(X + Y) = 7$. On the other hand, $E(X - Y) = 0$.

Example 2

An honest coin is thrown n times. There are 2^n sample points and each has probability $1/2^n$. (This is what is meant by the coin being "honest.") Define $X_1, \ldots, X_n$ by

$$X_i = \begin{cases} 1 \text{ if the } i\text{th toss is head} \\ 0 \text{ if the } i\text{th toss is tail} \end{cases} \qquad i = 1, \ldots, n$$

Then $Y = X_1 + \cdots + X_n$ is the total number of heads. For instance, if $n = 5$ and the sample point is $(H, H, T, T, H) = w$, then $X_1(w) = X_2(w) = X_5(w) = 1$, $X_3(w) = X_4(w) = 0$, and $Y(w) = 3$. Because 2^{n-1} sample points have H in the ith coordinate, it follows that $EX_i = 2^{n-1}/2^n = 1/2$. Hence, the expected number of heads in n throws is $EY = EX_1 + \cdots + EX_n = n/2$.

The following is an important special case of (2). Consider the random variable $aX + b$ (a and b are constants), which is the random variable assigning to a sample point w the value $aX(w) + b$. Then

$$E(aX + b) = aEX + b \tag{3}$$

assuming EX exists (see Exercise 26a).

2.6 INDICATOR RANDOM VARIABLES

The simplest of all random variables, X = constant, is usually too prosaic to be of any interest. A random variable having two possible values is considerably more interesting and useful. Let A be an event. Define a random variable X_A by

$$X_A(w) = \begin{cases} 1 \text{ if } w \text{ is in } A \\ 0 \text{ if } w \text{ is not in } A \end{cases} \tag{1}$$

The term X_A is called the *indicator random variable* of the set A. In a nonprobability setting, X_A is called the indicator function of the set A, or the set characteristic function of A. Since $EX_A = 1[P(A)] + 0[P(A')]$, it follows that

$$EX_A = P(A) \tag{2}$$

indicator random variable

A random variable whose only values are 0 and 1 is called an indicator random variable.

An indicator random variable X is always the indicator random variable of some event, namely the set of sample points w for which $X(w) = 1$. Thus, there is a one-to-one correspondence between events and indicator random variables. In particular, the random variable that is identically equal to 1 is $X_{\mathfrak{X}}$, the indicator random variable of the whole sample space, and it is usually denoted simply by 1. The random variable that is identically equal to 0 is $X_{\varnothing}$, the indicator random variable of the impossible event $\varnothing$, and it is usually denoted by 0.

Indicator random variables are often very useful in evaluating expected values when a direct evaluation might otherwise be rather difficult. Example 2, Section 2.5, provides one such instance. There follow several additional examples of the same type.

Example 1 Five cards marked 1, 2, 3, 4, 5 are shuffled and laid out against five fixed positions, also marked 1, 2, 3, 4, 5. Let X equal the number of matches—that is, the number of times a card marked i ends up against position i. Let the sample points be the $5! = 120$ permutations of the numbers 1, 2, 3, 4 and 5, each sample point having probability 1/120. Define the following indicator random variables.

$$X_i = \begin{cases} 1 \text{ if card } i \text{ is in position } i \\ 0 \text{ otherwise} \end{cases} \qquad i = 1, 2, 3, 4, 5$$

For example, if the sample point is $(5, 2, 3, 4, 1) = w$, then $X_1(w) = 0, X_2(w) = 1, X_3(w) = 1, X_4(w) = 1, X_5(w) = 0$, and $X(w) = 3$. The random variable of interest, X, is related to the X_i's by

$$X = X_1 + X_2 + X_3 + X_4 + X_5$$

Since there are $4! = 24$ sample points for which card i is in the ith position, it follows that $EX_i = P(X_i = 1) = 24/120 = 1/5$, and $EX = EX_1 + EX_2 + EX_3 + EX_4 + EX_5 = 5/5 = 1$. Actually the expected number of matches is 1 when the deck consists of any number of cards. (See Exercise 7a for a general version of this example.)

Example 2 Suppose n balls are randomly distributed among three cells. (Compare Exercises 10 and 11, Chapter 1.) Let X be the number of empty cells. Define

$$X_i = \begin{cases} 1 \text{ if the } i\text{th cell is empty} \\ 0 \text{ otherwise} \end{cases} \qquad i = 1, 2, 3$$

Then $X = X_1 + X_2 + X_3$ and $EX = EX_1 + EX_2 + EX_3$. Because $EX_i = P(X_i = 1) = 2^n/3^n$ (Exercise 11a, Chapter 1), then $EX = 2^n/3^{n-1}$ (see Exercise 5 for a general version of this example). This is an example of the occupancy problem, discussed in Sections 3.3 and 3.4.

2.7 INDICATOR RANDOM VARIABLES AND SET OPERATIONS

There are some useful relations between set operations with events and algebraic operations with the corresponding indicator random variables. Some basic relations are the following:

$$X_{A'} = 1 - X_A \tag{1}$$

$$X_{A_1 \cap A_2 \cap \cdots \cap A_n} = X_{A_1} \cdot X_{A_2} \cdots X_{A_n} \tag{2}$$

$$X_{A_1 \cup A_2 \cup \cdots \cup A_n} = 1 - (1 - X_{A_1})(1 - X_{A_2}) \cdots (1 - X_{A_n}) \tag{3}$$

Proof of (1). $1 - X_A(w) = 1$, if and only if w fails to be in A. ▮

Proof of (2). $X_{A_1 \cap A_2 \cap \cdots \cap A_n}(w) = 1$, if and only if w is in each of A_1, $A_2, \ldots, A_n$. ▮

Proof of (3). $X_{A_1 \cup A_2 \cup \cdots \cup A_n}(w) = 1$, if and only if w is in at least one of the events A_1 or A_2 or $\cdots$ or A_n. In particular, if w is in A_i, then $1 - X_{A_i}(w) = 0$ and the right side of (3) equals 1. If w fails to be in each of $A_1, A_2, \ldots, A_n$, then the right side of (3) is 0. ▮

Relations (1), (2), and (3) are useful in developing set relations among events. To show that two events, A and B, are the same, it is sufficient to show that $X_A = X_B$—that is, $X_A(w) = X_B(w)$ for all sample points w. For example, (1) and (3) together imply that

$$X_{(A_1 \cup A_2 \cup \cdots \cup A_n)'} = (1 - X_{A_1})(1 - X_{A_2}) \cdots (1 - X_{A_n})$$

Hence,

$$(A_1 \cup A_2 \cdots A_n)' = A_1' \cap A_2' \cap \cdots \cap A_n'$$

Example 1 Consider the distributive relationship

$$A \cup (B \cap C) = (A \cup B) \cap (A \cup C)$$

which is proved as follows:

$$X_{A \cup (B \cap C)} = X_A + X_B X_C - X_A X_B X_C \quad \text{by (2) and (3)}$$

$$\begin{aligned} X_{(A \cup B) \cap (A \cup C)} &= X_{A \cup B} X_{A \cup C} \quad \text{by (2)} \\ &= (X_A + X_B - X_A X_B)(X_A + X_C - X_A X_C) \quad \text{by (3)} \\ &= X_A + X_B X_C - X_A X_B X_C \quad \text{by using the fact that } X_A X_A = X_A \end{aligned}$$

(See Exercise 6a for an additional example.) Notice that the countability of the sample space has nothing to do with these methods. They make sense for arbitrary sets in an arbitrary space $\mathfrak{X}$.

2.8 THE INCLUSION-EXCLUSION FORMULA

Consider the relationship

$$P(A \cup B) = P(A) + P(B) - P(A \cap B) \tag{1}$$

which was first presented in Section 1.4. A proof of (1) is outlined in Figure 1.1. Here is another proof.

$$\begin{aligned} X_{A\cup B} &= 1 - (1 - X_A)(1 - X_B) \\ &= X_A + X_B - X_A X_B \\ &= X_A + X_B - X_{A\cap B} \end{aligned}$$

Hence, by Eq. (2) in Section 2.5, $EX_{A\cup B} = EX_A + EX_B - EX_{A\cap B}$, which is equivalent to (1). This method can be generalized to n events. Let $A_1, \ldots, A_n$ be events and introduce the following notation.

$$S_1 = \sum_i P(A_i)$$

$$S_2 = \sum_{i<j} P(A_i \cap A_j)$$

$$\vdots$$

$$S_k = \sum_{i_1<i_2<\cdots<i_k} P(A_{i_1} \cap A_{i_2} \cap \cdots \cap A_{i_k}), \qquad k = 1, \ldots, n$$

In other words, S_k is the sum of the probabilities of intersections of k of the events; it is summed out over all $\binom{n}{k}$ choices of k sets out of the available $A_1, \ldots, A_n$. The restriction $i_1 < \cdots < i_k$ means that each k-tuple appears just once. The generalization of Eq. (1) is Eq. 2:

inclusion-exclusion formula

$$P(A_1 \cup A_2 \cup \cdots \cup A_n) = S_1 - S_2 + \cdots + (-1)^{n+1} S_n \tag{2}$$

Proof of (2). Use (3), Section 2.7, and make the following expansion.

$$X_{A_1 \cup A_2 \cup \cdots \cup A_n} = 1 - (1 - X_{A_1})(1 - X_{A_2}) \cdots (1 - X_{A_n})$$
$$= \sum_{1 \leq i \leq n} X_{A_i} - \sum_{1 \leq i < j \leq n} X_{A_i} X_{A_j} + \cdots$$
$$+ (-1)^{n+1} X_{A_1} X_{A_2} \cdots X_{A_n}$$

Now take expected values of both sides to obtain Eq. (2) (Note 7). ∎

Example 1 For $n = 3$, the inclusion-exclusion formula becomes

$$P(A \cup B \cup C) = P(A) + P(B) + P(C)$$
$$- P(A \cap B) - P(A \cap C) - P(B \cap C)$$
$$+ P(A \cap B \cap C)$$
$$= S_1 - S_2 + S_3$$

Figure 2.3 suggests why the name "inclusion-exclusion" formula is appropriate.

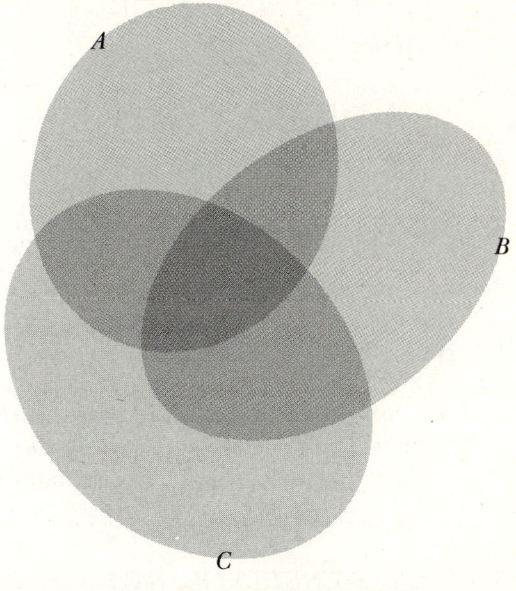

FIGURE 2.3 ***P(A ∪ B ∪ C)* is overestimated by S_1, because the double-shaded and triple-shaded regions are included too many times. The next estimate, $S_1 - S_2$, excludes the triple-shaded region altogether. Finally, $S_1 - S_2 + S_3$ gives the correct answer.**

Example 2 Ten balls are randomly distributed among four boxes. *What is the probability of at least one empty box?* Suppose that the sample space consists of the 4^{10} possible descriptions of assignments of the 10 balls, each sample point with probability $1/4^{10}$. (Compare with Exercises 10 and 11, Chapter 1.) Define the following events:

$A =$ event that box 1 is empty
$B =$ event that box 2 is empty
$C =$ event that box 3 is empty
$D =$ event that box 4 is empty

To compute $P(A \cup B \cup C \cup D)$, one needs the following preliminary computations.

$$P(A) = P(B) = P(C) = P(D) = (\tfrac{3}{4})^{10}$$

$$P(A \cap B) = P(A \cap C) = P(B \cap C) = \cdots = (\tfrac{2}{4})^{10}$$

$$P(A \cap B \cap C) = P(A \cap B \cap D) = \cdots = (\tfrac{1}{4})^{10}$$

$$P(A \cap B \cap C \cap D) = 0$$

Hence,

$$S_1 = \binom{4}{1}(\tfrac{3}{4})^{10}, \qquad S_2 = \binom{4}{2}(\tfrac{2}{4})^{10}, \qquad S_3 = \binom{4}{3}(\tfrac{1}{4})^{10}, \qquad S_4 = 0$$

and

$$P(A \cup B \cup C \cup D) = \frac{4(3)^{10} - 6(2)^{10} + 4}{4^{10}}$$

$$= \frac{230{,}056}{1{,}048{,}576} \approx 0.22$$

The following is another interpretation of this example. A maze consists of four possible paths, each of which ends in a separate cage. Ten mice are successively run through the maze. The probability that at least one cage remains empty is 0.22 . . . .

2.9 DENSITY FUNCTION OF A DISCRETE RANDOM VARIABLE

As pointed out in Section 2.2, a random variable defined over a discrete sample space can have at most a countable number of possible values.

density function

The function

$f_X(a) = P(X = a)$

where a varies over the possible values of X, is called the *density function of* X. Because f_X is defined over a countable set of real numbers, it is called a *discrete density function.* This name distinguishes it from other types that will be introduced later.

A useful way of describing a discrete density function is by means of a graph, as in Figure 2.4, in which $f_X(a)$ represents a "mass" assigned to the point a. The sum of the masses in a discrete density function equals 1. That is,

$f_X(a_1) + f_X(a_2) + \cdots = 1$

where $a_1, a_2, \ldots$ are the possible values of X. This is so because $(X = a_1), (X = a_2), \ldots$ are disjoint events whose union is $\mathfrak{X}$.

Example 1 An honest coin is thrown five times. A sample point is a sequence of five symbols each of which is H or T. There are $2^5 = 32$ sample points and suppose that each has probability 1/32. Let X be the number of heads. Thus, if $w = (H, H, T, T, H)$, then $X(w) = 3$. The event $(X = i)$ consists of $\binom{5}{i}$ sample points, and $P(X = i) = \binom{5}{i}\Big/32$,

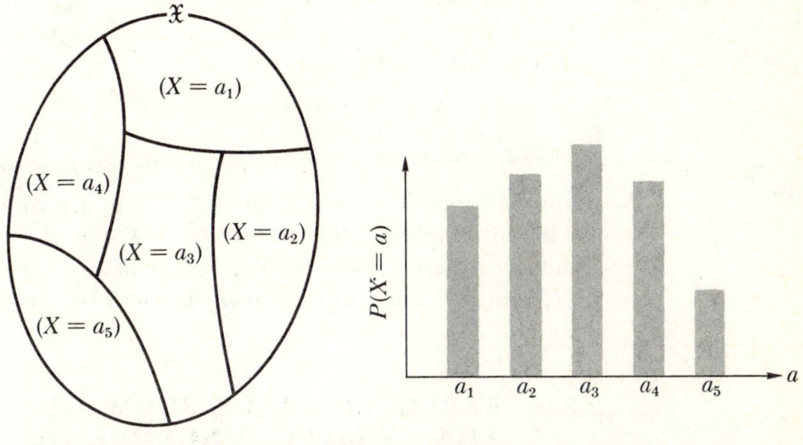

FIGURE 2.4 The random variable X partitions the sample space into mutually disjoint events $(X = a_1)$, $(X = a_2)$, ... The density function summarizes the amount of probability in each of these events.

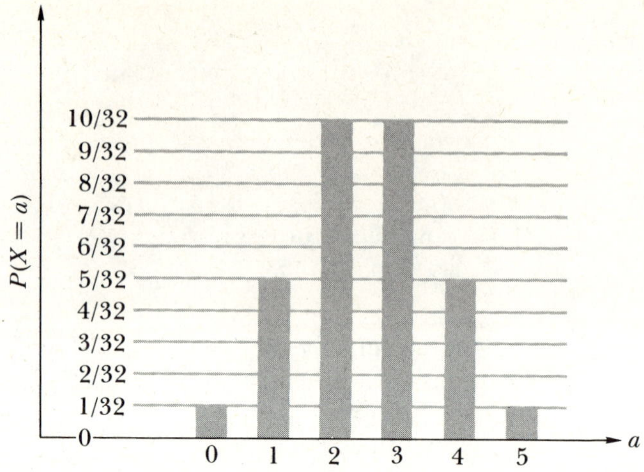

FIGURE 2.5 Density function of the number of heads in five throws of a coin.

$i = 0, 1, 2, 3, 4, 5$. The density function of X is shown in Figure 2.5. This density is a special case of the *binomial distribution,* which will be discussed in detail in Section 6.1.

The *density function* is often called by other names, such as *frequency function, frequency distribution,* and *distribution function.* There are various equivalent ways of describing the information inherent in the density function. Suppose that the possible values of X can be arranged in order as $a_1 < a_2 < a_3 < \cdots$. Then, if we know the probabilities $P(X > a_1)$, $P(X > a_2)$, . . . , it is equivalent to knowing the density function, since

$$P(X = a_1) = 1 - P(X > a_1)$$

$$P(X = a_2) = P(X > a_1) - P(X > a_2)$$

and so forth. The probabilities $P(X > a)$ are called *tail probabilities.* The probability information about X, whether described by the density function, tail probabilities, or in any other equivalent form, is generally called the *probability law of X* or the *distribution of X.*

2.10 EXPECTED VALUE FROM THE DENSITY FUNCTION

The definition of expected value of a random variable was given in Section 2.4. We now want to show that there is another way of computing expected value.

theorem for computing expected value from the density function

Let $a_1, a_2, \ldots$ be the possible values of a random variable X. Then EX exists if and only if $|a_1|P(X=a_1)+|a_2|P(X=a_2)+\cdots$ is finite, and its value is

$$EX = a_1P(X = a_1) + a_2P(X = a_2) + \cdots \quad (1)$$

Proof. By definition (Section 2.4), EX exists if $|X(w_1)|p_1+|X(w_2)|p_2+\cdots$ is finite and its value is

$$EX = X(w_1)p_1 + X(w_2)p_2 + \cdots \quad (2)$$

As pointed out in Section 2.4, the absolute convergence of (2) guarantees that any rearrangement of this series will converge to the same value. In particular, arrange the summation as follows. Group together all sample points w, such that $X(w) = a_i$. The contribution to the sum from these terms is $a_iP(X = a_i)$. Now sum over all possible values $a_1, a_2, \ldots$ to obtain (1). Conversely, if (1) converges absolutely, a particular arrangement of (2) converges absolutely; hence, any arrangement must converge, and, in particular, (2) itself converges to the same value as (1). ∎

If $EX = \infty$ in a well-defined way, then the right side of (1) also equals infinity. (See the discussion about infinite expected values in Section 2.4.) Of course, for random variables with a finite number of possible values, EX always exists.

Example 1

An honest coin is thrown five times. Let X be the number of heads obtained. As pointed out in Example 1, Section 2.9, $P(X = i) = \binom{5}{i}\Big/32$, $i = 0, 1, 2, 3, 4, 5$. Hence, $EX = [0(1) + 1(5) + 2(10) + 3(10) + 4(5) + 5(1)]/32 = 5/2$. Recall that the same expected value had already been obtained by a different method in Example 2, Section 2.5.

Example 2

Suppose n balls are randomly distributed among three cells. Let X be the number of empty cells. The density function of X is given as follows:

$$P(X = 0) = 1 - \frac{2^n - 1}{3^{n-1}}, \qquad P(X = 1) = \frac{2^n - 2}{3^{n-1}},$$

$$P(X = 2) = \frac{1}{3^{n-1}}$$

(Notice that these probabilities sum to 1. These computations appear in Exercise 11, Chapter 1.) Hence,

$$EX = \frac{1(2^n - 2) + 2(1)}{3^{n-1}} = \frac{2^n}{3^{n-1}}$$

Recall that this same expected value was obtained by a different method in Example 2, Section 2.6.

If the possible values of X are 0, 1, 2, . . . , then there is a useful way of computing EX in terms of tail probabilities. The result follows.

$$EX = P(X > 0) + P(X > 1) + P(X > 2) + \cdots \tag{3}$$

Proof of (3). The right side of (3) can be written as

$$\begin{aligned} P(X=1) + P(X=2) + P(X=3) + \cdots \\ + P(X=2) + P(X=3) + \cdots \\ + P(X=3) + \cdots \\ \vdots \end{aligned} \tag{4}$$

Sum (4) by adding first down the columns of the display, obtaining

$$P(X=1) + 2P(X=2) + 3P(X=3) + \cdots = \sum_{i=0}^{\infty} iP(X=i) = EX$$

by (1). If EX fails to exist, then both sides of (3) equal ∞. ∎

Example 3 An honest coin is thrown until a head is obtained. Let X equal the number of throws required. Because $P(X = k) = 1/2^k$, it follows that

$$\begin{aligned} P(X > k) &= \tfrac{1}{2}^{k+1} + \tfrac{1}{2}^{k+2} + \cdots \\ &= \tfrac{1}{2}^{k+1}(1 + \tfrac{1}{2} + \tfrac{1}{2}^2 + \cdots) \\ &= \tfrac{1}{2}^k, \qquad k = 1, 2, \ldots \end{aligned}$$

By (3), then,

$$EX = 1 + \tfrac{1}{2} + \tfrac{1}{2}^2 + \cdots = 2$$

This expected value was evaluated previously in a different way in Example 4, Section 2.4. (See Exercise 9 for a related problem.)

Equation (1) can be extended somewhat. Suppose that h is a real-valued function such that $h(X)$ is well defined. By $h(X)$ is meant the random variable that assigns to a sample point w the value $h[X(w)]$. The extension of (1) is the following:

$$E[h(X)] = h(a_1)P(X = a_1) + h(a_2)P(X = a_2) + \cdots \quad (5)$$

assuming the expected value exists.

The proof of (5) is almost the same as that of (1) and is left to the reader.

There are now available three distinct methods for computing $E[h(X)]$:

Method I. Summing over the sample space, from the original definition, Section 2.4.

$$E[h(X)] = h[X(w_1)]p_1 + h[X(w_2)]p_2 + \cdots$$

Method II. Using the density function of $h(X)$, Eq. (1), Section 2.10.

$$E[h(X)] = b_1P[h(X) = b_1] + b_2P[h(X) = b_2] + \cdots$$

[The possible values of $h(X)$ are $b_1, b_2, \ldots$.]

Method III. Using the density function of X, Eq. (5), Section 2.10.

$$E[h(X)] = h(a_1)P(X = a_1) + h(a_2)P(X = a_2) + \cdots$$

Example 4 Suppose that X has the following density function:

a	-2	-1	0	1	2
$P(X = a)$	1/16	4/16	6/16	4/16	1/16

The expected value of X^2 computed from this density function (Method III) is

$$\frac{(-2)^2 1}{16} + \frac{(-1)^2 4}{16} + \frac{(0)^2 6}{16} + \frac{(1)^2 4}{16} + \frac{(2)^2 1}{16} = 1$$

The density function of X^2 is the following:

b	0	1	4
$P(X^2 = b)$	3/8	4/8	1/8

The expected value of X computed from this density function (Method II) is

$$\frac{(0)3}{8} + \frac{(1)4}{8} + \frac{(4)1}{8} = 1$$

2.11 EXPECTED VALUE AS A CENTER OF GRAVITY

The probabilities $P(X = a_1)$, $P(X = a_2)$, . . . can be interpreted as a *mass distribution* on the points a_1, a_2, . . . on the real line, these masses summing to 1. An elementary physics formula tells how to compute the center of gravity of a number of masses dispersed on the line. Suppose that at points a_1, a_2, . . . on the line there are masses m_1, m_2, . . . , respectively. The center of gravity of this mass disposition is

$$\frac{a_1m_1 + a_2m_2 + \cdots}{m_1 + m_2 + \cdots}$$

In particular, when $m_i = P(X = a_i)$, then $m_1 + m_2 + \cdots = 1$, and the center of gravity is

$$a_1P(X = a_1) + a_2P(X = a_2) + \cdots$$

which equals EX by Eq. (1), Section 2.10 (see Figure 2.6).

A random variable X is said to be *symmetrically distributed* about the point c, if whenever $c + a$ is a possible value of X, then $c - a$ is also a possible value, and

$$P(X = c + a) = P(X = c - a)$$

It is the same as saying that

$$P(X - c = a) = P(c - X = a)$$

In other words, "X is symmetrically distributed about c" means that $X - c$ and $c - X$ have the same distribution. In particular, symmetry about 0 means that X and $-X$ have the same distribution.

Example 1 Suppose $P(X = i) = 1/n$, $i = 1, \ldots, n$ (see Figure 2.7). Then X is symmetrically distributed about the point $(n + 1)/2$, which is the midpoint of the two extreme possible values 1 and n. (See Exercise

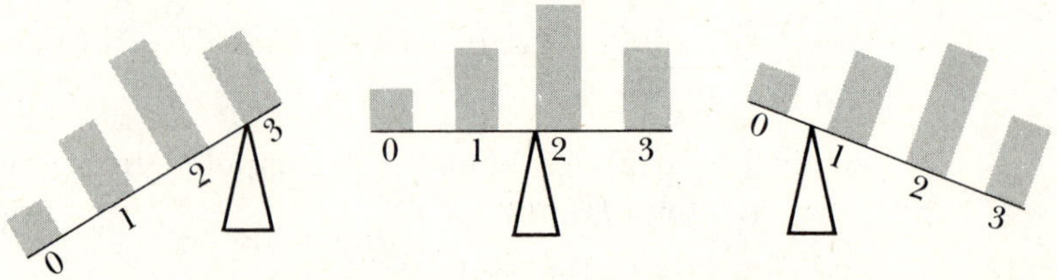

FIGURE 2.6 The graph of the density function will balance if a fulcrum is placed under the axis at the center of gravity.

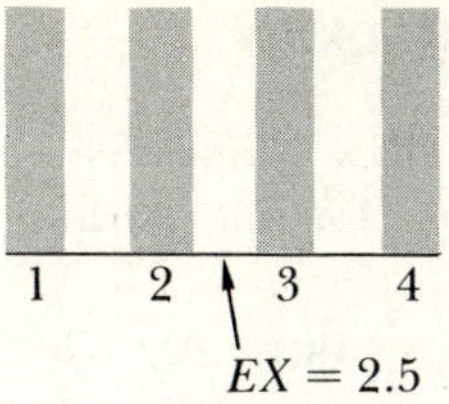

FIGURE 2.7 If X equals each of the values 1, 2, 3, 4 with probability 1/4 each, then X is symmetrically distributed about 2.5 = (1 + 4)/2.

20b for a generalization of this example.) According to (1), which follows, $EX = (n + 1)/2$.

point of symmetry equals expected value

> If X is symmetrically distributed about c and EX exists, then
>
> $$EX = c \qquad (1)$$

Proof. Inasmuch as $X - c$ and $c - X$ have the same density function, it follows from (1), Section 2.10, that

$$E(X - c) = E(c - X)$$

By (3), Section 2.5, it follows that $EX - c = c - EX$; hence, $EX = c$. ▌

The following example shows that the idea of symmetry can simplify the computation of expected value.

Example 2

Three numbers are selected without replacement from the 10 numbers 1, 2, . . . , 10. Let X be the middle valued number when the three numbers are arranged in order of magnitude. Thus,

$$X(9, 1, 5) = X(1, 5, 9) = X(5, 1, 9) = 5$$

It is easy to verify that the distribution of X is the following:

k	2	3	4	5	6	7	8	9
$P(X=k)$	4/60	7/60	9/60	10/60	10/60	9/60	7/60	4/60

[For instance, consider the computation of $P(X = 4)$. If we ignore order, there are $\binom{10}{3} = 120$ sample points. All triples of the form $i < 4 < j$ equal $(X = 4)$. There are three choices for i and six choices for j; therefore, there are $3 \cdot 6 = 18$ sample points in $(X = 4)$, and $P(X = 4) = 18/120 = 9/60$.] Since X is symmetrically distributed about the midpoint of 2 and 9, it follows without further computation that $EX = 5.5$.

SUMMARY

A *random variable* is a real-valued function defined on the sample space. If X is a random variable, the expected value of X is defined as

$$EX = \sum_i X(w_i)p_i$$

[The number assigned by X to the sample point w_i is $X(w_i)$, and p_i is the probability of that sample point.] The summation is over all sample points. If there are a countable infinity of sample points, the sum is required to be absolutely convergent—that is, $E|X| < \infty$—otherwise the expected value is said not to exist. In some cases, even if the expected value fails to exist, it makes sense to say that $EX = \infty$ or $EX = -\infty$.

If EX and EY exist, then $E(aX + bY) = aEX + bEY$.

The function $f_X(a) = P(X = a)$, where a varies over the possible values of X, is called the *density function* of X. We can compute EX from the density function by the relation

$$EX = a_1 f_X(a_1) + a_2 f_X(a_2) + \cdots$$

EX exists if and only if this sum is absolutely convergent. If $h(X)$ is a well-defined function of a random variable, then $E[h(X)]$ also can be computed from the density function of X by the relation

$$E[h(X)] = h(a_1)f_X(a_1) + h(a_2)f_X(a_2) + \cdots$$

Again, this expected value exists if and only if this sum is absolutely convergent.

Indicator random variables assume only two possible values, 0 and 1. Such random variables often are useful in evaluating expected values of more complicated random variables. They are also useful in proving combinatorial relationships, such as the inclusion-exclusion formula,

$$P(A_1 \cup A_2 \cup \cdots \cup A_n) = \sum_i P(A_i) - \sum_{i<j} P(A_i \cap A_j) + \cdots$$

EXERCISES

1 (a) Six cards are numbered 1, 2, 3, 4, 5, 6. A card is drawn at random. Show that the expected value of the number drawn is 3.5.

(b) An honest six-faced die is thrown. Show that the expected number of points to appear is 3.5.

(c) In (a), suppose that a card is drawn at random, is replaced, and a second card is drawn at random. Show that the expected value of the sum of the two numbers drawn is 7.

(d) An honest six-faced die is thrown twice. Show that the expected value of the total number of points to appear is 7.

(e) In (a), suppose that a card is drawn at random, is not replaced, and a second card is then drawn. Show that the expected value of the sum of the two numbers drawn is still 7. [Parts (a) and (b) are really identical and so are (c) and (d). The numbers of sample points for (a), (c), and (e) are 6, 36, and 30, respectively. The most direct way of computing the required expected values is simply to enumerate the sample points and to apply the definition of expected value in Section 2.4. However, a more sophisticated method will be needed for Exercise 2.]

2 Suppose n cards are numbered $1, 2, \ldots, n$. A card is drawn at random and is replaced, a second card is drawn, and so forth until k cards are drawn. Let X be the sum of the k numbers drawn. Show that $EX = k(n+1)/2$. [Hint: There are n^k sample points, each with the same probability, $1/n^k$. Let X_i be the number drawn on the ith drawing. The events $(X_i = 1), (X_i = 2), \ldots, (X_i = n)$ each have n^{k-1} sample points. Hence, $EX_i = 1n^{k-1}/n^k + \cdots + nn^{k-1}/n^k = (1+2+\cdots+n)/n = n(n+1)/2n = (n+1)/2$. $X = X_1 + \cdots + X_k$. $EX = EX_1 + \cdots + EX_k$ by (2), Section 2.5.]

3 Suppose k six-faced dice are thrown. Let X be the total number of points to appear. Show that $EX = 7k/2$.

4 (a) Show that if $E(X^2) = 0$, then $P(X = 0) = 1$.

(b) Suppose that f is a function such that $f(u) > 0$ if $u \neq 0$ and $f(0) = 0$. Show that if $E[f(X - a)] = 0$, then $P(X = a) = 1$. [Notice that (a) is a special case of (b).]

5 (a) Suppose n balls are randomly distributed among m cells. Show that the expected number of empty cells is $(m-1)^n/m^{n-1}$. (Hint: Use Example 2, Section 2.6 as a guide.)

(b) Show that the expected number of cells, which have two or more balls, is $[m^n - (m-1)^n - n(m-1)^{n-1}]/m^{n-1}$. [Hint: There are $(m-1)^n$ sample points for which cell i is empty. There are $n(m-1)^{n-1}$ sample points for which cell i has exactly 1 ball.]

(c) There are m telephone circuits available. Each of n users selects a circuit at random. (It is possible for several users to select the same circuit.) What is the expected number of circuits that are not used? [This problem is identical with (a).]

6 (a) Prove the set theoretic relation $A \cap (B \cup C) = (A \cap B) \cup (A \cap C)$ using indicator random variables. (Hint: Follow the method in Example 1, Section 2.7.)

(b) Show that $X_A \leqslant X_B$ if and only if $A \subset B$.

7 Suppose n cards marked $1, 2, \ldots, n$ are shuffled and laid out against n fixed positions, also marked $1, 2, \ldots, n$. Let X equal the number of matches—that is, the number of times a card marked i is in position i.

(a) Show that $EX = 1$ for any n. (Hint: Imitate the method using indicator random variables in Example 1, Section 2.6.)

(b) Let A_i be the event that card i is in position i, $i = 1, \ldots, n$. Let $S_1, S_2, \ldots$ be defined as in Section 2.8. Show that $S_1 = 1$, $S_2 = 1/2$, $S_3 = 1/3!, \ldots$. [Hint: $P(A_i) = 1/n$, $P(A_i \cap A_j) = (n-2)!/n! = 1/n(n-1)$ if $i \neq j$; $P(A_i \cap A_j \cap A_k) = 1/n(n-1)(n-2)$ if i, j, k are distinct; and so forth.]

(c) Use the inclusion-exclusion formula (Section 2.8) to show that the probability of no matches (which equals 1 − probability of at least one match) equals

$$\sum_{n=0}^{n} \frac{(-1)^n}{n!} = 1 - 1 + \frac{1}{2} - \frac{1}{6} + \cdots + \frac{(-1)^n}{n!}$$

8 Suppose n balls are randomly distributed among m boxes. Let A_i be the event that the ith box is empty, $i = 1, \ldots, m$.

(a) Show that

$$P(A_i) = \frac{(m-1)^n}{m^n}$$

$$P(A_i \cap A_j) = \frac{(m-2)^n}{m^n} \qquad (i \neq j) \text{ etc. and hence,}$$

$$S_k = \binom{m}{k}\frac{(m-k)^n}{m^n}, \qquad k = 1, \ldots, m$$

the S_k's being defined in Section 2.8.

(b) Use the inclusion-exclusion formula (Section 2.8) to show that the probability of no empty boxes is

$$\sum_{k=0}^{m} (-1)^k \binom{m}{k}\frac{(m-k)^n}{m^n}$$

(c) Deduce the following combinatorial formula from (b).

$$\sum_{k=0}^{m} (-1)^k \binom{m}{k}(m-k)^n = \begin{cases} m! & \text{if } n = m \\ 0 & \text{if } n < m \end{cases}$$

(Hint: If m balls are distributed among m boxes, then the probability of no empty boxes is $m!/m^m$. If fewer than m balls are distributed among m boxes, then the probability of no empty boxes is 0.)

9 An honest die is thrown until a 1 appears. Let X equal the number of throws required. Show that $EX = 6$. [Hint: *Method 1*. Imitate Example 4, Section 2.4. *Method 2*. Show that $P(X > k) = (5/6)^k$ and use Eq. (3), Section 2.10.]

10 (a) A drunk is searching for his house key among six loose keys in his pocket. (One of these keys is his house key, the other five are not.) Suppose he draws a key from his pocket, tries it, and if it is

not right returns it to the same pocket and tries again; he continues until he finds the right key. Show that the expected number of tries is 6 (compare with Exercise 9).

(b) Suppose that the drunk is sufficiently sober so that when he tries a wrong key he transfers it to another pocket and does not try it again. Show that the expected number of tries is now 7/2.

11 Consider the matching problem described in Exercise 7. Suppose that $n = 4$. In this case, there are 24 sample points.

(a) Show by direct enumeration of these sample points that the density function of X is the following:

k	0	1	2	3	4
$P(X = k)$	9/24	8/24	6/24	0	1/24

(b) Use the density function in (a) to show that $EX = 1$. (Compare this result with the general result in Exercise 7a.)

12 A deck of four cards numbered 1, 2, 3, 4, is shuffled and the cards laid out in a line. Suppose that the $4! = 24$ possible permutations each have the same probability, 1/24. Let the random variable X count the number of times an element of the permutation exceeds all of its preceding elements. Thus, if the permutation is (2, 1, 4, 3), then $X(2, 1, 4, 3) = 2$. [The convention is made that the first element exceeds all of its preceding elements because it has no preceding elements. The third element, 4, precedes everything before it. Hence, $X(2, 1, 4, 3) = 2$. Also, $X(1, 2, 4, 3) = 3$ and $X(1, 2, 3, 4) = 4$.]

(a) Show that the density function of X is given by

k	1	2	3	4
$P(X = k)$	6/24	11/24	6/24	1/24

(b) Notice that when EX equals 25/12, it can also be written as

$$EX = 1 + \tfrac{1}{2} + \tfrac{1}{3} + \tfrac{1}{4}$$

13 Suppose the same situation as in Exercise 12 but now there are five cards. Define indicator random variables.

$$X_i = \begin{cases} 1 & \text{if } i\text{th element of permutation exceeds all preceding elements} \\ 0 & \text{otherwise} \end{cases}$$

(X_1 is automatically 1 by convention.)

(a) Show that $P(X_i = 1) = 1/i$ ($i = 1, 2, 3, 4, 5$). [Hint: Consider those sample points the first i entries of which are the particular integers $b_1, \ldots, b_i$ in some order. For $1/i$th of these sample points the largest of these i integers is in the ith position. (Why?) Hence, for $1/i$th of all sample points, the largest of the first i entries appears in the ith position.]

(b) Conclude from (a) that $EX = 1 + 1/2 + 1/3 + 1/4 + 1/5$, where X is the total number of exceedances.

14 Suppose that a random variable X has the possible values 0, 1, 2, and $P(X=2)=p$, $EX=m$.
(a) Show that the density of X is given by

k	0	1	2
$P(X=k)$	$1-m+p$	$m-2p$	p

(b) Conclude from (a) that $2p \leq m \leq 1+p$.

15 Suppose n balls are randomly distributed among three boxes. Let X be the number of empty boxes. Show that the density function of X is given by

k	0	1	2
$P(X=k)$	$1-(2^n-1)/3^{n-1}$	$(2^n-2)/3^{n-1}$	$1/3^{n-1}$

[Hint: There are many ways to do this computation. One rather indirect method is to use Exercise 14 and the facts that $P(X=2) = p = 3/3^n = 1/3^{n-1}$, $EX = m = 2^n/3^{n-1}$. The expected value follows from Exercise 5 (a). Also, see Exercise 11, Chapter 1.]

16 A box contains n balls numbered 1, 2, . . . , n. Three balls are successively drawn without replacement. Assume that the probability model assigns probability $1/n^{(3)}$ to each of the $n^{(3)} = n(n-1)(n-2)$ possible outcomes. If the three selected numbers are arranged in order of magnitude, then X is the middle value. Thus, $X(9, 1, 5) = X(5, 1, 9) = 5$.
(a) Show that

$$P(X=k) = P(X=n-k+1) = \frac{6(k-1)(n-k)}{n^{(3)}}$$

$$k = 2, 3, \ldots, n-1$$

(b) Since (a) shows that X is symmetrically distributed, conclude that

$$EX = \frac{(n+1)}{2}$$

(This exercise generalizes Example 2, Section 2.11.)

There is a curious dual to the inclusion-exclusion formula. The formula remains correct if $\cap$ and $\cup$ are interchanged throughout. The following exercise asks you to prove some special cases.

17 Prove the following:
(a) $P(A \cap B) = P(A) + P(B) - P(A \cup B)$
(b) $P(A \cap B \cap C) = P(A) + P(B) + P(C) - P(A \cup B) - P(A \cup C) - P(B \cup C) + P(A \cup B \cup C)$

18 Suppose that A, B, C are events and X_A, X_B, X_C are their indicator random. Suppose also that $X = X_A + X_B + X_C$.
(a) Show that $EX = P(A) + P(B) + P(C)$.

(b) Establish the following expressions for the density function of X:

$$P(X=0) = 1 - P(A) - P(B) - P(C) + P(A \cap B) + P(A \cap C) + P(B \cap C) - P(A \cap B \cap C)$$

$$P(X=1) = P(A) + P(B) + P(C) - 2P(A \cap B) - 2P(A \cap C) - 2P(B \cap C) + 3P(A \cap B \cap C)$$

$$P(X=2) = P(A \cap B) + P(A \cap C) + P(B \cap C) - 3P(A \cap B \cap C)$$

$$P(X=3) = P(A \cap B \cap C)$$

(c) Show that $EX = P(A) + P(B) + P(C)$ by computing $\Sigma_{i=0}^{3} iP(X=i)$.

Since $E(X+Y) = EX + EY$, one might be misled into believing that $EXY = EXEY$ and $E(X/Y) = EX/EY$. Exercise 19 shows you that such facts are not generally true.

19 An honest coin is thrown three times. Let X equal the number of heads in the first two throws and let Y equal the number of heads in the last two throws.
(a) Show that $EX = EY = 1$.
(b) Show that $E(XY) = 5/4$.
(c) Show that $E[1/(1+Y)] = 7/12$.
(d) Show that $E[X/(1+Y)] = 1/2$.
(e) Show that $X/(1+Y)$ is symmetrically distributed about $1/2$. (The density assigns probabilities 2/8, 1/8, 2/8, 1/8, and 2/8 to the points 0, 1/3, 1/2, 2/3, and 1, respectively.)

20 (a) Suppose that a and b are constants and that the random variable X is symmetrically distributed about c. Show that $aX + b$ is symmetrically distributed about $ac + b$.
(b) Suppose that X is symmetrically distributed about c and X has a finite number of possible values, which, arranged in order, are $a_1 < a_2 < \cdots < a_n$. Show that $c = (a_1 + a_n)/2$. [Hint: By the definition of symmetry, $X - c$ and $-(X - c)$ have the same distributions, which implies that $X - c$ and $-(X - c)$ have the same possible values.]

21 An honest die with faces marked 1, 2, 3, 4, 5, 6 is thrown two times. Let X equal the number on the first throw and let Y equal the number on the second throw.
(a) Show that $EXY = EXEY$.
(b) Show that $E(X/Y) = (EX)E(1/Y)$. [Sections 4.7 and 4.8 will explain why (a) and (b) are true. In general, such relationships do not hold.]
(c) Determine the density function of $X + Y$. (Answer: Probabilities 1/36, 2/36, 3/36, 4/36, 5/36, 6/36, 5/36, 4/36, 3/36, 2/36, 1/36 are assigned to 2, 3, 4, 5, 6, 7, 8, 9, 10, 11, 12, respectively.)

(d) Show that the random variables Y and $7-Y$ are identically distributed [i.e., $P(Y=k)=P(7-Y=k)$].

(e) From (d), show that $X+Y$ and $7+(X-Y)$ have the same density functions. Now infer the density function of $X-Y$ without further computation using the result in (c). [Answer: The probabilities are the same as in (c) but are assigned to the points $-5, -4, -3, -2, -1, 0, 1, 2, 3, 4, 5$.]

(f) Argue without computations that $\log(X/Y)$ is symmetrically distributed about 0. (Hint: Argue, again without computations, that X/Y and Y/X have the same distributions.)

22 Suppose X and Y are nonnegative random variables such that $X \leqslant Y$ [which means that $X(w) \leqslant Y(w)$ for every sample point w]. Show that if EY exists then EX exists and $EX \leqslant EY$.

23 Suppose that $-\infty < a \leqslant X \leqslant b < \infty$ [which means that $a \leqslant X(w) \leqslant b$ for every sample point w]. Show that EX exists and $a \leqslant EX \leqslant b$.

24 (a) Show that if $E(X^2)$ exists then $E|X|$ exists, and hence, by definition, EX exists. [Hint: For any nonnegative real number a, it is true that $|a| \leqslant a^2+1$. (Check this separately for $|a| \leqslant 1$ and $|a| > 1$.) Hence, for any random variable, $|X| \leqslant X^2+1$. Now use Exercise 22 for the desired conclusion.]

(b) Generalize the preceding argument to show that if $E(X^m)$ exists then so does $E(X^n)$, if $0 \leqslant n \leqslant m$.

25 Suppose $E(X^2)$ exists.

(a) Show that $E[(X-t)^2]$ exists for any real number t and

$$E[(X-t)^2] = E(X^2) - 2tEX + t^2$$

(b) If $at^2+bt+c \geqslant 0$ for all t, then $b^2-4ac \leqslant 0$. Use this to conclude from (a) that

$$E(X^2) \geqslant [E(X)]^2$$

(c) If $b^2 = 4ac$, then there is a $t=t_0$ such that $at_0^2+bt_0+c=0$. Conclude that if $E(X^2)=[EX]^2$, then there is a t_0 such that $E[(X-t_0)^2]=0$. Now use Exercise 4a to conclude that X equals the constant t_0—that is, $P(X=t_0)=1$.

(d) The converse to (c) is easy. Show that if X equals a constant, then $E(X^2)=(EX)^2$.

(e) Show that $E[(X-t)^2]$ is minimized in t when $t=EX$.

26 (a) Suppose EX exists. Show that there exists a constant b such that $E(X-b)=0$. (Answer: $b=EX$.)

(b) Suppose $E(X^2)$ exists. Show that there are constants a and b such that $E[(X-b)/a]=0$ and $E\{[(X-b)/a]^2\}=1$. (Answer: $b=EX$, $a=\{E[(X-b)^2]\}^{1/2}$.)

27 (a) Suppose a and b are real numbers. Show that $(a+b)^2 \leqslant 2a^2+2b^2$. [Hint: $2a^2+2b^2-(a+b)^2=(a-b)^2$.]

(b) Use (a) to show that, if $E(X^2)$ and $E(Y^2)$ each exist, then $E[(X+Y)^2]$ also exists and $E[(X+Y)^2] \leqslant 2E(X^2) + 2E(Y^2)$.

(c) Suppose $E(X^2)$ and $E(Y^2)$ each exist. Use (b) to show that for any real number t, $E[(tX+Y)^2]$ exists. A direct calculation shows that

$$E[(tX + Y)^2] = t^2E(X^2) + 2tEXY + E(Y^2)$$

(d) If $at^2 + bt + c \geqslant 0$ for all t, then $b^2 - 4ac \leqslant 0$. Use this to conclude from (c) that

$$[E(XY)]^2 \leqslant E(X^2)E(Y^2)$$

This equation is known as the *Schwarz inequality*.

(e) Repeat parts (b), (c), and (d), with $|X|$ and $|Y|$ playing the roles of X and Y to conclude that if $E(X^2)$ and $E(Y^2)$ exist, then $E|XY|$ exists also and

$$[E(XY)]^2 \leqslant (E|XY|)^2 \leqslant E(X^2)E(Y^2)$$

NOTES

1 Random variables and expected values are discussed in Chapter IX of [9] and Chapter 2 of [10].

2 Cardano's *Liber de Ludo Aleae* had a formulation of the concept of expected value (see Note 1, Chapter 1).

3 The basic facts about absolute convergence of infinite series can be found in Sections 18.9 and 18.10 of [2] and in Sections 10.18 through 10.21 of [3]. In particular, the relation between rearrangements of series and absolute convergence is discussed in Section 10.21 of [3].

4 Let X be the number of tosses of a coin required to obtain the first head, as in Example 4, Section 2.4. Then $E(2^X) = 1 + 1 + 1 + \cdots = \infty$. This first known example of an infinite expected value was presented by Daniel Bernoulli (1700–1782) and was the source of much confusion.

5 The divergence of the harmonic series $1 + \frac{1}{2} + \frac{1}{3} + \cdots$ is shown in Section 18.2 of [2] and Section 10.1 of [3].

6 The triangle inequality is in Section I–4.8 of [3], and the proof of the absolute convergence of the termwise sum of two absolutely convergent series is in Section 10.18 of [3].

7 For a different proof of the inclusion-exclusion formula, see Chapter IV of [9].

Chapter Three

SOME COMBINATORIC PROBLEMS

In Chapter 3 we apply the ideas developed in the first two chapters to some combinatoric problems. These include matching, occupancy, and sampling from dichotomous populations. Special aspects of some of these problems have already appeared earlier in various examples and exercises. Combinatoric problems provide the main core of computational examples for discrete probability theory. A brute force approach to these problems is often quite forbidding; even the sparse amount of methodology developed thus far goes a long way in avoiding unpleasant computations (Note 1).

3.1 THE MATCHING PROBLEM

A deck of n cards are marked $1, 2, \ldots, n$. The cards are shuffled and laid out against n fixed positions also marked $1, 2, \ldots, n$. Suppose that the sample points are represented by the $n!$ permutations of $1, 2, \ldots, n$, each having probability $1/n!$ Let the random variable X denote the total number of matches throughout the n positions. A match occurs in position i if the card marked i is placed next to the fixed position marked i. For example, suppose $n = 4$ and the sample point is $(1, 3, 2, 4)$. Then $X(1, 3, 2, 4) = 2$ because 1 is in the first position, 4 is in the fourth position, and 3 and 2 are not in their corresponding positions. For $n = 4$, Table 3.1 shows the $4! = 24$ sample points and their X values. Figure 3.1 shows the density function.

TABLE 3.1

Sample points	*X values*	*Sample points*	*X values*	*Sample points*	*X values*
1, 2, 3, 4	4	2, 3, 1, 4	1	3, 4, 1, 2	0
1, 2, 4, 3	2	2, 3, 4, 1	0	3, 4, 2, 1	0
1, 3, 2, 4	2	2, 4, 1, 3	0	4, 1, 2, 3	0
1, 3, 4, 2	1	2, 4, 3, 1	1	4, 1, 3, 2	1
1, 4, 2, 3	1	3, 1, 2, 4	1	4, 2, 1, 3	1
1, 4, 3, 2	2	3, 1, 4, 2	0	4, 2, 3, 1	2
2, 1, 3, 4	2	3, 2, 1, 4	2	4, 3, 1, 2	0
2, 1, 4, 3	0	3, 2, 4, 1	1	4, 3, 2, 1	0

Example 1 At a dinner party, 10 persons seat themselves before noticing that there are place cards at every setting. The number of matches, X, is represented by the number of persons seated in their intended places.

Example 2 A person claims powers of extrasensory perception. He is tested as follows. An examiner draws cards successively from a deck. At each draw, the subject calls his guess of the card's designation. The subject is not informed of the sequence of cards actually drawn until the deck is exhausted. He keeps a record of his se-

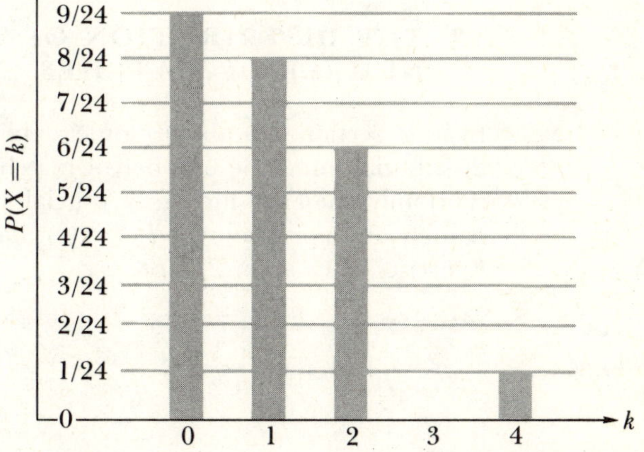

FIGURE 3.1 The density function of the number of matches, using a deck of four cards.

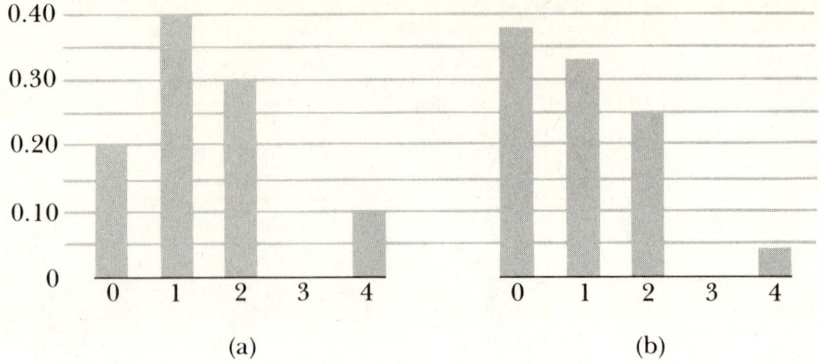

FIGURE 3.2 (a) Observed density of correct guesses in 100 trials, using a deck of four cards. (b) Theoretical density of matches, using a deck of four cards (the same density function as in Figure 3.1).

quence of guesses and does not use the same number twice. The number of matches, X—that is, the number of correct guesses—represents a measure of the subject's extrasensory ability. To appraise this ability, one should know the distribution of X when the subject's guesses are made purely at random. Suppose that this experiment is actually repeated 100 times using a deck of four cards. If the observed distribution of matches is the one shown in Figure 3.2a, it suggests that there is some nonchance phenomenon lying behind this particular subject's guesses. He seems to be guessing right too often.

3.2 THE DISTRIBUTION OF THE NUMBER OF MATCHES

In this section we obtain some explicit information about the distribution of X, the number of matches when n items are placed randomly in n positions. A crucial first piece of information, namely an expression for $P(X = 0)$, has been obtained earlier, in Exercise 7c, Chapter 2 (Note 2).

probability of no matches

$$p_n = P(X = 0) = 1 - 1 + \frac{1}{2} - \cdots + \frac{(-1)^n}{n!} \tag{1}$$

Since e^x has the power series expansion,

$$e^x = 1 + x + \frac{x^2}{2} + \cdots + \frac{x^n}{n!} + \cdots$$

and in particular, for $x = -1$,

$$e^{-1} = 1 - 1 + \frac{1}{2} - \cdots + \frac{(-1)^n}{n!} + \cdots$$

it follows that $p_n = P(X = 0)$ is for large n close in value to e^{-1}. In fact, because the series for e^{-1} has terms whose signs alternate and whose magnitudes decrease, it follows that $|p_n - e^{-1}| \leq 1/(n+1)!$ (Note 3). Table 3.2 shows some values of p_n. It is remarkable that the probability of no matches *almost does not depend on n.* Rounded off to the first two decimal places, the probability of no matches is 0.37 for all decks of five or more cards. As $n \to \infty$, the probability of no matches using n cards has the limit $1/e = 0.36787\ldots$ (see Table 3.2).

For fixed n, we now give an explicit expression for the density function of X in terms of the values of $p_1, p_2, \ldots, p_n$, as follows:

distribution of number of matches

$$P(X = k) = \frac{p_{n-k}}{k!}, \qquad k = 0, 1, \ldots, n \tag{2}$$

In the preceding expression, p_0 should be understood to equal 1. Then, for $k = n$, Eq. (2) says that $P(X = n) = 1/n!$ This statement is correct because there is only one permutation in which all the cards are matched. p_1 equals 0, since with a deck of only one card there cannot fail to be a match. Hence, $P(X = n-1) = p_1 = 0$; it is impossible for exactly $n - 1$ cards to be perfectly matched leaving one card mismatched. The general proof of (2) follows.

Proof. Let $N_{k,n}$ be the number of sample points in which there are

TABLE 3.2 Probabilities of no matches using n cards

n	$p_n = P(X = 0)$	$=$	$\sum_{i=0}^{n} (-1)^i/i!$
1	0		
2	1/2	=	0.5
3	2/6	=	0.33 ⋯
4	9/24	=	0.375
5	44/120	=	0.366 ⋯
6	265/720	=	0.36805 ⋯
7	1854/5040	=	0.367857 ⋯
⋮			
	$\lim_{n\to\infty} p_n = e^{-1}$	=	0.36787 ⋯

exactly k matches. Then $P(X=k)=N_{k,n}/n!$ We now want to evaluate $N_{k,n}$. Consider first the number of sample points in which there are exactly k matches in k specified positions. This means that in the remaining $n-k$ positions the $n-k$ numbers must be completely mismatched. The number of permutations of $n-k$ distinct integers that are completely mismatched is by definition $N_{0,n-k}$. There are $\binom{n}{k}$ ways of choosing the k specified positions. Hence, $N_{k,n} = \binom{n}{k}N_{0,n-k}$ and

$$P(X=k) = \binom{n}{k}\frac{N_{0,n-k}}{n!} = \frac{N_{0,n-k}}{(n-k)!k!}$$

But, since $N_{0,n-k}/(n-k)! = p_{n-k}$ = probability of no matches with a deck of $n-k$ cards, it follows that $P(X=k) = p_{n-k}/k!$ ∎

Example 1 Let us use (2) to determine the density function of X when $n=5$. We need the values $p_0=1, p_1=0, p_2=1/2, p_3=2/6, p_4=9/24$, which appear in Table 3.2. Then $P(X=1)=p_4=9/24=45/120$, $P(X=2)=p_3/2=20/120$, $P(X=3)=p_2/6=10/120$, $P(X=4)=0$, $P(X=5)=1/5!=1/120$. By subtraction, $P(X=0)=1-P(X=1)-P(X=2)-P(X=3)-P(X=4)-P(X=5)=44/120$. This density and the densities for $n=1, 2, \ldots, 5$ are summarized in Table 3.3.

It is interesting that the density function of X "stabilizes itself" as n goes to ∞.

TABLE 3.3 Density functions of *X*

$n=1$

k	$P(X=k)$
0	0
1	1

$n=2$

k	$P(X=k)$
0	1/2
1	0
2	1/2

$n=3$

k	$P(X=k)$
0	2/6
1	3/6
2	0
3	1/6

$n=4$

k	$P(X=k)$
0	9/24
1	8/24
2	6/24
3	0
4	1/24

$n=5$

k	$P(X=k)$
0	44/120
1	45/120
2	20/120
3	10/120
4	0
5	1/120

k	*Limiting probabilities of k matches as* $n \to \infty$
0	$e^{-1} = 0.3679 \cdots$
1	$e^{-1} = 0.3679 \cdots$
2	$e^{-1}/2 = 0.1839 \cdots$
3	$e^{-1}/3! = 0.0613 \cdots$
4	$e^{-1}/4! = 0.0153 \cdots$
5	$e^{-1}/5! = 0.0030 \cdots$
6	$e^{-1}/6! = 0.0006 \cdots$
⋮	
i	$e^{-1}/i!$
⋮	⋮

asymptotic distribution of number of matches

$$\lim_{n\to\infty} P(k \text{ matches using a deck of } n \text{ cards}) = \frac{e^{-1}}{k!} \qquad k = 0, 1, 2, \ldots \tag{3}$$

Proof. From (1), it follows that $\lim_{n\to\infty} p_n = e^{-1}$. Hence, for any fixed k, $\lim_{n\to\infty} p_{n-k} = e^{-1}$. From (2), the required limit is $\lim_{n\to\infty} p_{n-k}/k! = e^{-1}/k!$ ∎

The "limiting distribution" given by (3) is a special case of the important *Poisson distribution* with parameter $\lambda = 1$, to be presented in Section 6.7.

Figure 3.3 compares the graph of the density function for $n = 5$ with the limiting density function.

Finally we consider the expected number of matches.

expected number of matches

$$E(X) = 1 \quad \text{for any } n \tag{4}$$

One proof of (4), using indicator random variables, appears in Exercise 7a, Chapter 2. Here is a second evaluation of EX.

Second proof of (4). By (2), the probability of k matches using n cards is $p_{n-k}/k!$ Hence, for any n,

$$\sum_{k=0}^{n} \frac{p_{n-k}}{k!} = 1 \tag{5}$$

Evaluating EX from the density function of X [Eq. (1), Section 2.10],

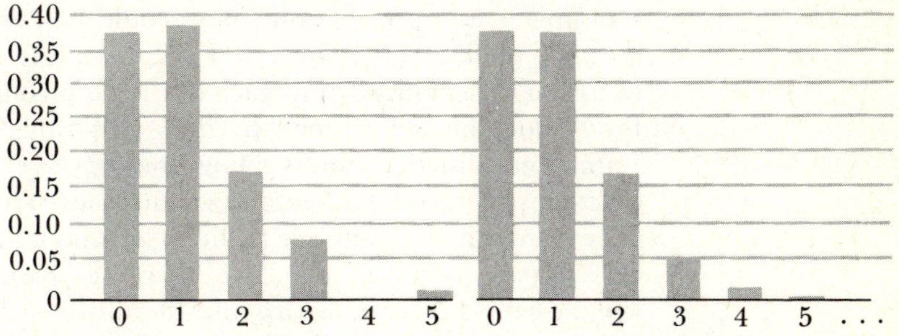

FIGURE 3.3 (a) Density of matches, using a deck of five cards. (b) Limiting density of matches.

$$EX = \sum_{k=0}^{n} \frac{kp_{n-k}}{k!} = \sum_{k=1}^{n} \frac{p_{n-k}}{(k-1)!} = \sum_{k-1=0}^{n-1} \frac{p_{n-1-(k-1)}}{(k-1)!}$$

This last expression is just (5) with n replaced by $n-1$—hence, $EX = 1$. ▮

3.3 BALLS RANDOMLY DISTRIBUTED IN CELLS

Suppose there are n balls numbered $1, 2, \ldots, n$, and m cells (or boxes) labeled $a_1, a_2, \ldots, a_m$. The balls are randomly distributed among the cells. An outcome of this experiment can be described by a sequence $a_{i_1}, a_{i_2}, \ldots, a_{i_n}$, which represents the cell assignments for balls $1, 2, \ldots, n$, respectively. Thus, if $n = 5$ and $m = 3$, then a_2, a_3, a_2, a_1, a_4 means that ball 1 entered cell a_2, ball 2 entered cell a_3, ball 3 entered cell a_2, ball 4 entered cell a_1 and ball 5 entered cell a_4. Since each of the m cell designations can appear in any of the n coordinates, there are m^n sample points altogether. The usual interpretation of "random distribution" is that each of these sample points has the same probability, $1/m^n$.

Example 1 Suppose two balls are randomly distributed among three cells. The following is the list of the $9 = 3^2$ equiprobable sample points: (a_1, a_1), (a_1, a_2), (a_1, a_3), (a_2, a_1), (a_2, a_2), (a_2, a_3), (a_3, a_1), (a_3, a_2), (a_3, a_3).

If the n balls are indistinguishable one from another (i.e., if the numbers $1, 2, \ldots, n$ are erased), how should an outcome of the experiment be described? The only reasonable possibility now is to give the inventory of how many balls are in each of the cells. For instance, in distributing two balls among three cells as in Example 1, the possible outcomes could be described as (2, 0, 0), (0, 2, 0), (0, 0, 2), (1, 1, 0), (1, 0, 1), (0, 1, 1). It is now not reasonable to assign probability 1/6 to each of these outcomes. The probabilities should be determined by first supposing that the balls are numbered and determining how the $9 = 3^2$ sample points listed in Example 1 "collapse" into new outcomes. Thus, (2, 0, 0) comes only from the single sample point (a_1, a_1) and is assigned probability 1/9. On the other hand, (1, 1, 0) comes from the two original sample points (a_1, a_2), (a_2, a_1) and therefore is assigned probability 2/9. (See the discussion in Example 4, Section 1.4.) Table 3.4 summarizes the probability assignments for this experiment. In general, in distributing n indistinguishable balls among m cells, an

TABLE 3.4 Two balls are randomly distributed among three cells

If the balls are distinguishable and numbered 1, 2, 3, *the sample points are the following, each with probability* 1/9	*If the balls are indistinguishable, the sample points are the following*	*Probability*
(a_1, a_1)	(2, 0, 0)	1/9
(a_2, a_2)	(0, 2, 0)	1/9
(a_3, a_3)	(0, 0, 2)	1/9
(a_2, a_1), (a_1, a_2)	(1, 1, 0)	2/9
(a_1, a_3), (a_3, a_1)	(1, 0, 1)	2/9
(a_3, a_2), (a_2, a_3)	(0, 1, 1)	2/9

outcome is a sequence of m integers, $(k_1, k_2, \ldots, k_m)$, giving the inventory of numbers of balls per cell. Since all n balls are being accounted for, $k_1 + k_2 + \cdots + k_m = n$.

The probability of the sequence $(k_1, k_2, \ldots, k_m)$ is

$$\left(\frac{n!}{k_1!k_2! \cdots k_m!}\right)\left(\frac{1}{m^n}\right) \tag{1}$$

Equation (1) is an example of a multinomial probability. The multinomial distribution will be discussed in Section 6.2.

Proof of (1). Suppose momentarily that the balls are numbered, and consider the sample points as originally designated in the form $(a_{j_1}, \ldots, a_{j_n})$, giving the cell assignments to the n balls. We need to know how many of these sample points collapse into the single designation $(k_1, k_2, \ldots, k_m)$. Any sequence $(a_{j_1}, \ldots, a_{j_n})$ in which there are k_1 a_1's, k_2 a_2's, $\ldots$, k_m a_m's has the required property. There are n a_i's of various types altogether. According to VIII, Section 1.5, there are $n!/k_1!k_2! \cdots k_m!$ sequences with the required property. Since each of the original sequences has probability $1/m^n$, the proof is complete. ∎

Example 2 A well-balanced pointer is mounted on a dial marked into four equal regions, a, b, c, d (see Figure 3.4). The pointer is spun three times. The outcome of the experiment describes the number of

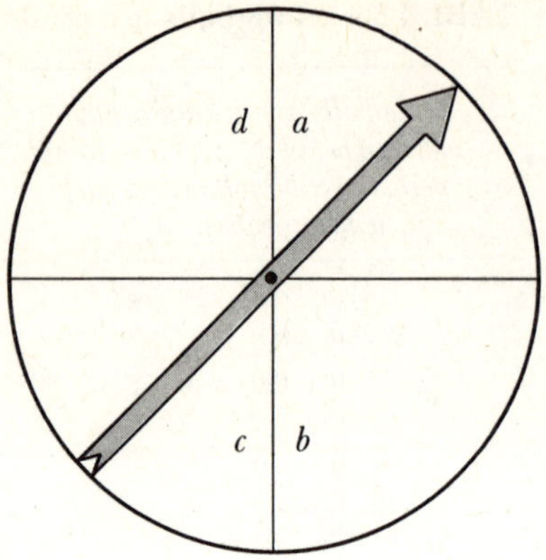

FIGURE 3.4

a's, b's, c's, and d's among the three spins. The sample points and probabilities, computed according to (1), are given in Table 3.5.

Example 3 The following are some alternative interpretations for the distribution of n balls among m cells.

(a) *Birthdays.* Among n persons there are various birthdays. If, for simplicity, we ignore leap years and say that $m = 365$, then there are $(365)^n$ patterns of birthdays for the n persons. These may reasonably be considered equally probable if human beings follow no seasonal mating patterns. Under this assumption, the probability that among n persons there are k_1 birthdays on January 1, k_2 on January 2, . . . , k_{365} on December 31, is given by (1). Some further discussion of birthdays appears in Section 3.6.

TABLE 3.5

Sample points	*Probabilities*	*Sample points*	*Probabilities*
3 0 0 0	1/64	0 2 1 0	3/64
0 3 0 0	1/64	0 1 2 0	3/64
0 0 3 0	1/64	0 2 0 1	3/64
0 0 0 3	1/64	0 1 0 2	3/64
2 1 0 0	3/64	0 0 2 1	3/64
1 2 0 0	3/64	0 0 1 2	3/64
2 0 1 0	3/64	1 1 1 0	6/64
1 0 2 0	3/64	1 1 0 1	6/64
2 0 0 1	3/64	1 0 1 1	6/64
1 0 0 2	3/64	0 1 1 1	6/64

(b) *School assignments.* In a proposed grade-school desegregation plan, pupils would be allowed to attend any city school of their choice. Suppose there are n pupils and m schools. If the choices are made randomly, then the probability model of assigning balls in cells is appropriate, and the probability of school 1 getting k_1 pupils, school 2 getting k_2 pupils, and so forth, is given by (1). A more realistic model would need modifications to recognize that friends and relatives influence each other in their choices.

3.4 SOME RANDOM VARIABLES RELATED TO OCCUPANCY

Suppose n balls are randomly distributed among m cells. Let U denote the number of balls that fall in the first cell. The possible values of U are 0, 1, . . . , n, and its distribution is given as follows.

distribution of number of balls in a cell

$$P(U=k)=\binom{n}{k}\frac{(m-1)^{n-k}}{m^n}, \qquad k=0, 1, \ldots, n \tag{1}$$

Proof. Suppose that the balls are numbered. There are m^n sample points, each with probability $1/m^n$. A sample point represents k balls in cell a_1 and $n-k$ balls elsewhere if k of its coordinates are a_1's and the remaining $n-k$ are not a_1's. Suppose first that k specified balls are the ones to fall in a_1. There are $(m-1)^{n-k}$ sample points in this event. This is simply the count of the number of $(n-k)$-tuples, each of whose entries can be any of the $m-1$ symbols $a_2, \ldots, a_m$. There are $\binom{n}{k}$ ways of specifying which k balls are the ones to fall in a_1; therefore, there are $\binom{n}{k}(m-1)^{n-k}$ sample points altogether. ∎

The distribution (1) is a special case of the *binomial distribution,* which will be discussed in Section 6.1. It will follow from properties of the binomial distribution that

$$EU=\frac{n}{m} \tag{2}$$

However, it is easy to give a direct proof of (2) as follows.

Proof of (2). Let U_i denote the number of balls in cell a_i, $i = 1, 2, \ldots, m$. It should be clear that each of $U_1, U_2, \ldots, U_m$ has the same density function, and hence, $EU_1 = EU_2 = \cdots = EU_m$. Because $U_1 + U_2 + \cdots + U_m = n$, the total number of balls, it follows that $EU_1 + \cdots + EU_m = mEU_1 = n$ and $EU_1 = n/m$. [See Exercise 7c for a generalization of (2).] ∎

Another random variable of interest is the *number of unoccupied cells.* Denote this random variable by X. A formula for $P(X = 0)$, which follows from the inclusion-exclusion formula, has already been given in Exercise 8b, Chapter 2.

probability of no unoccupied cells

$$P(X = 0) = \sum_{k=0}^{m} (-1)^k \binom{m}{k} \frac{(m-k)^n}{m^n} = p_{n,m} \tag{3}$$

The probability distribution of X can now be determined from these $p_{i,j}$'s.

distribution of number of unoccupied cells

Suppose n balls are randomly distributed among m cells. If X is the number of unoccupied cells, then

$$P(X = k) = \binom{m}{k}\left(\frac{m-k}{m}\right)^n p_{n,m-k}, \qquad k = 0, 1, \ldots, m-1 \tag{4}$$

Proof of (4). If i balls are distributed among j cells, let $N_{i,j}$ denote the number of sample points (i-tuples) representing no empty cells. Then $p_{i,j} = N_{i,j}/j^i$. To verify (4), we must count the number of sample points (n-tuples) representing exactly k empty cells. Suppose first that k specifically designated cells are empty. The count of these sample points is just $N_{n,m-k}$, the number of n-tuples representing the case where none of $m - k$ specified cells is unoccupied. There are $\binom{m}{k}$ ways of specifying the k unoccupied cells; hence, the event with the probability we want has $\binom{m}{k} N_{n,m-k}$ sample points. Its probability is

$$\binom{m}{k}\frac{N_{n,m-k}}{m^n} = \binom{m}{k}\left(\frac{m-k}{m}\right)^n \frac{N_{n,m-k}}{(m-k)^n}$$

which is the right-hand side of (4). ∎

Table 3.6 gives density functions of X for several values of m and n (see Exercise 8).

TABLE 3.6

$m=1, n=1$		$m=1, n=2$		$m=1, n=3$		$m=1, n=4$	
k	$P(X=k)$	k	$P(X=k)$	k	$P(X=k)$	k	$P(X=k)$
0	1	0	1	0	1	0	1

$m=2, n=1$		$m=2, n=2$		$m=2, n=3$		$m=2, n=4$	
k	$P(X=k)$	k	$P(X=k)$	k	$P(X=k)$	k	$P(X=k)$
0	0	0	1/2	0	3/4	0	7/8
1	1	1	1/2	1	1/4	1	1/8

$m=3, n=1$		$m=3, n=2$		$m=3, n=3$		$m=3, n=4$	
k	$P(X=k)$	k	$P(X=k)$	k	$P(X=k)$	k	$P(X=k)$
0	0	0	0	0	2/9	0	12/27
1	0	1	2/3	1	6/9	1	14/27
2	1	2	1/3	2	1/9	2	1/27

$m=4, n=1$		$m=4, n=2$		$m=4, n=3$		$m=4, n=4$	
k	$P(X=k)$	k	$P(X=k)$	k	$P(X=k)$	k	$P(X=k)$
0	0	0	0	0	0	0	6/64
1	0	1	0	1	6/16	1	36/64
2	0	2	3/4	2	9/16	2	21/64
3	1	3	1/4	3	1/16	3	1/64

Example 1 A meteorologist predicts that in four given counties there will be five tornadoes over the next six months. Assuming that nature will disperse these tornadoes according to the model of distributing five balls at random among four cells, what are the probabilities for the various number of counties that will suffer at least one tornado? By extending Table 3.6 (Exercise 8), we see that they are as follows:

k *number of counties suffering tornadoes*	*P(k counties suffer tornadoes)*
1	1/256
2	45/256
3	150/256
4	60/256

(If i counties are free of tornadoes, then $4-i=k$ counties suffer tornadoes.) The expected number of counties suffering tornadoes are $[1(1)+2(45)+3(150)+4(60)]/256 = 781/256 = 3.05\cdots$. This expected value also equals $4-3^5/4^4$, where $3^5/4^4$ is the expected number of unaffected counties. See Eq. (5) for the general expression that relates to this expected value.

Example 2 Suppose a hospital anticipates that 100 babies will be born in its delivery room over the next 90 days. What is the distribution of the number of these days in which the delivery room will not be used? The distribution is the same as that of the number of empty cells in distributing 100 balls among 90 cells. According to Eq. (5), the expected number of days that the delivery room is not used is $(89)^{100}/(90)^{99} \approx 30$.

Next, we give an expression for the expected number of empty cells.

expected number of empty cells

$$EX = \frac{(m-1)^n}{m^{n-1}} \tag{5}$$

The expected value (5) has already appeared earlier in Exercise 5a, Chapter 2.

First proof of (5). Define indicator variables $X_1, \ldots, X_m$ as follows: X_i is 1 if the ith cell is empty; it is 0 otherwise. Then $X = X_1 + \cdots + X_m$ and $EX = EX_1 + \cdots + EX_m$. Since $EX_i = P(X_i = 1) = (m-1)^n/m^n$, the result follows. This solution is the one that was intended for Exercise 5a in Chapter 2. ∎

Second proof of (5). Since $\Sigma_{k=0}^{m} P(X = k) = 1$, it follows from (4) that

$$\sum_{k=0}^{m} \binom{m}{k} \left(\frac{m-k}{m}\right)^n p_{n,m-k} = 1 \tag{6}$$

$$EX = \sum_{k=0}^{m} kP(X = k) = \sum_{k=0}^{m} k\binom{m}{k}\left(\frac{m-k}{m}\right)^n p_{n,m-k}$$

$$= m\left(\frac{m-1}{m}\right)^n \sum_{k-1=0}^{m-1} \binom{m-1}{k-1}\left[\frac{(m-1)-(k-1)}{m-1}\right]^n p_{n,(m-1)-(k-1)}$$

if we use the fact that $k\binom{m}{k} = m\binom{m-1}{k-1}$. Because the final summation equals 1 by replacing m by $m-1$ in (6), the result follows. ∎

3.5 WAITING FOR FULL OCCUPANCY

Balls are successively distributed among m boxes. How many balls must be distributed before all the cells are occupied? Clearly, the smallest possible value is m, but any integer greater than m is also a possible value. Let Z denote the random variable that counts the number of balls required to achieve a full occupancy. We want to get some information about the distribution of Z.

Example 1 In every box of a certain breakfast cereal is a picture of a baseball player. A full set is represented by m pictures. How many boxes of cereal must be purchased to obtain a full set of pictures? This version of the problem is sometimes called "the coupon collector's problem."

Example 2 (a) How many times must a die be thrown before every one of the six faces appears at least once? This is the same as asking how many balls must be distributed among six cells before full occupancy occurs.

(b) How many persons must one encounter before finding at least one with a January birthmonth, at least one with a February birthmonth, . . . , and at least one with a December birthmonth? The distribution is that of Z with $m = 12$.

The distribution of the "waiting time" Z can be completely described in terms of the numbers $p_{n,m}$ which were defined in (3), Section 3.4. Recall that $p_{n,m}$ is the probability of no unoccupied cells when n balls are randomly distributed among m cells. To say that $Z \leq n$ is exactly the same as saying that by the time the first n balls have been distributed, all m cells are occupied. Since the events are equivalent, we have

$$P(Z \leq n) = p_{n,m} \tag{1}$$

which is the only reasonable probability value to assign to the event $Z \leq n$. This value determines the density function of Z, as follows.

distribution of time until full occupancy

$$P(Z=n)=P(Z \leq n)-P(Z \leq n-1)=p_{n,m}-p_{n-1,m} \tag{2}$$

The first question of interest relating to Z is whether it is true that

$$\lim_{n\to\infty} P(Z \leq n) = \lim_{n\to\infty} p_{n,m} = 1 \tag{3}$$

It is certainly intuitively plausible that (3) holds, because if one continues distributing balls forever, full occupancy should eventually occur.

Proof of (3). The explicit expression for $p_{n,m}$ from (3), Section 3.4, is

$$p_{n,m} = 1 - \binom{m}{1}(m-1)^n/m^n + \binom{m}{2}(m-2)^n/m^n - \cdots + (-1)^{m-1}(1)^n/m^n \tag{4}$$

Since $\lim_{n\to\infty}(m-1)^n/m^n = \lim_{n\to\infty}(m-2)^n/m^n = \cdots = 0$, the result follows. ∎

Example 3 Suppose balls are successively distributed among 3 cells. The distribution of Z, the number of balls required to achieve full occupancy, is the following:

k	1	2	3	4	5	6	$\cdots$	n	$\cdots$
$P(Z=k)$	0	0	2/9	2/9	14/81	10/81	$\cdots$	$(2^{n-1}-2)/3^{n-1}$	$\cdots$

The general expression follows by evaluating (2) and (4) for $m = 3$.

Next we present, without proofs, several expressions for EZ.

$$EZ = m\left[\binom{m}{1}\Big/1 - \binom{m}{2}\Big/2 + \cdots + (-1)^{m+1}\binom{m}{m}\Big/m\right] \tag{5}$$

$$EZ = m\left(1 + \frac{1}{2} + \cdots + \frac{1}{m}\right) \tag{6}$$

Let us first give a few hints to the reader who may be anxious to prove (5) and (6). By using (3), Section 2.10, we have,

$$EZ = \sum_{n=0}^{\infty} P(Z > n) = \sum_{n=0}^{\infty} (1 - p_{n,m})$$

Application of this calculation to (4) will yield (5) after some messy algebra. That (6) is the same as (5) can be shown by induction (Note 4). Equation (6) is by far the more meaningful of the two expressions. We will give a direct and intuitive proof of (6) in Example 2, Section 6.4.

Example 4 (a) A coin is thrown until both head and tail appear at least once. By (6), the expected number of throws required is $2(1 + 1/2) = 3$ (see Exercise 10).

(b) A die is thrown until each of the six faces appears at least once. The expected number of throws required is $6(1 + 1/2 + \cdots + 1/6) = 6(49/20) = 14.7$. Thus, on the average, it takes almost 15 throws to obtain all six faces.

3.6 THE BIRTHDAY PROBLEM

What is the probability that among n persons, at least two have a common birthday? As is plausible, the greater the number of persons the greater is the probability of a multiple birthday. Of course, if there are more persons than there are days in the year, the probability is 1. It is surprising that for $n = 23$, the probability is already close to 1/2 (see Table 3.7). Suppose, for simplicity, that there are 365 days in any year and that a person's birthday is any one of these 365 days with equal likelihood. The model is just that of randomly distributing n balls among 365 cells.

The probability of a multiple birthdate is

$$1 - \frac{(365)^{(n)}}{(365)^n} \tag{1}$$

Proof of (1). There are 365^n sample points. Suppose that $n \leq 365$. The number of sample points in which the same cells is not entered twice is $(365)(364) \cdots (365 - n + 1) = 365^{(n)}$. Hence, the probability of no multiple birthdates is $(365)^{(n)}/365^n$. ∎

Notice that (1) is correct even if $n > 365$, for in that case $(365)^{(n)} =$

TABLE 3.7 The approximate probability for a multiple birthdate among n persons

n	$1 - [(365)^{(n)}/365^n]$	n	$1 - [(365)^{(n)}/365^n]$
1	0	17	0.32
2	0.00	18	0.35
3	0.01	19	0.38
4	0.02	20	0.41
5	0.03	21	0.45
6	0.05	22	0.48
7	0.06	23	0.51
8	0.08	24	0.54
9	0.10	25	0.57
10	0.12	26	0.60
11	0.15	27	0.63
12	0.17	28	0.66
13	0.20	29	0.68
14	0.23	30	0.70
15	0.26	31	0.73
16	0.29	32	0.75

0. Table 3.7 shows the probabilities of multiple birthdates for $n = 1, 2, \ldots, 32$.

The general version of the birthday problem is the following: n balls are distributed among m cells; what is the probability that some cell has more than one ball occupying it?

The probability of a multiple occupancy is

$$1 - \frac{m^{(n)}}{m^n} \tag{2}$$

The proof of (2) is exactly the same as that of (1).

Example 1 (a) The probability that among five persons there is a common birthday is

$$1 - \frac{(365)^{(5)}}{365^5} = 0.03 \cdots \qquad \text{(see Table 3.7).}$$

(b) The probability that among five persons there is a common birthmonth is

$$1 - \frac{(12)^{(5)}}{12^5} = 0.618 \cdots$$

(c) The probability that among five persons there is a common birthweek is

$$1 - \frac{(52)^{(5)}}{52^5} = 0.14 \cdots$$

3.7 BOSE–EINSTEIN OCCUPANCY

In the previous section, we assumed that when n balls are distributed among m cells, each of the m^n sample points has probability $1/m^n$. When the balls are indistinguishable, an outcome of the experiment is described by a sequence of m numbers $(k_1, \ldots, k_m)$, giving the numbers of balls in each of the m cells. As pointed out in (1), Section 3.3, the probability of such an outcome is conventionally $n!/(k_1!k_2! \cdots k_m!)m^n$. The model of distributing balls over cells is an appropriate one for various problems of statistical mechanics. The n balls are physical particles and the cells represent m equal parts into which a spatial region is divided. However, for many kinds of particles, the probability assumptions made earlier are not correct and other schemes are used. One such method says that if the balls are indistinguishable, then each of the outcomes $(k_1, \ldots, k_m)$ we have mentioned has *the same probability.* Such a system of particle allotment is called a *Bose–Einstein scheme* (Note 5).

Example 1

Suppose two balls are distributed among three cells. The sample space was described in Table 3.4, Section 3.3. Table 3.8 compares the ordinary probabilities with the Bose–Einstein probabilities.

The following describes the Bose–Einstein probabilities.

Bose–Einstein probabilities

If m balls are distributed among n cells, the probability of each outcome $(k_1, \ldots, k_m)$ under the Bose–Einstein scheme is $\binom{n+m-1}{n}^{-1}$.

Proof. Since each sequence of nonnegative integers $(k_1, \ldots, k_m)$, where $k_1 + \cdots + k_m = n$, has the same probability, we must show that there are $\binom{n+m-1}{n}$ such sequences. For simplicity, we present the proof in the special case $n = 3, m = 4$. An outcome can be represented by a sequence of five |'s and three X's. For instance, (2, 0, 1, 0) is represented by $|XX|\,|X|\,|$. The |'s can be thought of as the walls of the cells and the X's as the balls. Thus, $|XX|\,|X|\,|$ means that the first cell has two balls, the second cell is empty, the third cell has one ball, and the fourth cell is empty. Any arrangement of three balls in four cells is thus described by a sequence of five |'s and three X's, *provided that the sequence must start and end with a* |. There are then three |'s and three X's that are free to move, and

TABLE 3.8

Sample points	*Usual probabilities*	*Bose–Einstein probabilities*
2 0 0	1/9	1/6
0 2 0	1/9	1/6
0 0 2	1/9	1/6
1 1 0	2/9	1/6
1 0 1	2/9	1/6
0 1 1	2/9	1/6

<table>
<tr><th>outcomes</th><th>schematic representation used in proof of (1)</th></tr>
<tr><td>3 0 0 0</td><td>| X X X | | | |</td></tr>
<tr><td>0 3 0 0</td><td>| | X X X | | |</td></tr>
<tr><td>0 0 3 0</td><td>| | | X X X | |</td></tr>
<tr><td>0 0 0 3</td><td>| | | | X X X |</td></tr>
<tr><td>2 1 0 0</td><td>| X X | X | | |</td></tr>
<tr><td>2 0 1 0</td><td>| X X | | X | |</td></tr>
<tr><td>2 0 0 1</td><td>| X X | | | X |</td></tr>
<tr><td>1 2 0 0</td><td>| X | X X | | |</td></tr>
<tr><td>0 2 1 0</td><td>| | X X | X | |</td></tr>
<tr><td>0 2 0 1</td><td>| | X X | | X |</td></tr>
<tr><td>1 0 2 0</td><td>| X | | X X | |</td></tr>
<tr><td>0 1 2 0</td><td>| | X | X X | |</td></tr>
<tr><td>0 0 2 1</td><td>| | | X X | X |</td></tr>
<tr><td>1 0 0 2</td><td>| X | | | X X |</td></tr>
<tr><td>0 1 0 2</td><td>| | X | | X X |</td></tr>
<tr><td>0 0 1 2</td><td>| | | X | X X |</td></tr>
<tr><td>1 1 1 0</td><td>| X | X | X | |</td></tr>
<tr><td>1 1 0 1</td><td>| X | X | | X |</td></tr>
<tr><td>1 0 1 1</td><td>| X | | X | X |</td></tr>
<tr><td>0 1 1 1</td><td>| | X | X | X |</td></tr>
</table>

FIGURE 3.5 There are as many outcomes as there are arrangements of the six symbols | | | X X X, namely $\binom{6}{3} = 20$.

there are $\binom{6}{3}$ distinct patterns (see Figure 3.5). The general proof is similar and is left to the reader. ∎

The various aspects of occupancy presented in Sections 3.4 and 3.5 can be developed in a completely parallel way for the Bose–Einstein scheme. We shall now point out several of the facts, leaving the proofs to the reader.

Let X be the number of empty cells. Define $X_i = 1$ if the ith cell is empty and $X_i = 0$ otherwise. Then, $X = X_1 + \cdots + X_m$. To evaluate $EX = EX_1 + \cdots + EX_m$, we need $EX_i = P(i\text{th cell is empty}) = \binom{n+m-2}{n}$. Hence,

$$EX = m\frac{\binom{n+m-2}{n}}{\binom{n+m-1}{n}} = m\frac{(m-1)}{m+n-1} \tag{1}$$

Similarly, the probability of k specified cells being empty is $\binom{n+m-k-1}{n}\Big/\binom{n+m-1}{n}$. It follows, by the inclusion-exclusion formula, that

$$P(X=0) = 1 - P[(X_1 = 1) \cup \cdots \cup (X_m = 1)]$$

$$= \sum_{k=1}^{m} (-1)^{k-1} \binom{m}{k} \frac{\binom{n+m-k}{n}}{\binom{n+m-1}{n}} \tag{2}$$

(See Exercise 12.)

If the right side of (2) is denoted by p_m, the distribution of X can then be expressed as follows:

$$P(X=k) = \binom{m}{k} \frac{\binom{n+m-k-1}{n}}{\binom{n+m-1}{n}} p_{m-k}, \quad k = 0, 1, \ldots, m-1 \tag{3}$$

(See Exercise 13.)

3.8 SAMPLING FROM DICHOTOMOUS POPULATIONS

Many problems in probability and statistics are concerned with drawing samples from dichotomous populations. By a dichotomous population, we simply mean a collection of objects each of which can be classified as having one of two possible characteristics. For example, in a biological population each object is either male or female, above or below a given age, and so forth. A model that is often useful is that the population consists of N balls numbered $1, 2, \ldots, N$. Suppose the first M balls, numbers 1 through M, are white, and the remaining balls, numbers $M + 1$ through N, are black. A sample of n balls will be drawn from this population. There are two ways to sample.

sampling with replacement

Sampling with replacement assumes that when a ball is drawn, it is replaced before the next ball is drawn. In this case, there is no limitation on the size of n. If the outcome of the experiment is described by the n-tuple of numbers drawn, then there are N^n sample points. We suppose that the probability of each sample point is $1/N^n$.

sampling without replacement

Sampling without replacement assumes that when a ball is drawn, it is not replaced. In this case, n can be no greater than N, the population size. For some purposes, one may as well assume that "a handful" of n balls are drawn from the population, if the particular order in which the n balls are drawn is not relevant. However, often one does want to keep track of the order, and we shall abide by this convention. As in the previous case, the outcome of the experiment is described by the n-tuple of numbers drawn. There are then $N(N-1)\cdots(N-n+1)=N^{(n)}$ sample points, and each is assigned the probability $1/N^{(n)}$.

Example 1 Suppose $N=5$ and $n=2$ (the value of M is not relevant at this point). In sampling with replacement, the $5^2=25$ sample points are the following:

1, 1	2, 1	3, 1	4, 1	5, 1
1, 2	2, 2	3, 2	4, 2	5, 2
1, 3	2, 3	3, 3	4, 3	5, 3
1, 4	2, 4	3, 4	4, 4	5, 4
1, 5	2, 5	3, 5	4, 5	5, 5

In sampling without replacement, the diagonal entries are impossible, and the $5^{(2)}=20$ sample points are the following:

1, 2	2, 1	3, 1	4, 1	5, 1
1, 3	2, 3	3, 2	4, 2	5, 2
1, 4	2, 4	3, 4	4, 3	5, 3
1, 5	2, 5	3, 5	4, 5	5, 4

3.9 THE NUMBER OF WHITE BALLS DRAWN

The first random variable of interest is Y, the number of white balls obtained among the n balls drawn. Remember that there are M white balls in the population—namely, balls 1 through M.

distribution of white balls—sampling with replacement

$$P(Y=k)=\binom{n}{k}\frac{M^k(N-M)^{n-k}}{N^n}, \qquad k=0,1,\ldots,n \tag{1}$$

Equation (1) is an example of the important *binomial distribution* to be discussed in Section 6.1.

Proof of (1). Suppose first that k specified drawings result in white balls and the remaining $n - k$ drawings result in black balls. The number of sample points in this event is $M^k(N - M)^{n-k}$. (Each white ball can be any of the numbers $1, 2, \ldots, M$, and each black ball can be any of the numbers $M + 1, \ldots, N$.) There are $\binom{n}{k}$ choices for specifying the k positions of the white balls in the n-tuple. Hence, the number of sample points in the event $(Y = k)$ is $\binom{n}{k}M^k(N - M)^{n-k}$, which proves (1). ∎

distribution of white balls—sampling without replacement

$$P(Y = k) = \binom{n}{k}\frac{M^{(k)}(N - M)^{(n-k)}}{N^{(n)}}, \qquad k = 0, 1, \ldots, n \tag{2}$$

Equation (2) is an example of the *hypergeometric distribution,* which appears again later in Section 6.3. Notice the formal similarity between Eqs. (1) and (2), but make sure that you understand that the powers in (1) are conventional powers and those in (2) are factorial powers. If k is greater than M, then $P(Y = k)$ is automatically 0 because $M^{(k)} = 0$ (see Exercise 15, Chapter 1). Drawing k white balls is equivalent to drawing $n - k$ black ones. If $n - k$ is greater than $N - M$, then again $P(Y = k)$ is 0, which again is automatically taken care of by Eq. (2).

Proof of (2). The proof is practically the same as that of (1). First suppose that k specified drawings result in white balls. The number of sample points in this event is $M(M - 1) \cdots (M - k + 1)N(N - 1) \cdots (N - M - n - k + 1) = M^{(k)}(N - M)^{(n-k)}$. There are $\binom{n}{k}$ choices for specifying the k positions of the white balls in the n-tuple. Hence, the number of sample points in the event $(Y = k)$ is $\binom{n}{k}M^{(k)}(N - M)^{(n-k)}$, which proves (2). ∎

There is another expression for the distribution of Y in sampling without replacement, namely,

$$P(Y = k) = \frac{\binom{M}{k}\binom{N - M}{n - k}}{\binom{N}{n}}, \qquad k = 0, 1, \ldots, n \tag{3}$$

It is left to the reader to check that the right-hand sides of (2) and (3) are equal. It is important to give a direct interpretation of (3), however. Suppose that we do not keep track of the order in which the n balls are drawn. The sample point then is an *unordered* n-tuple of numbers drawn from the available numbers $1, 2, \ldots, N$. There are $\binom{N}{n}$ sample points each having probability $1\Big/\binom{N}{n}$. For how many of these sample points does Y equal k? There are $\binom{M}{k}$ possible unordered k-tuples of symbols from $1, 2, \ldots, M$ and $\binom{N-M}{n-k}$ possible unordered $(n-k)$-tuples of symbols from $M+1, \ldots, N$; $\binom{M}{k}\binom{N-M}{n-k}$ possible combinations of these form n-tuples. The required probability is, then, the right-hand side of (3).

Example 1 A sample of size three is drawn from an available group of four men and five women. The distributions of the number of men in the sample are shown in Table 3.9.

Notice that the expected values in both cases are 4/3. This fact will be explained next.

We now consider the expected number of white balls drawn in the sample. It is interesting that this expected value does not depend on whether the sampling is with or without replacement.

expected number of white balls drawn

$$EY = n\frac{M}{N} \tag{4}$$

TABLE 3.9

With replacement		*Without replacement*	
k	$P(Y=k)$	k	$P(Y=k)$
0	$\binom{3}{0}5^3/9^3 = 125/729$	0	$\binom{3}{0}5^{(3)}/9^{(3)} = 5/42$
1	$\binom{3}{1}(4)(5^2)/9^3 = 300/729$	1	$\binom{3}{1}(4)(5^{(2)})/9^{(3)} = 20/42$
2	$\binom{3}{2}(4^2)(5)/9^3 = 240/729$	2	$\binom{3}{2}(4^{(2)})(5)/9^{(3)} = 15/42$
3	$\binom{3}{3}4^3/9^3 = 64/729$	3	$\binom{3}{3}4^{(3)}/9^{(3)} = 2/42$

This result is certainly plausible. The proportion of whites in the population is M/N. Equation (4) says that the expected number of whites in the sample is the sample size times the proportion of whites in the population.

Proof of (4) (sampling with replacement). Introduce indicator random variables $Y_1, \ldots, Y_n$, where $Y_i = 1$ if the ith drawing is white and $Y_i = 0$ otherwise. Then $Y = Y_1 + \cdots + Y_n$ and $EY = EY_1 + \cdots + EY_n$. What we need is EY_i = probability that ith drawing is white. The number of sample points (n-tuples) in which the ith coordinate is one of $1, \ldots, M$ is $N^{n-1}M$. Hence, $EY_i = N^{n-1}M/N^n = M/N$. Thus, $EY = nM/N$. ▮

Proof of (4) (sampling without replacement). The proof is practically the same as in the previous case. Use the same indicator random variables $Y_1, \ldots, Y_n$ as before. The number of sample points in which the ith coordinate is one of $1, \ldots, M$ is $N^{(n-1)}M$. The particular position of the coordinate being considered is not relevant. Hence, $EY_i = N^{(n-1)}M/N^{(n)} = M/N$, and $EY = nM/N$. ▮

Example 2 Five cards are drawn from a conventional deck of playing cards.

(a) The expected number of black cards drawn is $5(26/52) = 2.5$.

(b) The expected number of spades drawn is $5(13/52) = 1.25$.

(c) The expected number of black queens drawn is $5(2/52) = 5/26 = 0.19 \cdots$.

Example 3 One-fifth of a biological population is female. Twenty specimens are drawn at random. The expected number of females among these is four.

3.10 WAITING FOR A WHITE

Consider a population of N balls, M of which are white and the remaining $N - M$ of which are black. Drawings are successively made until a white ball is obtained. Let X denote the number of drawings required to obtain a white ball. We suppose now that M is positive—in other words, that there actually are some white balls in the population.

waiting time for a white ball— sampling with replacement

If the sampling is with replacement, the distribution of X, the number of drawings required to obtain a white ball, is given by

$$P(X = k) = \frac{(N - M)^{k-1}M}{N^k}, \qquad k = 1, 2, \ldots \tag{1}$$

The distribution defined by (1) is a special case of the *geometric distribution,* which will be discussed in Section 6.4.

Proof of (1). That the first white ball appears on the kth drawing means that the first $k - 1$ drawings are black and the kth drawing is white. In making k drawings with replacement, there are N^k sample points. The number of these sample points with $k - 1$ black coordinates followed by a white is $(N - M)^{k-1}M$, from which (1) follows. ∎

An equivalent description of the distribution of X can be given in terms of the *tail probabilities.* These are

$$P(X > k) = \frac{(N - M)^k}{N^k} \tag{2}$$

There are several ways of seeing (2). If $X > k$, the first k drawings have all been black. Hence, the probability is $(N - M)^k/N^k$. A less direct way is to compute

$$\begin{aligned} P(X > k) &= P(X = k + 1) + P(X = k + 2) + \cdots \\ &= \frac{(N - M)^k M}{N^{k+1}} + \frac{(N - M)^{k+1} M}{N^{k+2}} + \cdots \\ &= \left[\frac{(N - M)^k M}{N^{k+1}}\right]\left[1 + \frac{(N - M)}{N} + \frac{(N - M)^2}{N^2} + \cdots\right] \\ &= \left[\frac{(N - M)^k M}{N^{k+1}}\right]\left[1 - \frac{(N - M)}{N}\right]^{-1} = \frac{(N - M)^k}{N^k} \end{aligned}$$

(The preceding infinite series is the geometric series $1 + r + r^2 + \cdots = 1/(1 - r)$, with $r = (N - M)/N = 1 - M/N$, which is less than 1 because M is positive. ∎

An important consequence of (2) is that the probabilities

$P(X = 1)$, $P(X = 2)$, . . . sum to 1 as long as M is not 0 [to prove this, let $k \to \infty$ in (2)]. The result implies that it is impossible to draw balls forever without obtaining a white one. No matter how small the proportion of white balls in the population may be, *eventually a white ball must be obtained.*

The easiest way to compute the expected value of X is to use Eq. (3), Section 2.10.

expected waiting time for a white ball

$$EX = \frac{N}{M} \tag{3}$$

Proof of (3).

$$\begin{aligned} EX &= P(X > 0) + P(X > 1) + P(X > 2) + \cdots \\ &= 1 + \frac{(N-M)}{N} + \frac{(N-M)^2}{N^2} + \cdots \\ &= \left[1 - \frac{(N-M)}{N}\right]^{-1} = \frac{N}{M} \quad \blacksquare \end{aligned}$$

Example 1 A box contains 999 black balls and 1 white ball. Balls are successively drawn with replacement until the white ball is obtained. The waiting time probabilities are as follows

k	1	2	3	. . .
$P(X = k)$	0.001	0.000999	0.000998001	. . .

The probability of the white ball showing up at an early stage is very small. Nevertheless, the probabilities sum to 1. Eventually, the white ball must appear! The expected number of draws required is 1000; therefore, on the average, one waits a long time.

Example 2 One-fourth of the birds on a certain island have been tagged. An experimenter posts himself at a feeding station and awaits the arrival of a tagged bird. On the average, the fourth bird to arrive will be tagged.

We next consider the waiting time for a white ball when the sampling is done without replacement. Again, let X denote the number of drawings required to obtain a white ball. The distribution is given as follows.

waiting time for a white ball—sampling without replacement

$$P(X = k) = \frac{(N - M)^{(k-1)}M}{N^{(k)}}, \qquad k = 1, 2, \ldots \tag{4}$$

Notice the formal similarity between this expression and (1). In (4), the powers are factorial, whereas in (1) they are conventional. For $k = 1$, $(N - M)^{(0)}$ should be interpreted as 1. Notice that (4) is automatically 0 when $k > N - M + 1$.

Proof of (4). The proof really depends upon to which sample space one chooses to refer. The most careful approach is to suppose that the drawings are repeatedly made until all N balls have been drawn. The sample space then consists of all $N!$ permutations of $1, 2, \ldots, N$, each of which has probability $1/N!$ The random variable X is the position number in which one of the symbols $1, 2, \ldots, M$ first appears. The event $(X = k)$ means that the first $k - 1$ entries represent black, the kth entry represents white, and the remaining entries can be anything. The number of sample points is

$$(N - M) \cdots (N - M - k + 2)M(N - k) \cdots (2)(1)$$
$$= (N - M)^{(k-1)}M(N - k)!$$

Hence, the probability is $(N - M)^{(k-1)}M(N - k)!/N! = (N - M)^{(k-1)}M/N^{(k)}$. ▮

Another approach to (4) is first to determine the tail probabilities, which are

$$P(X > k) = \frac{(N - M)^{(k)}}{N^{(k)}} \tag{5}$$

The proof of (5) is similar to that of (4) and is left to the reader.

Example 3

It is known that among 100 persons there are 3 persons having a certain unusual blood type. A search for the rare type is conducted by successively drawing persons at random and determining the blood type. The probability distribution of the number of persons that must be examined to find the right type is as follows:

k	1	2	$\cdots$	k	$\cdots$
$P(X = k)$	$\frac{3}{100}$	$\frac{(97)(3)}{(100)(99)}$	$\cdots$	$\frac{(97)^{(k-1)}(3)}{(100)^{(k)}}$	$\cdots$

It will be seen that the expected number of persons that must be examined is $101/4 = 25.25$. This expected value is of the form $(N + 1)/(M + 1)$, as will be explained.

Example 4 A deck of six cards numbered 1, 2, 3, 4, 5, 6 is shuffled and laid out in a line. What is the distribution of the first position in which an even integer appears? The distribution is the following:

k	1	2	3	4
$P(X=k)$	$\frac{10}{20}$	$\frac{6}{20}$	$\frac{3}{20}$	$\frac{1}{20}$

The expected position is $[1(10) + 2(6) + 3(3) + 4(1)]/20 = 7/4$.

To evaluate EX we first need a bit of algebraic machinery. Inasmuch as the sum of the probabilities in (4) equals 1,

$$\frac{M}{N} + \frac{(N-M)M}{N^{(2)}} + \frac{(N-M)^{(2)}M}{N^{(3)}} + \cdots + \frac{(N-M)^{(N-M)}M}{N^{(N-M+1)}} = 1 \quad (6)$$

Equation (6) is equivalent to

$$1 + \frac{(N-M)}{(N-1)} + \frac{(N-M)^{(2)}}{(N-1)^{(2)}} + \cdots + \frac{(N-M)^{(N-M)}}{(N-1)^{(N-M)}} = \frac{N}{M} \quad (7)$$

Because M and N are arbitrary positive integers, Eq. (7) is equivalent to

$$\frac{a^{(0)}}{b^{(0)}} + \frac{a^{(1)}}{b^{(1)}} + \frac{a^{(2)}}{b^{(2)}} + \cdots + \frac{a^{(a)}}{b^{(a)}} = \frac{(b+1)}{(b-a+1)} \quad (8)$$

whenever a and b are positive integers and $b > a - 1$. [Let $N - 1 = b$ and $N - M = a$ in (7).] Equation (8) now allows us to evaluate EX.

expected waiting time for a white ball—sampling without replacement

$$EX = \frac{N+1}{M+1} \quad (9)$$

Proof of (9). Again, we use Eq. (3), Section 2.10.

$$\begin{aligned} EX &= P(X > 0) + P(X > 1) + \cdots \\ &= 1 + \frac{(N-M)}{N} + \frac{(N-M)^{(2)}}{N^{(2)}} + \cdots + \frac{(N-M)^{(N-M)}}{(N)^{(N-M)}} \\ &= \frac{N+1}{M+1} \end{aligned}$$

[Set $N - M = a$, $N = b$ in (8).] ∎

It is interesting to compare the expected values for sampling with replacement and without replacement. These expected values are N/M and $(N + 1)/(M + 1)$, respectively. Since $N/M - (N + 1)/(M + 1) = (N - M)/(M + 1)M$, these expected values are equal only if $M = N$. In that case, the population consists of all white balls and a white ball is necessarily first drawn on the initial trial in both cases. Otherwise, the expected waiting time is greater when the sampling is with replacement. This latter is certainly reasonable—if the same black balls do not have to be examined again, as in sampling without replacement, the first white ball will be obtained sooner (see Exercise 15).

SUMMARY

Combinatorial problems are abundant in elementary probability theory. They are useful to illustrate techniques and as a source of examples. There is a tendency in first courses to identify all probability theory with combinatorial problems. This should not be the case, although a mastery of combinatorics can be very helpful.

The matching problem concerns the number of matches in a random permutation. Suppose n objects, marked $1, 2, \ldots, n$, are randomly laid out in a line against n fixed positions, also marked $1, 2, \ldots, n$. We count X, the total number of times that objects appear, against similarly marked fixed positions; X is called the number of matches. It is interesting that the expected number of matches equals 1, independent of n. By various combinatorial methods, we find the distribution of X and show that for large n, $P(X = k)$ is approximately equal to $e^{-1}/k!$

Other combinatorial problems are concerned with the random assignment of n balls among m cells. (There are various equivalent ways of formulating this situation.) We obtain information about the following:

the number of balls to land in a given cell,
the number of unoccupied cells,
waiting time to achieve full occupancy,
multiple occupancies (the birthday problem).

Finally, we consider combinatorial problems relating to sampling from dichotomous populations. (For simplicity, call these dichotomies black and white.) Such sampling can be with or without replacement. For either type of sampling, we find the distribution of the number of white elements in a sample and the number of drawings required to obtain a white.

EXERCISES

1 Example 1, Section 2.6, and Exercises 7 and 11, Chapter 2, relate to the matching problem. You might want to review these in connection with reading the material on matching in Chapter 3.

2 Verify the density functions for the number of matches given in Table 3.3 using Eq. (2), Section 3.2.

3 Show that $|P(X = 0) - P(X = 1)| = 1/n!$, $n = 1, 2, \ldots$, where X is the number of matches using n cards.

4 Suppose that $n > 2$.
(a) Show that the expected number of matches in the first two positions is $2/n$. (Hint: Use indicator random variables, as in Example 1, Section 2.6.)
(b) Show that the distribution of matches in the first two positions is the following:

k	0	1	2
probability of k matches	$\dfrac{(n^2 - n - 2n + 3)}{n(n-1)}$	$\dfrac{2(n-2)}{n(n-1)}$	$\dfrac{1}{n(n-1)}$

[Hint: Use Exercise 14a, Chapter 2.]

5 Let X be the number of matches using n cards. Equation (4), Section 3.2, says that $EX = 1$ for any n.
(a) Show also that if $n \geqslant 2$, $E[X(X-1)] = 1$; if $n \geqslant 3$, $E[X(X-1)(X-2)] = 1$; and, in general,

$$E(X^{(r)}) = 1, \qquad r = 1, 2, \ldots, n$$

The term $E[X^{(r)}]$ is called the *rth factorial moment* of X is 1 for r. [Hint: Imitate the second proof of Eq. (4), Section 3.2.]
(b) Show that $EX^{(r)} = 0$ if $r > n$. [Hint: For any positive integers k and r, $k^{(r)} = 0$ if $r > k$.]
(c) Verify (a) for several values of r and n using the density functions tabulated in Table 3.3. (For example, for $n = 5$ and $r = 2$, $E(X^{(2)}) = 2 \cdot 1(20/120) + 3 \cdot 2(10/120) + 4 \cdot 3(0) + 5 \cdot 4(1/120) = 120/120 = 1$.)

6 Example 2 in Section 2.6, Example 2 in Section 2.8, Example 2 in Section 2.10, and Exercises 5, 8, and 15 of Chapter 2 relate to the occupancy problem. You might want to review them while reading the material on occupancy in Chapter 3.

7 Suppose n balls are randomly distributed among m cells, and r is a positive integer, $r \leqslant m$. Let V be the number of balls to fall in r specified cells.
(a) Show that $P(V = k) = \binom{n}{k} r^k (m - r)^{n-k}/m^n$, $k = 0, 1, \ldots, n$. [Hint: Imitate the proof of (1), Section 3.4.]
(b) Show that $EV = rn/m$ by imitating the proof of (2), Section 3.4.
(c) Show that $EV = rn/m$ by evaluating

$$\sum_{k=0}^{n} \frac{k\binom{n}{k} r^k (m-r)^{n-k}}{m^n}$$

{Hint: $k\binom{n}{k} = n\binom{n-1}{k-1}$. Hence, the sum is the same as

$$rn \sum_{k-1=0}^{n-1} \binom{n-1}{k-1} \frac{r^{k-1}(m-r)^{n-1-(k-1)}}{m^n} = \frac{rn}{m}\left[\frac{r+(m-r)}{m}\right]^{n-1}$$

by the binomial expansion, IX, Section 1.5.}

8 Extend Table 3.6 to $n = 5, 6$ and $m = 5, 6$, using the iterative relationship, Eq. (4), Section 3.4.

9 (a) Verify the fact that $EX = (m-1)^n/m^{n-1}$ for several values of m and n, using the density functions tabulated in Table 3.6. [For instance, for $m = 4$, $n = 3$, $EX = 1(6/16) + 2(9/16) + 3(1/16) = 27/16 = 3^3/4^2$.]
(b) Show that $E[X(X-1)] = m(m-1)(m-2)^n/m^n$, and that, in general,

$$E(X^{(r)}) = \frac{m^{(r)}(m-r)^n}{m^n}, \qquad r = 1, 2, \ldots$$

[Hint: Imitate the second proof of (5), Section 3.4.]
(c) Verify that $E(X^{(2)}) = m(m-1)(m-2)^n/m^n$ computationally for several values of m and n, using the density functions in Table 3.6. {For instance, for $m = 4$, $n = 3$, $E[X(X-1)] = 2 \cdot 1(9/16) + 3 \cdot 2(1/16) = 24/16 = 4 \cdot 3(2^3/4^3)$.}

10 (a) A coin is thrown until both head and tail appear at least once. Let Z be the number of throws required. Show that

$$P(Z > n) = \frac{1}{2^{n-1}}, \qquad n = 1, 2, \ldots$$

[Hint: That $Z > n$ means the first n throws are all heads or all tails. Hence, the probability is $2(1/2^n)$.]
(b) Use Eq. (3), Section 2.10, to verify that $EZ = 3$. (See Example 4a, Section 3.5.) [Hint: $\Sigma_{n=0}^{\infty} P(Z > n) = 1 + 1 + 1/2 + 1/2^2 + \cdots = 3$.]

11 Suppose n balls are randomly distributed among m cells. Let Y be the number of cells occupied by two or more balls. Equation (2), Section 3.6, in effect says that $P(Y = 0) = m^{(n)}/m^n$.
(a) Determine the distribution of Y for $m = 12$, $n = 4$.
(b) Show that, in general,

$$P(Y = 1) = \frac{\binom{n}{2}m^{(n-1)} + \binom{n}{3}m^{(n-2)} + \cdots + \binom{n}{n}}{m^n}$$

12 Prove Eq. (2) for $P(X = 0)$ in Section 3.7.
$\left[\text{Hint: } S_k = \binom{m}{k}\binom{n+m-k-1}{n}\Big/\binom{n+m-1}{n}.\right]$

13 Prove Eq. (3) for $P(X = k)$ in Section 3.7. [Hint: Imitate the proof of (4), Section 3.4.]

14 Verify algebraically that the expressions for $P(Y = k)$ as given by (2) and (3), Section 3.9, are equivalent.

15 The intuitive idea that in sampling without replacement the first white ball appears earlier than in sampling with replacement is made rigorous by verifying that

$$\frac{(N-M)^k}{N^k} \geqslant \frac{(N-M)^{(k)}}{N^{(k)}} \quad \text{for} \quad k=1,2,\ldots$$

Explain why and verify the relationship [see Eqs. (2) and (5), Section 3.10.]

16 From a conventional deck of 52 playing cards, 13 are dealt.

(a) What is the probability of five hearts? [Answer: $\left[\binom{13}{5}\binom{39}{8}\right] \Big/ \binom{52}{13}$.]

(b) What is the probability of fewer than five black cards? [Answer: $\sum_{i=0}^{4}\left[\binom{26}{i}\binom{26}{13-i}\right] \Big/ \binom{52}{13}$.]

(c) Do (a) and (b) if the cards are drawn with replacement. [Answers: $\binom{13}{5}3^8/4^{13}$ and $\sum_{i=0}^{4}\binom{13}{i}(1/2)^{13}$.]

17 In Example 1, Section 3.9, suppose the sample size is four (instead of three); tabulate the two distributions. (Check that the probabilities sum to 1 and evaluate the expected values to check with the general formula.)

18 An electronic random-number generator produces triples of integers (i, j, k) repeatedly. (Each of i, j, and k can be $0, 1, \ldots, 9$.)

(a) What is the probability that a sequence of the form (i, i, i) appears among the first 10 sequences? [Answer: $1-(0.99)^{10}$.]

(b) What is the expected waiting time for such a sequence? (Answer: 100 sequences.) [Hint: As a model for this experiment, assume that the device is sampling with replacement from the population of the 10^3 possible triples (i, j, k).]

(c) What is the probability of exactly four sequences of the form (i, i, i) among the first 10 sequences? [Answer: $\binom{10}{4}(0.01)^4(0.99)^6$.]

19 A population contains five white objects and five black objects. Drawings are successively made *without replacement* until a black object is drawn. Tabulate the distribution of the number of drawings required. (Check that the probabilities sum to 1 and calculate the expected value to check with the general formula.)

20 Among N balls, M_1 are of type a, M_2 are of type b, and the remainder $N-M_1-M_2$ are of type c. Denote the proportions by

$$\frac{M_1}{N}=p_1 \qquad \frac{M_2}{N}=p_2$$

(a) Balls are successively drawn *with replacement.* Show that the probability of drawing a ball of type a before drawing one of type b is $p_1/(p_1+p_2)$.

(b) Balls are successively drawn *without replacement*. Show that the probability of drawing a ball of type a before drawing one of type b is still $p_1/(p_1 + p_2)$. [Hint: The required probability is

$$\sum_{k=1}^{N} \frac{M_1(N-M_1-M_2)^{(k-1)}}{N^{(k)}} = \frac{M_1}{N} \sum_{k=1}^{N} \frac{(N-M_1-M_2)^{(k-1)}}{(N-1)^{(k-1)}}$$

Use equation (8), Section 3.10, to evaluate this sum.)

21 From n balls numbered $1, 2, \ldots, n$, m balls are drawn. Let X be the largest number obtained.

(a) If the sampling is done *with replacement*, show that

$$P(X = k) = \frac{k^m - (k-1)^m}{n^m}, \qquad k = 1, 2, \ldots, n$$

[Hint: The event $(X \leq k)$ means that each of the m balls is no greater than k. There are n^m sample points, k^m of which make up the event $(X \leq k)$. $P(X = k) = P(X \leq k) - P(X \leq k - 1)$.]

(b) Still suppose the sampling is with replacement. Show that

$$EX = n - \sum_{k=0}^{n} \frac{k^m}{n^m}$$

Notice that for large n, EX/n is approximately equal to

$$1 - \int_0^1 x^m \, dx = \frac{m}{m+1}$$

(c) Now suppose the sampling is *without replacement*. (It is necessary to assume that $m \leq n$.) Show that

(i) $P(X \leq k) = \binom{k}{m} \Big/ \binom{n}{m}$

(ii) $P(X = k) = \binom{k-1}{m-1} \Big/ \binom{n}{m}$, $k = m, \ldots, n$

[Note that (i) and (ii) can be proved separately. Also, (ii) can be deduced from (i), since $P(X = k) = P(X \leq k) - P(X \leq k - 1)$.]

(d) Since $\Sigma_{k=m}^{n} P(X = k) = 1$, deduce the formula

$$\binom{a}{a} + \binom{a+1}{a} + \cdots + \binom{r}{a} = \binom{r+1}{a+1}$$

which holds whenever a and r are positive integers, $r > a$. [In (d) and (e), we still suppose that sampling is without replacement.]

(e) Show that $EX = m(n+1)/(m+1)$. [Hint: *Method 1.* $EX = P(X > 0) + P(X > 1) + \cdots + P(X > n - 1)$. Use the facts that $P(X > n) = P(X > n + 1) = \cdots = 0$, $P(X > 0) = \cdots = P(X > m - 1) = 1$. Now use (c), (i), and (d). *Method 2.* $EX = \Sigma_{k=1}^{n} k\binom{k-1}{m-1} \Big/ \binom{n}{m}$, $k\binom{k-1}{m-1} = m\binom{k}{m}$. Now use (d).]

22 A deck of $m+n$ cards marked $1, 2, \ldots, m, m+1, \ldots, m+n$ is randomly laid out against $m+n$ fixed positions, marked the same way. Let X be the number of matches in the first m positions and Y the number of matches in the next n positions.

(a) Show that $P[(X=k) \cap (Y=r)] = \binom{m}{k}\binom{n}{r}(m+n-k-r)!p_{m+n-k-r}/(m+n)!$ when $k=0, 1, \ldots, m$ and $r=0, 1, \ldots, n$. [Hint: Suppose first that the k and r matches are in specified positions. Then the remaining $m+n-k-r$ cards must be *completely mismatched.* The number of sample points having this property is $N_{0,m+n-k-r}$ [$N_{k,n}$ is defined in the proof of (2), Section 3.2]. Hence, the number of sample points in the event $(X=k) \cap (Y=r)$ is $\binom{m}{k}\binom{n}{r}N_{0,m+n-k-r}$.]

(b) Suppose that m and n approach ∞ is such a way that $m/(m+n)$ approaches ρ where $0 < \rho < 1$. Show that $P[(X=k) \cap (Y=r)]$ approaches

$$\frac{\rho^k e^{-\rho}}{k!} \frac{(1-\rho)^r e^{-(1-\rho)}}{r!}$$

{Hint: $P[(X=k) \cap (Y=r)]$ can be written as

$$\frac{m^{(k)}n^{(r)}}{(m+n)^{(k+r)}} \frac{p_{m+n-k-r}}{k!r!}$$

with a little algebra. As $m \to \infty$, $n \to \infty$ $[m^{(k)}n^{(r)}/(m+n)^{(k+r)}] \to \rho^k(1-\rho)^{k+r}$, and $p_{m+n-k-r} \to e^{-1} = e^{-\rho}e^{-(1-\rho)}$.}

Part (b) makes it plausible that $P(X=k) = P[(X=k) \cap (Y<\infty)]$ has the limit $\rho^k e^{-\rho}/k!$ because $\Sigma_{r=0}^{\infty} (1-\rho)^r e^{-(1-\rho)}/r! = e^{(1-\rho)}e^{-(1-\rho)} = 1$. Indeed, this statement is true, but a justification must be given for the interchange of limits. Similarly, it is plausible and true that $P(Y=r)$ has the limit $(1-\rho)^r e^{-(1-\rho)}/r!$ In other words, asymptotically X and Y are Poisson distributed with parameters ρ, $1-\rho$, respectively. In terms of the concept of independence to be introduced in Section 4.7, (b) says that X and Y are asymptotically independent.

23 A deck of n cards is marked $1, 2, \ldots, n$. There are n fixed positions also marked $1, 2, \ldots, n$. A card is drawn at random and its designation is entered in position 1. The card is replaced, a second drawing is made, and its designation is entered in position 2. The card is replaced, and so forth, until entries are made in all n positions. Let X be the number of matches. (If the cards were not successively replaced, then X would have the matching distribution described in Section 3.2.)

(a) Show that $EX = 1$, just as in the conventional matching problem.

(b) Show that

$$P(X=k) = \binom{n}{k}\frac{(n-1)^{n-k}}{n^n}, \qquad k = 0, 1, \ldots, n$$

(c) Use (b) to show that the probability of k matches using n cards has the limit $e^{-1}/k!$ just as in the conventional matching problem [see (3), Section 3.2]. [Hint: The probability in (b) can be written as $(n^{(k)}/n^k)(1 - 1/n)^n/(1 - 1/n)^k k!$ Use the facts that $\lim_{n\to\infty} n^{(r)}/n^r = 1$, $\lim_{n\to\infty}(1 - 1/n)^n = e^{-1}$, and $\lim_{n\to\infty}(1 - 1/n)^k = 1$, for fixed k.]

NOTES

1 For additional material on combinatoric analysis, see Chapter II of [9].

2 The matching problem has been discussed in probability theory since the eighteenth century. Montmort (1678–1719) determined Eq. (1), Section 3.2, for $n = 13$, attributing the proof to Nicolas Bernoulli (1687–1759). See [13].

3 For a review of the power series expansion of e^x, see Section 18.4 of [2] or Section 11.11 of [3]. For the assertion about the remainder in the expansion of e^{-1}, see the discussion of alternating series in Section 18.10 of [2] or Section 10.17 of [3].

4 The equality of the two expressions (5) and (6), Sections 3.5, appears as Exercise 18, Chapter II of [9].

5 A discussion of various occupancy models, including the Bose–Einstein scheme appears in Section 5, Chapter 2 of [14].

Chapter Four

CONDITIONAL PROBABILITY AND INDEPENDENCE

Empirically, $P(A)$ represents an appraisal of the likelihood that a chance experiment will produce an outcome in the set A. If the experimenter has prior information that the outcome must be in a set B, this information should be used to reappraise the likelihood that the outcome will also be in A. The reappraised probability is $P(A|B)$, the *conditional probability of A given B*. Possibly, knowing that the outcome must be in B does not change the likelihood that the outcome will also be in A, in which case A and B are *independent.* This chapter develops the concepts of independence and conditional probability, particularly as they relate to distributions of random variables (Note 1).

4.1 CONDITIONAL PROBABILITY

Suppose that B is an event the probability of which is not 0. Then for any event A, the *conditional probability* of A given B, denoted $P(A|B)$, is defined as follows.

conditional probability

$$P(A|B) = \frac{P(A \cap B)}{P(B)} \tag{1}$$

Since $(A \cap B) \subset B$ and $P(A \cap B) \leqslant P(B)$, it follows that

$$0 \leqslant P(A|B) \leqslant 1$$

Equally easy to check are the properties

$$P(\mathfrak{X}|B) = 1, \qquad P(\phi|B) = 0$$

and if A_1 and A_2 are disjoint, $P(A_1 \cup A_2|B) = P(A_1|B) + P(A_2|B)$. In fact, all the properties (a) through (e) of Section 1.4 are satisfied if the symbol $P(\ )$ is replaced by the symbol $P(\ |B)$ throughout (see Exercise 1).

We can interpret $P(A|B)$ as a reevaluation of the probability of A, *given that we restrict our interest only to sample points in the event* B. If B is to play the role of the whole sample space, then the reevaluated sample point probabilities, when summed over B, must sum to 1. One way of guaranteeing it is to divide all the original probabilities by $P(B)$. Now, to find the reevaluated probability of a set A, first find the original probability of that part of A that lies in B [namely, $P(A \cap B)$], and then divide by $P(B)$. Thus, the reevaluated probability is $P(A \cap B)/P(B) = P(A|B)$.

Insight as to the meaning of conditional probability can be gained from the frequency theory interpretation of probability (see Section 1.2). Suppose that the experiment in question is repeated N times. Let $N(B)$ count the *number* of repetitions among all N that are in the event B. According to the frequency theory, the observed *proportion* $N(B)/N$ will be close to the theoretical probability $P(B)$, if N is large. Now suppose that among those outcomes in B attention is restricted to outcomes also in event A, disregarding all other outcomes. What proportion of the retained outcomes will be in event A? This proportion will be $N(A \cap B)/N(B) = [N(A \cap B)/N]/[N(B)/N]$, which, according to the frequency theory, will be close to $P(A \cap B)/P(B) = P(A|B)$, the conditional probability of A given B.

A related interpretation can be made in a gaming context. Suppose that a gambler wants to bet that a game will turn out with one of the outcomes that make up the event A. His appraisal of the probability is $P(A)$, which may be based on his past experience in observing the proportion $N(A)/N$. Suppose now that by cheating or by buying some side information he learns that, in fact, the game has turned out with one of the outcomes making up B, but he does not know which particular outcome has actually turned up. How should he reevaluate the chance of observing an outcome in A? In his past experience over N trials, he has observed $N(B)$ outcomes in B, $N(A \cap B)$ of which were in A as well as B. Thus, his revised estimate of the chance of observing an outcome in A is $N(A \cap B)/N(B) = [N(A \cap B)/N]/[N(B)/N]$.

Example 1 A coin is thrown three times yielding a total of two heads. What is the conditional probability of a head on the first trial? Define the events,

$A =$ head on first trial $\qquad B =$ two heads in three trials

Then

$$P(A \cap B) = \frac{2}{8}, \qquad P(B) = \frac{3}{8}$$

since $A \cap B$ consists of the sample points (H, H, T), (H, T, H), and B consists of the sample points (T, H, H), (H, T, H), (H, H, T). Hence,

$$P(A|B) = \frac{2/8}{3/8} = \frac{2}{3}$$

Thus, with the prior information that two heads have been obtained, the reappraised probability of a head on the first trial, $P(A|B)$, is 2/3, instead of the original probability of a head, $P(A) = 1/2$ (see Exercise 4).

Example 2 An honest coin is thrown n times producing all tails. What is the conditional probability of a head on the next trial? Define the events,

$A =$ head on $(n + 1)$st trial $\qquad B = n$ tails on first n trials

Then,

$$P(A \cap B) = \frac{1}{2^{n+1}}, \qquad P(B) = \frac{1}{2^n}$$

and

$$P(A|B) = \frac{1/2^{n+1}}{1/2^n} = \frac{1}{2}$$

Hence, the conditional probability is the same as the original probability, $P(A) = 1/2$. This is related to the concept of *independence,* introduced in Section 4.6.

Equation (1) can be written as

$$P(A \cap B) = P(A|B)P(B) \tag{2}$$

Equation (2) is useful when it is possible to assign probabilities to $P(A|B)$ and to $P(B)$, but a plausible probability for $P(A \cap B)$ is not immediately evident. A version of (2) for three events A, B, C is the following:

$$P(A \cap B \cap C) = P(A|B \cap C)P(B|C)P(C) \tag{3}$$

Proof of (3). By (1), we have the two relations,

$$P(A|B \cap C) = \frac{P(A \cap B \cap C)}{P(B \cap C)}, \qquad P(B|C) = \frac{P(B \cap C)}{P(C)}$$

from which (3) follows. ▮

Example 3 A card is drawn at random from a deck of 10 cards numbered 1, 2, . . . , 10. Suppose that the number drawn is k. Then from among the cards 1 through k a second card is drawn at random. Suppose the number drawn on the second trial is r. Then from among the cards 1 through r a third card is drawn at random. What is a reasonable probability to assign to the sequence (9, 7, 5)? Define events A, B, and C as follows:

$A =$ (5 drawn on third trial)
$B =$ (7 drawn on second trial)
$C =$ (9 drawn on first trial)

The following probability assignments are reasonable:

$P(C) = 1/10$, because 1 card is being selected out of 10 cards
$P(B|C) = 1/9$, because, under the condition, 1 card is being selected out of 9 cards
$P(A|B \cap C) = 1/7$, because, under the condition, 1 card is being selected out of 7 cards

Once we accept these three probabilities, we must have, by (3),

$$P(A \cap B \cap C) = \left(\frac{1}{7}\right)\left(\frac{1}{9}\right)\left(\frac{1}{10}\right) = \frac{1}{630}$$

(See Exercise 5.)

Example 4 (Polya urn scheme, Note 2.) A box (or urn) contains two white balls and one black ball. A ball is drawn and then replaced along with an additional ball of the same color. The procedure is repeated—a second ball is drawn and is replaced together with an additional ball of the same color as drawn on the second drawing. Finally, a third ball is drawn and the experiment is ended. The problem is to set up a reasonable probability model. The sample points are represented by the triples (b, b, b), (b, b, w), (b, w, b), (b, w, w), (w, b, b), (w, b, w), (w, w, b), (w, w, w). Probabilities will be assigned to these sample points by first agreeing on plausible values for various conditional probabilities. Let B_i denote black on the ith trial and W_i white on the ith trial, $i = 1, 2, 3$. Thus, $B_1 \cap W_2 \cap B_3$ consists of the single sample point (b, w, b), $B_1 \cap B_2 \cap B_3$ is the sample point

TABLE 4.1

Sample points	*Probabilities*
(b, b, b)	6/60
(b, b, w)	4/60
(b, w, b)	4/60
(b, w, w)	6/60
(w, b, b)	4/60
(w, b, w)	6/60
(w, w, b)	6/60
(w, w, w)	24/60

(b, b, b), and so forth. The following probability assignments are intuitively plausible:

$P(B_1) = \frac{1}{3}$, $P(W_1) = \frac{2}{3}$ (initially, the box has 1 black and 2 white balls)

$P(W_2|W_1) = \frac{3}{4}$, $P(B_2|W_1) = \frac{1}{4}$ (under the condition W_1, the box now contains 1 black and 3 white balls)

$P(W_2|B_1) = \frac{2}{4}$, $P(B_2|B_1) = \frac{2}{4}$ (under the condition B_1, the box now contains 2 black and 2 white balls)

$P(W_3|W_1 \cap W_2) = \frac{4}{5}$, $P(B_3|W_1 \cap W_2 = \frac{1}{5}$ (under the condition $W_1 \cap W_2$, the box now contains 4 white and 1 black balls)

and so forth.

Thus, $P(w, w, w) = P(W_1)P(W_2|W_1)P(W_3|W_1 \cap W_2) = (\frac{2}{3})(\frac{3}{4})(\frac{4}{5}) = 24/60$, and $P(b, w, w) = P(B_1)P(W_2|B_1)P(W_3|B_1 \cap W_2) = (\frac{1}{3})(\frac{2}{4})(\frac{3}{5})$. The complete list of sample points and their probabilities appears in Table 4.1.

4.2 JOINT, MARGINAL, AND CONDITIONAL DISTRIBUTIONS

Suppose X and Y are two random variables, the possible values of which are $(a_1, a_2, \ldots)$, $(b_1, b_2, \ldots)$, respectively. The joint density function of X and Y is the function f defined as follows.

joint density function of two random variables

$$f(a, b) = P[(X = a) \cap (Y = b)], \qquad \begin{matrix}(a = a_1, a_2, \ldots)\\(b = b_1, b_2, \ldots)\end{matrix} \qquad (1)$$

Example 1 A coin is thrown three times. Let X be the number of heads in the first two throws and let Y be the number of heads in the last two throws. The joint density function of X and Y appears in Figure 4.1.

Since the events $(X = a_i) \cap (Y = b_j)$ are disjoint as (a_i, b_j) varies over all pairs of possible values of X and Y, and the union of these disjoint events is the whole sample space $\mathfrak{X}$, it follows that

$$\sum_{i,j} P[(X = a_i) \cap (Y = b_j)] = 1$$

(see Figure 4.2).

From the joint density function of X and Y, one can retrieve the individual density functions of X and Y. Since

$$[(X = a_i) \cap (Y = b_1)] \cup [(X = a_i) \cap (Y = b_2)] \cup \cdots = (X = a_i)$$

is a union of mutually disjoint events, it follows from (c), Section 1.4, that

$$P[(X = a_i) \cap (Y = b_1)] + P[(X = a_i) \cap (Y = b_2)] + \cdots$$
$$= P(X = a_i) \qquad (2)$$

Similarly,

$$P[(X = a_1) \cap (Y = b_j)] + P[(X = a_2) \cap (Y = b_j)] + \cdots$$
$$= P(Y = b_j) \qquad (3)$$

By way of emphasizing that the individual density functions of X and Y are obtained from the joint density using (2) and (3), these

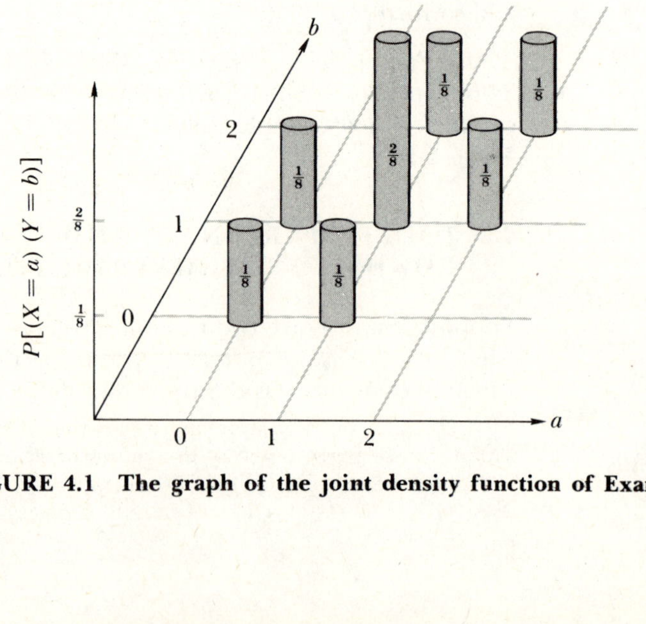

FIGURE 4.1 The graph of the joint density function of Example 1.

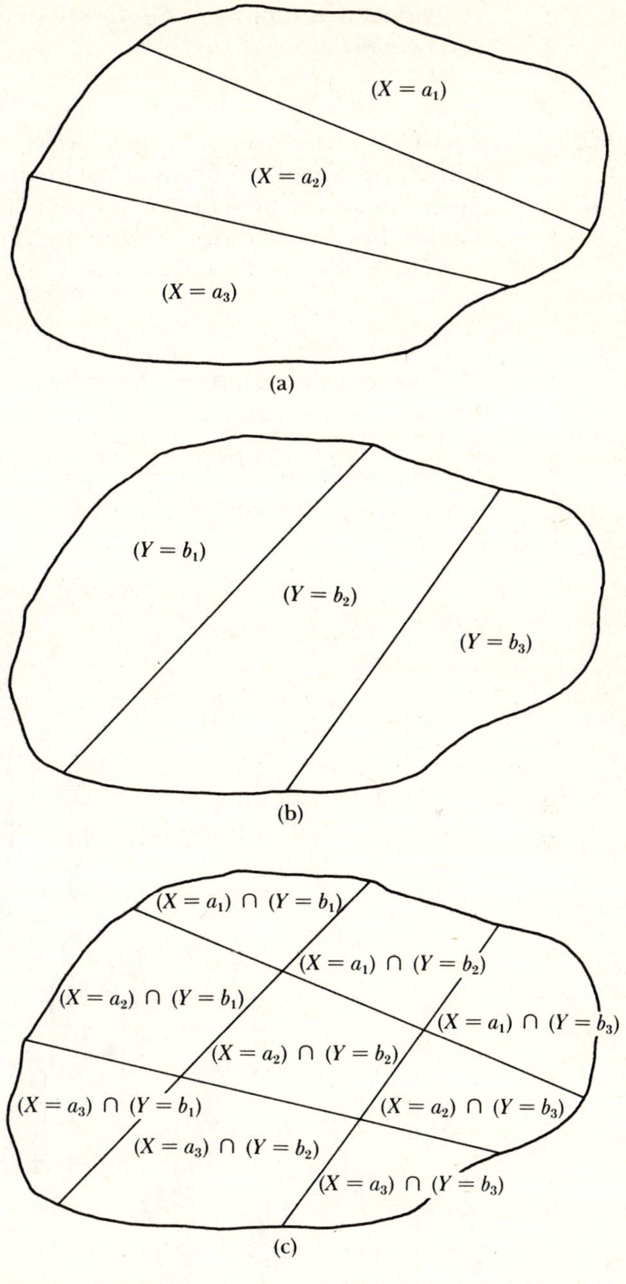

FIGURE 4.2 (a) A sample space partitioned by the values of *X*. (b) A sample space partitioned by the values of *Y*. (c) A sample space partitioned by the values of *X* and *Y* jointly.

individual densities may be referred to as *marginal density functions* (or *marginal distributions*).

Example 2 Consider again Example 1, in which a coin is thrown three times, with X the number of heads in the first two throws and Y the number of heads in the last two throws. Figure 4.3 repeats the joint density function shown in Figure 4.1. The marginal density functions of X and Y were obtained according to (2) and (3).

Consider now the events $(X = a)$ and $(Y = b)$ and the conditional probability $P(X = a|Y = b)$. According to the definition (1) of Section 4.1,

$$P(X = a|Y = b) = \frac{P[(X = a) \cap (Y = b)]}{P(Y = b)} \tag{4}$$

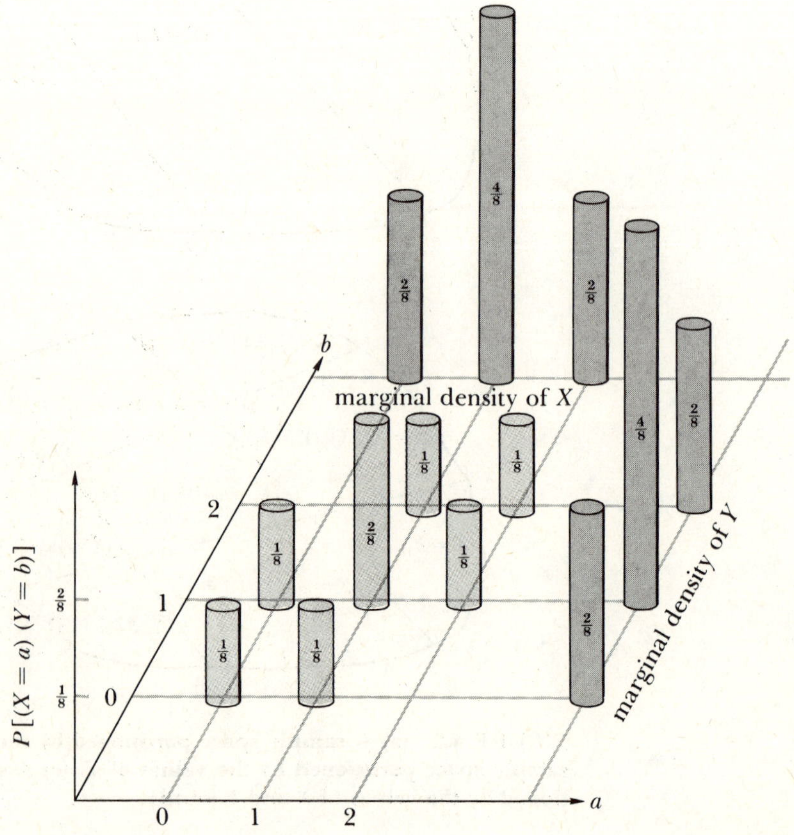

FIGURE 4.3 Marginal density functions.

This equation is well defined as long as $P(Y=b) > 0$. Suppose now that b is fixed and a varies over the possible values of X.

conditional distribution

The function f, defined by

$f(a) = P(X = a|Y = b)$

where a varies over $a_1, a_2, \ldots$, the possible values of X, is called the *conditional density function (or distribution) of X given that $Y = b$*, assuming $P(Y = b) \neq 0$.

The conditional density function is like an ordinary density function in that

$$0 \leq f(a) \leq 1 \quad \text{and} \quad f(a_1) + f(a_2) + \cdots = 1$$

The last assertion follows from (3), because

$$\sum_i P(X = a_i|Y = b) = \sum_i \frac{P[(X = a_i) \cap (Y = b)]}{P(Y = b)} = \frac{P(Y = b)}{P(Y = b)} = 1 \tag{5}$$

Example 3

Five balls are randomly distributed among three boxes. Let X be the number of balls in the first box and let Y be the number of balls in the first two boxes. The joint density function of X and Y is shown in Table 4.2, together with the conditional distributions of X, given the various Y values. (Verify the entries and see Exercise 8 for a generalization.)

Example 4 (Bayes' formula)

Suppose that X and Y are random variables with possible values $(a_1, a_2, \ldots)$, $(b_1, b_2, \ldots)$, respectively. We leave it to the reader to verify that

$$P(Y = a_i|X = b_j) = \frac{P(X = b_j|Y = a_i)P(Y = a_i)}{\sum_k P(X = b_j|Y = a_k)P(Y = a_k)} \tag{6}$$

The relation (6) is called Bayes' formula and has been used for some rather wild evaluations of the probability of life on Mars, or the probability of the sun rising tomorrow (Note 3).

The preceding ideas all extend in a natural way to more than two random variables. Thus, if X, Y, and Z are random variables, then the function f defined by

$$f(a, b, c) = P[(X = a) \cap (Y = b) \cap (Z = c)]$$

TABLE 4.2 $P[(X = a) \cap (Y = b)]$

b	a					
	0	1	2	3	4	5
5	$1/3^5$	$5/3^5$	$10/3^5$	$10/3^5$	$5/3^5$	$1/3^5$
4	$5/3^5$	$20/3^5$	$30/3^5$	$20/3^5$	$5/3^5$	
3	$10/3^5$	$30/3^5$	$30/3^5$	$10/3^5$		
2	$10/3^5$	$20/3^5$	$10/3^5$			
1	$5/3^5$	$5/3^5$				
0	$1/3^5$					

a	$P(X=a\|Y=0)$	$P(X=a\|Y=1)$	$P(X=a\|Y=2)$	$P(X=a\|Y=3)$	$P(X=a\|Y=4)$	$P(X=a\|Y=5)$
0	1	1/2	1/4	1/8	1/16	1/32
1	0	1/2	2/4	3/8	4/16	5/32
2			1/4	3/8	6/16	10/32
3				1/8	4/16	10/32
4					1/16	5/32
5						1/32

is the joint density function, and a, b, and c vary over the possible values of X, Y, and Z, respectively. From f, it is possible to retrieve the individual densities of X, Y, and Z, or the joint densities of (X, Y), (X, Z) and (Y, Z). The procedure is the same "marginal" summation described by (2) and (3). For instance,

$$P(X = a) = \sum_b \sum_c P[(X = a) \cap (Y = b) \cap (Z = c)]$$

$$P[(X = a) \cap (Y = b)] = \sum_c P[(X = a) \cap (Y = b) \cap (Z = c)]$$

Similarly, if $P[(Y = b) \cap (Z = c)] > 0$

$$f(a) = P[X = a|(Y = b) \cap (Z = c)] = \frac{P[(X = a) \cap (Y = b) \cap (Z = c)]}{P[(Y = b) \cap (Z = c)]}$$

defines the conditional density function of X given $Y = b$ and $Z = c$. If $P(Z = c) > 0$,

$$f(a, b) = P[(X = a) \cap (Y = b)|Z = c)]$$

$$= \frac{P[(X = a) \cap (Y = b) \cap (Z = c)]}{P(Z = c)}$$

defines the conditional density function of X and Y, given $Z = c$ (see Exercise 9).

4.3 CONDITIONAL EXPECTED VALUE

Conditional expected value is defined like ordinary expected value, except that the summation is over only that part of the sample space which describes the given condition, and the probabilities are the reappraised conditional probabilities. For instance, suppose one throws a die twice and the sum of the points is known to be five. In other words, one knows the outcome must be one of (1, 4), (2, 3), (3, 2), (4, 1). The conditional probabilities of these outcomes are 1/4 each. The conditional expected value of the number to appear on the first throw is, then, $(1+2+3+4)/4=2.5$. Of course, the expected value of the number on the first throw without any condition is 3.5. The definition of the conditional expected value of a random variable given a fixed value of another random variable is the following.

conditional expected value

Suppose X and Y are random variables and $P(Y=b)>0$. The *conditional expected value of X given $Y=b$*, denoted $E(X|Y=b)$ is defined as follows:

$$E(Y|Y=b) = \sum_{w \text{ in } (Y=b)} X(w)\frac{p(w)}{P(Y=b)} \tag{1}$$

assuming the sum to be absolutely convergent. The summation is over all sample points w of $\mathfrak{X}$ in the event $(Y=b)$, and $p(w)$ is the probability of a sample point w.

The definition of conditional expected value is similar to the definition of ordinary expected value (Section 2.4), except that in (1) the summation is not over all sample points w but only over those that make up $(Y=b)$. Dividing $p(w)$ by $P(Y=b)$ guarantees that

$$\sum_{w \text{ in } P(Y=b)} \frac{p(w)}{P(Y=b)} = \frac{P(Y=b)}{P(Y=b)} = 1$$

Thus, $E(X|Y=b)$ is like an ordinary expected value, in which $(Y=b)$ plays the role of the whole sample space and $p(w)/P(Y=b)$ behave like renormed probabilities over that new sample space. The reason for requiring the absolute convergence of (1) is exactly the same as with ordinary expected value. We refer the reader to the discussion about absolute convergence following the definition of EX in Section 2.4.

With ordinary expected value, EX can be computed from the density function of X instead of from the sample space [see (1),

Section 2.10]. In a similar way, one can compute $E(X|Y=b)$ from the conditional density function.

conditional expected value from conditional density function

$$E(X|Y=b)=\sum_i a_iP(X=a_i|Y=b) \tag{2}$$

The conditional expected value exists if and only if the right side is absolutely convergent.

The proof of (2) is exactly the same as that of (1), Section 2.10, and is left to the reader.

The linearity of expected value holds with conditional expectation just as in the ordinary case. Thus, if X, Y, Z are random variables and $E(X|Z = c)$ and $E(Y|Z = c)$ exist, then so does $E(aX+bY|Z=c)$, and

$$E(aX+bY|Z=c)=aE(X|Z=c)+bE(Y|Z=c) \tag{3}$$

The proof of (3) follows from the definition (1) and is left to the reader [use the proof of (1), Section 2.5, as a guide].

Finally, we point out that various other kinds of conditional expected values can be defined, such as

$$E[X|(Y=b)\cap(Z=c)], \qquad E(X|Y+Z=d)$$

and so forth. These can all be subsumed under the general definition, which follows.

$$E(X|A)=\sum_{w \text{ in } A} X(w)\frac{p(w)}{P(A)} \qquad \text{if } P(A)>0 \tag{4}$$

We again assume the sum to be absolutely convergent.

The formulation and proof of appropriate versions of (2) and (3) are left to the reader.

Example 1 Consider the joint distribution of X and Y described in Table 4.2. It is easy to see that

$$E(X|Y=5)=5/2, E(X|Y=4)=2, \quad E(X|Y=3)=3/2,$$

$$E(X|Y=2)=1, \quad E(X|Y=1)=1/2, E(X|Y=0)=0.$$

No computations are necessary because the conditional density functions are all symmetric (Section 2.11).

An iterative formula for EX in terms of conditional expected values is often useful.

$$EX=\sum_i E(X|Y=b_i)P(Y=b_i) \tag{5}$$

[We need only consider i's for which $P(Y=b_i)>0$.]

Implicit in (5) is that if EX exists, then each of $E(X|Y=b_i)$ exists. This fact follows immediately from the definition (1) (see Exercise 11).

Proof of (5). By (2),

$$E(X|Y=b_i)=\sum_j a_j \frac{P[(X=a_j)\cap(Y=b_i)]}{P(Y=b_i)}$$

The right side of (5) equals

$$\sum_i \left\{\sum_j a_j \frac{P[(X=a_j)\cap(Y=b_i)]}{P(Y=b_i)}\right\} P(Y=b_i)$$

$$=\sum_j a_j \sum_i P[(X=a_j)\cap(Y=b_i)] \qquad \text{(interchanging the order of summation)}$$

$$=\sum a_j P(X=a_j)=EX$$

The existence of EX guarantees the absolute convergence of all sums in the proof and allows the interchange in the order of summation. ∎

Example 2 A card is picked at random from among 10 cards numbered 1, 2, . . . , 10. If the card picked is k, then a second card is picked from among cards 1, . . . , k. Denote the first number selected by Y and the second number by X. Our object is to compute EX. First, $P(X=i|Y=k)=1/k$, because the second card is selected from among cards 1, . . . , k. The conditional distribution is symmetric about $(k+1)/2$, the midpoint of the extreme values 1 and k. Hence, $E(X|Y=k)=(k+1)/2$. Also, $P(Y=k)=1/10, k=1, \ldots, 10$. From (5), it follows that

$$EX = \sum_k E(X|Y = k)P(Y = k) = \sum_{k=1}^{10} \left(\frac{k+1}{2}\right)\left(\frac{1}{10}\right)$$

$$= \frac{(2 + 3 + \cdots + 11)}{20} = \frac{13}{4}$$

(See Exercise 12.)

4.4 MATRICES AND THEIR MULTIPLICATION

In this section, we define matrices and the formal aspects of matrix multiplication. The reader who is familiar with such matters may as well go directly to the next section where matrices are applied to conditional distributions.

An $m \times n$ matrix (of real numbers) means simply a rectangular array of real numbers, a_{ij}, arranged as follows:

$$\mathbf{A} = \begin{pmatrix} a_{11} & \cdots & a_{1n} \\ \vdots & & \\ a_{m1} & \cdots & a_{mn} \end{pmatrix}$$

An $m \times 1$ matrix is also called a *column vector*, and a $1 \times n$ matrix is also called a *row vector*.

Example 1

(a) (5) is a 1×1 matrix

(b) $(3, 4, 1/2)$ is a 1×3 matrix (row vector)

(c) $\begin{pmatrix} -2 \\ \pi \end{pmatrix}$ is a 2×1 matrix (column vector)

(d) $\begin{pmatrix} 7 & 4 \\ e & -1 \end{pmatrix}$ is a 2×2 matrix

If $\mathbf{A}$ is an $m \times n$ matrix with elements a_{ij}, and $\mathbf{B}$ is an $n \times p$ matrix with elements b_{ij}, then by $\mathbf{C} = \mathbf{AB}$ is meant the $m \times p$ matrix whose elements c_{ij} are defined as follows:

$$c_{ij} = a_{i1}b_{1j} + \cdots + a_{in}b_{nj}, \qquad i = 1, \ldots, m, \qquad j = 1, \ldots, p \quad (1)$$

If $\mathbf{A}$ is an $m \times n$ matrix, and $\mathbf{B}$ is an $r \times p$ matrix, then the product is only defined if $n = r$. The result is an $m \times p$ matrix.

Example 2

(a) $$\begin{pmatrix} a_{11} & a_{12} \\ a_{21} & a_{22} \end{pmatrix}\begin{pmatrix} b_{11} & b_{12} \\ b_{21} & b_{22} \end{pmatrix} = \begin{pmatrix} a_{11}b_{11} + a_{12}b_{21} & a_{11}b_{12} + a_{12}b_{22} \\ a_{21}b_{11} + a_{22}b_{21} & a_{21}b_{12} + a_{22}b_{22} \end{pmatrix}$$

(b) $(3,4,5)\begin{pmatrix}1\\0\\1\end{pmatrix}=(8)$

(c) $\begin{pmatrix}1\\0\\4\end{pmatrix}(3,4,5)=\begin{pmatrix}3&4&5\\0&0&0\\12&16&20\end{pmatrix}$

(d) $(1 \quad 0 \quad 1)\begin{pmatrix}4&5\\1&0\\1&2\end{pmatrix}=(5,7)$

(e) If $\mathbf{A}=\begin{pmatrix}a_{11}&a_{12}\\a_{21}&a_{22}\end{pmatrix}$ and $\mathbf{B}=\begin{pmatrix}b_{11}&b_{12}\\b_{21}&b_{22}\\b_{31}&b_{32}\end{pmatrix}$

then **AB** is not defined but **BA** is defined.

The preceding definition of **AB** may make sense for infinite matrices. Suppose

$$\mathbf{A}=\begin{pmatrix}a_{11}&a_{12}&\cdots\\a_{21}&a_{22}&\cdots\\&\vdots&\end{pmatrix},\qquad \mathbf{B}=\begin{pmatrix}b_{11}&b_{12}&\cdots\\b_{21}&b_{22}&\cdots\\&\vdots&\end{pmatrix}$$

are two rectangular arrays of numbers having sizes $m\times n$ and $p\times r$, respectively. If n and p are both infinite, then $\mathbf{C}=\mathbf{AB}$ can still be defined formally by (1). That is, c_{ij}, the (i, j)th element of C, is defined to be

$$c_{ij}=\sum_{k=1}^{\infty} a_{ik}b_{kj} \tag{2}$$

assuming that these infinite series converge for every i and j.

The only theoretical fact we need at this point relates to the associativity of matrix multiplication. That is, if **A**, **B**, **C** are matrices and (**AB**)**C** and **A**(**BC**) are well defined, then, under certain conditions,

$$(\mathbf{AB})\mathbf{C}=\mathbf{A}(\mathbf{BC}) \tag{3}$$

For finite matrices, (3) holds without further restriction. For infinite matrices, the truth of (3) is guaranteed when all sums that enter the multiplications [such as (2)] are convergent and all matrix elements are nonnegative. This is not the most general condition sufficient for associativity, but it will suffice for all our needs in this chapter. With associativity, the parentheses in (3) can be omitted because **ABC** is well defined. Similarly, under the same conditions, products of more than three matrices—such as **ABCD**—are also well defined (Note 4).

4.5 MATRICES AND CONDITIONAL PROBABILITIES

In this section, we show how relations involving conditional probabilities can often be succinctly expressed by using matrices. If X and Y are random variables with possible values $(a_1, a_2, \ldots)$, $(b_1, b_2, \ldots)$, respectively, then one matrix of interest is

$$\mathbf{B} = \begin{pmatrix} P(X = a_1 | Y = b_1) & P(X = a_2 | Y = b_1) & \cdots \\ P(X = a_1 | Y = b_2) & P(X = a_2 | Y = b_2) & \cdots \\ \vdots & \vdots & \end{pmatrix}$$

[We suppose that $P(Y = b_i) > 0$ for all i.] In effect, as pointed out in Eq. (5), Section 4.2, the elements of any row of **B** sum to 1.

Define row vectors **A** and **C** by

$$\mathbf{A} = [P(X = a_1), P(X = a_2), \ldots], \qquad \mathbf{C} = [P(Y = b_1), P(Y = b_2), \ldots]$$

Then the following relation holds

$$\mathbf{A} = \mathbf{CB} \tag{1}$$

Proof of (1).

$$\sum_j P(X = a_i | Y = b_j) P(Y = b_j) = \sum_j P[(X = a_i) \cap (Y = b_j)]$$

which is $P(X = a_i)$ by (2), Section 4.2. This is just what (1) asserts. ∎

Example 1 Consider Example 2, Section 4.3, once again. We can use (1) to determine the distribution of X as follows: **C** is a row vector with 10 elements all equal to 1/10; **B**, the matrix of conditional probabilities, is

$$\begin{pmatrix}
1 & 0 & 0 & 0 & 0 & 0 & 0 & 0 & 0 & 0 \\
1/2 & 1/2 & 0 & 0 & 0 & 0 & 0 & 0 & 0 & 0 \\
1/3 & 1/3 & 1/3 & 0 & 0 & 0 & 0 & 0 & 0 & 0 \\
1/4 & 1/4 & 1/4 & 1/4 & 0 & 0 & 0 & 0 & 0 & 0 \\
1/5 & 1/5 & 1/5 & 1/5 & 1/5 & 0 & 0 & 0 & 0 & 0 \\
1/6 & 1/6 & 1/6 & 1/6 & 1/6 & 1/6 & 0 & 0 & 0 & 0 \\
1/7 & 1/7 & 1/7 & 1/7 & 1/7 & 1/7 & 1/7 & 0 & 0 & 0 \\
1/8 & 1/8 & 1/8 & 1/8 & 1/8 & 1/8 & 1/8 & 1/8 & 0 & 0 \\
1/9 & 1/9 & 1/9 & 1/9 & 1/9 & 1/9 & 1/9 & 1/9 & 1/9 & 0 \\
1/10 & 1/10 & 1/10 & 1/10 & 1/10 & 1/10 & 1/10 & 1/10 & 1/10 & 1/10
\end{pmatrix}$$

Then (1) says that $[P(X = 1), P(X = 2), \ldots] = \mathbf{CB}$. Thus, $P(X = 10) = 1/100$, $P(X = 9) = 1/90 + 1/100$, and so forth.

Relation (1) can be extended to more than two random variables.

Consider the case of three random variables, X, Y, and Z. Define

$$\mathbf{A} = [P(X = a_1), P(X = a_2), \ldots]$$

$$\mathbf{B} = [P(Z = c_1), P(Z = c_2), \ldots]$$

$$\mathbf{C} = \begin{pmatrix} P(Y = b_1|Z = c_1) & P(Y = b_2|Z = c_1) & \cdots \\ P(Y = b_1|Z = c_2) & P(Y = b_2|Z = c_2) & \cdots \\ \vdots & \vdots & \end{pmatrix}$$

$$\mathbf{D} = \begin{pmatrix} P(X = a_1|Y = b_1) & P(X = a_2|Y = b_1) & \cdots \\ P(X = a_1|Y = b_2) & P(X = a_2|Y = b_2) & \cdots \\ \vdots & \vdots & \end{pmatrix}$$

Then we have

$$\mathbf{A} = \mathbf{BCD} \tag{2}$$

Proof of (2). Define $\mathbf{E} = [P(Y = b_1), P(Y = b_2), \ldots]$. Then, by (1), $\mathbf{A} = \mathbf{ED}$. Applying (1) again, we have $\mathbf{E} = \mathbf{BC}$. Hence, $\mathbf{A} = \mathbf{BCD}$. Associativity follows from the fact that all sums in these computations are *convergent* and all elements of the matrices are nonnegative (Note 4). ▮

Relations (1) and (2) are extended in a natural way for any number of random variables, which is illustrated in the following examples.

Example 2 A box contains two white balls and one black ball. A ball is drawn and is replaced and an additional ball of the same color is also placed in the box. The procedure is repeated—a second ball is drawn and is replaced together with an additional ball of the same color that was drawn on the second drawing, and so forth. Let X_n be the number of black balls in the box after the nth drawing, $n = 1, 2, \ldots$. The reader should verify the following relationships (compare with Example 4, Section 4.1):

$$[P(X_1 = 1), P(X_1 = 2)] = (2/3, 1/3)$$

$$[P(X_2 = 1), P(X_2 = 2), P(X_2 = 3)]$$

$$= [P(X_1 = 1), P(X_1 = 2)]\begin{pmatrix} 3/4 & 1/4 & 0 \\ 0 & 2/4 & 2/4 \end{pmatrix} = (3/6, 2/6, 1/6)$$

$$[P(X_3 = 1), P(X_3 = 2), P(X_3 = 3), P(X_3 = 4)]$$

$$= [P(X_2 = 1), P(X_2 = 2), P(X_2 = 3)]\begin{pmatrix} 4/5 & 1/5 & 0 & 0 \\ 0 & 3/5 & 2/5 & 0 \\ 0 & 0 & 2/5 & 3/5 \end{pmatrix}$$

$$= (4/10, 3/10, 2/10, 1/10)$$

It is interesting to notice that each of X_1, X_2, X_3 has a density function of the same "triangular" shape. These are depicted in Figure 4.4 (see Exercise 13).

Example 3 Balls are successively distributed among four cells. Let X_n be the number of empty cells remaining after the nth ball has been distributed. It is easy to determine the conditional probability relationships between X_n and X_{n-1}. Consider the matrix whose (i, j)th entry is $P(X_n = i|X_{n-1} = j)$ $(i, j = 0, 1, 2, 3)$. This matrix equals

$$\begin{pmatrix} 1 & 0 & 0 & 0 \\ 1/4 & 3/4 & 0 & 0 \\ 0 & 2/4 & 2/4 & 0 \\ 0 & 0 & 3/4 & 1/4 \end{pmatrix}$$

For instance, $P(X_n = 1|X_{n-1} = 1) = 3/4$ and $P(X_n = 0|X_{n-1} = 1) = 1/4$ for the following reason: $X_{n-1} = 1$ means that after the $(n-1)$th ball has been distributed, three cells are occupied and one is unoccupied. If the nth ball enters one of the already occupied cells, then X_n still has the value 1; if it enters the single unoccupied cell, then X_n becomes 0. The conditional probabilities of these outcomes are then 3/4 and 1/4. The distribution of X_1 assigns probability 1 to the single possible value 3. Hence,

$$[P(X_2 = 0), P(X_2 = 1), P(X_2 = 2), P(X_2 = 3)]$$

$$= (0, 0, 0, 1)\begin{pmatrix} 1 & 0 & 0 & 0 \\ 1/4 & 3/4 & 0 & 0 \\ 0 & 2/4 & 2/4 & 0 \\ 0 & 0 & 3/4 & 1/4 \end{pmatrix}$$

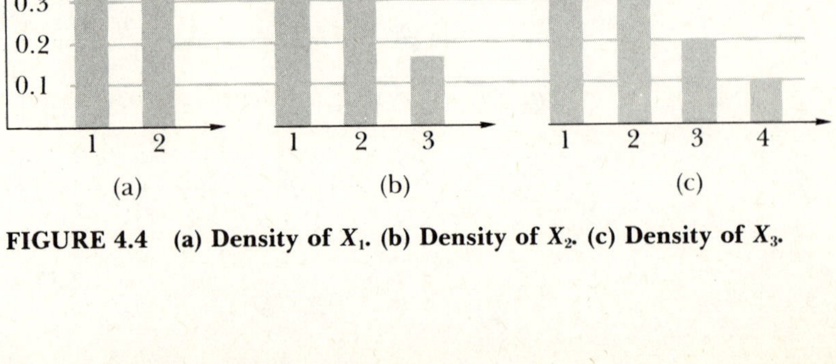

FIGURE 4.4 **(a) Density of X_1. (b) Density of X_2. (c) Density of X_3.**

and, in general,

$$[P(X_n = 0), P(X_n = 1), P(X_n = 2), P(X_n = 3)]$$

$$= (0, 0, 0, 1)\begin{pmatrix} 1 & 0 & 0 & 0 \\ 1/4 & 3/4 & 0 & 0 \\ 0 & 2/4 & 2/4 & 0 \\ 0 & 0 & 3/4 & 1/4 \end{pmatrix}^{n-1}$$

The computed values of the distributions are shown in Table 4.3. Notice that the entries agree with some of those obtained earlier in Table 3.6 (see Exercise 14).

Example 4 The following is a simple example of what is technically called a "pure death process." A population initially consists of four living organisms. At successive stages, a scourge enters the scene and falls upon one of the members of the population. The scourge, being stupid, cannot distinguish between a living member of the population and one it has already eliminated. At every stage, he selects his one victim out of the available four with equal likelihood. Let X_n be the number of organisms still living after the nth visit of the scourge. This situation is exactly the same as the one described in Example 3. The nth ball looking for its place among the four cells, some of which are occupied and some not, is the same as the scourge during its nth visit looking for a victim among the four candidates, some of which are already dead and some not. The conditional probabilities for the X_n's in this death process coincide with those already given in Example 3; the distributions are the same as those in Table 4.3. According to (6), Section 3.5, the expected number of visits required to decimate the population completely is

$$4\left(1 + \frac{1}{2} + \frac{1}{3} + \frac{1}{4}\right) = \frac{25}{3}$$

TABLE 4.3

$n =$	1	2	3	4	5
$P(X_n = 0)$	0	0	0	6/64	60/256
$P(X_n = 1)$	0	0	6/16	36/64	150/256
$P(X_n = 2)$	0	3/4	9/16	21/64	45/256
$P(X_n = 3)$	1	1/4	1/16	1/64	1/256

4.6 INDEPENDENCE AND CONDITIONAL PROBABILITY

According to the discussion in Section 4.1, $P(A|B)$ can be interpreted as a reappraisal of the likelihood that the experiment will turn up with an outcome in A *if* one has the prior information that an outcome somewhere in B has already turned up. *Suppose that this prior information does not make any difference.* That is, suppose that the observed proportion $N(A \cap B)/N(B)$ is about the same as the observed proportion $N(A)/N$, over N actual trials, using the frequency theory interpretation as in Section 4.1. This state of affairs can be described by saying that "A is not influenced by B" or that "A is independent of B." In terms of the theoretical probabilities, the property being asserted, assuming $P(B) > 0$, is that

$$P(A|B) = P(A) \tag{1}$$

or, equivalently, that

$$P(A \cap B) = P(A)P(B) \tag{2}$$

When (2) holds, A and B are called *independent*.

definition of independence of two events

Two events, A and B, are said to be independent if

$$P(A \cap B) = P(A)P(B) \tag{3}$$

Suppose that $P(A) > 0$, $P(B) > 0$, and that A and B are independent. Then

$$P(A|B) = \frac{P(A \cap B)}{P(B)} = \frac{P(A)P(B)}{P(B)} = P(A)$$

and, by the same argument, $P(B|A) = P(B)$. In other words, if the events are independent, then the conditional probability is the same as the "ordinary" or "unconditional" probability. Notice also that if either $P(A) = 0$ or $P(B) = 0$, then both sides of (2) are zero and A and B are automatically independent.

For more than two events, the definition of independence is slightly more complicated. For instance, three events A, B, and C are said to be mutually independent if all four of the following conditions are satisfied:

$$\begin{aligned} P(A \cap B) &= P(A)P(B) \\ P(A \cap C) &= P(A)P(C) \\ P(B \cap C) &= P(B)P(C) \\ P(A \cap B \cap C) &= P(A)P(B)P(C) \end{aligned} \tag{4}$$

definition of independence of n events

In general, n events $A_1, \ldots, A_n$ are said to be mutually independent if

$$P(A_i \cap A_j) = P(A_i)P(A_j) \qquad 1 \leq i < j \leq n$$

$$P(A_i \cap A_j \cap A_k) = P(A_i)P(A_j)P(A_k) \qquad 1 \leq i < j < k \leq n \tag{5}$$

$$\vdots$$

$$P(A_1 \cap A_2 \cap \cdots \cap A_n) = P(A_1)P(A_2) \cdots P(A_n)$$

Since $\binom{n}{0} + \binom{n}{1} + \cdots + \binom{n}{n} = 2^n$ (Exercise 5, Chapter 1), it follows that there are $2^n - n - 1$ conditions specified in (5). For instance, for $n = 3$, there are $2^3 - 3 - 1 = 4$ conditions listed in (4). Notice that if $A_1, \ldots, A_n$ are mutually independent, then any subcollection of two or more of these events are also mutually independent. For instance, if A, B, and C are mutually independent, then A and B are independent as well (see Exercise 21).

Example 1 Suppose that a sample space consists of eight sample points $w_1, w_2, \ldots, w_8$, each of which has probability 1/8. Define the following events:

$$A = (w_1, w_2, w_3, w_4)$$

$$B = (w_1, w_2, w_5, w_6)$$

$$C = (w_1, w_3, w_5, w_7)$$

Then A, B, and C are mutually independent. For instance, since $A \cap B = (w_1, w_2)$, $P(A \cap B) = 2/8 = P(A)P(B)$. The verification of the other conditions in (4) is left to the reader.

4.7 INDEPENDENCE OF RANDOM VARIABLES

definition of independence of two random variables

Two random variables X and Y are said to be *independent* if the events (X in E) and (Y in F) are independent for all sets of real numbers E and F. By definition, this statement means that

$$P[(X \text{ in } E) \cap (Y \text{ in } F)] = P(X \text{ in } E)P(Y \text{ in } F) \tag{1}$$

for all sets of real numbers E and F. It is not enough for (1) to hold for *some* sets E, F; it must hold for *all* sets.

We can make the same intuitive interpretation about independence of random variables as was made for independence of events in Section 4.6. The independence of two random variables suggests that having prior information about the value of the one should not help in making probability statements about the other. If (1) holds, then the conditional distribution of X given any value of Y is the same as the distribution of X itself. That is, whenever $P(Y = b_i) > 0$,

$$P(X = a_i|Y = b_j) = P(X = a_i) \tag{2}$$

The proof of (2) follows immediately from (1), because

$$P(X = a_i|Y = b_j) = \frac{P[(X = a_i) \cap (Y = b_j)]}{P(Y = b_j)} = \frac{P(X = a_i)P(Y = b_j)}{P(Y = b_j)}$$

$$= P(X = a_i)$$

The situation is quite symmetric in X and Y. The conditions (1) also imply that the conditional distribution of Y given any value of X is the same as the distribution of Y. In fact, the same calculation shows that the converse also holds; that is, (2) implies (1).

Example 1 (a) A person is selected at random from a given population. Let X be the day of the week on which the person is born and let Y be his age. In making a mathematical model for the joint distribution of X and Y, a reasonable assumption is that X and Y are independent.

(b) A biologist is observing the numbers of progeny in successive generations of hamsters. Let X be the number of hamsters born in the first generation and Y the number born in the second generation. In a probability model for this process, it would not be reasonable to suppose that X and Y are independent.

Example 2 A coin is thrown four times. Let X be the number of heads in the first two throws and let Y be the number of heads in the last two throws. Suppose that each of the 16 sample points is assigned the probability 1/16. *Then X and Y are independent.* Intuitively, of course, this assertion is very plausible, for why should the first two throws influence the third and fourth throws? Consider what is involved in the formal verification of (1), however. Among the conditions that must be checked are the following:

$$P[(X = 0 \text{ or } X = 1) \cap (Y = 2)] = \frac{3}{16}$$

$$= P(X = 0 \text{ or } X = 1)P(Y = 2)$$

$$P[(X = 1) \cap (Y < 2)] = \frac{3}{8} = P(X = 1)P(Y < 2)$$

and so forth. If one lets E and F vary over all possible subsets of (0, 1, 2), then 64 conditions altogether have to be checked! Fortunately, there will be an easier method of recognizing this independence than the brutal verification of (1) for all sets E and F. This method is presented in Section 4.9.

The definition of independence for two random variables is extended to more than two random variables in a natural way.

independence of n random variables

Random variables $X_1, \ldots, X_n$ are said to be mutually independent if for all sets of real numbers $E_1, \ldots, E_n$,

$$P[(X_1 \text{ in } E_1) \cap \cdots \cap (X_n \text{ in } E_n)] = P(X_1 \text{ in } E_1) \cdots P(X_n \text{ in } E_n) \qquad (3)$$

Example 3 A coin is thrown six times. Let X be the number of heads in the first two throws, Y the number of heads in the next two throws, Z the number of heads in the last two throws. Suppose that each of the 64 sample points is assigned the probability 1/64. Then, just as in Example 2, X, Y, Z are mutually independent, for the same plausible reasons. Verifying condition (3) is even more forbidding here than in Example 2. Again, we remind the reader there will be a relatively painless criterion for recognizing independence presented in Section 4.9. Meanwhile, you might want to check (3) for at least a few sets. For instance,

$$P[(X = 0) \cap (Y = 2) \cap (Z = 1)] = \frac{1}{32} = P(X = 0)P(Y = 2)P(Z = 1)$$

4.8 SOME CONSEQUENCES OF INDEPENDENCE

If $X_1, \ldots, X_n$ are mutually independent random variables, then any two or more of the X_i's are also mutually independent. For instance, if X_1, X_2, X_3 are mutually independent, then X_1, X_2 are independent also. The proof in this special case proceeds as follows:

$$\begin{aligned} P[(X_1 \text{ in } E) \cap (X_2 \text{ in } F)] &= P[(X_1 \text{ in } E) \cap (X_2 \text{ in } F) \cap (X_3 \text{ in } R)] \\ &= P(X_1 \text{ in } E)P(X_2 \text{ in } F)P(X_3 \text{ in } R) \\ &= P(X_1 \text{ in } E)P(X_2 \text{ in } F) \end{aligned}$$

where R is the whole real line, and, hence, $P(X_3 \text{ in } R) = P(\mathfrak{X}) = 1$. The proof for general n is similar and is left to the reader.

Another fact about independent random variables is that functions of individual independent random variables are still independent. For example, if X_1, X_2, X_3 are mutually independent, then $Y_1 = \phi_1(X_1)$, $Y_2 = \phi_2(X_2)$, and $Y_3 = \phi_3(X_3)$ are also mutually independent. To be even more specific, if X_1, X_2, X_3 are mutually independent, then $(X_1)^2, (X_2)^4, |X_3| + 5$ are also mutually independent. The proof of the general assertion follows directly from the definition of independence and is left to the reader as an exercise.

Here is another important consequence of independence. Suppose that X and Y are random variables the expected values of which exist; assume $E(XY)$ also exists. Then

$$E(XY) = (EX)(EY) \tag{1}$$

Proof of (1). By definition,

$$E(XY) = \sum_w X(w)Y(w)p(w) \tag{2}$$

where $p(w)$ is the probability of a sample point w and the summation is over all sample points. First sum over the sample points in $(X = a_i) \cap (Y = b_j)$. The contribution to the sum is

$$a_i b_j P[(X = a_i) \cap (Y = b_j)] = a_i b_j P(X = a_i)P(Y = b_j)$$

Now hold i fixed and sum over all possible values b_j, which gives

$$a_i P(X = a_i) \sum_j b_j P(Y = b_j) = a_i P(X = a_i)EX$$

Now complete the summation over all a_i, giving the final sum as

$$(EX) \sum_i a_i P(X = a_i) = (EX)(EY)$$

Remember that the fact that $E(XY)$ exists means (2) is absolutely convergent; hence, the order of summing (2) does not affect the sum. ∎

Example 1 Suppose that X_A and X_B are the indicator random variables of the sets A and B. If X_A and X_B are independent, then

$$E(X_A X_B) = (EX_A)(EX_B)$$

Since $X_A X_B = X_{A \cap B}$,

$$P(A \cap B) = P(A)P(B)$$

Hence, the independence of X_A, X_B implies the independence of the events A, B. Conversely, if A and B are independent, then X_A and X_B are independent random variables. (Prove this!) A general version of this example is given in Exercise 26.

Example 2 Consider the following probability space and random variables X and Y.

Sample points	*Probabilities*	X	Y
w_1	1/4	0	0
w_2	1/4	−1	1
w_3	1/4	1	1
w_4	1/4	0	2

We know X and Y are not independent, because

$$P[(X = 0) \cap (Y = 0)] = \frac{1}{4} \neq P(X = 0)P(Y = 0) = \left(\frac{1}{2}\right)\left(\frac{1}{4}\right)$$

However, $E(XY) = 0 = (EX)(EY)$. Hence, the truth of (1) does not necessarily imply independence.

Equation (1) holds for any number of independent random variables. For instance, if X, Y, Z are mutually independent, then

$$E(XYZ) = (EX)(EY)(EZ) \tag{3}$$

assuming that all the expected values referred to exist. The proof of (3) is similar to that of (1) and is left to the reader.

Example 3 Consider X, Y, and Z of Example 3, Section 4.7. For the time being, accept the independence of these random variables. By enumerating the 64 sample points one sees that $E(XYZ) = 1$. (Can you think of an easier way of doing this computation?) Also, $EX = EY = EZ = 1$, so (3) holds.

4.9 A CRITERION FOR INDEPENDENCE

In this section, we present a criterion for independence that is easier than conditions (1) or (3) of Section 4.7. Consider first only two random variables.

Theorem

X and Y are independent random variables if, and only if, for all numbers a and b,

$$P[(X = a) \cap (Y = b)] = P(X = a)P(Y = b) \tag{1}$$

Proof. In condition (1) of Section 4.7, take E and F to be the one-point sets a and b; thus, independence implies that (1) holds. Now we want to show that (1) implies condition (1) of 4.7. Suppose E is made up of the sequence of numbers $a_1, a_2, \ldots$, and F is made up of $b_1, b_2, \ldots$. (Remember that discrete random variables cannot have more than countably many possible values.)

$$P[(X \text{ in } E) \cap (Y \text{ in } F)] = \sum_{i,j} P[(X = a_i) \cap (Y = b_j)] \tag{2}$$

The sum is over all a_i in E and b_j in F. By (1),

$$\sum_{i,j} P[(X = a_i) \cap (Y = b_j)] = \sum_{i,j} P(X = a_i)P(Y = b_j) \tag{3}$$

Now hold a_i fixed and sum over j. The right side of (3) becomes $P(X = a_i)P(Y \text{ in } F)$. Now complete the summation over i, giving $P(X \text{ in } E)P(Y \text{ in } F)$. ∎

The same theorem holds for any number of random variables. As the proof is practically the same, we leave it for the reader.

Theorem

$X_1, \ldots, X_n$ are mutually independent if, and only if, for all numbers $a_1, \ldots, a_n$,

$$P[(X_1 = a_1) \cap \cdots \cap (X_n = a_n)] = P(X_1 = a_1) \cdots P(X_n = a_n) \tag{4}$$

A succinct way of describing the last two theorems is as follows: *Random variables are independent if, and only if, their joint density function is the product of their marginal density functions.* Notice that when a is not a possible value of X_i, both sides of (4) are zero. Hence, (4) need only be checked for possible values of the random variables.

Example 1

A coin is thrown six times. Suppose that each of the $2^6 = 64$ outcomes has the same probability 1/64. Let

$$X_i = \begin{cases} 1 \text{ if } i\text{th throw is a head} \\ 0 \text{ otherwise} \end{cases} \qquad i = 1, 2, \ldots, 6$$

It is easy to check that $P(X_i = 1) = P(X_i = 0) = 1/2$. Let $e_1, e_2, \ldots, e_6$ be any sequence of 0's and 1's. Then

$$P[(X_1 = e_1) \cap (X_2 = e_2) \cap \cdots \cap (X_6 = e_6)] = \frac{1}{64} = \frac{1}{2^6}$$

$$= P(X_1 = e_1)P(X_2 = e_2) \cdots P(X_6 = e_6)$$

Hence, $X_1, X_2, \ldots, X_6$ are mutually independent. It follows that the three random variables

$$X_1 + X_2, \qquad X_3 + X_4, \qquad X_5 + X_6$$

are also mutually independent (see Exercise 20), which proves that X, Y, and Z of Example 3, Section 4.7, are independent.

Often one knows that random variables are independent because one defines them to be so. That is, one proposes that the correct mathematical model to describe the joint distribution of a set of random variables is one in which the conditions of independence are satisfied. Suppose, specifically, that a model is posed in which X, Y, Z are random variables with individual density functions f, g, and h. If it is assumed that X, Y, and Z are mutually independent, then, according to (4), the joint density must be specified as follows:

$$P[(X = a_i) \cap (Y = b_j) \cap (Z = c_k)] = f(a_i)g(b_j)h(c_k) \tag{5}$$

where a_i, b_j, c_k vary over all possible values for each of the three variables. In other words, the random variables are independent because the independence has been "built into the system." Notice that nothing is being said about a background sample space $\mathfrak{X}$ over which the random variables are being defined, and usually this will not be necessary. The purist who may be concerned over the absence of a sample space should know that it is always possible to construct an appropriate space. For instance, one can take for the sample space the set of all triples of real numbers (a_i, b_j, c_k) and define

$$X(a_i, b_j, c_k) = a_i, \; Y(a_i, b_j, c_k) = b_j, \; Z(a_i, b_j, c_k) = c_k$$

and the probability space structure will then be complete. When the joint density is prescribed by (5), the individual densities of X, Y, and Z will indeed turn out to be f, g, and h. To check the last assertion for X, for example, we must do the following calculation:

$$P(X = a_i) = \sum_{j,k} P[(X = a_i) \cap (Y = b_j) \cap (Z = c_k)]$$

(In other words, hold a_i fixed and sum out over all b_j and c_k.) By (5), then,

$$P(X = a_i) = f(a_i) \sum_{j,k} g(b_j)h(c_k)$$
$$= f(a_i) \sum_{j} g(b_j) \sum_{k} h(c_k)$$
$$= f(a_i) \cdot 1 \cdot 1 = f(a_i)$$

Example 2 Suppose we postulate that X, Y, and Z are three indicator random variables that are independent and for which

$$P(X = 1) = \tfrac{1}{2}, \qquad P(Y = 1) = \tfrac{1}{3}, \qquad P(Z = 1) = \tfrac{1}{4}$$

The joint density of X, Y, Z must then be given according to the tabulation in Table 4.4. The independence of X, Y, and Z is guaranteed by the proper prescription of the joint probabilities according to (5).

Despite the fact that independence is often found in a probability model because joint distributions are defined that way, sometimes random variables turn out to be independent when there is no intuitive reason to expect them to be so. The following example illustrates such a situation.

Example 3 Consider the $3! = 6$ permutations of 1, 2, and 3, and suppose that each has probability 1/6. Define random variables X and Y as follows:

$$X(i, j, k) = \begin{cases} 1 \text{ if } j > i \\ 0 \text{ otherwise} \end{cases} \qquad Y(i, j, k) = \begin{cases} 1 \text{ if } k > i \text{ and } k > j \\ 0 \text{ otherwise} \end{cases}$$

By simply listing the six sample points (this is left for the reader),

TABLE 4.4

a_i	b_j	c_k	$P[(X = a_i) \cap (Y = b_j) \cap (Z = c_k)]$
1	1	1	(1/2)(1/3)(1/4) = 1/24
1	1	0	(1/2)(1/3)(3/4) = 3/24
1	0	1	(1/2)(2/3)(1/4) = 2/24
1	0	0	(1/2)(2/3)(3/4) = 6/24
0	1	1	(1/2)(1/3)(1/4) = 1/24
0	1	0	(1/2)(1/3)(3/4) = 3/24
0	0	1	(1/2)(2/3)(1/4) = 2/24
0	0	0	(1/2)(2/3)(3/4) = 6/24

we find the joint distribution of X and Y. The entries for $P[(X=i) \cap (Y=j)]$ are the following:

		$i = 0$	$i = 1$
j	0	2/6	2/6
	1	1/6	1/6

Hence, X and Y are independent. Can you give any intuitive reason for this independence? (See Exercises 22 and 27.)

SUMMARY

The *conditional probability of A given B* is, by definition, $P(A|B) = P(A \cap B)/P(B)$, assuming $P(B) > 0$. This represents a reappraisal of the likelihood that the experiment will produce an outcome in A when one has the prior information that it definitely must produce an outcome in B.

If X and Y are random variables, then, for fixed b, the function f given by

$$f(a) = P[(X=a)|(Y=b)] = \frac{P[(Y=a) \cap (Y=b)]}{P(Y=b)}$$

with a varying over the possible values of X, is the *conditional density function of X given Y = b*. The roles of X and Y can each be played by more than one random variable.

Events A and B are independent if $P(A \cap B) = P(A)P(B)$, with similar definitions for more than two events. If $P(A) > 0$ and $P(B) > 0$, then A and B are independent if, and only if, $P(A|B) = P(A)$ and $P(B|A) = P(B)$.

Random variables X and Y are, by definition, independent if the events (X in A) and (Y in B) are independent for all sets of real numbers A and B. It turns out that equivalently X and Y are independent if, and only if, $(X=a)$ and $(Y=b)$ are independent for all pairs of possible values a and b. For n random variables, this criterion says that $X_1, \ldots, X_n$ are independent if, and only if,

$$P[(X_1 = a_1) \cap \cdots \cap (X_n = a_n)] = P(X_1 = a_1) \cdots P(X_n = a_n)$$

for all n-tuples of possible values $a_1, \ldots, a_n$. In other words, to say that random variables are independent means that their joint density function is the product of their marginal density functions.

The *conditional expected value* of X given an event A is

$$E(X|A) = \sum_{w \text{ in } A} X(w)\frac{p(w)}{P(A)}$$

This is just the ordinary expected value of X with the summation restricted to the set A, using the conditional probabilities given A. Conditional expected value can be computed from the conditional density function as well as from a sample space summation, just as with ordinary expected value.

We have the useful formula,

$$EX = \sum_b E(X|Y=b)P(Y=b)$$

summing over the possible values of Y.

The vector of probabilities $P(X=a)$ can be nicely expressed in terms of the vector of probabilities $P(Y=b)$ and the matrix of conditional probabilities, $P(X=a|Y=b)$.

EXERCISES

1 Establish the assertions in the first paragraph of Section 4.1; that is, show that

(a) $0 \leq P(A|B) \leq 1$

(b) $P(\mathfrak{X}|B) = P(B|B) = 1$

(c) If $A_1, A_2, \ldots$ is a sequence of mutually disjoint events, then $P(A_1 \cup A_2 \cup \cdots |B) = P(A_1|B) + P(A_2|B) + \cdots$.

(d) $P(A'|B) = 1 - P(A|B)$

(e) $P(C \cup D|B) = P(C|B) + P(D|B) - P(C \cap D|B)$

2 (a) Show that $P(A \cap B|B) = P(A|B)$.

(b) Show that $P(A \cup B|B) = 1$.

3 Formulate a general version of (2) and (3), Section 4.1, for n events.

4 An honest coin is thrown n times.

(a) Show that the conditional probability of a head on any specified trial, given a total of k heads over the n trials, is k/n ($k > 0$; otherwise the conditional probability is not defined).

(b) Suppose $0 \leq r \leq k$ and $0 < m < n$ are integers. Show that the conditional probability of r heads over the first m trials given a total of k heads over all n trials is

$$\frac{\binom{m}{r}\binom{n-m}{k-r}}{\binom{n}{k}}$$

5 From five cards numbered 1, 2, 3, 4, 5, a card is drawn at random. Suppose that the number drawn is k. Then, from among the cards

1 through k, a second card is drawn at random. Suppose that the number drawn on the second trial is r. Then, from among the cards 1 through r, a third card is drawn at random.

(a) List all 35 sample points (why 35?), and determine their probabilities. Make sure that these probabilities add to 1. [Selected answers: The probabilities of (5, 5, 5), (4, 4, 1), and (1, 1, 1) are 1/125, 1/80, and 1/5.]

(b) Evaluate the conditional probability of drawing a 1 on the first trial, given the drawing of a 5 on the third trial. (Answer: Zero.)

(c) Evaluate the conditional probability of drawing a 5 on the first trial, given the drawing of a 1 on the third trial. (Answer: 0.287.)

6 Consider the process described in Example 4, Section 4.1. A box contains one black ball and two white balls. Each time a ball is drawn it is replaced together with an additional one of the same color. Let X_n denote the number of black balls in the box after n trials have been completed. Verify the following density functions:

k	$P(X_1 = k)$	k	$P(X_2 = k)$	k	$P(X_3 = k)$	k	$P(X_4 = k)$
1	2/3	1	3/6	1	4/10	1	5/15
2	1/3	2	2/6	2	3/10	2	4/15
		3	1/6	3	2/10	3	3/15
				4	1/10	4	2/15
						5	1/15

[Hint: The density functions for X_1, X_2, X_3 can be read off from Table 4.1. For instance, $(X_2 = 2) = (b, w, b), (b, w, w), (w, b, b), (w, b, w)$; hence, $P(X_2 = 2) = (4/60) + (6/60) + (4/60) + (6/60) = 2/6$. The density function for X_4 can be obtained by first constructing the sample space of 16 sample points, describing a sequence of four drawings.]

7 (a) Repeat Example 4, Section 4.1, and Exercise 6, assuming that the box initially has one white and one black ball.

(b) Verify the results of Example 5, Section 2.4, using conditional probabilities.

8 (a) Verify the conditional distribution in Table 4.2.

(b) If n balls are randomly distributed among m cells, let X be the number that fall in the first cell and let Y be the number that fall in the first two cells. Show that the conditional distributions of X, given the various values of Y, are the following:

$$P(X = k|Y = r) = \binom{r}{k}(\tfrac{1}{2})^r, \qquad k = 0, 1, \ldots, r$$

[Hint: $P[(X = k) \cap (Y = r)] = \dfrac{n!}{k!(r-k)!(n-r)!}\left(\dfrac{1}{m}\right)^r\left(1 - \dfrac{2}{m}\right)^{n-r}$.]

(There is a direct intuitive interpretation of the result. If r balls are randomly distributed among two cells, the distribution of the number of balls in the first cell is exactly the same as the preceding conditional distribution. Is this interpretation reasonable?)

9 Define

$$f(a, b) = P[(X = a) \cap (Y = b)|Z = c)]$$

Suppose $P(Z = c) > 0$. Show that

$$\sum_i f(a_i, b) = P(Y = b|Z = c)$$

$$\sum_j f(a, b_j) = P(X = a|Z = c)$$

[Compare with (2) and (3), Section 4.2.]

10 Prove (2) and (3) of Section 4.3. [Use as guides the proofs of (1), Section 2.10, and (1), Section 2.5.]

11 Prove that if EX exists, then $E(X|A)$ exists also, if $P(A) > 0$. (Hint: The fact that EX exists means that $\sum_w |X(w)|p(w) < \infty$. What if w ranges only over A?)

12 A card is picked at random from among n cards numbered $1, 2, \ldots, n$. If the card is k, then a second card is picked from among cards $1, \ldots, k$. Denote the first number selected by Y and the second by X. Show that

$$EX = \frac{n+3}{4}$$

(Hint: Imitate Example 2, Section 4.3. You need to know that $1 + 2 + \cdots + k = k(k+1)/2$.)

13 Follow the method of Example 2, Section 4.5, to find the distributions of X_4 and X_5. (These follow the "triangular" pattern shown in Figure 4.4. Also see Exercise 6.)

14 Balls are successively distributed among five cells. Let X_n be the number of empty cells remaining after the nth ball has been distributed. Follow the method used in Example 3, Section 4.5, to determine the distribution of X_1, X_2, X_3, X_4, and X_5.

15 When n balls are distributed among m cells, the expected number of empty cells is $(m-1)^n/m^{n-1}$ (Eq. 5, Section 3.4). Use this fact to show that

$$(0\ \ 0\ \ 0\ \ 1)\begin{pmatrix} 1 & 0 & 0 & 0 \\ 1/4 & 3/4 & 0 & 0 \\ 0 & 2/4 & 2/4 & 0 \\ 0 & 0 & 3/4 & 1/4 \end{pmatrix}^{n-1}\begin{pmatrix} 0 \\ 1 \\ 2 \\ 3 \end{pmatrix} = \frac{3^n}{4^{n-1}}, \qquad n = 1, 2, \ldots$$

16 Show that the following is true:

$$P(X = a_i|Z = c_j) = \sum_k P[(X = a_i|(Y = b_k) \cap (Z = c_j)]P(Y = b_k)|Z = c_j)$$

[Compare with (2), Section 4.5.]

17 (a) Show that $\mathfrak{X}$ and A are independent for any event A.

(b) Show that $\varnothing$ and A are independent for any event A.

(c) Show that A and B are independent if, and only if, A and B' are independent. [Hint: $P(A' \cap B') + P(A' \cap B) = P(A'), P(A \cap B') + P(A \cap B) = P(A)$.]

(d) Show that any of the following four statements are equivalent:

(i) A and B are independent
(ii) A and B' are independent
(iii) A' and B are independent
(iv) A' and B' are independent

[Hint: Part (c) already shows that (1) and (2) are equivalent. A second application of (c) shows that (2) and (4) are equivalent and a third application of (c) shows that (4) and (3) are equivalent.]

18 Suppose A, B, and C are mutually independent. Show that

(a) $P(A|B \cap C) = P(A)$
(b) $P(A|B \cup C) = P(A)$
(c) $P(A \cap B|C) = P(A \cap B)$
(d) $P(A \cup B|C) = P(A \cup B)$

assuming these conditional probabilities to be well defined. [Hint for (a): $P(A|B \cap C) = P(A \cap B \cap C)/P(B \cap C) = P(A)P(B)P(C)/P(B)P(C) = P(A)$.] Give a nonmathematical plausibility argument for (a), (b), (c), (d).

19 Notice that (1), Section 4.8, includes the definition of independence itself in the following sense. Suppose that X and Y are independent. Then $\phi_1(X)$, $\phi_2(Y)$ are also independent, and

$$E[\phi_1(X)\phi_2(Y)] = E[\phi_1(X)]E[\phi_2(Y)]$$

assuming all expected values exist. Show that by choosing ϕ_1 and ϕ_2 properly, it implies that

$$P[(X \text{ in } E) \cap (Y \text{ in } F)] = P(X \text{ in } E)P(Y \text{ in } F)$$

20 (a) Suppose $X_1, X_2, X_3, X_4, X_5, X_6$ are mutually independent. Show that $\phi_1(X_1, X_2)$, $\phi_2(X_3, X_4)$, and $\phi_3(X_5, X_6)$ are also mutually independent.

(b) Formulate a general version of (a).

21 Suppose X, Y, and Z have the following joint distribution:

a	b	c	$P[(X = a) \cap (Y = b) \cap (Z = c)]$
1	0	0	1/4
0	1	0	1/4
0	0	1	1/4
1	1	1	1/4

Show that each of (X, Y), (X, Z), (Y, Z) is a pair of independent random variables, but X, Y, Z are not mutually independent.

22 Suppose each of the four permutations of 1, 2, 3, 4 has the same probability, $1/4! = 1/24$. Define random variables X_1, X_2, X_3, X_4 as follows:

$$X_i = \begin{cases} 1 \text{ if } i\text{th element of permutation exceeds} \\ \quad \text{all preceding elements} \\ 0 \text{ otherwise} \end{cases}$$

and $X_1 = 1$ (compare Exercises 12 and 13, Chapter 2). Show that X_1, X_2, X_3, X_4 are mutually independent. (Hint: The easiest approach may be a brute-force one. Enumerate all 24 sample points and tabulate the joint distribution. A theoretical approach to the general version of this problem appears in Exercise 27.)

23 (a) Suppose that each of X and Y has two possible values. (You may as well suppose that these possible values are 0 and 1.) Show that the single condition

$$P[(X = 1) \cap (Y = 1)] = P(X = 1)P(Y = 1)$$

implies that X and Y are independent. {Hint: $P[(X = 0) \cap (Y = 1)] = P(Y = 1) - P[(X - 1) \cap (Y = 1)]$, and so forth. See Exercise 17d.}

(b) Show that if X, Y, and Z are indicator random variables, then the four conditions

$$P[(X = 1) \cap (Y = 1) \cap (Z = 1)] = P(X = 1)P(Y = 1)P(Z = 1)$$
$$P[(X = 1) \cap (Y = 1)] = P(X = 1)P(Y = 1)$$
$$P[(X = 1) \cap (Z = 1)] = P(X = 1)P(Z = 1)$$
$$P[(Y = 1) \cap (Z = 1)] = P(Y = 1)P(Z = 1)$$

imply that X, Y, and Z are mutually independent.

(c) Formulate a general version of (a) and (b).

24 Suppose that events $A_1, \ldots, A_n$ are mutually independent. Show that

$$P(A_1 \cup A_2 \cup \cdots \cup A_n) = 1 - [1 - P(A_1)][1 - P(A_2)] \cdots [1 - P(A_n)]$$

[Hint: Use the fact that $(A_1 \cup A_2 \cup \cdots \cup A_n)' = A_1' \cap A_2' \cap \cdots \cap A_n'$ (Section 2.7). Alternate method: Take expected values of both sides of (3), Section 2.7, and use the fact that $X_{A_1}, X_{A_2}, \ldots, X_{A_n}$ are mutually independent.]

25 Suppose in the matrix of joint probabilities $P[(X = a_i) \cap (Y = b_j)] = p_{ij}$,

$$\begin{pmatrix} p_{11} & p_{12} & \cdots \\ p_{21} & p_{22} & \cdots \\ \vdots & \vdots & \end{pmatrix}$$

any two rows are proportional to each other. That is, for every i, j,

$$\frac{p_{i1}}{p_{j1}} = \frac{p_{i2}}{p_{j2}} = \cdots$$

(Numerators and denominators must both be zero if either one is zero.) Show that X and Y are independent. [Hint: $p_{ik} = c_{ij}p_{jk}$ for each k. Sum over k to obtain

$$P(X = a_i) = c_{ij}P(X = a_j) \tag{i}$$

and sum over i to obtain

$$P(Y = b_k) = \sum_i c_{ij}p_{jk} \tag{ii}$$

Sum (i) over i to obtain

$$1 = \sum_i c_{ij}P(X = a_j) \tag{iii}$$

Now combine (ii) and (iii) to obtain

$$p_{jk} = P(X = a_j)P(Y = b_k)$$

(What happens if $p_{jk} = 0$?)]

26 Suppose that $A_1, \ldots, A_n$ are events and $X_{A_1}, \ldots, X_{A_n}$ are their indicator random variables. Show that the events $A_1, \ldots, A_n$ are mutually independent if, and only if, the random variables $X_{A_1}, \ldots, X_{A_n}$ are mutually independent.

27 Suppose the sample space consists of all $n!$ permutations of $1, 2, \ldots, n$, each having probability $1/n!$ Let $(i_1, \ldots, i_n)$ denote a typical sample point and define random variables $R_1, R_2, \ldots, R_n$ as follows:

R_k = number of indices among $i_1, \ldots, i_k$ that are less than or equal to i_k

(R_1 is automatically 1). For instance, for $n = 5$, the following illustrates the definition:

Sample points	R_1	R_2	R_3	R_4	R_5
1 2 3 4 5	1	2	3	4	5
4 3 1 5 2	1	1	1	4	2
2 3 5 4 1	1	2	3	3	1
⋮			⋮		

(a) Show that there is a one-to-one correspondence between the n-tuples of possible values $(r_1, r_2, \ldots, r_n)$ such that

$$P[(R_1 = r_1) \cap (R_2 = r_2) \cap \cdots \cap (R_n = r_n)] > 0$$

and sample points $(i_1, \ldots, i_n)$. [*Hint:* Sample points $(i_1, \ldots, i_n)$

completely determine the values of $R_1, \ldots, R_n$. Conversely, if $(r_1, \ldots, r_n)$ are values of $R_1, \ldots, R_n$, they completely determine the sample point $i_1, \ldots, i_n$. Work backward: r_n determines the value of i_n, r_{n-1} determines the value of i_{n-1}, and so forth. For instance, for $n = 5$, suppose $(r_1, \ldots, r_5) = (1, 2, 3, 3, 1)$. $r_5 = 1$ says that all numbers in positions 1 through 4 exceed the number in position 5; therefore, that number is 1. $r_4 = 3$ says that $4 - 3 = 1$ of the numbers in positions 1 through 3 exceeds the number in position 4; therefore, that number must be 4; and so forth. The sample point is (2, 3, 5, 4, 1), finally.]

(b) Show that for every $k = 1, \ldots, n$,

$$P(R_k = i) = \frac{1}{k}, \qquad i = 1, \ldots, k$$

(c) Conclude from (a) that

$$P[(R_1 = r_1) \cap \cdots \cap (R_n = r_n)] = \frac{1}{n!}$$

and, hence, $R_1, \ldots, R_n$ are mutually independent.

(d) Define

$$X_k = \begin{cases} 1 \text{ if } R_k = k \\ 0 \text{ otherwise} \end{cases}$$

Conclude from (c) that $X_1, \ldots, X_n$ are independent. (Notice that Exercise 22 is a special case for $n = 4$.)

NOTES

1 For some additional discussions of conditional probability and independence, see Chapter V of [9] and Chapters 4 and 5 of [10].

2 G. Polya (1887–) is a contemporary American mathematician.

3 Bayes' formula is the work of Thomas Bayes (1702–1761), an English clergyman who worked on the "doctrine of chances," as he called probability theory. The calculation of the probability of the sun rising was done under Bayes' influence by Laplace (1749–1827), who should have known better (see p. 124 of [9]).

4 Matrices are discussed in Sections 13.2 and 13.3 of [2], and in Chapter 16 of [3]. In particular, the associativity of matrix multiplication for *finite* matrices is proved in Theorem 16.17 of [3]. The fact that associativity holds for products of infinite matrices, with nonnegative elements, is less elementary, but is proved in Proposition 1-4 of [15]. The first 10 pages of [15] provides a useful review of formal matrix operations.

Chapter Five

GENERATING FUNCTIONS

The probability distribution of a nonnegative, integer-valued random variable is concisely summarized by a generating function. The generating function is simply a power series in a dummy variable t, with the coefficient of t^k being the probability at k. This chapter shows how generating functions are useful for evaluating moments and for determining the distribution of a sum of independent random variables (Notes 1 and 2).

5.1 GENERATING FUNCTIONS

Suppose that X has the possible values 0, 1, 2, ... It is possible to summarize the probabilities $P(X = 0)$, $P(X = 1)$, . . . in one function, called the generating function.

generating function

By the *generating function of X* is meant the following function of t:

$$\phi(t) = P(X = 0) + P(X = 1)t + P(X = 2)t^2 + \cdots, \qquad |t| \leq 1$$

(Note 3.) (1)

The only role that the t plays is as a dummy variable in constructing the function ϕ. If X has only a finite number of possible values, then ϕ is a polynomial in t and is defined for all real numbers t. Otherwise, ϕ is a power series in t that converges for $|t| \leq 1$, and possibly for no other values of t. Notice that $\phi(1) = 1$ (Note 4).

In general, if $\phi(t) = p_0 + p_1t + p_2t^2 + \cdots$, and the p_i's are all nonnegative, and $p_0 + p_1 + p_2 + \cdots = 1$, then ϕ will be called a *probability generating function,* without necessarily referring to an underlying random variable. A polynomial in t or a power series in t completely determines, and is completely determined by, the sequence of coefficients of $t^0, t^1, t^2, \ldots$. By differentiating both sides of (1) with respect to t and setting $t = 0$, we conclude that

$$\frac{\phi^{(k)}(0)}{k!} = p_k, \qquad k = 1, 2, \ldots$$

(We also have $\phi(0) = p_0$.) We shall see a number of contexts in which it will be easier to determine the generating function than to determine the probabilities directly. Meanwhile, we give several examples of functions that are probability generating functions, and one example of a function that is not.

Example 1 Suppose X is uniformly distributed over $0, 1, \ldots, n$. That is, $P(X = k) = 1/(n + 1)$, $k = 0, 1, \ldots, n$. Then,

$$\phi(t) = \frac{(1 + t + t^2 + \cdots + t^n)}{(n + 1)}$$

$$= \begin{cases} \dfrac{(1 - t^{n+1})}{(1 - t)(n + 1)} & \text{if } t \neq 1 \\ 1 & \text{if } t = 1 \end{cases}$$

Example 2 Suppose X is the waiting time to obtain a head in tossing an honest coin. That is, $P(X = 1) = 1/2$, $P(X = 2) = 1/2^2, \ldots, P(X = k) = 1/2^k, \ldots$. Then,

$$\phi(t) = \left(\frac{1}{2}\right)t + \left(\frac{1}{2}\right)^2 t^2 + \cdots = \frac{t}{2 - t}, \qquad |t| \leq 1$$

Example 3 From calculus, recall the power-series expansion

$$-\log(1 - u) = u + \frac{u^2}{2} + \frac{u^3}{3} + \cdots$$

if $|u| < 1$ (Note 5). Now define

$$\phi(t) = \frac{\log(1 - pt)}{\log(1 - p)}$$

with $0 < p < 1$. Now ϕ has the expansion

$$\phi(t) = cpt + c\frac{p^2}{2}t^2 + c\frac{p^3}{3}t^3 + \cdots$$

where $c = [-\log(1-p)]^{-1}$. This expansion is valid for $|t| \leqslant 1$ (actually even for $|t| < 1/p$), and

$$cp + c\left(\frac{p^2}{2}\right) + \cdots = \phi(1) = \frac{\log(1-p)}{\log(1-p)} = 1$$

Hence, ϕ is a probability generating function.

Example 4 $\phi(t) = 2/(1+t)$ is *not* a probability generating function. Even though $\phi(1) = 1$, the coefficients are not all nonnegative.

The point brought out in the preceding example deserves emphasis.

criterion for a probability generating function

Suppose $\phi(t) = p_0 + p_1 t + p_2 t^2 + \cdots$ is a power series that converges when $|t| \leqslant 1$. Then ϕ is a probability generating function if, and only if,

$$p_k \geqslant 0, \qquad k = 0, 1, \ldots$$

and

$$\phi(1) = 1$$

5.2 GENERATING FUNCTION OF A SUM OF INDICATOR RANDOM VARIABLES

In this section, we obtain an explicit and useful expression for the generating function of a sum of indicator random variables. Suppose that $X = X_1 + \cdots + X_n$, where *each X_i is an indicator random variable.* Let $A_1, \ldots, A_n$ be the events for which the X_i's are indicator random variables. That is,

$$(X_i = 1) = A_i \quad \text{and} \quad P(X_i = 1) = P(A_i)$$

Recall the notation introduced earlier in Section 2.8.

$$S_1 = \sum_i P(A_i), \qquad S_2 = \sum_{i<j} P(A_i \cap A_j), \qquad \ldots$$

The generating function of X is

$$\phi(t) = 1 + (t-1)S_1 + (t-1)^2 S_2 + \cdots + (t-1)^n S_n \tag{1}$$

Notice that ϕ is not explicitly expressed by (1) as a polynomial in t, but rather as a polynomial in $(t-1)$; but, of course, the one determines the other. In any case, ϕ is a polynomial of at most nth degree, because n is the largest possible value of X.

Proof of (1). The proof depends on examining the following "random functions" of t,

$$Y = [1 + (t-1)X_1] \cdots [1 + (t-1)X_n] \qquad (2)$$

Consider the value of $Y(w)$ for any sample point w where $X(w) = k$. For such a w, $Y(w) = t^k$. This is so because k of the $X_i(w)$'s are equal to 1 and the remaining $X_i(w)$'s are equal to 0, and therefore, the right-hand side of (2) is $[1 + (t-1)]^k 1^{n-k} = t^k$. Hence,

$$P(Y = t^k) = P(X = k), \qquad k = 0, 1, \ldots, n$$

and

$$EY = t^0 P(Y = t^0) + \cdots + t^n P(Y = t^n)$$
$$= t^0 P(X = 0) + \cdots + t^n P(X = n)$$

Thus, EY, considered as a function of the dummy variable t, is just the generating function of X. Now let us examine the expected value of Y from another point of view. Expand the right-hand side of (2) in powers of $(t-1)$, giving,

$$Y = 1 + (t-1) \sum_i X_i + (t-1)^2 \sum_{i<j} X_i X_j + \cdots$$
$$+ (t-1)^n X_1 \cdots X_n \qquad (3)$$

Now take the expected value of the right-hand side of (3), giving

$$EY = 1 + (t-1)S_1 + (t-1)^2 S_2 + \cdots + (t-1)^n S_n \quad \blacksquare$$

Example 1 Consider the special case of $n = 2$. Then

$$\phi(t) = 1 + (t-1)[P(A_1) + P(A_2)] + (t-1)^2 P(A_1 \cap A_2)$$
$$= [1 - P(A_1) - P(A_2) + P(A_1 \cap A_2)]$$
$$+ [P(A_1) + P(A_2) - 2P(A_1 \cap A_2)]t + P(A_1 \cap A_2)t^2$$

Hence,

$$P(X = 0) = 1 - P(A_1) - P(A_2) + P(A_1 \cap A_2)$$
$$P(X = 1) = P(A_1) + P(A_2) - 2P(A_1 \cap A_2)$$
$$P(X = 2) = P(A_1 \cap A_2)$$

(See Figure 5.1.)

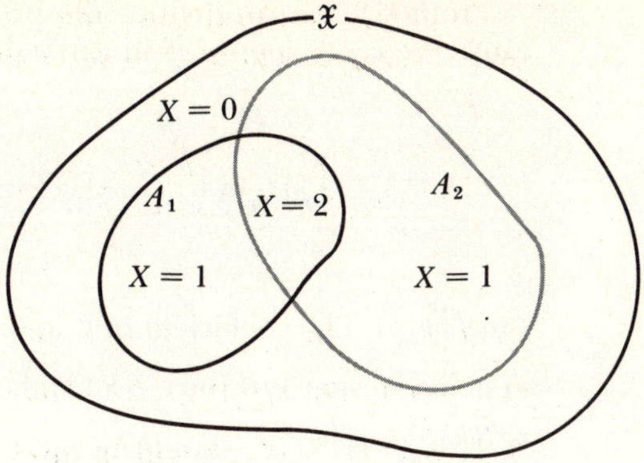

FIGURE 5.1

Example 2 An honest coin is thrown n times. Let X_i be 1 if the ith throw is head, and 0 otherwise. Then $X = X_1 + \cdots + X_n$ is the total number of heads. Inasmuch as $S_k = \binom{n}{k}\Big/2^k$ (verify this!),

$$\phi(t) = \sum_{k=0}^{n} \binom{n}{k}(t-1)^k/2^k = \left[\frac{(t-1)}{2} + 1\right]^n = \left[\frac{1+t}{2}\right]^n$$

Because $[(1+t)/2]^n = \Sigma_{k=0}^{n} \binom{n}{k} t^k/2^n$, it follows that

$$P(X = k) = \frac{\binom{n}{k}}{2^n}, \qquad k = 0, 1, \ldots, n$$

Of course, this distribution of X is easily verified directly.

Example 3 Let X be the number of matches in the matching problem, using a deck of n cards (Section 3.2). Let X_i be 1 if there is a match in the ith position and 0 otherwise, and let $X = X_1 + \cdots + X_n$. It has already been pointed out that

$$S_k = \frac{1}{k!}, \qquad k = 1, 2, \ldots, n$$

Hence,

$$\phi(t) = \sum_{k=0}^{n} \frac{(t-1)^k}{k!} \tag{4}$$

Notice that (4) is the partial sum of the first n terms of the expansion of $\exp(t-1)$ in powers of $(t-1)$.

From (1), we can deduce the probability distribution of X, because the coefficient of t^k in $\phi(t)$ equals $P(X = k)$.

$$P(X = k) = \binom{k}{k}S_k - \binom{k+1}{k}S_{k+1} + \cdots + (-1)^{n-k}\binom{n}{k}S_n \qquad (5)$$

Proof of (5). The coefficient of t^k in $(t-1)^i$ is $\binom{i}{k}(-1)^{i-k}$, $i=k, \ldots, n$. (The coefficient is 0 for $i < k$.) Hence, the coefficient of t^k in $\phi(t)$ is $\sum_{i=k}^{n}\binom{i}{k}S_k(-1)^{i-k}$. ($S_0$ should be understood to equal 0.) (See Note 6.) ∎

Notice that the special case of $k = 0$ in (5) is

$$P(X = 0) = 1 - S_1 + S_2 - \cdots + (-1)^n S_n$$

which is equivalent to the *inclusion-exclusion formula*, (Section 2.8).

Example 4 Again suppose that X is the number of matches with n cards, as in Example 3. According to (5), the distribution of X is given by

$$P(X = k) = \binom{k}{k}\frac{1}{k!} - \binom{k+1}{k}\frac{1}{(k+1)!} + \cdots + (-1)^{n-k}\binom{n}{k}\frac{1}{n!}$$

$$= \frac{1}{k!}\sum_{i=0}^{n-k}\frac{(-1)^i}{i!} \qquad (6)$$

since

$$\frac{\binom{k+i}{k}}{(k+i)!} = \frac{1}{i!k!}$$

Notice that the right side of (6) is a partial sum in the expansion of $e^{-1}/k!$ As pointed out in (3), Section 3.2, the limit of $P(X = k)$ as $n \to \infty$ equals $e^{-1}/k!$

5.3 GENERATING FUNCTIONS AND MOMENTS

moments

By the kth moment of a random variable is meant

$$E(X^k), \qquad k = 1, 2, \ldots$$

assuming that this expected value exists [that is, $E(|X|^k) < \infty$.]

If reference to the 0th moment is ever required, it is understood to equal 1. The first moment, EX, is just the expected value of X. If the moment for $k = m$ exists, then all moments for $k < m$ also exist (see Exercise 24, Chapter 2).

A variant of the moment that is also of interest is the factorial moment.

factorial moments

By the kth factorial moment of a random variable X is meant

$$E(X^{(k)}) = E[X(X-1)\cdots(X-k+1)], \qquad k = 1, 2, \ldots$$

again assuming that this expected value exists. Here $E(X^{(0)})$ is understood to equal 1.

The ordinary moments are determined by the factorial moments, and vice versa. For instance,

$$E(X^1) = E(X^{(1)})$$

$$E(X^2) = E(X^{(2)}) + E(X^{(1)})$$

$$E(X^3) = E(X^{(3)}) + 3E(X^{(2)}) + E(X^{(1)})$$

These equations follow from the immediately verified relationships that

$$a^2 = a(a-1) + a$$

$$a^3 = a(a-1)(a-2) + 3a(a-1) + a$$

and, conversely,

$$E(X^{(2)}) = E(X^2) - EX$$

$$E(X^{(3)}) = E(X^3) - 3E(X^2) + 2EX$$

which are just as easy to verify.

In general, one can develop such relationships in the form

$$\left.\begin{aligned} &E(X^k) = \sum_{i=0}^{k} a_i E(X^{(i)}) \\ &\text{and} \\ &E(X^{(k)}) = \sum_{i=0}^{k} b_i E(X^i) \end{aligned}\right\} \qquad (1)$$

where the a_i's and the b_i's are uniquely determined constants. The explicit forms of these relationships are of no particular interest to us now, however. It should be pointed out that $E(X^k)$ exists if, and only if, $E(X^{(k)})$ exists. This fact follows from (1), and from the fact that if $E(X^k)$ exists then $E(X^i)$ also exists for $i < k$.

Still suppose that X is a nonnegative, integer-valued random variable, as in Section 5.1. Suppose also that ϕ is its generating function. The factorial moments of X are determined from the derivatives of the generating function at $t = 1$, as follows.

$$\phi^{(k)}(1) = E(X^{(k)}) \tag{2}$$

By $\phi^{(k)}(t)$ we mean $[d^k\phi(t)]/dt^k$, and by $\phi^{(k)}(1)$ we mean $\lim_{t\uparrow 1} \phi^{(k)}(t)$. Since $X \geqslant 0$, if $E(X^{(k)})$ does not exist, it equals ∞ in a well-defined way, and then both sides of (2) equal ∞. By $\lim_{t\uparrow 1}$ is meant that t approaches 1 from the left, because $\phi(t)$ may not even be defined if $t > 1$.

Heuristic proof of (2). By successive differentiations of ϕ, we have

$$\phi(t) = p_0 + p_1 t + p_2 t^2 + p_3 t^3 + \cdots$$

$$\phi^{(1)}(t) = p_1 + 2p_2 t + 3p_3 t^2 + \cdots$$

$$\phi^{(2)}(t) = 2p_2 + 3 \cdot 2p_3 t + \cdots$$

and, in general,

$$\phi^{(k)}(t) = \sum_{i=k}^{\infty} i^{(k)} p_i t^{i-k} = \sum_{i=0}^{\infty} i^{(k)} p_i t^{i-k} \qquad (i^{(k)} = 0 \quad \text{if} \quad 0 \leqslant i < k)$$

Now let t approach 1 to obtain

$$\phi^{(k)}(1) = \sum_{i=0}^{\infty} i^{(k)} p_i = E(X^{(k)})$$

This proof is heuristic and not rigorous because it involves interchange of limit operations, which although plausible, must still be justified. We shall not do so here. ▮

Example 1 Suppose that X is the number of throws required to obtain the first head when throwing an honest coin. Then $p_i = 1/2^i$, $i = 1, 2, \ldots$, and $\phi(t) = \Sigma_{i=1}^{\infty} t^i/2^i = t/(2-t)$. Successive differentiations give

$$\phi^{(1)}(t) = \frac{2}{(2-t)^2}, \ \phi^{(2)}(t) = \frac{4}{(2-t)^3}, \ldots, \ \phi^{(k)}(t) = \frac{2k!}{(2-t)^{k+1}}$$

Hence, all moments exist and

$$E(X^{(k)}) = (k+1)!, \qquad k = 1, 2, \ldots$$

It was observed in Example 3, Section 2.10, that $EX = 2$.

Example 2 Consider the Poisson distribution given by $p_i = e^{-1}/i!$, $i = 0, 1, \ldots$. The generating function is

$$\phi(t) = e^{-1} + e^{-1}t + \frac{e^{-1}t^2}{2} + \frac{e^{-1}t^3}{3!} + \cdots$$
$$= e^{-1}e^t = e^{t-1}$$

Because $\phi^{(k)}(t) = e^{t-1}$, it follows that $E(X^{(k)}) = 1$, $k = 1, 2, \ldots$.

Example 3 Let X be the number of matches using n cards. As shown in Example 3, Section 5.2,

$$\phi(t) = \sum_{i=0}^{n} \frac{(t-1)^i}{i!}$$

It follows immediately that

$$E(X^{(k)}) = \begin{cases} 1 & \text{if } k = 1, \ldots, n \\ 0 & \text{if } k > n \end{cases}$$

Now consider the generating function of $X = X_1 + \cdots + X_n$, where the X_i's are indicator random variables. According to (1), Section 5.2,

$$\phi(t) = 1 + (t-1)S_1 + (t-1)^2 S_2 + \cdots + (t-1)^n S_n$$

It follows that

$$E(X^{(k)}) = \phi^{(k)}(1) = \begin{cases} k!S_k & \text{if } k = 1, 2, \ldots, n \\ 0 & \text{if } k > n \end{cases}$$

Hence, the generating function of X is of the form

$$\phi(t) = \sum_{k=0}^{n} \frac{E(X^{(k)})(t-1)^k}{k!} \tag{3}$$

Actually, (3) generalizes to any nonnegative integer-valued random variable.

Suppose that X is nonnegative and integer valued, and that all moments of X exist. Then

$$\phi(t) = \sum_{k=0}^{\infty} \frac{E(X^{(k)})(t-1)^k}{k!}, \qquad |t| \leq 1 \tag{4}$$

Notice that (3) is a special case of (4), since in (3) $E(X^{(k)}) = 0$, if $k > n$.

Proof of (4).

$$\phi(t) = p_0 + p_1 t + p_2 t^2 + \cdots \tag{5}$$

Now write $t^i = (t - 1 + 1)^i = \sum_{k=0}^{i} \binom{i}{k}(t-1)^k$

Hence,

$$\phi(t) = \sum_{k=0}^{\infty} (t-1)^k \sum_{i=k}^{\infty} \binom{i}{k} p_i$$

The rearrangement in the order of summing (5) is valid, because (5) is absolutely convergent for $|t| \leq 1$. The result now follows from the facts that

$$\binom{i}{k} = \frac{i^{(k)}}{k!}, \qquad \sum_{i=k}^{\infty} \binom{i}{k} p_i = \frac{E(X^{(k)})}{k!} \qquad \blacksquare$$

An important corollary of (4) is the following.

> If X is nonnegative and integer valued and all its moments exist, then the distribution of X is completely determined by the moments. In other words, if two distributions over 0, 1, 2, . . . have the same moments and all moments are finite, then the distributions are identical. "Moments" can be interpreted to mean either regular or factorial moments (Note 7).

Proof. All moments existing is equivalent to all factorial moments existing. Moreover, the factorial moments completely determine and are determined by the ordinary moments. According to (4), ϕ, the generating function, is completely determined by the factorial moments. On the other hand, the distribution is completely determined by ϕ. $\blacksquare$

Example 4 Suppose all factorial moments of X equal 1. Then

$$\phi(t) = \sum_{k=0}^{\infty} \frac{(t-1)^k}{k!} = e^{t-1}$$

and X has the Poisson distribution described in Example 2.

Despite (4), a distribution over the nonnegative integers is not determined by its existing moments if all moments do not exist.

5.4 SUMS OF INDEPENDENT RANDOM VARIABLES

The most important use of generating functions is in determining the distributions of sums of independent random variables. Although the material in this section is restricted to nonnegative, integer-valued random variables, general versions of the same techniques exist for arbitrary random variables. These are discussed in Sections 8.5 through 8.7.

It is convenient to introduce an alternate way of describing the generating function. Namely,

$$\phi(t) = E(t^X), \qquad |t| < 1 \tag{1}$$

Proof of (1). Because X is a random variable with values $0, 1, 2, \ldots$, t^X is a well-defined random variable with values $t^0, t^1, t^2, \ldots$, and $E(t^X) = t^0P(X=0) + t^1P(X=1) + \cdots = \phi(t)$. ∎

Now suppose that X and Y are independent random variables. The following theorem relates the generating functions of X and Y with that of $X + Y$.

generating function of sum of independent random variables is product of generating functions

> Suppose that X and Y are independent, nonnegative, integer-valued random variables with generating functions ϕ_1 and ϕ_2 respectively. Let ϕ be the generating function of $X + Y$. Then
>
> $$\phi(t) = \phi_1(t)\phi_2(t) \tag{2}$$

Proof of (2). We want to evaluate $\phi(t) = E(t^{X+Y}) = E(t^X t^Y)$. Since X and Y are independent, then so are t^X and t^Y (Section 4.8). For independent random variables, the expected value of a product is the product of the expected values [(1), Section 4.8]. Hence,

$$\phi(t) = E(t^X t^Y) = [E(t^X)][E(t^Y)] = \phi_1(t)\phi_2(t) \quad ∎$$

By a similar argument, one shows that if $X_1, X_2, \ldots, X_n$ are mutually independent, with generating functions $\phi_1, \phi_2, \ldots, \phi_n$, respectively, then ϕ, the generating function of $X_1 + \cdots + X_n$, is given as follows.

> $$\phi(t) = \phi_1(t)\phi_2(t) \cdots \phi_n(t) \tag{3}$$

Example 1 Suppose $X_1, X_2, \ldots, X_n$ are mutually independent, Poisson distributed random variables, with $P(X_i = k) = e^{-1}/k!$, $k = 0, 1, \ldots$. Then $E(t^{X_i}) = e^{t-1}$ and $E(t^{X_1 + \cdots + X_n}) = e^{n(t-1)}$

Since

$$e^{n(t-1)} = \sum_{k=0}^{\infty} e^{-n}\frac{n^k t^k}{k!}$$

it follows that

$$P(X_1 + \cdots + X_n = k) = e^{-n}\frac{n^k}{k!}, \qquad k = 0, 1, \ldots$$

It will be seen in Section 6.7 that this example is also of a Poisson distribution, with parameter equal to n.

Example 2 Consider the distribution of the number of throws required to obtain a first head in tossing an honest coin. That is,

$$p_k = \tfrac{1}{2}^k, \qquad k = 1, 2, \ldots$$

The probability generating function of this distribution is $t/(2-t)$ (Example 2, Section 5.1). Now suppose that $X_1, X_2, \ldots, X_n$ are mutually independent, each with this same distribution. Let $X = X_1 + \cdots + X_n$. We can interpret X as the number of throws required to accumulate first a total of n heads. Then ϕ, the generating function of X, is given by $\phi(t) = t^n/(2-t)^n$. The power series expansion of ϕ is

$$\phi(t) = \sum_{k=n}^{\infty} \binom{k-1}{n-1}\left(\frac{1}{2}\right)^k t^k \tag{4}$$

(Series of this sort are discussed in Section 6.5. This section can be read now if you wish, for it does not depend on the preceding material.) Hence,

$$P(X = k) = \binom{k-1}{n-1}\left(\frac{1}{2}\right)^k, \qquad k = n, n+1, \ldots$$

Example 3 Suppose X_i equals 1 and 0, both with probability 1/2, 1/2, and $X = X_1 + \cdots + X_n$. Suppose that X_i's are independent. We can interpret X as the number of heads obtained in n throws of an honest coin. The generating function of each X_i is $(1/2)t^0 + (1/2)t^1 = (1+t)/2$. Hence, the generating function of X is $(1+t)^n/2^n$. Because

$$(1+t)^n/2^n = \sum_{k=0}^{n} \binom{n}{k} \frac{1}{2^n} t^n$$

it follows that

$$P(X=k) = \binom{n}{k} \frac{1}{2^n}$$

(Compare Example 2, Section 5.2.)

5.5 THE CONVOLUTION FORMULA

In the last section, we saw that the generating function for a sum of independent random variables is the product of the generating functions of the individual variables. We now want to examine the explicit relation between the probabilities themselves. Suppose, as before, that X and Y are independent random variables that are nonnegative and integer valued, and define,

$$P(X=k) = p_k, \qquad P(Y=k) = q_k, \qquad k = 0, 1, 2, \ldots$$

Let us now obtain the distribution of $X+Y$.

convolution formula

$$P(X+Y=k) = p_0q_k + p_1q_{k-1} + \cdots + p_kq_0 \qquad k = 0, 1, \ldots \quad (1)$$

Equation (1) says that the probability of the sum being equal to k is the sum of all products p_iq_j for all indices i and j that sum to k. This relation is called a *convolution formula.*

First proof of (1). The event $(X+Y=k)$ can be decomposed into the disjoint union

$$[(X=0) \cap (Y=k)] \cup \cdots \cup [(X=k) \cap (Y=0)]$$

Since $P[(X=i) \cap (Y=j)] = P(X=i)P(X=j)$, by independence, it follows that $P(X=k) = \sum_{i=0}^{k} P(X=i)P(X=k-i)$, which is what (1) asserts. ▮

Second proof of (1). By (2), Section 5.4, the generating function of $X+Y$ is

$$(p_0 + p_1t + p_2t^2 + \cdots)(q_0 + q_1t + q_2t^2 + \cdots)$$

The coefficient of t^k in this product is just the right side of (1). This coefficient is $P(X+Y=k)$. ▮

Example 1 A six-faced die with faces marked 1, 2, 3, 4, 5, 6 is thrown repeatedly. The mathematical model is that $X_1, X_2, \ldots, X_n$ are independent random variables, with

$$P(X_i = k) = \tfrac{1}{6}, \qquad k = 1, \ldots, 6$$

First, let us find the distribution of $X_1 + X_2$ using the convolution formula. In (1), we make the identification

$$p_1 = p_2 = \cdots = p_6 = \tfrac{1}{6}, \qquad \text{all other } p_i = 0$$

$$q_1 = q_2 = \cdots = q_6 = \tfrac{1}{6}, \qquad \text{all other } q_i = 0$$

The distribution of $X_1 + X_2$ then turns out as follows:

k	2	3	4	5	6	7	8	9	10	11	12
$P(X+Y=k)$	1/36	2/36	3/36	4/36	5/36	6/36	5/36	4/36	3/36	2/36	1/36

To find the distribution of $X_1 + X_2 + X_3$, treat $X_1 + X_2$ as X in the convolution formula and X_3 as Y. Variables $X_1 + X_2$ and X_3 are independent because X_1, X_2, X_3 are mutually independent. Some of the probabilities for $X_1 + X_2 + X_3$ are as follows:

$$P(X_1 + X_2 + X_3 = 3) = (1/36)(1/6) = 1/216$$

$$P(X_1 + X_2 + X_3 = 4) = (1/36)(1/6) + (2/36)(1/6) = 3/216$$

$$P(X_1 + X_2 + X_3 = 5) = (1/36)(1/6) + (2/36)(1/6) + (3/36)(1/6) = 6/216$$

The full distributions of $X_1 + \cdots + X_n$ for several values of n appear in Table 5.1.

Example 2 A coin is thrown $m + n$ times. Let X be the number of heads on the first m throws and Y the number of heads on the next n throws. Then $X + Y$ is the total number of heads. We know that

$$P(X = k) = \binom{m}{k}\frac{1}{2^m}, \qquad k = 0, 1, \ldots, m$$

$$P(Y = k) = \binom{n}{k}\frac{1}{2^n}, \qquad k = 0, 1, \ldots, n$$

$$P(X + Y = k) = \binom{m+n}{k}\frac{1}{2^{m+n}}, \qquad k = 0, 1, \ldots, m + n$$

In this case, (1), the convolution formula, asserts that

$$\sum_{i=0}^{k} \binom{m}{i}\binom{n}{k-i} = \binom{m+n}{k}, \qquad k = 0, 1, \ldots$$

TABLE 5.1

k	$P(X_1 = k)$	$P(X_1 + X_2 = k)$	$P(X_1 + X_2 + X_3 = k)$	$P(X_1 + X_2 + X_3 + X_4 = k)$
1	1/6			
2	1/6	1/36		
3	1/6	2/36	1/216	
4	1/6	3/36	3/216	1/1296
5	1/6	4/36	6/216	4/1296
6	1/6	5/36	10/216	10/1296
7		6/36	15/216	20/1296
8		5/36	21/216	35/1296
9		4/36	25/216	56/1296
10		3/36	27/216	80/1296
11		2/36	27/216	104/1296
12		1/36	25/216	125/1296
13			21/216	140/1296
14			15/216	146/1296
15			10/216	140/1296
16			6/216	125/1296
17			3/216	104/1296
18			1/216	80/1296
19				56/1296
20				35/1296
21				20/1296
22				10/1296
23				4/1296
24				1/1296

Notice in particular that if $m = n$ and $k = n$, we have the interesting relation

$$\sum_{i=0}^{n} \binom{n}{i}\binom{n}{n-i} = \sum_{i=0}^{n} \binom{n}{i}^2 = \binom{2n}{n}$$

Example 3 Suppose that $X_1, X_2, \ldots$ are independent indicator random variables with

$$P(X_n = 0) = 1 - \frac{1}{n}, \qquad P(X_n = 1) = \frac{1}{n}, \qquad n = 1, 2, \ldots$$

If the convolution formula is successively applied to determine the distributions of $X_1 + X_2$, $(X_1 + X_2) + X_3$, $(X_1 + X_2 + X_3) + X_4$, and so forth, they turn out as tabulated in Table 5.2. Notice that the distribution of $X_1 + X_2 + X_3 + X_4$ is exactly the one that appeared

TABLE 5.2

k	$P(X_1=k)$	$P(X_1+X_2=k)$	$P(X_1+X_2+X_3=k)$	$P(X_1+X_2+X_3+X_4=k)$	$P(X_1+X_2+X_3+X_4+X_5=k)$
1	1	1/2	2/6	6/24	24/120
2		1/2	3/6	11/24	50/120
3			1/6	6/24	35/120
4				1/24	10/120
5					1/120

previously in Exercise 12a Chapter 2 and in Exercise 22, Chapter 4. Exercise 27, Chapter 4, explains this fact.

SUMMARY

If a random variable X has possible values 0, 1, 2, . . . , then its *generating function* is defined to be

$$\phi(t) = E(t^X) = P(X=0) + P(X=1)t + P(X=2)t^2 + \cdots, \qquad |t| \leqslant 1$$

The generating function ϕ completely determines the probabilities. In fact,

$$\frac{\phi^{(k)}(0)}{k!} = P(X=k), \qquad k = 0, 1, \ldots$$

The *generating function of a sum of indicator random variables* $X = X_1 + \cdots + X_n$ has a particularly simple form, from which it is possible to determine explicit, useful expressions for $P(X = k)$. In particular, the expression for $P(X = 0)$ is equivalent to the inclusion-exclusion formula.

The generating function can be used to evaluate *moments* when they exist. The *kth factorial moment* is given by $E(X^{(k)}) = \phi^{(k)}(1)$. If all moments of a nonnegative, integer-valued random variable exist, then the generating function is determined in terms of the factorial moments as follows:

$$\phi(t) = \sum_{k=0}^{\infty} \frac{E(X^{(k)})(t-1)^k}{k!}, \qquad |t| \leqslant 1$$

If $X_1, \ldots, X_n$ are independent, nonnegative, integer-valued random variables with generating functions $\phi_1, \ldots, \phi_n$, then the generating function of $X_1 + \cdots + X_n$ is given by

$$\phi(t) = \phi_1(t) \cdots \phi_n(t)$$

If $\Sigma_{n=0}^{\infty} p_n t^n = \phi_1(t)$ and $\Sigma_{n=0}^{\infty} q_n t^n = \phi_2(t)$ are two probability generating functions, then $\phi(t) = \phi_1(t)\phi_2(t) = \Sigma_{n=0}^{\infty} r_n t^n$ is a probability generating function. The sequence $\{r_n\}$ is called the *convolution* of the sequences $\{p_n\}$ and $\{q_n\}$. Explicitly,

$$r_n = \sum_{i=0}^{n} p_i q_{n-i}$$

EXERCISES

1 (a) A die is thrown repeatedly until a 2 is obtained. Let X be the number of throws required. Show that the generating function of X is $t/(6-5t)$.
(b) A die is thrown repeatedly until either a 2 or a 3 is obtained. Show that the generating function of the number of throws required is $t/(3-2t)$.

2 (a) Show that $\phi(t) = 1$ is the generating function of a random variable which is 0 with probability 1. If $P(X = k) = 1$, then the generating function is $\phi(t) = t^k$, $k = 0, 1, \ldots$.
(b) If the generating function of X is $\phi(t)$, show that the generating function of $X + k$ is $t^k\phi(t)$, where k is a positive integer.

3 (a) If ϕ is a probability generating function, show that $|\phi(t)| < 1$ if $|t| < 1$, assuming $P(X = 0) \neq 1$.
(b) Show that $1/[2 - \phi(t)]$ is a probability generating function. {Hint: Expand $(1/2)[1 - \phi(t)/2]^{-1}$ as a geometric series in $\phi(t)/2$.}

4 (a) Suppose n balls are randomly distributed among m cells. Let X be the number of unoccupied cells. Show that the generating function of X is

$$\sum_{i=0}^{m} \binom{m}{i}\left(\frac{m-i}{m}\right)^n (t-1)^i$$

[Hint: Use (1), Section 5.2.

$$S_i = \binom{m}{i}\left(\frac{m-i}{m}\right)^n]$$

(b) Show that the probability distribution of X is given by

$$P(X = k) = \sum_{i=0}^{m-k} (-1)^i \binom{k+i}{k}\binom{m}{k+i}\left(\frac{m-k-i}{m}\right)^n$$

[Hint: Use (5), Section 5.2.]

5 (a) Let X be a random variable having the possible values $0, 1, \ldots, n$. Show that there exist indicator random variables $X_1, \ldots, X_n$ such that $X = X_1 + \cdots + X_n$. [Hint: Let A_1 be the event $(X > 0)$, A_2 the event $(x > 1) \ldots$, A_n the event $(x > n - 1)$, and let

$X_1, X_2, \ldots, X_n$ be the indicator random variables of $A_1, \ldots, A_n$, respectively. Draw a picture.]

(b) Consider the events A_i defined in the hint for (a). Notice that $A_1 \supset A_2 \supset \cdots \supset A_n$. Let $S_1, S_2, \ldots, S_n$ be the quantities in (1), Section 5.2. Show that

$$S_1 = \sum_{i=0}^{n-1} P(X > i)$$

$$S_2 = \sum_{i=1}^{n-1} iP(X > i)$$

$$\vdots$$

$$S_k = \sum_{i=k-1}^{n-1} \binom{i}{k-1} P(X > i)$$

(c) Conclude from (b) [(1), Section 5.2] and (4), Section 5.3, that

$$E(X^{(k)}) = k! \sum_{i=k-1}^{n-1} \binom{i}{k-1} P(X > i) \qquad (*)$$

Inasmuch as $P(X > n) = P(X > n+1) = \cdots = 0$, (*) can be written as

$$E(X^{(k)}) = k! \sum_{i=k-1}^{\infty} \binom{i}{k-1} P(X > i) \qquad (**)$$

Notice that when $k = 1$, (**) corresponds to Eq. (3), Section 2.10.

(d) Part (c) suggests that (**) should be true even without the restriction of a finite number of possible values. Show that this fact is indeed true by imitating the proof of (3), Section 2.10. [Hint: The right side of (**) is

$$k! \sum_{i=k-1}^{\infty} \binom{i}{k-1} \Big(\sum_{j>i} p_j \Big) = k! \sum_{j=k}^{\infty} p_j \sum_{i=k-1}^{j-1} \binom{i}{k-1}$$

$$= \sum_{j=k}^{\infty} j^{(k)} p_j = E(X^{(k)})$$

which uses the fact that

$$\sum_{i=k-1}^{j-1} \binom{i}{k-1} = \binom{j}{k}$$

which is proved in Exercise 21d, Chapter 3.

6 Suppose that X has generating function ϕ and that $E(X^3)$ exists. Show that

$$E(X^3) = \phi^{(3)}(1) + 3\phi^{(2)}(1) + \phi^{(1)}(1)$$

$$E(X^2) = \phi^{(2)}(1) + \phi^{(1)}(1)$$

7 Suppose that for a positive integer n,

$$\frac{E(X^{(k)})}{k!} = \begin{cases} \binom{n}{k} & \text{if } k = 0, 1, \ldots, n \\ 0 & \text{if } k > n \end{cases}$$

Show that $P(X = n) = 1$. [Hint: Use (4), Section 5.3, to show that $\phi(t) = t^n$.]

8 (a) If $\phi(t) = Et^X$ is the generating function of X, show that the generating function of the "tail probabilities" is the following:

$$\sum_{k=0}^{\infty} P(X > k)t^k = \frac{1 - \phi(t)}{1 - t}, \qquad |t| < 1$$

{Hint: Let $p_k = P(X = k)$. Then sum

$$\begin{aligned} & t^0\ [1] \\ + & t^1\ [1 - p_1] \\ + & t^2\ [1 - p_1 - p_2] \\ + & t^3\ [1 - p_1 - p_2 - p_3] \\ & \vdots \end{aligned}$$

down the columns of the triangular display, which gives

$$\frac{1}{1-t} - \frac{p_1 t}{1-t} - \frac{p_2 t^2}{1-t} - \cdots\}$$

(b) Show that

$$\lim_{t \uparrow 1} \frac{1 - \phi(t)}{1 - t} = EX$$

even if $EX = \infty$. (The term $t \uparrow 1$ means t approaches 1 from the left.) [Hint: Use part (a) of this exercise and Eq. (3), Section 2.10.]

(c) Suppose that $0 < EX < \infty$. Show that $[1 - \phi(t)]/[(1 - t)EX]$ is a probability generating function.

9 Suppose that $X_1, \ldots, X_n$ are independent indicator random variables. Define $p_i = P(X_i = 1)$, $i = 1, \ldots, n$. Show that the generating function of $X_1 + \cdots + X_n$ is $[p_1 t + (1 - p_1)] \cdots [p_n t + (1 - p_n)]$.

10 A box contains four balls, numbered 0, 1, 1, and 2. Suppose n balls are successively drawn, *with replacement* between drawings. Let X be the sum of the n numbers drawn. Show that

$$P(X = k) = \binom{2n}{k} \frac{1}{2^{2n}}, \qquad k = 0, 1, \ldots, 2n$$

{Hint: The generating function for any one drawing is $(1 + 2t + t^2)/4 = [(1 + t)/2]^2$. The generating function of X is $[(1 + t)/2]^{2n}$.}

11 Suppose A will throw a die repeatedly until he obtains a 2. Independently, B will throw a die repeatedly until he obtains a 2 or a 3. Let X be the number of throws required by both A and B together. Show that

$$P(X=k)=\frac{5}{18}\left(\frac{5}{6}\right)^{k-2}-\frac{2}{9}\left(\frac{2}{3}\right)^{k-2}, \qquad k=2, 3, \ldots$$

(Hint: From Exercise 1, the generating function is $[t/(6-5t)][t/(3-2t)]$. Express this as

$$\frac{5}{18}\frac{t^2}{1-(5/6)t}-\frac{2}{9}\frac{t^2}{1-(2/3)t}$$

and expand each part as a geometric series.)

12 (a) Balls are successively distributed among three cells. Let Z be the number of balls that must be distributed to occupy all three cells. Show that the generating function of Z is

$$t^3\frac{1/3}{1-(2/3)t}\frac{2/3}{1-(1/3)t}$$

(Hint: Use Example 3, Section 3.5.)

(b) Notice that each of

$$\phi_1(t)=\frac{1/3}{[1-(2/3)t]} \quad \text{and} \quad \phi_2(t)=\frac{2/3}{[1-(1/3)t]}$$

is a probability generating function. Conclude that Z is distributed like $X+Y+3$, where X and Y are independent and have generating functions ϕ_1 and ϕ_2, respectively.

13 (a) An honest die is thrown six times. Let X be the total number of points. Verify the following entries in the distribution of X:

k	6	7	8	9	10
$P(X=k)$	$1/6^6$	$6/6^6$	$21/6^6$	$56/6^6$	$126/6^6$

(Hint: Use Table 5.1. Find the first five terms in the distribution of a sum of two independent random variables, each of which has the distribution of $X_1+X_2+X_3$.)

(b) An honest die is thrown n times. Let X be the total number of points. Verify the following entries in the distribution of X.

k	n	$n+1$	$n+2$	$n+3$	$n+4$
$P(X=k)$	$1/6^n$	$n/6^n$	$\binom{n+1}{2}\Big/6^n$	$\binom{n+2}{3}\Big/6^n$	$\binom{n+3}{4}\Big/6^n$

(Unfortunately, this pattern does not continue for all k.) {Hint: The generating function for X is

$$\frac{(t + t^2 + \cdots + t^6)^n}{6^n} = \frac{t^n}{6^n}\frac{(1 - t^6)^n}{(1 - t)^n}$$

$$= \frac{t^n}{6^n}(1 - nt^6 + \cdots)\left[1 + nt + \binom{n+1}{2}t^2 + \binom{n+2}{3}t^3 + \cdots\right]\}$$

14 (a) Suppose that $X_1, X_2, \ldots$ are independent random variables, each with the same distribution,

$$P(X_i = 0) = P(X_i = 1) = P(X_i = 2) = \tfrac{1}{3}$$

Tabulate the distributions of X_1, $X_1 + X_2$, $X_1 + X_2 + X_3$, $X_1 + X_2 + X_3 + X_4$.

(b) Verify the following general expressions.

k	$P(X_1 + \cdots X_n = k)$
0	$1/3^n$
1	$n/3^n$
2	$\binom{n+1}{2}\Big/3^n$
3	$\left[\binom{n+2}{3} - n\right]\Big/3^n$
4	$\left[\binom{n+3}{4} - n^2\right]\Big/3^n$
5	$\left[\binom{n+4}{5} - n\binom{n+1}{2}\right]\Big/3^n$

(The general pattern is more complicated!) (Hint: Follow the hint to Exercise 13b.)

(c) A box of Squiffles is supposed to contain 100 Squiffles, but it typically contains 99, 100, or 101 with equal probabilities. What is the distribution of the total number of Squiffles in four boxes? [Hint: All necessary computations will have been done in part (a).]

15 Show that the only way that ϕ and $1/\phi$ can both be probability generating functions is for $\phi(t)$ to be identically 1. [Hint: Condition $\phi(1/\phi) = 1$ means it is possible for a sum $X + Y$ of two independent, nonnegative, integer-valued random variables to equal 0. Show that it can only happen if $P(X = 0) = P(Y = 0) = 1$.]

16 Consider the following version of the matching problem using a deck with a *random number* of cards. First a preliminary experiment is done, which produces a random variable Y such that

$$P(Y = k) = \frac{1}{2^{k+1}}, \qquad k = 0, 1, 2, \ldots$$

If $Y = k$, then a deck of k cards marked $1, \ldots, k$ is laid out against fixed positions also marked $1, \ldots, k$. Let X be the total number of matches. (If $Y = 0$, then $X = 0$.) Show that the generating function of X is $e^{(t-1)/2}$. According to Section 6.7, this fact says that X is Poisson distribution with parameter 1/2. [Hint: Given that $Y = k$, the conditional generating function of X is $1 + (t-1) + \cdots + (t-1)^k/k!$ See Example 3, Section 5.2.]

17 Suppose ϕ_1 and ϕ_2 are probability generating functions. Show that ϕ defined by

$$\phi(t) = \phi_1[\phi_2(t)]$$

is a probability generating function also.

18 The nth generation of a certain population contains k organisms with probability p_k, $k = 0, 1, 2, \ldots$. Each organism of the nth generation gives rise to r new organisms of the $(n+1)$st generation with probability q_r, $r = 0, 1, \ldots$. Show that the generating function of the size of the $(n+1)$st generation is $\phi_1[\phi_2(t)]$, where

$$\phi_1(t) = p_0 + p_1 t + p_2 t^2 + \cdots, \qquad \phi_2(t) = q_0 + q_1 t + q_2 t^2 + \cdots$$

What assumptions are used to obtain this result? {Hint: Assuming that the members of the nth generation produce their progeny independently one of another, the generating function of the size of the $(n+1)$st generation, given k elements in the nth generation, is $[\phi_2(t)]^k$. Hence, the unconditional generating function is

$$\sum_{k=0}^{\infty} [\phi_2(t)]^k p_k = \phi_1[\phi_2(t)]\}$$

NOTES

1 For additional reading on probability generating functions, see Chapter XI of [9].

2 Probability generating functions were already used by the English mathematician Abraham DeMoivre (1667–1754) and by the French mathematician Pierre Simon Laplace (1749–1827).

3 The term *generating function* should not be confused with *moment generating function.* The generating function is $\phi(t) = E(t^X)$, whereas the moment generating function is $E(e^{tX}) = \phi(e^t)$. The latter is related to the Laplace and Fourier transforms, to be discussed in Section 8.5.

4 See Section 18.3 of [2] and Chapter 7 of [3] for a review of power series expansions.

5 For the power series expansion of $\log(1-u)$, see Formula (8), Section 18.6, of [2], or Formula (10.32), Section 10.8, of [3].

6 For a different proof of (5), Section 5.2, see Section 5, Chapter 4, of [9].

7 In general, the distributions of nondiscrete random variables (which will be introduced in Chapter 7) need not be uniquely determined by their moments, even when all moments exist. The problem of whether a sequence of numbers can be the moments of some distribution, and if so, whether that distribution is uniquely determined, is an important chapter of classical analysis, called the *moment problem.*

Chapter Six

SOME SPECIAL DISCRETE DISTRIBUTIONS

A number of important discrete probability distributions arise in both theoretical and practical settings. This chapter reviews the most important of these—the binomial, multinomial, hypergeometric, negative binomial and Poisson distributions.

6.1 THE BINOMIAL DISTRIBUTION

The binomial distribution is one of the most useful distributions, both in theoretical probability as well as in its applications. It can be described in terms of the number of *successes* in a sequence of success-failure experiments. Suppose that the same experiment is repeated n times and these are "independent" repetitions (that is, the result of any one trial does not influence, nor is it influenced by, the result of any other trial). Each trial produces *success* or *failure*, with probabilities p and $1-p$, respectively, where p is the same for each of the n trials. What is the probability distribution of the number of successes? A formal way of posing the same question is as follows: Suppose that $X_1, X_2, \ldots, X_n$ are n *independent* indicator random variables with

$$X_i = \begin{cases} 1 \text{ with probability } p, & (0 \leq p \leq 1 \\ 0 \text{ with probability } 1-p \end{cases}$$

What is the distribution of $X = X_1 + \cdots + X_n$?

binomial distribution

The distribution of X is given as follows:

$$P(X=k) = \binom{n}{k} p^k (1-p)^{n-k}, \qquad k = 0, 1, \ldots, n \tag{1}$$

This distribution is called the *binomial distribution with parameters n and p* (Note 1).

Even though the result is very simple, the binomial distribution is so important it is worth giving several proofs.

First proof of (1). Let $e_1, \ldots, e_n$ be a sequence of k 1's and $n-k$ 0's. Inasmuch as $X_1, \ldots, X_n$ are independent,

$$P[(X_1 = e_1) \cap \cdots \cap (X_n = e_n)] = P(X_1 = e_1) \cdots P(X_n = e_n) = p^k(1-p)^{n-k}$$

There are $\binom{n}{k}$ distinguishable sequences $(e_1, \ldots, e_n)$ consisting of k 1's and $n-k$ 0's. Therefore, $P(X = k) = \binom{n}{k}p^k(1-p)^{n-k}$. ∎

Second proof of (1). The generating function of any X_i is $1 - p + pt$. Inasmuch as the X_i's are independent, then, by (3), Section 5.4, the generating function of $X_1 + \cdots + X_n$ is $(1-p+pt)^n = \sum_{k=0}^{n} \binom{n}{k} p^k(1-p)^{n-k}t^k$. The coefficient of t^k, $P(X = k)$, is $\binom{n}{k}p^k(1-p)^{n-k}$. ∎

You have already met the binomial distribution several times before. Some of the examples that follow will remind you of some of these encounters.

Example 1

(a) A coin is tossed n times. The number of heads is binomially distributed with parameters n, 1/2. (See Example 2, Section 5.2.)

(b) A die is tossed n times. The number of aces is binomially distributed with parameters n, 1/6.

(c) A pair of dice is tossed n times. The number of tosses with point values exceeding 10 is binomially distributed with parameters n, 1/12.

(d) Suppose n balls are randomly distributed among five cells. The number of balls landing in cell 1 is binomially distributed with parameters n, 1/5. The number of balls landing in any of the cells 2, 3, 4, or 5 is binomially distributed with parameters n, 4/5. The number of balls landing in cells 1, 3, or 4 is binomial with parameters n, 3/5. (See Exercise 7, Chapter 3.)

(e) Whenever a certain typist strikes a typewriter key, she will make a mistake with probability 1/100. The total number of mistakes in copy that requires a total number of 10,000 key strikes is binomially distributed with parameters $n = 10{,}000$ and $p = 1/100$.

(f) The total number of boys in a family with 10 children is binomially distributed with parameters $n = 10$ and $p = 1/2$. What assumptions are required for this fact to be correct?

Example 2 (a) The binomial distribution for $n = 5$, $p = 1/4$ is the following:

k	0	1	2	3	4	5
$P(X = k)$	$\frac{243}{1024}$	$\frac{405}{1024}$	$\frac{270}{1024}$	$\frac{90}{1024}$	$\frac{15}{1024}$	$\frac{1}{1024}$

(b) The binomial distribution for $n = 5$, $p = 3/4$ is the following:

k	0	1	2	3	4	5
$P(X = k)$	$\frac{1}{1024}$	$\frac{15}{1024}$	$\frac{90}{1024}$	$\frac{270}{1024}$	$\frac{405}{1024}$	$\frac{243}{1024}$

Notice that the probabilities in (a) and (b) are reversals of the same sequence (see Figure 6.1).

By examining the ratio $P(X = k + 1)/P(X = k)$, one can get information about the graph of the binomial distribution. The ratio is

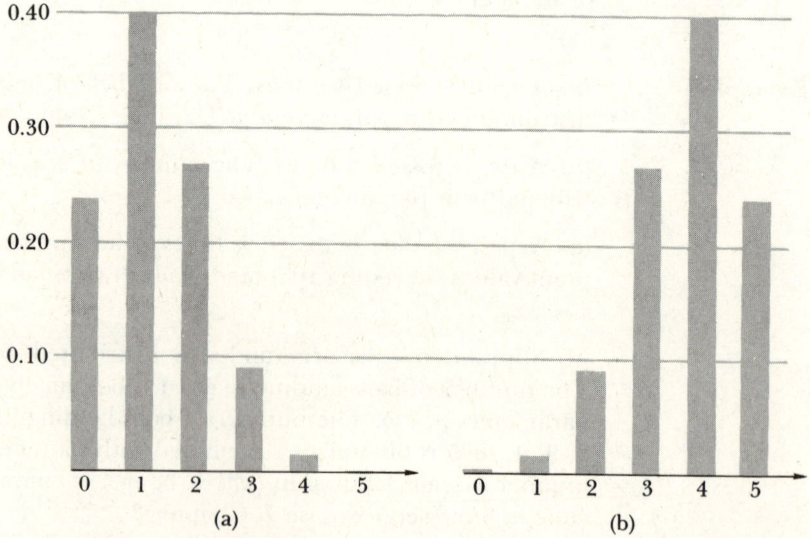

FIGURE 6.1 (a) Binomial distribution for $n = 5$, $p = 1/4$. (b) Binomial distribution for $n = 5$, $p = 3/4$.

$$\frac{\binom{n}{k+1}p^{k+1}(1-p)^{n-k-1}}{\binom{n}{k}p^k(1-p)^{n-k}} = 1 + \frac{(n+1)p-(k+1)}{(k+1)(1-p)} \tag{2}$$

From (2), one determines when the probabilities are increasing or decreasing by seeing when this ratio is greater or less than 1. The criterion follows.

The probabilities increase up to the point $(n+1)p$ and decrease past that point. If $(n+1)p$ happens to be a possible value of X (that is, an integer between 0 and n), then there is a double maximum at $(n+1)p-1$ and $(n+1)p$.

If $0 < (n+1)p < 1$, the maximum probability is at 0 with monotone decrease beyond 0. What happens if $(n+1)p = 1$?

Example 3 Consider the two distributions in Example 2. In (a), $(n+1)p = 1.5$—hence, the maximum probability is at 1; the probabilities decrease monotonically beyond it. In (b), $(n+1)p = 4.5$; therefore, the maximum probability is at 4.

Because the binomial distribution is so widely used in applications, it has been extensively tabulated (Note 2). Table 6.1 shows a typical section of a table of the binomial distribution.

The moments for the binomial distribution are easily computed from the generating function,

$$\phi(t) = (1-p+pt)^n$$

by computing $\phi^{(k)}(1)$ [see (2), Section 5.3].

factorial moments of binomial distribution

If X is binomially distributed with parameters n and p, then

$$E(X^{(k)}) = \begin{cases} n^{(k)}p^k & \text{if } k = 1, \ldots, n \\ 0 & \text{if } k > n \end{cases}$$

In particular, $EX = np$.

Example 4 (a) In throwing a coin 10 times, the expected number of heads is 5.

(b) In throwing a pair of dice 36 times, the expected number of double aces is 1.

TABLE 6.1 Table of several binomial distributions, for $n = 85$ and $p = 0.46$, and $n = 85$ and $p = 0.47$*

$p = 0.46$			$p = 0.47$		
x	*Individual term*	*Cumulative (x or less)*	x	*Individual term*	*Cumulative (x or less)*
25	0.000710	0.001317	26	0.000811	0.001523
26	0.001397	0.002713	27	0.001571	0.003094
27	0.002600	0.005313	28	0.002887	0.005981
28	0.004588	0.009901	29	0.005031	0.011012
29	0.007681	0.017582	30	0.008329	0.019341
30	0.012214	0.029796	31	0.013104	0.032444
31	0.018459	0.048255	32	0.019609	0.052053
32	0.026535	0.074790	33	0.027928	0.079982
33	0.036304	0.111094	34	0.037878	0.117860
34	0.047298	0.158392	35	0.048945	0.166805
35	0.058709	0.217101	36	0.060284	0.227089
36	0.069460	0.286561	37	0.070798	0.297887
37	0.078360	0.364921	38	0.079304	0.377191
38	0.084317	0.449239	39	0.084753	0.461944
39	0.086559	0.535798	40	0.086432	0.548375
40	0.084796	0.620594	41	0.084125	0.632500
41	0.079281	0.699875	42	0.078153	0.710653
42	0.070752	0.770627	43	0.069306	0.779959
43	0.060270	0.830897	44	0.058666	0.838626
44	0.049007	0.879904	45	0.047400	0.886026
45	0.038036	0.917940	46	0.036552	0.922578
46	0.028175	0.946115	47	0.026896	0.949474
47	0.019916	0.966031	48	0.018882	0.968357
48	0.013431	0.979461	49	0.012644	0.981001
49	0.008639	0.988100	50	0.008073	0.989074
50	0.005299	0.993399	51	0.004913	0.993987
51	0.003098	0.996497	52	0.002849	0.996836
52	0.001725	0.998222	53	0.001573	0.998409
53	0.000915	0.999137	54	0.000827	0.999235

*Reproduced with kind permission of Bell Telephone Laboratories Incorporated, from 50–100 Binomial Tables by Harry G. Romig, 1947, 1953.

(c) One hundred triples of integers are selected from a table of random numbers. The expected number of triples of the form (i, i, i) $(i = 0, 1, \ldots, 9)$ is 1.

Finally, we want to show that it is possible to express the binomial tail probabilities,

$$P(X \geq k) = \binom{n}{k} p^k (1-p)^{n-k} + \cdots + \binom{n}{n} p^n, \qquad k = 1, \ldots, n$$

in a concise way by an integral, as follows:

$$P(X \geq k) = n\binom{n-1}{k-1} \int_0^p x^{k-1}(1-x)^{n-k}\,dx, \qquad k = 1, 2, \ldots, n \quad (3)$$

The right side of (3), as a function of p, is called an *incomplete beta function,* for which there exist many tables, providing an alternate tabulation of the binomial distribution (Note 3). Equation (3) may seem strange because it expresses a sum as an integral. Its real meaning will be apparent in Section 9.6; meanwhile, it can be proved in a purely formal way by differentiating both sides of (3).

Proof of (3).

$$\frac{d}{dp}\binom{n}{i} p^i (1-p)^{n-i}$$

$$= n\binom{n-1}{i-1} p^{i-1}(1-p)^{n-1-(i-1)} - n\binom{n-1}{i} p^i (1-p)^{n-1-i}$$

Denote $n\binom{n-1}{i} p^i (1-p)^{n-1-i}$ by a_i, and in terms of the a_i's, we have

$$\frac{dP(X \geq k)}{dp} = (a_{k-1} - a_k) + (a_k - a_{k+1}) + \cdots + (a_n - 0)$$

$$= a_{k-1} = n\binom{n-1}{k-1} p^{k-1}(1-p)^{n-k}$$

Because a_{k-1} equals the derivative of the right side of (3), by the *fundamental theorem* of integral calculus (Note 4), both sides of (3) can differ only by a constant (with respect to p). Set $p = 0$ to see that this difference is actually 0. ∎

6.2 THE MULTINOMIAL DISTRIBUTION

The binomial distribution concerns n independent trials, where on each trial one of two outcomes are possible with probabilities p and $1 - p$. Consider now the generalization to n independent trials, where on each trial any one of m *outcomes* are possible, whose probabilities are $p_1, \ldots, p_m$ with $p_1 + \cdots + p_m = 1$. We shall be interested in the vector $(Y_1, \ldots, Y_m)$ where Y_i equals the total number of outcomes of type i. The joint distribution of $Y_1, \ldots, Y_m$, derived as follows, is called the *multinomial distribution.*

multinomial distribution

Let $k_1, k_2, \ldots, k_m$ be m nonnegative integers the sum of which is n. Then

$$P[(Y_1 = k_1) \cap \cdots \cap (Y_m = k_m)] = \binom{n}{k_1, \ldots, k_m} p_1^{k_1} \cdots p_m^{k_m} \quad (1)$$

This equation is called the multinomial distribution with parameters $n; p_1, \ldots, p_m$.

Recall from VIII, Section 1.5, that $\binom{n}{k_1, \ldots, k_m}$ denotes the multinomial coefficient $n!/k_1! \cdots k_m!$ Each of the n trials must be one of the m possible types; thus, $k_1 + \cdots + k_m = n$, for this sum accounts for all n trials.

Proof of (1). Suppose that among the n trials, k_1 specified trials are of type 1, k_2 are of type 2, and so forth. By independence, the probability of such a sequence is $p_1{}^{k_1} p_2{}^{k_2} \cdots p_m{}^{k_m}$. The event $(Y_1 = k_1) \cap \cdots \cap (Y_m = k_m)$ consists of *all* sequences where there are k_i elements of type i, $i = 1, \ldots, m$. The number of such sequences is $\binom{n}{k_1, \ldots, k_m}$, the number of arrangements of n objects, k_1 of one type, k_2 of a second type, $\ldots$, k_m of an mth type (see VIII, Section 1.5). ∎

The sum of the probabilities on the right of (1), over all m-tuples of nonnegative integers that sum to n, must, of course, equal 1. This sum is the multinomial expansion of $(p_1 + \cdots + p_m)^n = 1^n = 1$ (X, Section 1.5; see Exercise 12).

Example 1

(a) Ten balls are randomly distributed among three cells. Let Y_i denote the number of balls in cell i, $i = 1, 2, 3$. Then Y_1, Y_2, Y_3 is multinomially distributed with parameters 10; 1/3, 1/3, 1/3 [see (1), Section 3.3].

(b) A die is thrown n times. The 6-tuple, which counts the number of times each of the six faces appears, is multinomially distributed with parameters n; $p_1 = \cdots = p_6 = 1/6$.

(c) A target is divided into three concentric rings. Suppose that the probabilities of a single shot entering these rings are 1/8, 4/8, and 3/8. The triple, which counts the disposition of n shots among these three rings, is multinomially distributed with parameters n; 1/8, 4/8, 3/8.

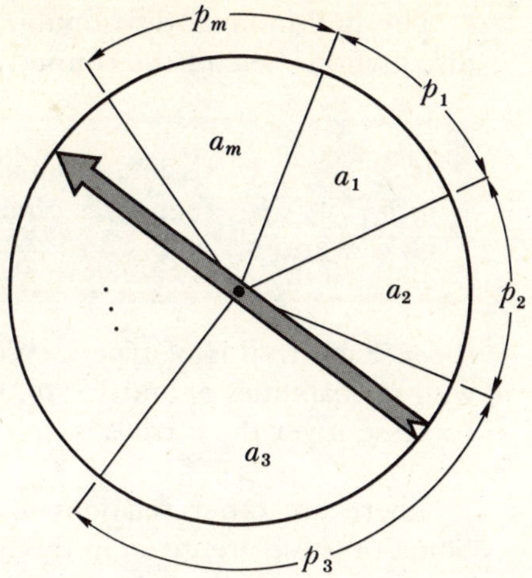

FIGURE 6.2 The probability is p_i that the pointer will stop in region a_i.

(d) A dial is marked into m regions $a_1, \ldots, a_m$, as in Figure 6.2. A well-balanced pointer is spun n times. Suppose Y_i counts the number of these spins that stop in a_i, $i = 1, \ldots, m$. Then $(Y_1, \ldots, Y_m)$ is multinomially distributed with parameters n; $p_1, \ldots, p_m$.

Example 2 Suppose $n = 4$, $p_1 = 2/9$, $p_2 = 3/9$, and $p_3 = 4/9$. The multinomial distribution for these parameters is tabulated in Table 6.2.

TABLE 6.2

k_1	k_2	k_3	$P[(Y_1 = k_1) \cap (Y_2 = k_2) \cap (Y_3 = k_3)]$
4	0	0	16/6561
0	4	0	81/6561
0	0	4	256/6561
3	1	0	96/6561
1	3	0	216/6561
3	0	1	128/6561
1	0	3	512/6561
0	3	1	432/6561
0	1	3	768/6561
2	2	0	216/6561
2	0	2	384/6561
0	2	2	864/6561
2	1	1	576/6561
1	2	1	864/6561
1	1	2	1152/6561

The multinomial distribution is closely related to the binomial distribution. One of the connections is the following.

> If $(Y_1, \ldots, Y_m)$ is multinomially distributed with parameters $n; p_1, \ldots, p_m$, then Y_1 is binomially distributed with parameters n and p_1. (2)

Proof. Each trial is of type 1 (success), or something else (failure) with probabilities p_1 and $1-p_1$, respectively. The total number of successes over the n trials is Y_1. ▮

There are other relations like (2), which are equally obvious. Some of these are given in the following example.

Example 3 (a) Suppose (Y_1, Y_2) is multinomially distributed with parameters $n; p, 1-p$. Then Y_1 is binomially distributed with parameters n and p.

(b) Suppose (Y_1, Y_2, Y_3, Y_4) is multinomially distributed with parameters $n; p_1, p_2, p_3, p_4$. Then (Y_1+Y_2, Y_3, Y_4) is multinomially distributed with parameters $n; p_1+p_2, p_3, p_4$ (see Exercise 13).

(c) Suppose (Y_1, Y_2, Y_3) is multinomially distributed with parameters $n; p_1, p_2, p_3$. The conditional distribution of Y_1, given that $Y_1+Y_2=r$, is binomial with parameters r and $p=p_1/(p_1+p_2)$ (see Exercise 14).

6.3 THE HYPERGEOMETRIC DISTRIBUTION

This distribution is the counterpart of the binomial distribution when sampling is done from a dichotomous population *without replacement.* The binomial cannot apply here because the drawings are now dependent. Suppose a population consists of N objects, M of type 1 and the remaining $N-M$ of type 2. If n random drawings are made without replacement, let X count the number of objects of type 1 drawn. The distribution of X is as follows.

hypergeometric distribution

$$P(X=k) = \binom{n}{k}\frac{M^{(k)}(N-M)^{(n-k)}}{N^{(n)}} = \frac{\binom{M}{k}\binom{N-M}{n-k}}{\binom{N}{n}},$$

$k=0, 1, \ldots, n$

This distribution is called the hypergeometric distribution, with parameters n, M, and N (n is the sample size, N the population size, and M the number of objects of type 1 in the population).

This distribution was derived and fully discussed in Section 3.9. The only points that we want to add here concern the relation between the hypergeometric distribution and the binomial distribution. If the sampling is with replacement, then X is binomially distributed with parameters n and $p = M/N$. This fact is true because, in sampling with replacement, the successive trials are independent, and each trial produces an object of type 1 with probability p. From a practical point of view, if N is very large compared to n, then it really makes no difference whether the sampling is with or without replacement; the point is that failing to replace the drawn item hardly changes the composition of the population for the next trial. The formal fact follows.

$$\lim_{\substack{M\to\infty \\ N\to\infty \\ M/N\to p}} \binom{n}{k}\frac{M^{(k)}(N-M)^{(n-k)}}{N^{(n)}} = \binom{n}{k}p^k(1-p)^{n-k} \qquad (1)$$

That is, for large M and N, the hypergeometric probability almost equals the binomial probability, if the proportion M/N remains close to a fixed p (see Exercise 21).

Proof of (1). The proof is quite routine. Since

$$\lim \frac{M}{N} = \lim \frac{M-1}{N-1} = \cdots = \lim \frac{M-k+1}{N-k+1} = p$$

and

$$\lim \frac{N-M}{N-k} = \lim \frac{N-M-1}{N-k-1} = \cdots = \lim \frac{N-M-n+k+1}{N-n+1}$$
$$= 1-p$$

it follows that

$$\lim \frac{M^{(k)}(N-M)^{(n-k)}}{N^{(n)}} = p^k(1-p)^{n-k} \qquad \blacksquare$$

Example 1 Suppose $n = 4$, $M = 25$, and $N = 100$. Table 6.3 shows the related binomial and hypergeometric distributions (Note 5).

TABLE 6.3

Binomial distribution $n = 4, p = 1/4$		*Hypergeometric distribution* $n = 4, M = 25, N = 100$	
k	$P(X = k)$	k	$P(X = k)$
0	0.316	0	0.310
1	0.422	1	0.431
2	0.211	2	0.212
3	0.047	3	0.044
4	0.004	4	0.003

6.4 THE GEOMETRIC DISTRIBUTION

Consider a sequence of independent trials, each of which produces "success" or "failure" with probabilities p and $1 - p$. How many trials are required to achieve a first success? If X is the trial number on which a success first takes place, then its distribution is as follows.

geometric distribution

$$P(X = k) = (1 - p)^{k-1}p, \qquad k = 1, 2, \ldots \tag{1}$$

This distribution is called the geometric distribution with parameter p.

We should assume that $0 < p \leq 1$. If $p = 0$, a success never occurs, and the only reasonable distribution assigns probability 1 to the possible value $k = \infty$. At the other extreme, if $p = 1$, then the first success automatically occurs on the first trial; therefore, $P(X = 1) = 1$. The probability given by (1) is reasonable, because it represents the probability of a sequence of $k - 1$ failures followed by a success. There is a concise expression for the "tail probabilities" $P(X > k)$—namely,

$$P(X > k) = (1 - p)^k, \qquad k = 0, 1, 2, \ldots \tag{2}$$

If $p > 0$, then $\lim_{k \to \infty}(1 - p)^k = 0$, which implies that

$$\sum_{k=1}^{\infty} P(X = k) = 1$$

This fact is important because it guarantees that no matter how small the probability of a success may be, as long as it is positive *a success will eventually occur within a finite waiting time.* Intuitively, (2) is plausible because it is the probability of a sequence of k

failures. To say that the first k trials are failures is the same as saying that $(X > k)$. A more formal proof follows.

Proof of (2).

$$\begin{aligned} P(X > k) &= (1-p)^k p + (1-p)^{k+1} p + (1-p)^{k+2} p + \cdots \\ &= (1-p)^k p[1 + (1-p) + (1-p)^2 + \cdots] \\ &= (1-p)^k p \frac{1}{1-(1-p)} = (1-p)^k \quad \blacksquare \end{aligned}$$

The generating function of X is

$$\phi(t) = pt + (1-p)pt^2 + \cdots = \frac{pt}{[1-(1-p)t]} \tag{3}$$

Since $\phi'(1) = 1/p$, it follows from (2), Section 5.3, that

$$EX = \frac{1}{p} \tag{4}$$

That $EX = 1/p$ also follows from Eq. (3), Section 2.10, because

$$\begin{aligned} P(X > 0) + P(X > 1) + \cdots &= 1 + (1-p) + (1-p)^2 + \cdots \\ &= \frac{1}{1-(1-p)} = \frac{1}{p} \end{aligned}$$

The general factorial moments are obtained just as easily by using (2), Section 5.3.

$$E(X^{(k)}) = k! \frac{(1-p)^{k-1}}{p^k}, \qquad k = 1, 2, \ldots \tag{5}$$

Proof of (5). First rewrite the generating function as

$$\phi(t) = \frac{pt}{1-(1-p)t} = \frac{p}{(1-p)[1-(1-p)t]} - \frac{p}{1-p}$$

and evaluate $\phi^{(k)}(1)$. $\blacksquare$

Example 1 (a) A coin is thrown until a head is obtained. The number of throws required is geometrically distributed with $p = 1/2$. The expected number of throws is $1/p = 2$ (see Example 4, Section 2.4).

(b) In a dichotomous population of size N, there are M objects of type 1 and $N - M$ objects of type 2. If drawings are made with successive replacement, the number of drawings required to obtain a type 1 object is geometrically distributed with parameter $p = M/N$. The expected number of drawings is $1/p = N/M$ (Section 3.10).

(c) Consider a continuing one-line history of generations of an organism (*A* begets *B*, *B* begets *C*, and so forth). In any organism, a certain genetic mutation may take place with probability $1/10^4$. The number of generations that must evolve before a mutation takes place is geometrically distributed with $p = 1/10^4$. The expected waiting time is 10^4 generations. Despite the rareness of the mutation, it will definitely take place eventually if the generations continue, which is true because $P(X = 1) + P(X = 2) + \cdots = 1$. Even though the expected waiting time is large, the probabilities do sum to 1.

Example 2 Using the geometric distribution, we can now give an explanation of an expected value that arose earlier. Suppose that balls are successively distributed among m cells. Recall that it was asserted in (6), Section 3.5, that the expected number of balls required to occupy all m cells equals

$$m\left(1 + \frac{1}{2} + \frac{1}{3} + \cdots + \frac{1}{m}\right)$$

Define random variables $Z_1, Z_2, \ldots, Z_m$ as follows:

$Z_1 =$ number of balls that must be distributed for any one cell to be occupied (of course, $Z_1 = 1$, because the first ball automatically occupies one cell)

$Z_2 =$ additional number of balls that must be distributed for any two cells to be occupied

$Z_3 =$ additional number of balls that must be distributed for any three cells to be occupied

and so forth. We know Z_1 is identically 1. Z_2 is geometrically distributed with parameter $p = (m - 1)/m$, for the following reason. Once one cell is occupied, as successive balls are distributed, failures occur when these balls enter the already occupied cell. A success occurs when one of the available $m - 1$ unoccupied cells is finally entered. The probability of success is $(m - 1)/m$. Once two cells are occupied, as additional balls are distributed, failures occur when they enter the already occupied cells. A success occurs when one of the available $m - 2$ unoccupied cells is entered. Thus, Z_3 is geometrically distributed with parameter $p = (m - 2)/m$. Similarly, each Z_1 is geometrically distributed with the sequence of parameters being

$$\frac{m}{m}, \frac{m-1}{m}, \ldots, \frac{1}{m}$$

Let $Z = Z_1 + Z_2 + \cdots + Z_m$, which is the total number of balls required to fully occupy all the cells. It follows that

$$EZ = EZ_1 + EZ_2 + \cdots + EZ_m = m\left(\frac{1}{m} + \frac{1}{m-1} + \cdots + \frac{1}{2} + 1\right)$$

We now describe an interesting property of a geometrically distributed random variable, which says that its "excess over a boundary" is itself geometrically distributed. That is, suppose that it is known that X is greater than k. Then, under this condition, the amount by which X exceeds k is still geometrically distributed. This property can also be called one of "lack of memory" or "agelessness" (see Example 3). The formal statement follows.

excess over boundary is geometrically distributed

Suppose that X is geometrically distributed with parameter p. Then, for every positive integer k,

$$P(X = k + r | X > k) = (1-p)^{r-1}p = P(X = r), \qquad r = 1, 2, \ldots \tag{6}$$

Proof of (6).

$$P(X = k + r | X > k) = \frac{P(X = k + r)}{P(X > k)} = \frac{(1-p)^{k+r-1}p}{(1-p)^k} = (1-p)^{r-1}p \qquad \blacksquare$$

For a converse to (6), see Exercise 26.

Example 3 Suppose that in a population, the life span is given by the geometric distribution. In other words, at birth the projected life span of an individual is k units of time with probability $(1-p)^{k-1}p$, $k = 1, 2, \ldots$, where p is an appropriately chosen constant. Property (6) says that in the population of individuals that have already achieved age k, the projected future life span is the same as at birth. The life expectancy tables for 100-year-olds is the same as for 1-year-olds! If this model were appropriate, everyone should pay the same life insurance rates.

Another interesting property of the geometric distribution says that the minimum of a set of independent geometrically distributed random variables is still geometrically distributed.

minimum of geometric random variables is geometric

Suppose that $X_1, \ldots, X_n$ are mutually independent, geometrically distributed random variables with parameters $p_1, \ldots, p_n$, respectively. Let $Y = \min(X_1, \ldots, X_n)$. Then Y is geometrically distributed with parameter $p = 1 - (1-p_1) \cdots (1-p_n)$.

(7)

Proof. If we succeed in showing that $P(Y > k) = a^k$ for some positive a, then Y must be geometrically distributed with parameter $1 - a$, since

$$P(Y = k) = P(X > k - 1) - P(Y > k) = a^{k-1} - a^k = a^{k-1}(1 - a)$$

Now

$$\begin{aligned} P(Y > k) &= P[(X_1 > k) \cap \cdots \cap (X_n > k)] \\ &= (1 - p_1)^k \cdots (1 - p_n)^k = [(1 - p_1) \cdots (1 - p_n)]^k \quad \blacksquare \end{aligned}$$

Example 4 The geometric distribution sometimes describes the life span of a physical particle. If p is the parameter of the distribution, then $1/p$ is the expected lifetime. If the expected life spans of n different particles are $1/p_1, \ldots, 1/p_n$, then the first extinction will also occur according to the geometric distribution, and the expected waiting time for the first extinction is $[1 - (1 - p_1) \cdots (1 - p_n)]^{-1}$.

Finally, we mention a slight extension in the use of the term *geometric distribution.* If X is geometrically distributed according to our original definition, then its possible values are 1, 2, ... On the other hand, the random variable $Y = X + k - 1$ has possible values $k, k + 1, \ldots$, and we will say that Y is geometrically distributed over the values $k, k + 1, \ldots$. For example, Y is geometrically distributed over 0, 1, 2, . . . means that

$$P(Y = k) = (1 - p)^k p, \qquad k = 0, 1, 2, \ldots$$

Often, the use of the term geometric distribution is restricted to this case.

6.5 NEWTON'S BINOMIAL EXPANSION

To discuss the negative binomial distribution in Section 6.6, and for various other applications, we review the power series expansion of $(1 + t)^a$. First, we extend the definition of the binomial coefficient. Until now, $\binom{n}{k}$ has been interpreted as

$$\binom{n}{k} = \frac{n^{(k)}}{k!} = \frac{n!}{k!(n-k)!} \tag{1}$$

with both n and k restricted to nonnegative integers (Section 1.5). Although the right-hand side of (1) makes sense only for nonnegative integers, the quantity $n^{(k)}/k!$ makes sense if n is *any real number.* This fact allows for a general definition of $\binom{n}{k}$ as follows:

$$\binom{n}{k} = \frac{n^{(k)}}{k!} = \frac{n(n-1)\cdots(n-k+1)}{k!} \tag{2}$$

In (2), k is still a nonnegative integer but n can be any real number. In this sense, $\binom{n}{k}$ is called a *generalized binomial coefficient.* By convention, $\binom{n}{0}$ still equals 1.

Example 1

(a) $\binom{1/2}{3} = \frac{(1/2)(1/2-1)(1/2-2)}{3!} = 1/16$

(b) $\binom{1}{2} = \frac{(1)(0)}{2} = 0$ (c) $\binom{-1}{2} = \frac{(-1)(-2)}{2} = 1$

There exists a relationship between binomial coefficients that we need occasionally.

$$\binom{n}{k} = (-1)^k\binom{-n+k-1}{k} \tag{3}$$

Proof of (3).

$$\binom{n}{k} = \frac{n(n-1)\cdots(n-k+1)}{k!}$$

$$= \frac{(-1)^k(-n+k-1)\cdots(-n+1)(-n)}{k!}$$

$$= (-1)^k\binom{-n+k-1}{k} \quad \blacksquare$$

When n is a positive integer, then the binomial expansion of $(1+t)^n$ is

$$(1+t)^n = \sum_{k=0}^{n} \binom{n}{k} t^k = \sum_{k=0}^{\infty} \binom{n}{k} t^k \tag{4'}$$

(Since $\binom{n}{k} = 0$ when k is an integer greater than n, the sum in (4′) can be written $\Sigma_{k=0}^{\infty}$.) It is interesting that (4′) is still correct when n is *any real* number.

Newton's binomial expansion

If a is any real number, then, for $|t| < 1$,

$$(1+t)^a = \sum_{k=0}^{\infty} \binom{a}{k} t^k \tag{4}$$

(Note 6).

Example 2

(a) Suppose $a = -1$. Then

$$(1+t)^{-1} = \binom{-1}{0} + \binom{-1}{1} t + \binom{-1}{2} t^2 + \cdots = 1 - t + t^2 - t^3 + \cdots$$

because $\binom{-1}{k} = (-1)^k$. This series is the familiar geometric series. For its conventional form, replace t by $-t$ and we have

$$\frac{1}{1-t} = 1 + t + t^2 + \cdots, \qquad |t| < 1$$

A useful form of (4) follows.

$$(1-t)^{-a} = \sum_{k=0}^{\infty} \binom{a+k-1}{k} t^k, \qquad |t| < 1 \tag{5}$$

Proof of (5). By (4),

$$(1-t)^{-a} = \sum_{k=0}^{\infty} \binom{-a}{k} (-t)^k$$

By (3), $\binom{-a}{k}(-1)^k = \binom{a+k-1}{k}$, from which (5) follows. ∎

When a is a positive integer, then (5) can be proved in a different way. By successively differentiating

$$\frac{1}{1-t} = 1 + t + t^2 + \cdots$$

we obtain

$$\frac{1}{(1-t)^2} = 1 + 2t + 3t^2 + \cdots$$

$$\frac{2}{(1-t)^3} = 2^{(2)} + 3^{(2)}t + 4^{(2)}t^2 + \cdots$$

$$\vdots$$

$$\frac{n!}{(1-t)^{n+1}} = n^{(n)} + (n+1)^{(n)}t + (n+2)^{(n)}t^2 + \cdots$$

or, equivalently, because $(n+k)^{(n)}/n! = \binom{n+k}{n} = \binom{n+k}{k}$, we have

$$\frac{1}{(1-t)^{n+1}} = \sum_{k=0}^{\infty} \binom{n+k}{k} t^k \tag{6}$$

Notice that (6) is a special case of (5), with $n + 1 = a$.

6.6 THE NEGATIVE BINOMIAL DISTRIBUTION

We next want to describe a distribution that is closely related to the geometric distribution.

negative binomial distribution

A random variable X is said to have the negative binomial distribution with parameters a and p ($a > 0$, $0 < p \leq 1$), if

$$P(X = k) = \binom{a+k-1}{k}(1-p)^k p^a, \qquad k = 0, 1, 2, \ldots \tag{1}$$

(Notes 7 and 8).

Since

$$\binom{a+k-1}{k} = \frac{(a+k-1)(a+k-2)\cdots(a)}{k!}$$

and a is positive, it follows that the terms on the right side of (1) are positive. It must be checked that their sum is 1. By (5), Section 6.5,

$$p^a \sum_{k=0}^{\infty} \binom{a+k-1}{k}(1-p)^k = \frac{p^a}{[1-(1-p)]^a} = 1$$

Thus, (1) does correctly describe a probability distribution.

Example 1 The negative binomial distribution is often used to describe the number of accidents or epidemic victims in a population (Note 9).

Example 2 (a) If $a = 1$, then (1) is the geometric distribution over the values $0, 1, 2, \ldots$.

(b) Suppose that repeated, independent success-failure experiments are performed until n successes have been obtained. Let p be the probability of a success on any trial. If Y is the number of trials required to obtain n successes, then

$$P(Y = n + k) = \binom{n+k-1}{k} p^n (1-p)^k, \qquad k = 0, 1, 2, \ldots \tag{2}$$

is the probability of $n - 1$ successes in the first $n + k - 1$ trials followed by a success in the $(n + k)$th trial. The probability of $n - 1$ successes in the first $n + k - 1$ trials is the binomial probability

$$\binom{n+k-1}{n-1} p^{n-1}(1-p)^{n+k-1-(n-1)} = \binom{n+k-1}{k} p^{n-1}(1-p)^k$$

When this probability is multiplied by p, the probability of a success in the $(n+k)$th trial, the result is just the right side of (2). In effect, (2) says that $Y - n$ has the negative binomial distribution with parameters n and p. A minimum of n trials is required to obtain n successes; therefore, $Y - n$ is the additional number of trials required beyond this minimum to obtain n successes.

(c) A coin is thrown until 10 heads are obtained. The number of throws beyond 10 that are required is negative binomially distributed with parameters $a = 10$, $p = 1/2$.

generating function of negative binomial distribution

The generating function of the negative binomial distribution is

$$\phi(t) = \frac{p^a}{[1-(1-p)t]^a}, \qquad |t| \leq 1 \tag{3}$$

Proof of (3). The proof follows immediately from (5), Section 6.5.

$$p^a \sum_{k=0}^{\infty} \binom{a+k-1}{k} (1-p)^k t^k = \frac{p^a}{[1-(1-p)t]^a} \qquad \blacksquare$$

By computing $\phi^{(k)}(1)$ from Eq. (3), we have the factorial moments.

$$E(X^{(k)}) = (-a)^{(k)}\left(\frac{1-p}{p}\right)^k = k!\binom{a+k-1}{k}\left(\frac{1-p}{p}\right)^k \qquad (4)$$

Example 3 If X is geometric over 1, 2, . . . , then, by (4), Section 6.4, $EX = 1/p$. Because $X - 1$ has the negative binomial distribution with $a = 1$, then $E(X - 1) = (1 - p)/p$, which agrees with (4) when $a = 1$ and $k = 1$.

From the generating function follows an important *closure property* of the negative binomial distribution.

a sum of independent negative binomial variables has a negative binomial distribution

Suppose that $X_1, \ldots, X_n$ are mutually independent, negative binomially distributed variables with parameters $(a_1, p), \ldots, (a_n, p)$. (The p is the same for each X_i.) Then

$$X = X_1 + \cdots + X_n$$

has the negative binomial distribution with parameters $a = a_1 + \cdots + a_n$ and p.

Proof. The proof follows immediately from the fact that

$$\frac{p^{a_1}}{[1-(1-p)t]^{a_1}} \cdot \cdots \cdot \frac{p^{a_n}}{[1-(1-p)t]^{a_1}} = \frac{p^{a_1+\cdots+a_n}}{[1-(1-p)t]^{a_1+\cdots+a_n}} \quad ∎$$

Example 4 Let X be the total number of times that a coin must be thrown to obtain a total of n heads. Let X_1 be the number of throws required to obtain the first head, X_2 the number of additional throws required to obtain the second head, and so forth. Then

$$X = X_1 + \cdots + X_n$$

It is certainly plausible that $X_1, \ldots, X_n$ are mutually independent, although we shall not verify this fact here. Since each X_i is geometrically distributed over 1, 2, . . . , it follows that

$$(X_1 - 1) + \cdots + (X_n - 1) = X - n$$

is a sum of n independent, negative binomial variables, each having the parameters $a = 1$ and $p = 1/2$. Hence, $X - n$ has the negative binomial distribution with parameter $a = n$ and $p = 1/2$, which agrees with the result in Example 2b.

6.7 THE POISSON DISTRIBUTION

The Poisson distribution is an important distribution with many applications. The definition of the Poisson distribution follows.

Poisson distribution

A random variable X is Poisson distributed with parameter λ ($\lambda > 0$) if

$$P(X = k) = \frac{\lambda^k e^{-\lambda}}{k!}, \qquad k = 0, 1, \ldots \tag{1}$$

Since e^x is not negative for any real number x, and because $e^\lambda = \sum_{k=0}^{\infty} \lambda^k/k!$, it follows that (1) does really define a probability distribution over the nonnegative integers.

Example 1

(a) The number of particles picked up by a Geiger counter in a unit of time is usually assumed to be Poisson distributed. We shall see that λ is related to the rate at which particles are arriving.

(b) The number of calls coming through a telephone switchboard per unit time is also usually assumed to be Poisson distributed. The parameter λ depends on the intensity of calls and varies with the time of day (Note 10).

Example 2

Suppose n cards marked 1, 2, . . . , n are matched against n similarly marked fixed positions. It was shown in (3), Section 3.2, that the number of matches is approximately Poisson distributed with parameter $\lambda = 1$. The larger that n is, the better is the approximation (see Exercise 33).

The Poisson distribution is often used to approximate the binomial distribution for n "large" and p "small." The following limit relationship is the basis of this approximation.

Poisson approximation to binomial

$$\lim_{\substack{n \to \infty \\ p = \lambda/n}} \binom{n}{k} p^k (1-p)^{n-k} = \frac{\lambda^k e^{-\lambda}}{k!}, \qquad k = 0, 1, \ldots \tag{2}$$

What (2) asserts is that if n is large and p is small, of the order of λ/n, then the binomial probability distribution is approximated by the Poisson distribution with parameter $\lambda = np$ (Note 11).

Proof of (2). The limit follows from the following facts:

$$\lim_{\substack{n\to\infty \\ p=\lambda/n}} \binom{n}{k} p^k = \lim_{n\to\infty} \frac{n^{(k)}}{n^k} \frac{\lambda^k}{k!} = \frac{\lambda^k}{k!}$$

$$\lim_{n\to\infty}\left(1 - \frac{\lambda}{n}\right)^n = e^{-\lambda}$$

$$\lim_{n\to\infty}\left(1 - \frac{\lambda}{n}\right)^{-k} = 1 \quad \blacksquare$$

Example 3 Suppose $n = 100$, $p = 1/100$. Table 6.4 compares the binomial distribution with parameters n, p and the Poisson distribution with parameter $\lambda = np = 1$ (Note 12).

We now evaluate the probability generating function for the Poisson distribution.

Poisson generating function

The generating function of a random variable that is Poisson distributed with parameter λ is

$$\phi(t) = e^{\lambda(t-1)} \tag{3}$$

TABLE 6.4

Binomial distribution		*Poisson distribution*	
k	$P(X = k)$	k	$P(X = k)$
0	0.3660	0	0.3679
1	0.3697	1	0.3679
2	0.1849	2	0.1839
3	0.0610	3	0.0613
4	0.0149	4	0.0153
5	0.0029	5	0.0031
6	0.0005	6	0.0005
7	0.0001	7	0.0001
8	0.0000	8	0.0000
.		.	
.		.	
.		.	

Proof.

$$\phi(t) = \sum_{k=0}^{\infty} t^k \frac{\lambda^k e^{-\lambda}}{k!} = e^{\lambda t} e^{-\lambda} = e^{\lambda(t-1)} \quad \blacksquare$$

It follows immediately from (3) that

$$\phi^{(k)}(1) = E(X^{(k)}) = \lambda^k, \qquad k = 1, 2, \ldots \tag{4}$$

In particular, we see that the parameter λ equals EX.

The Poisson distribution also has an *important closure property.*

a sum of independent Poisson variables is Poisson

Suppose $X_1, \ldots, X_n$ are mutually independent, Poisson distributed, random variables with parameters $\lambda_1, \ldots, \lambda_n$. Then $X = X_1 + \cdots + X_n$ is Poisson distributed with parameter $\lambda = \lambda_1 + \cdots + \lambda_n$.

Proof. The generating function of $X_1 + \cdots + X_n$ is

$$\exp[\lambda_1(t-1)] \cdots \exp[\lambda_n(t-1)] \\ = \exp\,[(\lambda_1 + \cdots + \lambda_n)(t-1)] \quad \blacksquare$$

Example 4 Radioactive particles from two different sources are being picked up by one Geiger counter. Suppose that the model prescribes that the numbers of particles arriving in a unit of time from each of these sources are independent random variables, Poisson distributed with parameters λ_1 and λ_2, respectively. Then the total number of particles recorded by the counter in a unit of time is Poisson distributed with parameter $\lambda = \lambda_1 + \lambda_2$ (compare Example 1).

An expression for cumulative Poisson probabilities in terms of an integral follows.

$$\sum_{i=0}^{k} \frac{\lambda^i e^{-\lambda}}{i!} = \int_{\lambda}^{\infty} \frac{x^{k-1}}{k!} e^{-x}\, dx \tag{5}$$

The proof of this relation will be given in (5), Section 9.5. The integral on the right side of (5) is a so-called *incomplete gamma integral* (Note 12).

Some important distributions related to the Poisson are the *compound Poisson distributions* (Note 13).

compound Poisson distribution

Suppose that the joint distribution of a pair of random variables X and Y is described as follows:

1. Y is Poisson distributed.
2. The conditional distribution of X, given that $Y = k$, is that of

$$X_1 + \cdots + X_k$$

The X_i's are independent random variables, each having the same distribution. We assume that the distribution of the X_i's does not depend on k, and if $k = 0$, then $X = 0$. The distribution of X is said to be a *compound Poisson distribution.*

Notice that the preceding definition does not describe the compound Poisson probabilities explicitly; in general, they are difficult to describe. When the distribution of the X_i's happens to have possible values 0, 1, 2, . . . , then a nice expression for the probability generating function of the compound Poisson distribution can be given.

compound Poisson generating function

Suppose that in the preceding definition of the compound Poisson distribution, the X_i's are nonnegative integer valued and have a generating function of $\phi(t)$. Then the generating function of the compound Poisson distribution is

$$\psi(t) = \exp \lambda[\phi(t) - 1] \tag{6}$$

Proof. The assumptions for the compound Poisson distribution assert that

$$E(t^X | Y = k) = [\phi(t)]^k, \qquad k = 0, 1, 2, \ldots$$

Now use the iterative formula for conditional expected values, (5), Section 4.3, to conclude that

$$\begin{aligned} E(t^X) &= \sum_{k=0}^{\infty} E(t^X | Y = k) P(Y = k) = \sum_{k=0}^{\infty} \frac{[\phi(t)]^k \lambda^k e^{-\lambda}}{k!} \\ &= \exp \lambda[\phi(t) - 1] \quad \blacksquare \end{aligned}$$

Example 5

If $\phi(t) = t$ (in other words, X_i is 1 with probability 1), then $\psi(t) = \exp \lambda(t - 1)$, and the distribution is the ordinary Poisson distribution.

Example 6 Suppose that the X_i's in the definition of the compound Poisson distribution are indicator random variables and $\phi(t) = 1 - p + pt$. Then

$$\psi(t) = \exp \lambda[1 - p + pt - 1] = \exp \lambda p(t - 1)$$

which is the generating function of a Poisson distribution with parameter λp. A meaningful interpretation can be made in terms of Example 1. Radioactive particles are picked up by a Geiger counter and the number arriving per unit time is Poisson distributed with parameter λ. Suppose that the counter may skip the counting of particles. When a particle enters, it is recorded with probability p or skipped with probability $1 - p$, and the particles are treated independently. The number of particles actually recorded by the counter fits the definition of a compound Poisson distribution. Variable Y represents the number of particles that would be recorded if none were skipped. Given that $Y = k$,

$$X_1 + \cdots + X_k$$

represents the number among these k particles that are actually counted. The result is that X, the total number of particles recorded, is still Poisson distributed, but now the parameter is λp. Since $EX = \lambda p$, and $EY = \lambda$, the expected number of particles counted is p times the number of particles that arrive.

Example 7 Suppose that a box contains three balls numbered 0, 1, 2. Balls are successively drawn, *with replacement*, a random number of times according to the following scheme. First, an experiment is performed producing a Poisson random variable Y, the parameter of which is λ; if $Y = k$, then k drawings are made. Let X be the sum of the numbers drawn. X is compound Poisson distributed with generating function

$$\psi(t) = \exp \lambda\left[\frac{1 + t + t^2}{3} - 1\right] = \exp \frac{\lambda}{3}(t - 1) \exp \frac{\lambda}{3}(t^2 - 1)$$

(see Exercise 31).

SUMMARY

A random variable X is binomially distributed with parameters n and p if

$$P(X = k) = \binom{n}{k} p^k (1 - p)^{n-k}, \qquad k = 0, 1, \ldots, n$$

Variable X is distributed like a sum of n independent indicator random variables, each of which equals 1 with probability p. An equivalent description is that X is distributed like the number of successes in n independent success-failure trials, where p is the probability of success on any single trial. The expected number of successes, EX, equals p.

If any one of m alternatives can occur on a single trial with probabilities $p_1, \ldots, p_m$, this situation gives rise to the multinomial distribution. Specifically, let $(Y_1, \ldots, Y_m)$ count the number of times these alternatives have appeared over n independent trials. Then

$$P[(Y_1 = k_1) \cap \cdots \cap (Y_m = k_m)] = \binom{n}{k_1, \ldots, k_m} p_1^{k_1} \cdots p_m^{k_m}$$

whenever $(k_1, \ldots, k_m)$ is an m-tuple of nonnegative integers with sum n. The marginal distribution of Y_i is binomial with parameters n and p_i.

If n drawings are made with replacement from a dichotomous population consisting of M elements of one kind (type 1) and $N - M$ elements of a second kind (type 2), then X, the number of type 1 elements in the sample, is binomially distributed with parameters n and $p = M/N$. On the other hand, if the n drawings are made without replacement, then X has the hypergeometric distribution given by

$$P(X = k) = \binom{n}{k} \frac{M^{(k)}(N-M)^{(n-k)}}{N^{(n)}} = \frac{\binom{M}{k}\binom{N-M}{n-k}}{\binom{N}{n}},$$

$$k = 0, 1, \ldots, n$$

Whether the sampling is with or without replacement, $EX = np = nM/N$.

If X is the number of trials required to obtain a first success in repeated independent success-failure trials, then

$$P(X = k) = (1-p)^{k-1}p, \qquad k = 1, 2, \ldots$$

This is the geometric distribution with parameter p. The expected number of trials required for a success, EX, equals $1/p$.

A random variable X has the negative binomial distribution with parameters a and p if

$$P(X = k) = \binom{a+k-1}{k}(1-p)^k p^a,$$

$$k = 0, 1, \ldots, \qquad a > 0, \qquad 0 < p \leqslant 1$$

If Y is the number of trials required to obtain n successes in repeated independent success-failure experiments, then $Y - n$ has the negative binomial distribution with parameters $a = n$ and p.

A random variable X has the Poisson distribution with parameter λ if

$$P(X = k) = \frac{\lambda^k e^{-\lambda}}{k!}, \qquad k = 0, 1, 2, \ldots$$

The expected value of X is λ. A sum of n independent Poisson random variables with parameters $\lambda_1, \ldots, \lambda_n$ is also Poisson, with parameter $\lambda_1 + \cdots + \lambda_n$. The number of matches in the matching problem is approximately Poisson with parameter $\lambda = 1$. The Poisson distribution approximates the binomial distribution for n large and p small with $\lambda = np$.

EXERCISES

1 If X is binomially distributed with parameters n, p, show that $n - X$ is binomially distributed with parameters $n, 1 - p$.

2 If X and Y are *independent* binomially distributed random variables with parameters n_1, p and n_2, p, respectively, show that $X + Y$ is binomially distributed with parameters $n_1 + n_2, p$. Find three different proofs! (What happens if the two p's are different?)

3 Suppose X is binomially distributed with parameters n, p. Show that X is symmetrically distributed about c if, and only if, $p = 1/2$ and $c = n/2$. [Hint: $p^n = (1 - p)^n$ must hold.]

4 Suppose X is binomially distributed with parameters n, p. Show that if $p = 0$, then X is the constant 0, and if $p = 1$, then X is the constant n.

5 Two dice are thrown n times. Show that the number of throws in which the number on the first die exceeds the number on the second die is binomially distributed with parameters n and $p = 5/12$.

6 Suppose n balls numbered $1, 2, \ldots, n$ are randomly distributed among n cells, also numbered $1, 2, \ldots, n$. Let X be the number of balls that fall in cells marked with the same number as the balls. Show that X is binomially distributed with parameters n and $p = 1/n$. (In effect, this problem is a variant of the matching problem that appeared in Exercise 23, Chapter 3.)

7 A particle performs a random walk over the positions $0, \pm 1, \pm 2, \ldots$ in the following way. The particle starts at 0. It makes successive one-unit steps that are mutually independent; each step is to the right with probability p, or to the left with probability $1 - p$. Let X be the position of the particle after n steps.

(a) Show that $(X+n)/2$ is binomially distributed with parameters n, p.
(b) Show that the expected position of the particle after n steps is $n(2p-1)$.

8 Suppose X and Y are independent, binomially distributed, random variables with parameters n, p and n, $1-p$, respectively. Show that $X+Y$ is distributed like $U_1+\cdots+U_n$, a sum of n independent random variables, each U_i with the following distribution:

k	0	1	2
$P(U_i=k)$	$p(1-p)$	$(1-p)^2+p^2$	$p(1-p)$

(Hint: What is the distribution of a sum of two independent indicator random variables that equal one with probabilities p and $1-p$, respectively?)

9 Use (3), Section 6.1, to show that if $p_1 < p_2$, then

$$\sum_{i=0}^{k} \binom{n}{i} p_1^i (1-p_1)^{n-i} \geq \sum_{i=0}^{k} \binom{n}{i} p_2^i (1-p_2)^{n-i}$$

for every $k=0, 1, \ldots, n$. In effect, this equation says that the bigger p is, the more the distribution is "shifted to the right."

10 (a) Four balls are randomly distributed among nine cells. Let Y_1 be the number of balls in the first two cells, Y_2 the number in the next three cells, and Y_3 the number in the last four cells. Show that the distribution of (Y_1, Y_2, Y_3) is the one tabulated in Table 6.2.
(b) Let $Z=\max(Y_1, Y_2, Y_3)$. Show that Z has the following distribution:

k	2	3	4
$P(Z=k)$	4056/6561	2152/6561	353/6561

11 A six-faced die has two red faces, two white faces, one black face, and one green face. The die is thrown n times. Let $Y_1=$ number of reds, $Y_2=$ number of whites, $Y_3=$ number of blacks, and $Y_4=$ number of greens. Show the following:

(a) $P[(Y_1=k_1) \cap (Y_2=k_2) \cap (Y_3=k_3) \cap (Y_4=k_4)]$

$$=\binom{n}{k_1, k_2, k_3, k_4} \frac{2^{k_1+k_2}}{6^n}$$

(b) $P(Y_1+Y_3=k)=\dfrac{\binom{n}{k}}{2^n}$

(c) $E(Y_1+Y_4)=\dfrac{n}{2}$

12 (a) Show that

$$\sum_{k_1 \geq 1, \ldots, k_m \geq 1} \binom{n}{k_1, \ldots, k_m} = \sum_{k=0}^{m} (-1)^k \binom{m}{k} (m-k)^n$$

[Hint: Consider Eq. (3), Section 3.4.]

(b) Check (a) directly for $m = 2$.

(c) Define

$$c(x) = \begin{cases} 1 & \text{if } x = 0 \\ 0 & \text{if } x \neq 0 \end{cases}$$

Show that

$$\sum \binom{n}{k_1, \ldots, k_m} [c(k_1) + \cdots + c(k_m)] = m(m-1)^n$$

The summation is over all m-tuples of nonnegative integers that sum to n. [Hint: Consider Eq. (5), Section 3.4.]

13 Suppose $(Y_1, Y_2, Y_3, Y_4, Y_5)$ is multinomially distributed with parameters $n; p_1, p_2, p_3, p_4, p_5$. What is the distribution of $(Y_1 + Y_2, Y_3, Y_4 + Y_5)$? (Answer: Multinomial, n; $p_1 + p_2, p_3, p_4 + p_5$.)

14 Prove the assertion in Example 3c, Section 6.2. {Hint:

$$P(Y_1 = k | Y_1 + Y_2 = r) = \frac{P[(Y_1 = k) \cap (Y_2 = r - k) \cap (Y_3 = n - r)]}{P(Y_1 + Y_2 = r)}$$

Use the fact that $Y_1 + Y_2$ is binomial n, $p_1 + p_2$.)

15 Exercises 16 and 17 in Chapter 3 relate to the hypergeometric distribution. You might want to review them at this point.

16 Suppose that 100 cards marked 1, 2, . . . , 100 are randomly arranged in a line. Show that the number of even integers in the first 20 positions is hypergeometrically distributed with parameters $n = 20$, $M = 50$, $N = 100$.

17 A lot of factory produced goods consists of N items; M of these are defective and $N - M$ are acceptable. Items are successively drawn from the lot, inspected, and returned to the lot. Inspected items are marked so they do not have to be inspected again. Drawings are made until a total of n distinct items have been inspected. Show that the number of defectives among the inspected items has the hypergeometric distribution with parameters n, M, and N.

18 Suppose that $X_1, \ldots, X_N$ are mutually independent indicator random variables, with $P(X_i = 1) = p$, $0 < p < 1$. Show that for $1 \leq M \leq N$,

$$P\left(\sum_{i=1}^{M} X_i = k \,\middle|\, \sum_{i=1}^{N} X_i = n\right) = \frac{\binom{M}{k}\binom{N-M}{n-k}}{\binom{N}{n}}$$

(Hint: $\Sigma_{i=1}^{M} X_i$, $\Sigma_{i=M+1}^{N} X_i$ are independent and binomial.)

19 A deck of $m+n$ cards marked $1, 2, \ldots, m, \ldots, m+n$ is dealt out against $m+n$ similarly marked fixed positions. Let X be the number of matches in the first m positions and let Y be the number of matches in the next n positions. Show that the conditional distribution of X given $X+Y$ is hypergeometric. Specifically,

$$P(X=k|X+Y=r)=\frac{\binom{m}{k}\binom{n}{r-k}}{\binom{m+n}{r}}$$

[Hint: Consult Exercise 22a, Chapter 3, and Eq. (2), Section 3.2.]

20 Suppose N balls are randomly distributed among $m+n$ cells. Let X be the number of unoccupied cells among the first m cells, and let Y be the number of unoccupied cells among the next n cells. Show that the conditional distribution of X given $X+Y$ is exactly the same hypergeometric distribution as in Exercise 19.

21 (a) Suppose that a population consists of m types of objects, M_1 of type 1, M_2 of type 2, and so forth. Then $N=M_1+\cdots+M_m$ is the total population size. Suppose that a sample of size n is drawn *without replacement.* Let Y_1 be the number drawn of type 1, Y_2 the number drawn of type 2, and so forth. Show that

$$\begin{aligned}P[(Y_1=k_1)\cap(Y_2=k_2)\cap\cdots\cap(Y_m=k_m)] &= \frac{\binom{M_1}{k_1}\binom{M_2}{k_2}\cdots\binom{M_m}{k_m}}{\binom{N}{n}}\\ &= \binom{n}{k_1, k_2, \ldots, k_m}\frac{M_1^{(k_1)}M_2^{(k_2)}\cdots M_m^{(k_m)}}{N^{(n)}}\end{aligned}$$

for every sequence of nonnegative integers $k_1, \ldots, k_m$, the sum of which is n.

(b) Suppose that $N, M_1, \ldots, M_m$ all approach infinity in such a way that

$$\frac{M_1}{N}\to p_1, \frac{M_2}{N}\to p_2, \ldots, \frac{M_m}{N}\to p_m$$

$$(p_1+p_2+\cdots+p_m=1)$$

Show that the distribution in (a) approaches the multinomial distribution with parameters $n; p_1, p_2, \ldots, p_m$. [Hint: Imitate the proof of (1), Section 6.3.]

22 Suppose that X and Y have the following joint distribution:

(a) X is hypergeometric with parameter

$$n=10, \quad M=50, \quad N=200$$

(b) The conditional distribution of Y, given that $X = k$, is hypergeometric with parameter

$$n = 10, \qquad M = 50 - k, \qquad N = 190$$

Show that $X + Y$ is hypergeometric with parameters $n = 20$, $M = 50$, $N = 200$. (Intuitive explanation: A first sample of size 10 is drawn, and X is the number of type 1 objects in the sample. Then, without this sample being replaced, a second sample of size 10 is drawn, and Y is the number of type 1 objects in the second sample.)

23 (a) A pair of coins are thrown together. What is the distribution of the number of throws required for both to show head simultaneously? (Answer: Geometric $p = 1/4$.)

(b) What is the distribution of the number of throws required for at least one of the coins to show a head? [Answer: Geometric, $p = 3/4$. Use (7), Section 6.4, or do directly.]

24 Suppose that X and Y are independent random variables that are geometrically distributed with parameters p_1 and p_2, where $p_1 \neq p_2$. Show that

$$P(X + Y = k) = \frac{p_1 p_2}{p_2 - p_1}[(1 - p_1)^{k-2} - (1 - p_2)^{k-2}], \qquad k = 2, 3, \ldots$$

{Hint: The generating function of $X + Y$ is

$$\left[\frac{p_1 t}{1 - (1 - p_1)t}\right]\left[\frac{p_2 t}{1 - (1 - p_2)t}\right] = \frac{p_1 p_2 t^2}{p_2 - p_1}\left[\frac{1 - p_1}{1 - (1 - p_1)t} - \frac{1 - p_2}{1 - (1 - p_2)t}\right]\}$$

25 Show that the generating function of the number of balls that must be distributed to occupy fully m cells is

$$[t]\left[\frac{m-1}{m}\left(\frac{t}{1 - (1/m)t}\right)\right]\left[\frac{m-2}{m}\left(\frac{t}{1 - (2/m)t}\right)\right] \cdots$$
$$\left[\frac{1}{m}\left(\frac{t}{1 - [(m-1)/m]t}\right)\right]$$
$$= \frac{m!}{m^m} t^m \left(\frac{1}{1 - (1/m)t}\right)\left(\frac{1}{1 - (2/m)t}\right) \cdots \left(\frac{1}{1 - [(m-1)/m]t}\right)$$

(Hint: Use Example 2, Section 6.4. You will have to accept the fact that $Z_1, Z_2, \ldots, Z_m$ are mutually independent. Can you prove this independence?)

26 Suppose that X has possible values $1, 2, \ldots$, and $P(X = 1) > 0$. Suppose also that for every positive integer k, X has the following property:

$$P(X = k + r \mid X > k) = P(X = r), \qquad r = 1, 2, \ldots$$

Show that X must be geometrically distributed with parameter $p =$

$P(X = 1)$. [Notice that this statement is the converse to (6), Section 6.4.] [Hint: Actually, the asserted condition need only hold for $k = 1$. The following is typical of the computations that are needed:

$$P(X = 2|X > 1) = \frac{P(X = 2)}{P(X > 1)} = P(X = 1)$$

Hence, $P(X = 2) = p(1 - p)$.]

27 (a) Suppose that X and Y are independent, each geometrically distributed over 1, 2, . . . with parameter p. Show that

$$P(\max X, Y = k) = 2p(1-p)^{k-1} - p(2-p)(1-p)^{2k-2}, \qquad k = 1, 2, \ldots$$

(b) Each of two persons independently throws a coin until he obtains a head. Show that the maximum number of throws has the following distribution:

$$p_k = \left(\frac{1}{2}\right)^{k-1} - \frac{3}{4}\left(\frac{1}{4}\right)^{k-1}, \qquad k = 1, 2, \ldots$$

28 Suppose that X and Y are independent, each geometrically distributed over 0, 1, 2, . . . with parameter p. Let $U = \min(X, Y)$, $V = \max(X, Y)$.

(a) Show that

$$P[(U = k) \cap (V = k + r)] = \begin{cases} (1-p)^{2k}p^2 & \text{if } r = 0 \\ 2(1-p)^{2k+r}p^2 & \text{if } r = 1, 2, \ldots \end{cases}$$

(b) Conclude from (a) that

$$\begin{aligned} P(U = k) &= (1-p)^{2k}p(2-p) \\ &= (1-a)^k a, \qquad k = 0, 1, 2, \ldots \end{aligned}$$

where $a = 1 - (1 - p)^2$. [In other words, U is geometric with parameter $1 - (1 - p)^2$. Compare (7), Section 6.4.]

(c) Conclude from (a) that

$$P(V - U = r) = \begin{cases} \dfrac{p}{2-p} & \text{if } r = 0 \\ \dfrac{2p}{2-p}(1-p)^r & \text{if } r = 1, 2, \ldots \end{cases}$$

(d) Conclude from (a), (b), and (c) that U and $V - U$ are independent.

29 (a) Show that

$$(-1)^n\binom{-1/2}{n} = \binom{2n}{n}\frac{1}{2^{2n}}$$

[For $n = 3$,

$$\binom{-1/2}{3} = \frac{(-1/2)(-3/2)(-5/2)}{3!} = \frac{1 \cdot 2 \cdot 3 \cdot 4 \cdot 5 \cdot 6}{3!3!2^6}(-1)^3]$$

(b) Show that

$$\sum_{i=0}^{n} \frac{\binom{2n-2i}{n-i}\binom{2i}{i}}{2^{2n}} = 1$$

[Hint: Use $(1-t)^{-1/2}(1-t)^{-1/2} = 1 + t + t^2 + \cdots$.]

30 (a) Suppose that X has possible values 0, 1, 2, . . . , and the generating function of X is $\phi(t)$. If k is a positive integer, show that the generating function of kX is $\phi(t^k)$. (The possible values of kX are $0, k, 2k, 3k, \ldots$.)

(b) If X is Poisson distributed with parameter λ and k is a positive integer, show that the generating function of kX is $\exp \lambda(t^k - 1)$.

31 (a) Show that the compound Poisson random variable X of Example 7, Section 6.7, is distributed like $U + 2V$, where U and V are independent and each is Poisson with parameter $\lambda/3$. (Hint: Use Exercise 30b.)

(b) Suppose that X is compound Poisson with $\phi(t) = p_0 + p_1 t + \cdots + p_n t^n$. Show that X is distributed like

$$U_1 + 2U_2 + \cdots + nU_n$$

where $U_1, U_2, \ldots, U_n$ are mutually independent, Poisson distributed, with parameters $\lambda p_1, \lambda p_2, \ldots, \lambda p_n$, respectively. {Hint:

$$E(t^X) = \exp \lambda\left[\sum_{i=1}^{n} p_i t^i - 1\right] = \exp[\lambda p_1(t-1) + \cdots + \lambda p_n(t^n - 1]\}$$

32 (a) Suppose that X and Y are independent, Poisson distributed with parameters λ_1 and λ_2. Show that the conditional distribution of X, given that $X + Y = n$, is binomial with parameters n and $p = \lambda_1/(\lambda_1 + \lambda_2)$ $(n > 0)$.

(b) Suppose that $X_1, \ldots, X_m$ are mutually independent, Poisson distributed variables with parameters $\lambda_1, \ldots, \lambda_m$. Let $\lambda = \lambda_1 + \cdots + \lambda_m$. Show that the joint conditional distribution of $X_1, \ldots, X_m$, given that $X_1 + \cdots + X_m = n$, is multinomial with parameters n; $\lambda_1/\lambda, \ldots, \lambda_m/\lambda$ $(n > 0)$.

33 Repeat Exercise 16, Chapter 5, supposing now that

$$P(Y = k) = (1-p)^k p, \qquad k = 0, 1, 2, \ldots$$

Conclude that X is Poisson distributed with parameter $\lambda = 1 - p$.

NOTES

1 The first expositions of the binomial distribution were made by Abraham DeMoivre, and by Jacob Bernoulli (1654–1705), of the family of eminent Swiss mathematicians. Bernoulli studied what we call a sequence of independent and identically distributed indicator

random variables. In recognition, such a sequence is still called a *Bernoullian sequence of trials.* Also, the expression $\binom{n}{k}p^k(1-p)^{n-k}$ for k successes in n Bernoullian trials is sometimes called Bernoulli's formula.

2 For several useful tables of the binomial distribution, see [16], [17], and [18]. Also see [19] for a guide to such tables. The introductory chapters to [16], [17], and [18] describe various applications of the binomial distribution to practical problems.

3 Tables of the incomplete beta function appear in [20]. See [19] for additional references.

4 For the fundamental theorem of integral calculus, see page 176 of [2], or Section 5.1 of [3].

5 Reference [21] is a table of the hypergeometric distribution. Chapter 2 describes some applications.

6 For some additional material on Newton's binomial expansion, see Problem 2, Section 18.3 of [2], and Section 11.15 of [3]. The discovery by Isaac Newton (1642–1727) that the ordinary binomial expansion can be extended to $(1 + t)^a$ with a arbitrary, was one of the great discoveries of seventeenth-century mathematics.

7 The negative binomial distribution is also known as Pascal's distribution, after the French scientist, Blaise Pascal (1623–1662).

8 For a table of the negative binomial distribution, see [22]. Also see [19] for references to other tables.

9 The introductory chapter to [22] describes the application of the negative binomial distribution referred to in Example 1, Section 6.6, as well as other applications.

10 For applications of the Poisson distribution to telephone system engineering, see [23] and [24].

11 The Poisson distribution is named after the French mathematician Simeon Denia Poisson (1781–1840). The approximation described by (2), Section 6.7, is the work of Poisson; hence, the distribution is sometimes called *Poisson's exponential binomial limit.*

12 Reference [25] is a table of the Poisson distribution. See [19] for references to other tables, as well as tables of the incomplete gamma integral.

13 For additional reading on the compound Poisson distribution, see Section 2, Chapter XII, of [9].

Chapter Seven

INTEGRATING DENSITY FUNCTIONS

The density function of a discrete random variable X was earlier defined by $f(a) = P(X = a)$, a varying over the possible values of X. It follows that

$$P(X \text{ in } A) = \sum_{a \text{ in } A} f(a)$$

There are many probability situations where a reasonable model prescribes that

$$P(X \text{ in } A) = \int_A f(x)\, dx$$

In such a case, we say that X has an *integrating density function f.* The values of X can lie in one or higher dimensional space. A discrete probability space does not provide a sufficient setting for such random variables. We first describe a broader setting—*the general probability space*—to be able to introduce integrating density functions.

7.1 GENERAL PROBABILITY SPACES

Chapters 1 through 6 are about discrete probability spaces; these provide a mathematical model for chance experiments with at most a countable number of outcomes. What should the model be for experiments with more than a countable number of outcomes? (For instance, consider Example 3, Section 1.1.) Suppose one starts with the naïve hope that a theory can be based on the same axioms (a), (b), and (c) of Section 1.4, which formed the basis for discrete probability spaces. Remember these axioms were the following.

review of axioms for the discrete case

(a) For any event A, $0 \leq P(A) \leq 1$.
(b) $P(\mathfrak{X}) = 1, P(\varnothing) = 0$.
(c) If $A_1, A_2, \ldots,$ is a finite or countable sequence of mutually disjoint events, then

$$P(A_1 \cup A_2 \cup \cdots) = P(A_1) + P(A_2) + \cdots$$

When $\mathfrak{X}$ is not discrete, it is in general *not possible* to assign probability values to *every* subset of $\mathfrak{X}$ in such a way that (a), (b), and (c) are satisfied. (This assertion is not at all obvious, nor will it be proved here.) The way out of the difficulty is to become more selective about which subsets of $\mathfrak{X}$ one calls *events*, the sets to which probability values can be assigned. A rigorous approach to probability theory begins with a collection of events that remains closed under the basic set operations and such that (a), (b), and (c) hold. This method will now be outlined in greater detail (Note 1).

axioms for general probability spaces

I. There is a sample space $\mathfrak{X}$ that is the set of basic sample points.

II. There is a collection $\mathscr{F}$ of subsets of $\mathfrak{X}$, called *events*, having the following properties.

(a) The whole sample space, $\mathfrak{X}$, and $\varnothing$, the empty set, are members of $\mathscr{F}$. (These are the *certain* event and the *impossible* event.)
(b) If A is in $\mathscr{F}$, then A' is also in $\mathscr{F}$. (In other words, complements of events are events.)
(c) If $A_1, A_2, \ldots$ is a finite or countable sequence of sets in $\mathscr{F}$, then $A_1 \cup A_2 \cup \cdots$ is in $\mathscr{F}$ and $A_1 \cap A_2 \cap \cdots$ is in $\mathscr{F}$. (In other words, countable unions and intersections of events are events.) $\mathscr{F}$ is called an *algebra of events* (Note 2).

III. There exists a probability measure P, which is a set function that assigns values to every event and satisfies the following requirements.

(a) For any event A (i.e., for any element of the collection $\mathscr{F}$), $0 \leq P(A) \leq 1$
(b) $P(\mathfrak{X}) = 1, P(\varnothing) = 0$.
(c) If $A_1, A_2, \ldots$ is a finite or countable sequence of mutually disjoint events, then

$$P(A_1 \cup A_2 \cup \cdots) = P(A_1) + P(A_2) + \cdots$$

The sample space $\mathfrak{X}$, equipped with its algebra of events $\mathcal{F}$ and its probability measure P, is called a *general probability space.*

The preceding is almost the same as the original development in Section 1.4 for the discrete case. The main difference is that in the discrete case *any* subset of $\mathfrak{X}$ is an event; in the general case, not every subset of $\mathfrak{X}$ may be an event. There is a collection $\mathcal{F}$ of subsets of $\mathfrak{X}$; these, and only these, are the events. Except for restricting oneself only to these sets, the axioms are really the same as before.

The following are several important "continuity" consequences of the axioms. These are presented without proof (see Exercise 25).

continuity properties

(a) Suppose that $A_1 \subset A_2 \subset A_3 \subset \cdots$ and $A = A_1 \cup A_2 \cup A_3 \cup \cdots$. Then $\lim_{n\to\infty} P(A_n) = P(A)$.

(b) Suppose $B_1 \supset B_2 \supset B_3 \supset \cdots$ and $B = B_1 \cap B_2 \cap B_3 \cap \cdots$. Then $\lim_{n\to\infty} P(B_n) = P(B)$.

(In other words, for monotone sequences of events, the probability of the limit is the limit of the probabilities. The limit sets A and B are events by axiom IIc.)

Example 1

Consider a probability model for the experiment of selecting a point at random from the unit interval $[0, 1]$. At a minimum, one should be able to make probability statements about the selected point lying in any given interval. The interpretation of the phrase "at random" is that the probability of the selected point lying between a and b, where $0 \leqslant a \leqslant b \leqslant 1$, should be $b - a$. What is a suitable algebra of events for the probability model? The conventional approach to this problem will now be described in an informal, nonrigorous way. Consider all the sets that can be obtained from intervals by any succession of finite or countable sequences of the set operations of union, intersection, and complement. For example, a countable union of intervals is such a set; the intersection of countably many countable unions of intervals is such a set, and so forth. Call the collection of sets obtained in this way by the name $\mathcal{F}$. Hopefully, it should be clear, at least at this informal level of presentation, that $\mathcal{F}$ satisfies the requirements (a), (b), and (c) of axiom II. For instance, consider (b). If A is a set obtained by a countable sequence of set operations on intervals, then A' involves just one more operation and is also in $\mathcal{F}$. The sets that make up $\mathcal{F}$ are called *Borel sets,* and $\mathcal{F}$ is called the algebra *generated* by the intervals. Of course, the construction of $\mathscr{F}$ must be done in a more rigorous way; the description of $\mathscr{F}$ as the collection of "all sets that can be obtained from intervals by any succession of finite or countable sequences of the set operations of union, intersection, and complement" is imprecise and clumsy (Note 3). Now it is a fact that there is a probability measure P that assigns values to the sets making up the collection $\mathscr{F}$, and that satisfies (a), (b), and (c) of axiom III. It also assigns to any interval in $[0, 1]$ just the length of that interval. In fact, there is *only one* probability measure satisfying these requirements (Note 4). Of course, a consequence of these facts is that if A is a union of mutually disjoint intervals in $[0, 1]$, then $P(A)$ is the sum of the lengths of these intervals.

Example 2 A point is selected at random from the unit interval $[0, 1]$.

(a) The probability that the point selected lies between 0 and 1/2 is 1/2.

(b) The probability of any one point set is 0.

(c) The probability of the set of rational numbers is 0, because there are a countable number of rational numbers, each having probability 0.

(d) The probability of the union of disjoint intervals,

$$\left[\frac{1}{2}, \frac{1}{2} + \frac{1}{10}\right] \cup \left[\frac{2}{3}, \frac{2}{3} + \frac{1}{10^2}\right] \cup \left[\frac{3}{4}, \frac{3}{4} + \frac{1}{10^3}\right] \cup \cdots$$

equals $(1/10) + (1/10)^2 + (1/10)^3 + \cdots = 1/9$.

It should be pointed out that the various relationships among sets and indicator random variables, discussed in Section 2.7, did not depend on the discreteness of the sample space; these are still correct for general probability spaces. In particular, the inclusion-exclusion formula (Section 2.8) is correct.

7.2 RANDOM VARIABLES

On a discrete sample space, $\mathfrak{X}$, a random variable was defined to be *any* real-valued function defined on $\mathfrak{X}$ (Section 2.1). For general probability spaces, however, some restrictions are necessary to ensure that the functions used are "nice" enough. Let X be a function over $\mathfrak{X}$. Then if A is a set of real numbers, as before, (X in A) denotes the set of sample points w such that $X(w)$ is in A. In particular,

$(a \leq X \leq b)$ denotes the set of sample points w such that $a \leq X(w) \leq b$, with similar definitions for $(a < X < b)$, $(X < b)$, and so forth. Now what is desired is that (X in A) should be an event (i.e., an element of the algebra $\mathscr{F}$) whenever A is an interval. (An interval may or may not contain either of its endpoints or extend to ∞ or $-\infty$.) This fact will guarantee that the probabilities $P(a \leq X \leq b)$, $P(X \leq b)$, $P(X = a)$, and so forth, are all defined.

definition of a random variable

> A real-valued function X defined over a sample space $\mathfrak{X}$ is a random variable if (X in A) is an event whenever A is an interval.

If X has a finite or countable number of possible values, then X is called a *discrete* random variable. That is, X is discrete if there is a finite or countable sequence of numbers $a_1, a_2, \ldots$ such that $P(X = a_1) + P(X = a_2) + \cdots = 1$ ($a_1, a_2, \ldots$ are called the *possible values* of X). The only random variables that can possibly arise in discrete sample spaces are discrete random variables. However, discrete random variables also arise in nondiscrete sample spaces, as the following examples show.

Example 1

Suppose the sample space is the unit interval $[0, 1]$, and the probability of any interval in the unit interval is just the length of that interval. (This topic is discussed in Examples 1 and 2, Section 7.1.) Define a random variable as follows:

$$X(w) = 0 \quad \text{if } 0 \leq w < \tfrac{1}{4}$$

$$X(w) = 1 \quad \text{if } \tfrac{1}{4} \leq w \leq 1$$

Then $P(X = 0) = 1/4$, $P(X = 1) = 3/4$, and X is a discrete random variable.

Example 2

Consider the same situation as in Example 1. Define a random variable Y as follows:

$$Y(w) = n \quad \text{if } \frac{1}{n+1} < w \leq \frac{1}{n}, \qquad n = 1, 2, \ldots$$

Then

$$P(Y = n) = \frac{1}{n} - \frac{1}{n+1} = \frac{1}{n(n+1)}$$

and

$$P(Y = 1) + P(Y = 2) + \cdots = 1$$

Therefore, Y is a discrete random variable.

Example 3 Again, we have the same situation as in Example 1. Define Z by $Z(w) = w^2$; Z is not a discrete random variable.

7.3 CUMULATIVE DISTRIBUTION FUNCTIONS

Suppose that X is a random variable. Define a function F as follows.

definition of a cdf

$$F(t) = P(X \leq t), \qquad -\infty < t < \infty \tag{1}$$

(Notice that since X is a random variable, then $(X \leq t)$ is an event.) Function F is called the *cumulative distribution function* (or cdf for short) of the random variable X (Note 5).

The function F has the following properties.

properties of a cdf

(a) $0 \leq F(t) \leq 1$ for any t.
(b) If $t_1 < t_2$, then $F(t_1) \leq F(t_2)$.
(c) F is right continuous. That is,

$$\lim_{\epsilon_n \to 0} F(t + \epsilon_n) = F(t)$$

whenever $\epsilon_1 > \epsilon_2 > \cdots$ is any sequence of positive numbers that tends to 0.

(d) $\lim_{t \to -\infty} F(t) = 0$ and $\lim_{t \to \infty} F(t) = 1$.

Proof. Because $F(t)$ is the probability of an event, (a) is true. Because $(X \leq t_1) \subset (X \leq t_2)$, (b) is true. To prove (c), we notice that the events $(X \leq t + \epsilon_1)$, $(X \leq t + \epsilon_2)$, . . . form a monotone decreasing sequence of events the intersection of which equals $(X \leq t)$. By the *continuity properties* described in Section 7.1, it follows that $\lim_{n \to \infty} P(X \leq t + \epsilon_n) = P(X \leq t)$. To prove (d), let t_n be a sequence of real numbers that converge monotonically to $-\infty$, and let u_n be a sequence of real numbers that converge monotonically to ∞. Then $(X \leq t_1)$, $(X \leq t_2)$, . . . is a monotone decreasing sequence of events the intersection of which is the empty set $\varnothing$, and $(X \leq u_1)$, $(X \leq u_2)$, . . . is a monotone increasing sequence of events having as a union the whole sample space $\mathfrak{X}$. Again, by the continuity properties,

$$\lim_{n\to\infty} P(X \leq t_n) = P(\varnothing) = 0$$

$$\lim_{n\to\infty} P(X \leq u_n) = P(\mathfrak{X}) = 1 \quad \blacksquare$$

Example 1 Suppose that X is an indicator random variable. That is, X equals 1 with probability p and X equals 0 with probability $1 - p$, where $0 \leq p \leq 1$. Then the cdf of X is given by

$$F(t) = \begin{cases} 0, & t < 0 \\ 1 - p, & 0 \leq t < 1 \\ 1, & 1 \leq t \end{cases}$$

and F is continuous everywhere except at the two points 0 and 1. At these points, F has a jump discontinuity and is continuous from the right but not from the left. For instance, the limit of $F(t)$ as t approaches 0 from the left is 0, whereas $F(0) = 1 - p$. Figure 7.1 shows the graph of F.

Example 2 A six-faced die is thrown. As usual, suppose the faces are marked 1, 2, 3, 4, 5, 6, and let X be the number to appear. The cdf of X is the following:

$$F(t) = \begin{cases} 0, & t < 1 \\ 1/6, & 1 \leq t < 2 \\ 2/6, & 2 \leq t < 3 \\ \vdots & \\ 5/6, & 5 \leq t < 6 \\ 6/6, & 6 \leq t \end{cases}$$

Figure 7.2 shows the graph of F.

Example 3 A point is selected at random from the unit interval [0, 1]. Suppose Z is the square of the number selected (the random variable defined in Example 3 of Section 7.2). When t is between 0 and 1, the event that the square of the number selected is less than or equal to t is just the event that the original number selected is less than or equal to $\sqrt{t}$. Hence, the cdf is

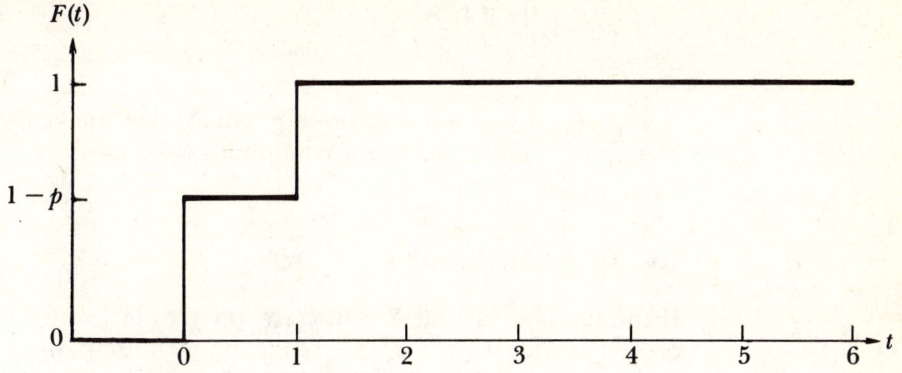

FIGURE 7.1 The cdf of an indicator random variable.

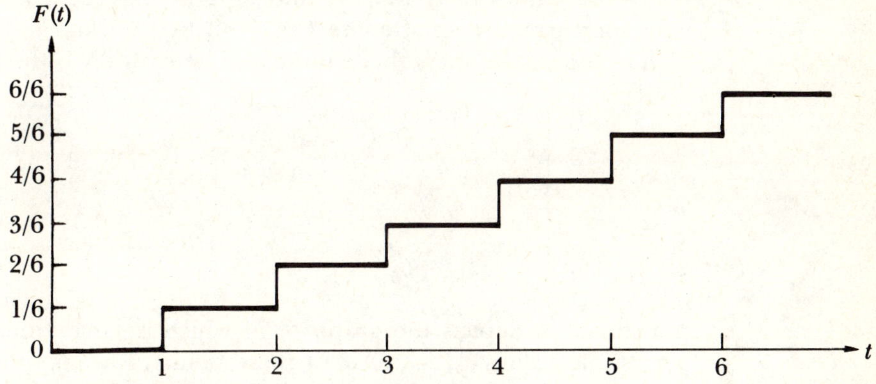

FIGURE 7.2 The cdf for a die throw.

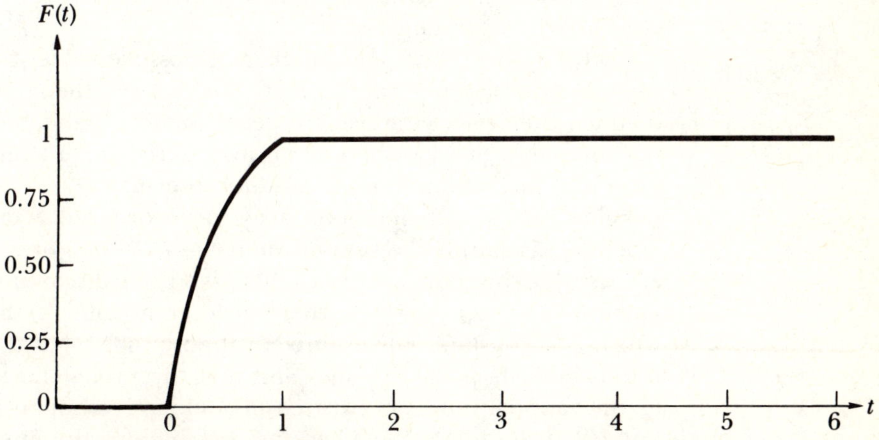

FIGURE 7.3 The cdf of the square of a number selected at random from [0, 1].

$$F(t) = \begin{cases} 0 & \text{if } t < 0 \\ \sqrt{t} & \text{if } 0 \leq t \leq 1 \\ 1 & \text{if } 1 < t \end{cases}$$

Figure 7.3 shows the graph of F. Notice that unlike the cdf's in Examples 1 and 2, this cdf is continuous for all t.

7.4 DISCRETE cdf's

If the random variable X is discrete, then, as in Examples 1 and 2 of Section 7.3, the cdf is a step function with jump discontinuities at the possible values of X. This fact will now be examined in detail. Consider first the simple case that X has only a finite number of possible values $a_1, a_2, \ldots, a_n$, and suppose that these are arranged in increasing order. Denote $P(X = a_i)$ by p_i; thus, $p_1, p_2, \ldots, p_n$ are positive numbers that sum to 1. The cdf of X is the following:

$$F(t) = \begin{cases} 0 & \text{if } t < a_1 \\ p_1 & \text{if } a_1 \leq t < a_2 \\ p_1 + p_2 & \text{if } a_2 \leq t < a_3 \\ \vdots & \\ p_1 + \cdots + p_{n-1} & \text{if } a_{n-1} \leq t < a_n \\ p_1 + \cdots + p_n = 1 & \text{if } a_n \leq t \end{cases}$$

Figure 7.4 depicts the graph of F, which is discontinuous at the possible values $a_1, \ldots, a_n$. At these points, F is continuous from the right but discontinuous from the left. At the possible value a_i, the graph jumps by the amount p_i. It increases from height 0 to height 1 only through these jumps at $a_1, a_2, \ldots, a_n$. Everywhere else the cdf is constant.

If there are a countable number of possible values that can be written in increasing order $a_1 < a_2 < a_3 < \cdots$, then the situation is much the same. However, in general, a countable sequence of real numbers may not be written in ordered form (consider, for example, the rational numbers). Suppose that $a_1, a_2, \ldots$ are the possible values, but not necessarily in an ordered arrangement. Then $F(t)$ is simply the sum of the terms $P(X = a_i)$ over all a_i that are less than or equal to t. At a_i, there is a jump discontinuity in the amount $P(X = a_i)$; however, there need be no interval throughout which F is constant. For example, suppose that X has the rational numbers as its possible values and that the probabilities assigned to the rational numbers (written in some agreed-upon sequence) are $1/2, 1/2^2, 1/2^3, \ldots$. There are infinitely many jumps in any positive length interval. At any rational number, F is discontinu-

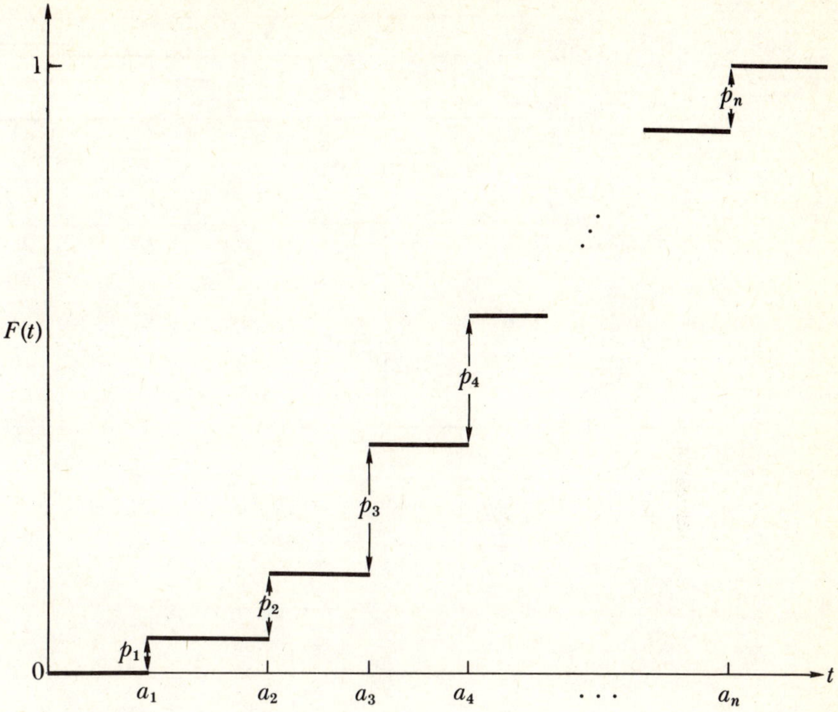

FIGURE 7.4 A discrete cdf.

ous; at any irrational number, it is continuous. What does the graph of this cdf look like?

Fortunately, most of the examples of discrete cdf's of interest will not be this complicated; the step function graphs shown in Figures 7.4 and 7.5 are typical of the usual ones that will be met.

Example 1 Let X be the number of throws required to obtain a head using an honest coin. The following is the cdf:

$$F(t) = \begin{cases} 0 & \text{if } t < 1 \\ 1 - 1/2 & \text{if } 1 \leq t < 2 \\ 1 - 1/2^2 & \text{if } 2 \leq t < 3 \\ \vdots & \\ 1 - 1/2^k & \text{if } k \leq t < k + 1 \\ \vdots & \end{cases}$$

The graph appears in Figure 7.5.

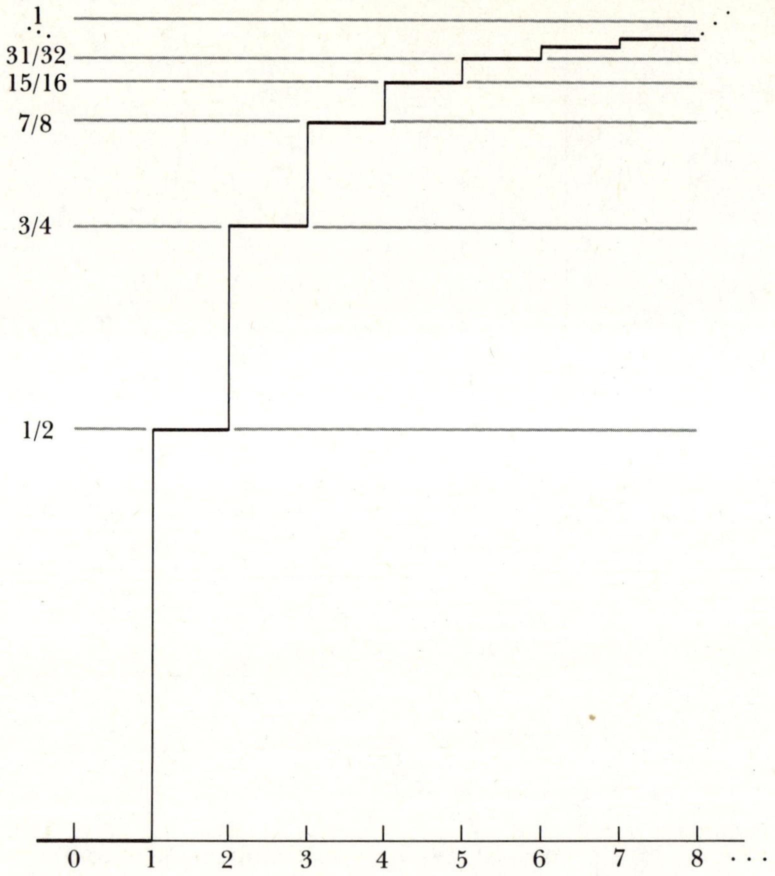

FIGURE 7.5 The cdf of the number of throws required to obtain a head.

7.5 INTEGRATING DENSITY FUNCTIONS

If X is a discrete random variable then $P(X \text{ in } A)$ is evaluated by summing $P(X = a)$ over all possible values a that make up the set A. There exists an important special case in which X is not discrete and probabilities for X are obtained by integrating a function that we shall call the integrating density function.

definition of integrating density function

There is a nonnegative function f, and F is expressed in terms of f by the relationship

$$F(t) = \int_{-\infty}^{t} f(x)\, dx, \qquad -\infty < t < \infty \tag{1}$$

Function f is called an integrating density function for the random variable x (Note 6).

Throughout this book, f will be a Riemann integrable function (Note 7).

There is a strong analogy between the integrating density function and the discrete density function (Section 2.9). For a discrete random variable X, the density function is $f(a) = P(X = a)$. If a_1, $a_2, \ldots$ are the possible values of X, then the relationship between the cdf F and the discrete density function f is

$$F(t) = \sum_{a_i \le t} f(a_i)$$

In other words, $F(t)$ is the sum of values of the density function for all possible values a_i less than or equal to t. When there is an integrating density function, the operation of summing is replaced by integrating. In this case, $F(t)$ is the integral of values of the density function for all arguments x less than or equal to t. Also, (1) implies that $\int_{-\infty}^{\infty} f(x)\, dx = 1$, just as in the discrete case $\Sigma_i\, f(a_i) = 1$. These relationships are illustrated in Figure 7.6.

It must be pointed out that an integrating density function is not unique. If f is an integrating density of X, and if g differs from f only at a finite number of points, then, for every t,

$$P(X \le t) = \int_{-\infty}^{t} f(x)\, dx = \int_{-\infty}^{t} g(x)\, dx$$

and g is also an integrating density of X. [Actually, g can differ from f on a more complicated set, namely, a Borel set of measure 0 (Note 3).] This nonuniqueness should never cause us any trouble. In every example, we shall choose some one version of the integrating density and stick with it. We allow ourselves the slight abuse of language in speaking of "the integrating density function" when we should really say "one of the equivalent versions of an integrating density function."

From (1) follow several important properties of the cdf.

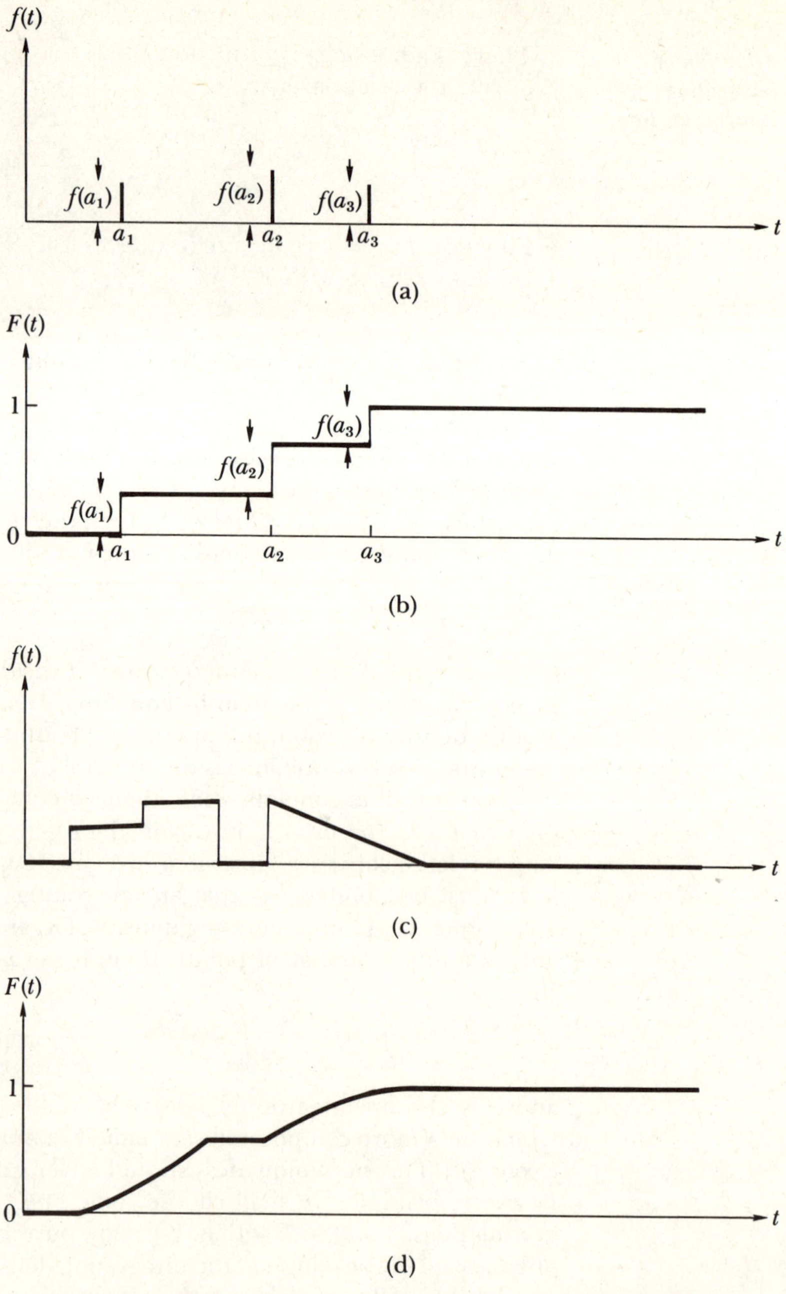

FIGURE 7.6 (a) Discrete density function. (b) Discrete cdf. (c) Integrating density function. (d) The cdf determined from the integrating density, $F'(t) = f(t)$.

Errata for Probability and Statistics by Dwass. (Errata listed for pages 1 through 402 are applicable to Statistics by Dwass.)

Page	Line	Correction
8	21	The display starting P(A) should be labelled as equation (1).
13		last four entries of Table 1.7 should be:
		7 5,040
		8 40,320
		9 362,880
		10 3628.800
15	5	495 should be 4,950
15	11	$n = 2$ should be $r = 2$.
25	2	Lower case x should be upper case X.
48	Exercise 18, lines 2 and 3 from bottom	random should be random variables
69	11	m balls should be n balls n cells should be m cells
70	2 from bottom	should read: empty) = $\binom{n+m-2}{n}/\binom{n+m-1}{n}$. Hence,
71	Equation 2	should appear as: $$= \sum_{k=0}^{m} (-1)^k \binom{m}{k-1} \frac{\binom{n+m-k-1}{n}}{\binom{n+m-1}{n}}$$
73	21 and 22 (turnover)	N(N-1) should be (N-M)
75	15 and 17	$N^{(n-1)}$ should be $(N-1)^{(n-1)}$
82	Exercise 11	In the right side of the display $\binom{n}{n}$ should be $\binom{n}{n}m$
84	2 of Exercise 21(b)	The summation should be $$\sum_{k=0}^{n-1}$$
85	8 of Exercise 22(b)	$\rho^k(1-\rho)^{k+r}$ should be $\rho^k(1-\rho)^r$
91	11	$P(B_3 \mid W_1 \cap W_2$ should be $P(B_3 \mid W_1 \cap W_2)$
97	18	$E(Y \mid Y=b)$ should be $E(X \mid X=b)$
110	10 from bottom	EX should be EY
110	8 from bottom	First (EX) should be (EY)
115	8 of Summary	The numerator $P[(Y=a) \cap (Y=b)]$ should be $P[(X=a) \cap (Y=b)]$

SHENANDOAH COLLEGE
LIBRARY
WINCHESTER, VA.

QA
273
.D887
1970

Page	Line	Correction
117	Exercise 5(c)	0.287 should be 0.137
128	6	should read: $\cdots S_i(-1)^{i-k}$. (S_0 should be understood to equal 1.)$\cdots$
130	2 from bottom Example 1	$(k+1)!$ should be $2\,k!$
131	3 and 4 from bottom	should be: Suppose that X is nonnegative and integer valued, that all moments of X exist, and $\phi(t)$ exists for $\lvert t\rvert < 2$.
	2 from bottom	$\lvert t\rvert < 1$ should be $\lvert t-1\rvert < 1$
134	3	$E(t^{X}i)$ should be $E(t^{X_i})$
135	1	The final t^n should be t^k
	3	$\binom{n}{k}\frac{1}{2}n$ should be $\binom{n}{k}\frac{1}{2}k$
	8 from bottom	$(X=j)$ should be $(Y=j)$
	7 from bottom	$(X=k)$ should be $(X+Y=k)$
138	5 from bottom	$\lvert t\rvert \le 1$ should be $\lvert t-1\rvert \le 1$
141	5 through 8 Exercise 8(a)	replace 1 by $1-p_0$ (four places)
	10	$\frac{1}{1-t}$ should be $\frac{1-p_0}{1-t}$
146	5 from bottom	add closing parenthesis after $(0 \le p \le 1$
151	15	$(a_n - 0)$ should be $(a_{n-1} - 0)$
158	2 from bottom	Z_1 should be Z_i
160	9	X should be Y
165	1	$(-a)^{(k)}$ should be replaced by $(-1)^k(-a)^{(k)}$
168	4 from bottom	x^{k-1} should be x^k
171	6	p should be np
176	4 of Exercise 24	Exponent k-2 should be k-1 (two places)
207	2	Exponent -at should be -2at
214	2 of Example 2	$-\surd a$ should be $-\sqrt{a}$ (upper bar of radical is missing)
215	8 of text material	$(X-Y<a)$ should be $X+Y<a)$
218	3	$\int\cdots\int$ should be $\int_{-\infty}^{\infty}\cdots\int_{-\infty}^{\infty}$
	3 from bottom	$\lvert b\lvert u_2\rvert$ should be $\lvert b\rvert\lvert u_2\rvert$
222	2	Add ∞ to upper limits of integral (two places).

Page	Line	Correction
231	13	In right side of display interchange u and v
233	Exercise 16	Answers for Exercise 5 should be 1/c, c+1, -1
234	6	mx^{n-1} should be mx^{m-1}
	Exercise 20(d)	Integral should be $\int_{-1}^{1}$
	2 from bottom	$\frac{\log 2}{2}$ should be $\frac{3 \log 2}{2}$
239	7 from bottom	J(x,y;u,r) should be J(x,y;u,v)
245	5 from bottom	$(X=u-v_i)$ should be $(Y=u-v_i)$
248	bottom display	Exponent -(u+v) should be -(u-v)
250	4 from bottom	$\Theta < e^{-\Theta u} \leq 1$ should be $0 < e^{-\Theta u} \leq 1$
251	15 from bottom	For no value of Θ except $\Theta = 0$ does the should read Only if Θ satisfies $\lvert\Theta\rvert < 1$ does the
260	4 of Example 3	(6) should read (7)
263	last display	if $0 \leq x \leq 1$ should be if $0 < x \leq 1$
265	last line of Exercise 9	x+y=r should be x+y=s
269	last line of Exercise 20	Replace arrow under double integral by >
275	Example 3	Display should be $$h(x) = \begin{cases} 0 & \text{if } \frac{1}{2} \leq x < 1 \\ 1 & \text{if } (\frac{1}{2})^2 \leq x < \frac{1}{2} \\ 2 & \text{if } (\frac{1}{2})^3 \leq x < (\frac{1}{2})^2 \end{cases}$$
279	Third section of Table 9.3	$0 \leq x < 2$ should be $0 \leq x < 1$
	last line	Delete second (n-1)! in denominator
280	9	$\int_{\infty}^{\infty}$ should be $\int_{-\infty}^{\infty}$
281	3 from bottom	$\frac{[c(x-k)]^n + (-1)^n [c(x-n-1)]^n}{n!}$ should be $\frac{[c(x-k)]^n}{n!} + \frac{(-1)^n [c(x-n-1)]^n}{n!}$
	last line	This last sum equals (3)... should read This is (3) with n replaced by n+1...
286	10	delete dt
288	last display	J(x,y;r,s) should be J(x,y;s,r)

Page	Line	Correction
289	last line	Insert λ in front of integral. That is, the middle expression should be $\lambda \int_t^{\infty} e^{-\lambda x}\,dx$
296	last line	replace (sin Θx) by $\left(\frac{\sin \Theta x}{\sqrt{2\pi}}\right)$
		replace (cos Θ x) by $\left(\frac{\cos \Theta x}{\sqrt{2\pi}}\right)$
298	Table 9.4	In the column headings (two places) $(2\pi)^{-1}$ should be $(2\pi)^{-1/2}$
299	5	$N(m, \sigma)$ should be $N(m, \sigma^2)$
	6	Right side of Equation (9) should read $\exp(i\Theta m - \frac{\Theta^2\sigma^2}{2})$
300	10	$N(m, \sigma_1^2)$ should be $N(m_1, \sigma_3^2)$
302	13	should read $\alpha = n/2$, $\lambda = 1/2$
304	7 from bottom	The expression in brackets should read $[(x-m)^2/2\sigma^2]$
306	5 of Exercise 8	$2e^{-\lambda(u+v)}$ should be $2\lambda^2 e^{-\lambda(u+v)}$
307	3	$2e^{-\lambda(u+v)}$ should be $2\lambda^2 e^{-\lambda(u+v)}$
	4 of Exercise 10	$\lambda = 1.$ should be λ .
309	Exercise 18(b)	$\|a\|$ should be a^2
320	8	First summation should read $\sum_{i=1}^{n}$
		(X_iX_j) should be (X_i, X_j)
	7 from bottom	Last − should be +
322	2 from bottom	Proof of (1) should read Proof of (2)
323	8	Proof of (2) should read Proof of (1)
	last line	Under lim, $n > \infty$ should be $n \to \infty$
329	6 from bottom	$(n\sigma)^{1/2}$ should be $\sqrt{n}\,\sigma$
333	3	$\int_{\infty}^{a}$ should be $\int_{-\infty}^{a}$
334	16	Add dx to integral
338	4 of Exercise 12	$\leq$ should be $>$
341	4, Exercise 21(d)	$(\frac{c-m}{2})$ should be $2(c-m)$

Page	Line	Correction
341	6, Exercise 21(d)	$(\frac{m-c}{2})$ should be $2(m-c)$
346	2 from bottom	time 0 should be time t
356	11 from bottom	limit in (7) may not exist, but... should read $\lim_{n \to \infty} U_n$ may not exist, but...
366	5	$(Y_1 = e_n)$ should be $(Y_n = e_n)$
374	14	In the matrix expression for P_n, X_n should be Y_n (three places)
376	8	$\pi_0 = \pi_3 = \pi_6 = \ldots$ should be $\pi_1 = \pi_3 = \pi_5 \ldots$
	9	$\pi_2 = \pi_5 = \pi_8 = \ldots$ should be $\pi_2 = \pi_4 = \pi_6 \ldots$
	12	$[P(Y_0=0), P(Y_0=1), P(Y_0=1)]$ should be $[P(Y_0=0), P(Y_0=1), P(Y_0=2)]$
385	Equation 1	should read $Y(t) = \begin{cases} 0 & 0 \leq t < W_1 \\ X_1 & W_1 \leq t < W_1 + W_2 \\ X_1+X_2 & W_1+W_2 \leq t < W_1 + W_2 + W_3 \\ \vdots & \end{cases}$
400	Exercise 20	In the matrix expression for P, the third line should appear as $p \quad 0 \quad 0 \quad 1-p \quad 0 \quad 0 \quad \ldots$
401	4 from bottom	$E(Y_{n+1}^2)$ should be $E(Y_n^2)$
406	14	$e^{-\lambda tx}$ should be $e^{-\lambda t}$
413	13 from bottom	h(X) should be u(X)
418	6 from bottom	i balls should be i black balls
428	13	$u_s(X)$ should be $u_2(X)$
	3 from bottom	X should be $\bar{X}$
430	10 from bottom	$\alpha^{-1/n} - 1$ should be $u(X)(\alpha^{-1/n} - 1)$
447	5 from bottom	$\psi(X)$ should be $\phi(X)$
448	8	ψ should be ϕ
454	caption for Fig. 13.7	(10) should be (9)
	Fig. 13.7	ART... On horizontal axis, $1 - \frac{1}{2}^n$ should be $(1-\frac{1}{2}^n)/n+1$ and $-1+\frac{1}{2}^n$ should be $(-1+\frac{1}{2}^n)/n+1$

Page	Line	Correction
455	3	should read $D(x) = \frac{1}{2\pi}\left\{\int_{-\infty}^{0} \exp\left[-\frac{\Theta^2+(x-\Theta)^2}{2}\right]d\Theta - \int_{0}^{\infty} \exp\left[-\frac{\Theta^2+(x-\Theta)^2}{2}\right]d\Theta\right\}$
462	1	Add closing parenthesis in denominator; $f(X, \Theta_2$ should be $f(X, \Theta_2)$
463	15	a should be α
	3 from bottom	$P_1(X<d)=1-e^{-d}$ should be $P_1(X<d)=e^{-d}$
	5 from bottom	$X<d$ should be $X>d$
	6 from bottom	$X>d$ should be $X<d$
464	4	$e^{-\Theta d}$ should be $1 = e^{-\Theta d}$
	7	same correction
477	last line	add closing parenthesis following $\}$
485	18	formula (2) should read formula (5)
486	1	Insert minus sign; $u(X)\ h(\Theta)$ should be $u(X) - h(\Theta)$
487	2 from bottom	(2) should be (5)
488	8	$P(X_i=p)$ should be $P(X_i=1) = P$
493	4 of Exercise 7	should read: succession of n numbers drawn, assuming the balls are numbered 1, 2, . . . , N.
506	6 from bottom	true $\hat{\hat{p}}$ should read true p (delete second carat)
510	2	F should be ϕ
516	1	$\psi(^{M}_{N})$ should be $\psi(\frac{M}{N})$
520	3 of Exercise 6	M/n should be M/N
	10 of Exercise 6	$a(0)=1$ should be $a(0)=0$
527	13	x_i should be x_1
	9 from bottom	In middle of line, the x_i in $a_i x_i$ should be boldface.
529	6 from bottom	Y_2 should be Y_3
531	5	y should be boldface
533	4 from bottom	Add closing parenthesis; $\|(x,u\|$ should be $\|(x,u)\|$
541	last line	$X=$ should be $\bar{X}=$

Page	Line	Correction
541	4 of shaded block	Y in Y_1 should not be boldface
544	14	Θ should be boldface
	5 from bottom	Θ and a_1 should both be boldface
546	Exercise 7 (first display)	Insert large closing parenthesis after $\begin{matrix} a_{n1} \\ \vdots \\ a_{nn} \end{matrix}$
		Insert large opening parenthesis before $\begin{matrix} U_1 \\ \vdots \\ U_n \end{matrix}$
551	16	$(0, \ldots, 0, 1, \ldots, 0)$ should be $(0, \ldots, 0, 1, \ldots, 1)$
555	4	second β should be boldface
556	1	u and X should be boldface (2 places each)
557	5	α should be boldface (four places)
	9	α_1 and α_2 should be a_1 and a_2
560	7 of Example 3	Boldface t_n should not be boldface
562	10, 11, and 12	$X_{i.} - X_{j.}$ should be $\bar{X}_{i.} - \bar{X}_{j.}$ (bars missing in 4 places)
	first shaded block	Right parenthesis missing $(\beta\ \partial_i$ should be $(\beta\ \partial_i)$
563	1	$\sum_{i=1}^{n}$ should be $\sum_{i=1}^{k}$
565	12	$(Xi)^2$ should be $(\bar{X}i)^2$ (bar missing)
	15	$-m\sum_{i=1}^{m}$ should be $-m\sum_{i=1}^{k}$
	8 from bottom	Right side of display should read $\sum_{i=1}^{n} X_i^2 - n(\bar{X})^2 - \sum_{i=1}^{n} (ti = \bar{t})^2(\beta_2)^2$
	4 from bottom	β_1, β_2 (first appearance) should be $\hat{\beta}_1, \hat{\beta}_2$
566	5 and 6 from bottom	In EX the X should be boldface (2 places)
567	2 and 3 of Example 2	m should be k

Page	Line	Correction
568	9	W_1 should be V_1
570	Example 2	Change first sentence to read: Following Example 3, Section 17.5, we now discuss the hypothesis $\delta_1 = \cdots = \delta m$, that there is no difference among row effects in the two-way layout model.
	4 from bottom	y_{m-1} should be z_{m-1}
571	2	$(X^2_{\cdot}m)$ should be $(\overline{X}^2_{\cdot}m)$
	7	$V(z_1, \ldots, z_{m-1})$ should be $V(y_1, \ldots, y_{k-1})$ (y's boldface)
	9	the first X_{ij} should be X^2_{ij}
572	19	X on W_2 should be EX on W_2
573	2 from bottom	$<c$ should be $>c$.
	last line of text	accepting should be rejecting
574	5 of Example 1	X should be boldface
	Fig. 17.4	In caption change 0.95 to 0.05 0.99 to 0.01 0.95 to 0.05
575	1	X should be boldface
	2	X in EX_i should not be boldface
	3	Right side of display β_1 should be β_i
578	Fig. 17.5	Bottom display within figure: $\hat{\beta}_2$ should be $\hat{\beta}_2 t$
580	4	E in EX should not be boldface
	7 from bottom	β should be $\overline{\beta}$
589-591		In page heading MEDIONS should be MEDIANS

properties of cdf with integrating density

If X has an integrating density function f, then,
(a) $P(X=t)=0$, for any number t.
(b) F is everywhere continuous.
(c) $P(a<X\leq b)=P(a\leq X\leq b)=P(a<X<b)=P(a\leq X<b)=F(b)-F(a)$, $(a<b)$.
(d) F is differentiable at any t where f is continuous. At such a point,

$$F'(t)=f(t) \tag{2}$$

Proof. That F is continuous is a basic property of the Riemann integral. Suppose $a<b$. Then, since $(X\leq b)=(X\leq a)\cup(a<X\leq b)$ is a union of disjoint events, it follows that

$$P(X\leq b)=P(X\leq a)+P(a<X\leq b)$$

and, hence

$$P(a<X\leq b)=P(X\leq b)-P(X\leq a)=F(b)-F(a) \tag{3}$$
$$=\int_a^b f(x)\,dx$$

If t is contained in $[a, b]$, then

$$P(X=t)\leq F(b)-F(a) \tag{4}$$

By continuity, the right side of (4) can be made as close to zero as we please by making $b-a$ sufficiently small; hence, $P(X=t)$ is less than any arbitrarily small number and must be zero. Notice that (a) and (b) are really asserting the same thing. The relation (3) proves the first part of (c). The remaining parts follow from (a). For instance,

$$P(a\leq X\leq b)=P(a<X\leq b)+P(X=a)=P(a<X\leq b)$$

Part (d) is part of the so-called *fundamental theorem* of integral calculus (Note 4, Chapter 6). ▮

All the cases described by (c) can be unified by the assertion that

$$P(X \text{ in } A)=\int_A f(x)\,dx \tag{5}$$

whenever A is an interval. (This interval may include or exclude either of its endpoints, or extend to $-\infty$ or ∞.) It follows that if A

is a *disjoint union of intervals,* (5) is still correct. Actually, (5) holds if A is any set on the real line over which the integral of f is defined.

Example 1 A number is selected at random from the unit interval [0, 1]. Denoting the number by X, its cdf is

$$F(t) = \begin{cases} 0 & \text{if } t < 0 \\ t & \text{if } 0 \leqslant t \leqslant 1 \\ 1 & \text{if } 1 < t \end{cases}$$

The integrating density function is

$$f(t) = \begin{cases} 1 & \text{if } 0 < t < 1 \\ 0 & \text{otherwise} \end{cases}$$

Notice that F is everywhere continuous and is differentiable except at the points 0 and 1, where f fails to be continuous. Except at these points, $F'(t) = f(t)$. For any number t, $F(t) = \int_{-\infty}^{t} f(x)\, dx$ (see Figure 7.7).

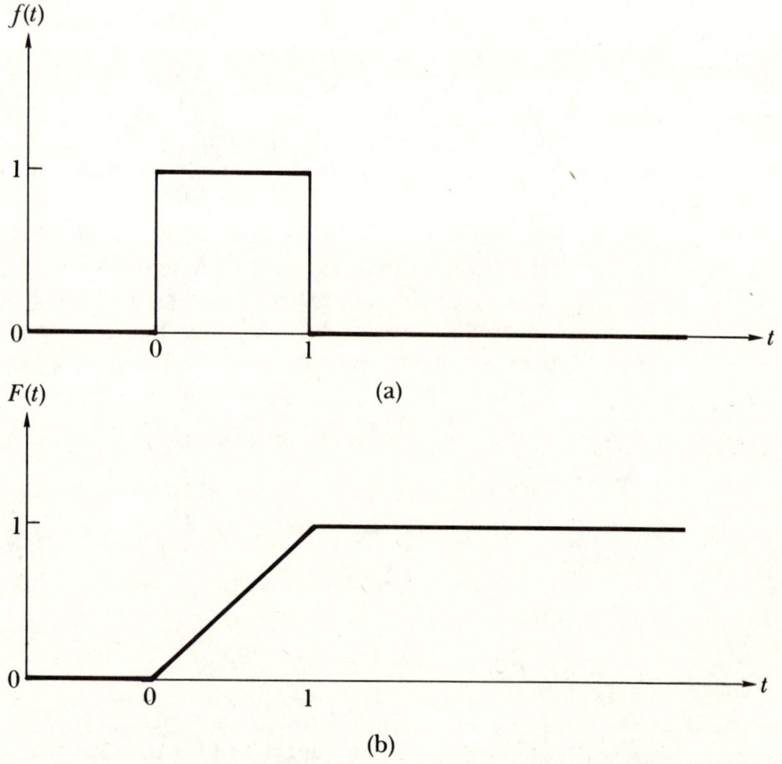

FIGURE 7.7 (a) Integrating density of a number selected at random from [0, 1]. (b) The cdf of a number selected at random from [0, 1].

Example 2 A number is selected at random from the unit interval [0, 1]. Suppose Z is the square of the number selected. Then, as shown in Example 3 of Section 7.3, the cdf of Z is

$$F(t) = \begin{cases} 0 & \text{if } t < 0 \\ \sqrt{t} & \text{if } 0 \leq t \leq 1 \\ 1 & \text{if } 1 < t \end{cases}$$

Because $\sqrt{t} = \int_0^t (1/2\sqrt{u})\, du$ when $0 \leq t \leq 1$, it follows that F has the integrating density function

$$f(t) = \begin{cases} 1/2\sqrt{t} & \text{if } 0 \leq t \leq 1 \\ 0 & \text{otherwise} \end{cases}$$

Notice, as asserted by properties (b) and (d), that F is everywhere continuous and is differentiable everywhere except at $t = 0$ and $t = 1$. This example shows that an integrating density function can exceed 1 in value, or it can even be unbounded, unlike a discrete density function, which cannot exceed 1 because $f(a) = P(X = a)$. To evaluate $P(1/4 < Z < 1/2)$, we compute

$$\int_{1/4}^{1/2} \frac{1}{2\sqrt{x}}\, dx = \left(\frac{1}{2}\right)^{1/2} - \left(\frac{1}{4}\right)^{1/2}$$

which is the *area under the density function* between 1/4 and 1/2. In general,

$$P(a < Z < b) = \int_a^b \frac{1}{2\sqrt{x}}\, dx = \sqrt{b} - \sqrt{a}, \qquad 0 \leq a < b \leq 1$$

which is the area under the density function between a and b.

Example 3 A particle has a lifetime described by the following integrating density function:

$$f(x) = \begin{cases} ae^{-ax} & \text{if } x \geq 0 \\ 0 & \text{otherwise} \end{cases}$$

where a is a positive constant. The probability that the particle survives beyond age t is, for positive t, equal to

$$\int_t^\infty ae^{-ax}\, dx = e^{-at}$$

If $0 < t_1 < t_2$, the probability that the particle survives beyond age t_1 but not beyond age t_2 is

$$\int_{t_1}^{t_2} ae^{-ax}\, dx = \exp(-at_1) - \exp(-at_2)$$

7.6 INTEGRATING DENSITY FUNCTIONS FOR SEVERAL RANDOM VARIABLES

The concept of integrating density function is also meaningful for several random variables. The definition for two random variables follows.

joint integrating density for two random variables

Suppose that X and Y are two random variables defined on the same sample space. They have a joint integrating density function f, if f is a nonnegative function such that

$$P[(X, Y) \text{ in } A] = \iint_A f(u, v)\, du\, dv \tag{1}$$

whenever A is any set in the (u, v) plane over which the integral of f is well defined.

As with the integrating density function for one random variable, it will suffice for our purposes to assume that f is Riemann integrable, and that the sets A are sets over which the Riemann integral of f is defined. The reader should be familiar with the theory and techniques of integrating functions of several variables (Note 8). If we take A to be the whole (u, v) plane, (1) implies that

$$\int_{-\infty}^{\infty}\int_{-\infty}^{\infty} f(u, v)\, du\, dv = 1$$

This is so because $(-\infty < X < \infty) \cap (-\infty < Y < \infty) = \mathfrak{X}$, the whole sample space. Thus, the integrating density function for two random variables is in spirit the same as that for one random variable. It is a nonnegative function having a total integral of 1. The probability that the random variables jointly take on values in a set is obtained by integrating the joint density function over that set (see Figure 7.8).

Example 1 Consider the experiment of selecting a point at random from the unit square, $0 \leq u \leq 1$, $0 \leq v \leq 1$. Suppose X is the u coordinate

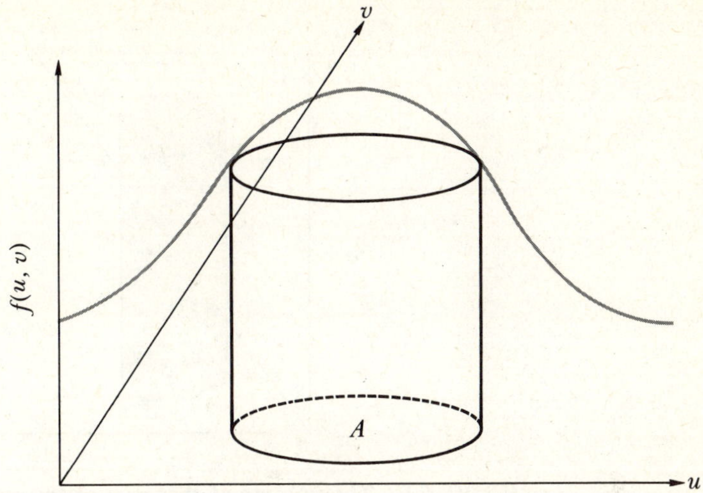

FIGURE 7.8 The volume beneath the surface that projects onto the set A is $\int\int_A f(u, v)\,du\,dv = P[(X, Y) \text{ in } A]$.

of the point selected and Y is the v coordinate. The usual model for this experiment is that X and Y have a joint integrating density function

$$f(u, v) = \begin{cases} 1 & \text{if } 0 \leqslant u \leqslant 1 \text{ and } 0 \leqslant v \leqslant 1 \\ 0 & \text{otherwise} \end{cases}$$

If A is a set within the unit square,

$$P[(X, Y) \text{ in } A] = \int\int_A f(u, v)\,du\,dv$$

is simply the area of A. If A is an arbitrary set in the plane, then $P[(X, Y) \text{ in } A]$ is the area of that part of A lying within the unit square (see Figure 7.9). The following are some probability evaluations relating to this experiment:

$$P(X \leqslant Y) = P(X < Y) = P(X > Y) = \tfrac{1}{2}$$

$$P(X^2 + Y^2 < 1) = \frac{\pi}{4}$$

$$P[(X < \tfrac{1}{2}) \cap (Y < \tfrac{1}{2})] = \tfrac{1}{4}$$

Example 2 A point is selected at random from the circle of radius 1, the center of which is at the origin; X and Y represent the coordinates of the point selected. The interpretation of "at random" is that the

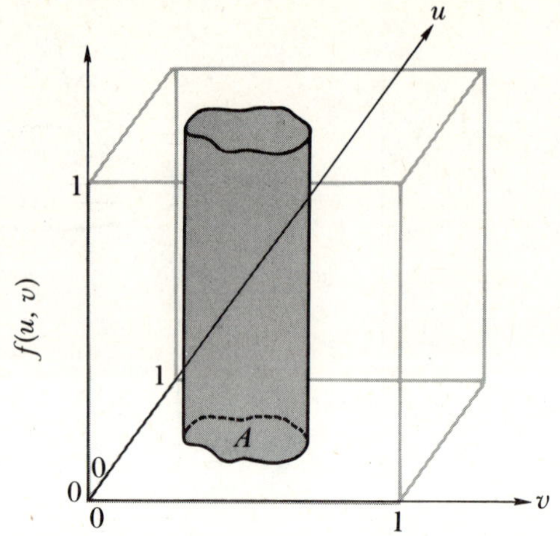

FIGURE 7.9 The volume of that part of the cube that projects down on *A* is *P*(*A*).

density function has the same value throughout the circle. That is,

$$f(u, v) = \begin{cases} \dfrac{1}{\pi} & \text{if } 0 \leq u^2 + v^2 \leq 1 \\ 0 & \text{otherwise} \end{cases}$$

(The constant must be $1/\pi$, in order that $\int\int f(u, v)\, du\, dv = 1$, where the integration is over the circle.) The following is another interpretation of the experiment. A marksman fires at the previously described circle. His projectile hits the target at random, meaning that if A is a region in the target, the probability of hitting somewhere in A is

$$\int\int_A \frac{1}{\pi}\, du\, dv = \frac{(\text{area of } A)}{\pi}$$

For instance, the probability of hitting within one half-unit of the origin is

$$\int\int_{(u^2+v^2)^{1/2} \leq 1/2} \frac{1}{\pi}\, du\, dv = \frac{\pi(1/2)^2}{\pi} = \frac{1}{4}$$

Example 3 Consider the target model described in Example 2, but we now deal with a more skillful marksman. The integrating density function describing his marksmanship assigns more mass toward the center of the target. Specifically, suppose the integrating density is the following:

$$f(u, v) = \begin{cases} \dfrac{2}{\pi}[1 - (u^2 + v^2)] & \text{if } 0 \leq u^2 + v^2 \leq 1 \\ 0 & \text{otherwise} \end{cases}$$

Using polar coordinates $\int\int (2/\pi)[1 - (u^2 + v^2)]\, du\, dv$ over the circle equals

$$\int_0^{2\pi} d\theta \int_0^1 \frac{2}{\pi}(1 - r^2)r\, dr = 1$$

Similarly, the probability that the hit is within c of the origin is

$$\iint_{(u^2+v^2)^{1/2} \leq c} f(u, v)\, du\, dr = \int_0^{2\pi} d\theta \int_0^c \frac{2}{\pi}(1 - r^2)r\, dr = c^2(2 - c^2)$$

In particular, the probability of hitting within a half-unit of the origin is

$$\left(\frac{1}{2}\right)^2\left[2 - \left(\frac{1}{2}\right)^2\right] = \frac{7}{16}$$

We define a joint density for n random variables in the same way as for two random variables.

definition of integrating density for n random variables

Suppose that $X_1, \ldots, X_n$ are n random variables defined over the same sample space. If f is a nonnegative function of n variables $u_1, \ldots, u_n$ such that

$$P[(X_1, \ldots, X_n) \text{ in } A] = \int \cdots \int_A f(u_1, \ldots, u_n)\, du_1 \cdots du_n$$

for all sets A in n-space over which the integral is defined, then f is called a *joint integrating density function* for the n random variables.

Taking A to be all of n space, we have

$$\int_{-\infty}^{\infty} \cdots \int_{-\infty}^{\infty} f(u_1, \ldots, u_n)\, du_1 \cdots du_n = 1$$

because $(-\infty < X_1 < \infty) \cap \cdots \cap (-\infty < X_n < \infty) = \mathfrak{X}$, the whole sample space. Thus, as in the other cases, an integrating density function for n random variables is a nonnegative function of n variables, whose total integral over all of n-space equals 1. The probability that $(X_1, \ldots, X_n)$ jointly lie in a set A equals the integral of the density function over that set.

Example 4 A point is picked at random from the unit cube in 3-space, by which we mean the set of coordinates u_1, u_2, u_3 satisfying $0 \leq u_i \leq 1$, $i = 1, 2, 3$. The interpretation of random selection is that the integrating density is

$$f(u_1, u_2, u_3) = \begin{cases} 1 & \text{if } (u_1, u_2, u_3) \text{ lies in the unit cube} \\ 0 & \text{otherwise} \end{cases}$$

The integral of f over the whole space is simply the volume of the cube, which is 1. In general, $P[(X_1, X_2, X_3) \text{ in } A]$ is the volume of that part of A which lies in the unit cube; X_1, X_2, X_3 represent the coordinates of the point selected. (The computation reduces to a volume only because the density is constant within the cube. For other densities, the integration is more complicated.) The following are some probabilities computed in terms of this density function.

(a) $P(X = 1/2) = 0$.

(b) $P(X_1 = X_2) = 0$ (the *volume* of a plane in 3-space is 0).

(c) $P(X_1 = X_2 = X_3) = 0$ (the *volume* of a line in 3-space is 0).

(d) $P(X_1 < X_2 < X_3) = 1/6$ [by symmetry, each of the six sets $(X_1 < X_2 < X_3)$, $(X_1 < X_3 < X_2)$, $(X_2 < X_1 < X_3)$, $(X_2 < X_3 < X_1)$, $(X_3 < X_1 < X_2)$, $(X_3 < X_2 < X_1)$ has the same probability. The union of these six sets falls short of being the whole sample space $\mathfrak{X}$ only by sets of the type described in (b) and (c), whose probabilities are zero].

Next we show how the integrating density functions of individual random variables are obtained from the joint integrating density function.

obtaining marginal densities from a joint density

Suppose that X and Y have a joint integrating density function. Then

$$\int_{-\infty}^{\infty} f(u, v)\, dv = g(u) \quad \text{is an integrating density function for } X$$

$$\int_{-\infty}^{\infty} f(u, v)\, du = h(v) \quad \text{is an integrating density function for } Y \tag{2}$$

Proof of (2). It will suffice to prove the first statement. Inasmuch as $P[(X, Y) \text{ in } A] = \iint_A f(u, v)\, du\, dv$, then, in particular,

$$P(X \leq t) = P[(X \leq t) \cap (Y < \infty)] = \int_{-\infty}^{t} du \int_{-\infty}^{\infty} f(u, v) dv$$

Hence, $g(u) = \int_{-\infty}^{\infty} f(u, v)\, dv$ is a nonnegative function such that $P(X \leq t) = \int_{-\infty}^{t} g(u)\, du$, and g is an integrating density function for X. ∎

As in the discrete case, the individual densities may be called *marginal densities* to emphasize that they have been obtained from the joint density function. Notice the counterpart of (2) in the discrete case [see (2) and (3) Section 4.2]. Suppose that $u_1, u_2, \ldots$ are the possible values of X, and $v_1, v_2, \ldots$ are the possible values of Y. Then

$$\sum_j P[(X = u_i) \cap (Y = v_j)] = P(X = u_i)$$

$$\sum_i P[(X = u_i) \cap (Y = v_j)] = P(Y = v_j)$$

In the same way, one obtains the integrating density function for X_1 when f is a joint integrating density function for n random variables $X_1, \ldots, X_n$. Hold u_1, the dummy variable relating to X_1, fixed and integrate out over all the remaining variables $u_2, \ldots, u_n$. In other words, set $u_1 = u$, $g(u) = \int f(u, u_2, \ldots, u_n)\, du_2 \cdots du_n$. The proof is the same as for two variables and will not be repeated.

More generally, suppose $X_1, \ldots, X_m, Y_1, \ldots, Y_n$ have a joint integrating density function $f(u_1, \ldots, u_m, v_1, \ldots, v_n)$. Then the joint integrating density of $X_1, \ldots, X_m$ is

$$g(u_1, \ldots, u_m) = \int_{-\infty}^{\infty} \cdots \int_{-\infty}^{\infty} f(u_1, \ldots, u_m, v_1, \ldots, v_n) dv_1 \cdots dv_n$$

That is, the dummy variables referring to $X_1, \ldots, X_m$ are held fixed and all other variables are integrated out. Again, it is worth pointing out that the counterpart of (3) in the discrete case is

$$P[(X_1 = u_1) \cap \cdots \cap (X_m = u_m)]$$

$$= \sum_{v_1} \cdots \sum_{v_n} P[(X_1 = u_1) \cap \cdots \cap (X_m = u_m) \cap (Y_1 = v_1) \cap \cdots \cap (Y_n = v_n)]$$

Example 5 A point is picked at random from the circle of radius 1 centered at the origin. As in Example 2, Let X and Y be the coordinates of the point selected. The joint density of X and Y equals $1/\pi$ within the circle and 0 outside. Let us find the marginal density function of X. If $-1 \leqslant u \leqslant 1$, then

$$\int_{-\infty}^{\infty} f(u, v)\, dv = \int_{-(1-u^2)^{1/2}}^{(1-u^2)^{1/2}} \frac{1}{\pi}\, dv = \frac{2}{\pi}(1 - u^2)^{1/2}$$

Hence, g, the marginal density of X, is

$$g(u) = \begin{cases} \dfrac{2}{\pi}(1 - u^2)^{1/2} & \text{if } -1 \leqslant u \leqslant 1 \\ 0 & \text{otherwise} \end{cases}$$

By a similar argument, the density function of Y has the same functional form. Incidentally, because g is an integrating density, we must have

$$\int_{-\infty}^{\infty} g(u)\, du = \int_{-1}^{1} \frac{2}{\pi}(1 - u^2)^{1/2}\, du = 1$$

which can also be verified directly.

7.7 INDEPENDENCE

Recall that two discrete random variables X and Y are called independent if

$$P[(X \text{ in } A) \cap (Y \text{ in } B)] = P(X \text{ in } A)P(Y \text{ in } B)$$

for arbitrary sets of real numbers A and B [(1), Section 4.7]. The definition of independence of X and Y on a general probability space will be the same, except for the restriction that the sets (X in A) and (Y in B) must be events—that is, they must be members of $\mathfrak{F}$. According to the definition in Section 7.2, X and Y are random variables only if (X in A) and (Y in B) are events whenever A and B are intervals. Hence, we give the following definition of independence.

independence of two random variables

Random variables X and Y, defined on a general probability space, are said to be independent if

$$P[(X \text{ in } A) \cap (Y \text{ in } B)] = P(X \text{ in } A)P(Y \text{ in } B) \qquad (1)$$

whenever A and B are intervals.

(As usual, an interval may or may not include its endpoints, or may extend to ∞ or $-\infty$.) It is easy to check that if (1) holds for intervals then it also holds when A and B are each unions of intervals. (Prove this.) As a consequence, (1) also holds when A and B are Borel sets, which, roughly speaking, are sets obtained from intervals by countably many set operations. We shall not prove these facts here.

The definition of independence of n random variables is similar.

independence of n random variables

Random variables $X_1, \ldots, X_n$, defined on a general probability space, are said to be mutually independent if

$$P[(X_1 \text{ in } A_1) \cap \cdots \cap (X_n \text{ in } A_n)] = P(X_1 \text{ in } A_1) \cdots P(X_n \text{ in } A_n) \quad (2)$$

whenever $A_1, \ldots, A_n$ are intervals.

In particular, if we choose $A_1 = (-\infty, t_1], \ldots, A_n = (-\infty, t_n]$, then (2) becomes,

$$P[(X_1 \leq t_1) \cap \cdots \cap (X_n \leq t_n)] = P(X_1 \leq t_1) \cdots P(X_n \leq t_n) \quad (3)$$

The left side of (3), as a function of the variables $t_1, \ldots, t_n$, is called the joint cdf of $X_1, \ldots, X_n$. Thus, (3) asserts that mutual independence of $X_1, \ldots, X_n$ implies that the joint cdf is the product of the individual cdf's. Actually, (3) is equivalent to (2), but we shall not prove it here.

In the discrete case, a useful criterion for independence of X and Y is that for all possible values u and v,

$$P[(X = u) \cap (Y = v)] = P(X = u)P(Y = v)$$

which says that the joint density function is the product of the individual density functions [(1), Section 4.9]. With integrating density functions, a similar criterion holds.

criterion for independence with integrating density functions

Suppose that $f_1, \ldots, f_n$ are integrating density functions for $X_1, \ldots, X_n$, respectively. Suppose also that f is a version of the joint integrating density of $X_1, \ldots, X_n$, such that

$$f(u_1, \ldots, u_n) = f_1(u_1) \cdots f_n(u_n) \quad \text{for all } u_1, \ldots, u_n \quad (4)$$

Then $X_1, \ldots, X_n$ are mutually independent.

Proof. Let $A_1, \ldots, A_n$ be intervals. Using (4), we have

$$P[(X_1 \text{ in } A_1) \cap \cdots \cap (X_n \text{ in } A_n)]$$

$$= \int \cdots \int_{\substack{u_1 \text{ in } A_1 \\ \vdots \\ u_n \text{ in } A_n}} f_1(u_1) \cdots f_n(u_n)\, du_1 \cdots du_n$$

$$= \left[\int_{A_1} f_1(u_1)\, du_1\right] \cdots \left[\int_{A_n} f_n(u_n)\, du_n\right]$$

$$= P(X_1 \text{ in } A_1) \cdots P(X_n \text{ in } A_n) \quad \blacksquare$$

There is a converse to the above criterion. Namely, if $X_1, \ldots, X_n$ are independent and have a joint integrating density f and individual integrating densities $f_1, \ldots, f_n$, then (4) must hold. This statement should be properly interpreted, however. Inasmuch as integrating densities are not uniquely determined, what is meant is that there are equivalent versions of each of the densities such that (4) holds. We shall not prove this assertion here, but we shall use it on several occasions.

We should point out that in most problems random variables will be independent because we shall stipulate that they are independent. In particular, with joint densities we shall stipulate that the densities satisfy (4). For applications, one should, of course, be careful that such an imposition of independence actually describes the "real world" situation.

Example 1 Two points are selected independently and at random from the unit interval [0, 1]. The mathematical model is that the points selected, X and Y, are independent random variables, each with the same integrating density,

$$g(u) = h(u) = \begin{cases} 1 & \text{if } 0 \leq u \leq 1 \\ 0 & \text{otherwise} \end{cases}$$

The joint density must then be

$$f(u, v) = \begin{cases} 1 & \text{if } 0 \leq u \leq 1 \text{ and } 0 \leq v \leq 1 \\ 0 & \text{otherwise} \end{cases}$$

Notice that f can be interpreted as the integrating density for the selection of a point at random from the unit square, $0 \leq u \leq 1$, $0 \leq v \leq 1$, as in Example 1, Section 7.6.

Example 2 (Example 2 is a continuation of Example 1.) Let $0 \leq t \leq 1$. What is the probability that the two points selected are within t of each other? That is, what is $P(|X - Y| \leq t)$? This probability is

$$\iint\limits_{|u-v| \leq t} f(u, v)\, du\, dv$$

which equals the area of the set depicted in Figure 7.10a. This area is $1 - (1 - t)^2$. Hence,

$$P(|X - Y| \leq t) = 1 - (1 - t)^2, \qquad 0 \leq t \leq 1$$

$$\frac{d}{dt}[1 - (1 - t)^2] = 2(1 - t)$$

$$1 - (1 - t)^2 = \int_0^t 2(1 - x)\, dx$$

Thus, $|X - Y|$, the distance between two points independently

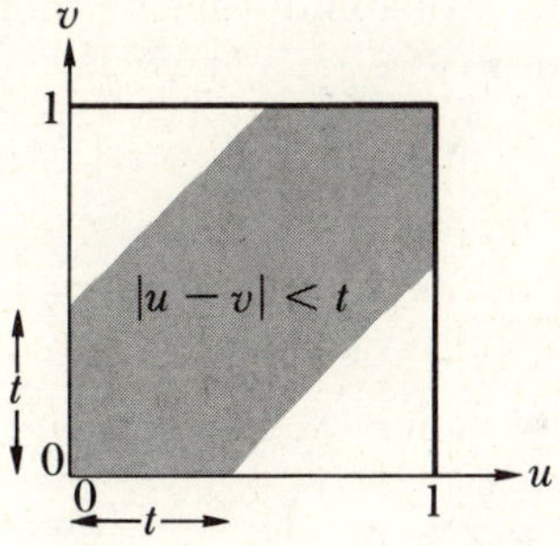

(a)

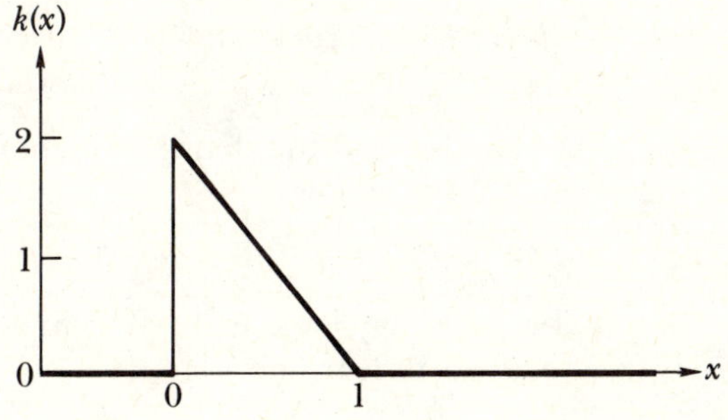

(b)

FIGURE 7.10 (a) The set in which $|X - Y| < t$. (b) The integrating density function of $|X - Y|$.

selected at random from [0, 1], has the integrating density function k given by

$$k(x) = \begin{cases} 2(1-x) & \text{if } 0 \leqslant x \leqslant 1 \\ 0 & \text{otherwise} \end{cases}$$

(See Figure 7.10b.)

Example 3 Two particles have lifetimes described by the following mathematical model. The two lifetimes are independent random variables, X and Y, each of which has the same integrating density function,

$$g(x) = h(x) = \begin{cases} ae^{-ax} & \text{if } x \geqslant 0 \\ 0 & \text{otherwise} \end{cases} \qquad a > 0$$

$$f(x, y) = \begin{cases} a^2e^{-a(x+y)} & \text{if } x \geqslant 0 \text{ and } y \geqslant 0 \\ 0 & \text{otherwise} \end{cases}$$

Let us now consider the probabilities of several events. For $t > 0$, the probability that neither particle has a lifetime of more than t units is

$$P[(X \leqslant t) \cap (Y \leqslant t)] = P(X \leqslant t)P(Y \leqslant t) \qquad \text{[by Eq. (3)]}$$

$$= \left(\int_0^t ae^{-ax}\, dx\right)^2 = (1 - e^{-at})^2$$

If $Z = \max(X, Y)$, then, in effect, the preceding relation shows that

$$P(Z \leqslant t) = \begin{cases} 0 & \text{if } t < 0 \\ (1 - e^{-at})^2 & \text{if } t \geqslant 0 \end{cases}$$

The density function of Z is then given by

$$\frac{dP(Z \leqslant t)}{dt} = \begin{cases} 2a(1 - e^{-at})e^{-at} & \text{if } t \geqslant 0 \\ 0 & \text{otherwise} \end{cases}$$

We can also find the density function of the *minimum* lifetime $W = \min(X, Y)$ as follows:

$$P(W \leqslant t) = 1 - P(W > t) = 1 - P[(X > t) \cap (Y > t)]$$

(This trick may look artificial at first, but you will be using it often so it is worth learning.) Hence,

$$P(W \leqslant t) = 1 - \left(\int_t^\infty ae^{-ax}\, dx\right)^2$$

$$= \begin{cases} 1 - (e^{-at})^2 = 1 - e^{-2at} & \text{if } t \geqslant 0 \\ 0 & \text{otherwise} \end{cases}$$

Hence, the density function of W is given by

$$\frac{dP(W \leq t)}{dt} = \begin{cases} 2ae^{-at} & \text{if } t \geq 0 \\ 0 & \text{if } t < 0 \end{cases}$$

All the consequences of independence described for discrete random variables in Section 4.8 hold for random variables on general probability spaces. For instance, consider the assertion that *subsets of mutually independent random variables are also mutually independent.* This statement holds in the present general setting. Also, if $X_1, \ldots, X_n$ are independent, and if $Y_1 = \phi_1(X_1), \ldots, Y_n = \phi_n(X_n)$ are well-defined random variables, then $Y_1, \ldots, Y_n$ are mutually independent.

Example 4 Suppose that X and Y are independent, each with the same integrating density,

$$f(x) = g(x) = \begin{cases} 1 & 0 \leq x \leq 1 \\ 0 & \text{otherwise} \end{cases}$$

Then X^2 and Y^2 are also independent and have the joint density,

$$f(x, y) = \begin{cases} \dfrac{1}{4x^{1/2}y^{1/2}} & \text{if } 0 \leq x \leq 1 \text{ and } 0 \leq y \leq 1 \\ 0 & \text{otherwise} \end{cases}$$

(see Example 2, Section 7.5).

7.8 MIXTURES OF DISCRETE AND INTEGRATING DENSITY FUNCTIONS

Up to now, two basic kinds of random variables have been considered—random variables with discrete density functions and random variables with integrating density functions. Although these types are those most commonly arising in practice, by no means do they exhaust all the possibilities, as the following example shows.

Example 1 Consider the following experiment. A coin is thrown. If it shows head, then a second coin with faces marked 1 and 2 is thrown. On the other hand, if the first coin shows tail, then a point is selected at random from the unit interval [0, 1]. Let X denote the number that appears as the final result of the experiment. What is the cdf of X? A simple calculation shows it to be

$$F(t) = P(X \le t) = \begin{cases} 0 & \text{if } t < 0 \\ t/2 & \text{if } 0 \le t < 1 \\ \frac{3}{4} & \text{if } 1 \le t < 2 \\ 1 & \text{if } 2 \le t \end{cases}$$

(see Figure 7.11). Then F is a "mixture" or "convex combination" of two cdf's G and H in the following sense. Let G be the cdf of the number that arises in throwing the second coin and let H be the cdf of the number selected from [0, 1]. Then,

$$G(t) = \begin{cases} 0 & \text{if } t < 1 \\ \frac{1}{2} & \text{if } 1 \le t < 2 \\ 1 & \text{if } 2 < t \end{cases} \qquad \text{and} \qquad H(t) = \begin{cases} 0 & \text{if } t < 0 \\ t & \text{if } 0 \le t \le 1 \\ 1 & \text{if } 1 < t \end{cases}$$

The cdf G is discrete, with density function given by

$$g(1) = g(2) = \tfrac{1}{2}$$

There is an integrating density function for H given by

$$h(t) = \begin{cases} 1 & \text{if } 0 \le t \le 1 \\ 0 & \text{otherwise} \end{cases}$$

Now $F(t) = (1/2)G(t) + (1/2)H(t)$, $-\infty < t < \infty$, which can be interpreted in the following way. When the original coin shows head, the cdf of the number that finally shows is G; when the coin shows tail, the cdf is H. Each of these cdf's arises with equal probabilities 1/2, 1/2, and the cdf for the composite experiment is the average of these cdf's.

In general, suppose that g is a discrete density function and h is an integrating density function. Suppose that $a_1, a_2, \ldots$ are the possible values on which g concentrates its probabilities. The cdf's determined by g and h are

$$G(t) = \sum_{a_i \le t} g(a_i) \qquad \text{and} \qquad H(t) = \int_{-\infty}^{t} h(x)\,dx$$

definition of mixture of cdf's

Let $0 \le p \le 1$. Then F, defined by

$$F(t) = pG(t) + (1 - p)H(t), \qquad -\infty < t < \infty \tag{1}$$

is also a cdf, which is called a mixture of G and H.

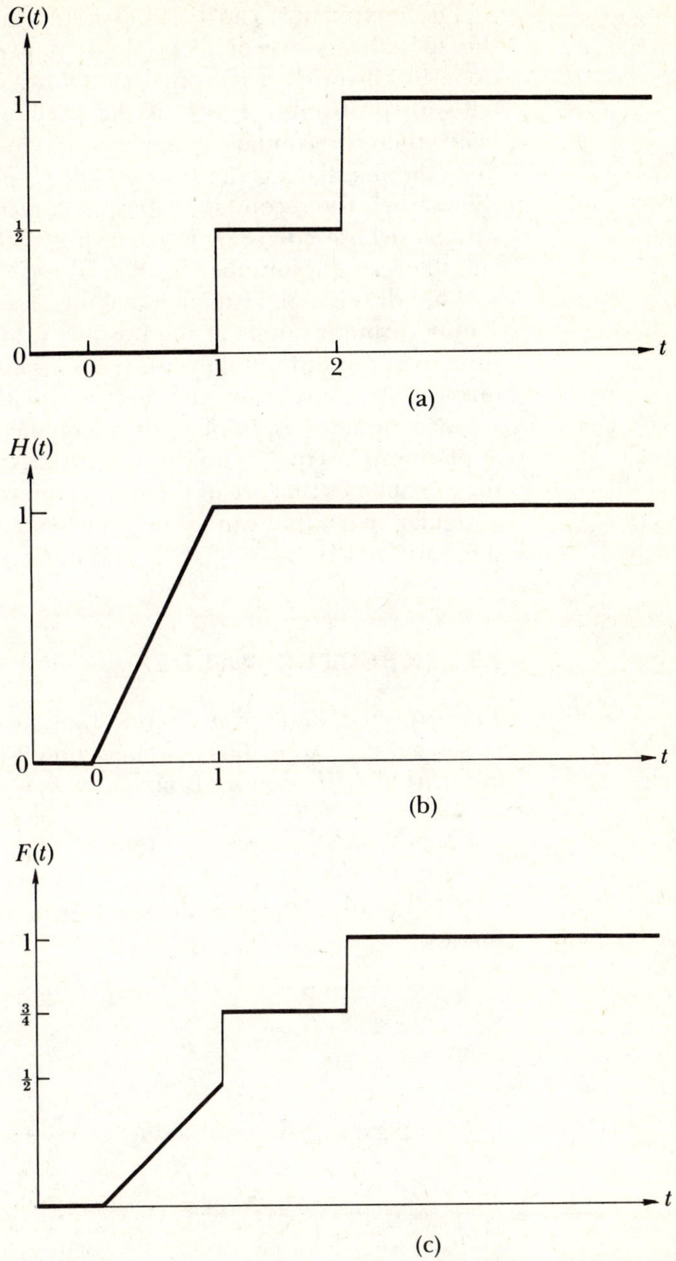

FIGURE 7.11 (a) G is the cdf of a random variable that equals 1 and 2 with probabilities 1/2, 1/2. (b) H is the cdf of a random variable that is uniformly distributed on [0, 1]. (c) The average of G and H is F.

The mixture (1) can be interpreted in much the same way as the special mixture in Example 1. A preliminary experiment is done by throwing a coin that shows head with probability p and tail with probability $1 - p$. If the preliminary experiment shows head, then a secondary experiment is done that produces a random variable having the discrete cdf G; on the other hand, if tail shows, then the secondary experiment is one producing a random variable whose cdf is H. Considering both possibilities head and tail, the resulting number has the cdf $pG + (1 - p)H$.

The cdf G is a step function that increases from height 0 to height 1 only through jumps at the possible values $a_1, a_2, \ldots$; H is continuous and is differentiable with $H'(t) = h(t)$ whenever h is continuous at t. By considering the mixtures of the form $pG + (1 - p)H$, a much broader class of cdf's is obtained that suffices for most applications. Actually, however, such mixtures do not exhaust all the possibilities that are of theoretical interest. There exist so-called "singular" cdf's that cannot be represented as such a mixture, but these will not be considered in this book.

7.9 EXPECTED VALUES

The expected value of a discrete random variable X was defined in Section 2.4. According to (1), Section 2.10, the evaluation of EX in terms of f, the density function of X, is

$$EX = \sum_i a_i P(X = a_i) = \sum_i a_i f(a_i)$$

If X has an integrating density function f, then EX is defined as follows.

definition of expected value for integrating density

$$EX = \int_{-\infty}^{\infty} x f(x)\, dx \tag{1}$$

This expected value is said to exist only if

$$\int_{-\infty}^{\infty} |x|\, f(x)\, dx < \infty$$

Notice the formal similarity between the sum $\Sigma_i\, a_i f(a_i)$ in the discrete case and the integral $\int_{-\infty}^{\infty} x f(x)\, dx$ in the integrating density case. The interpretation of EX as a center of gravity in the dis-

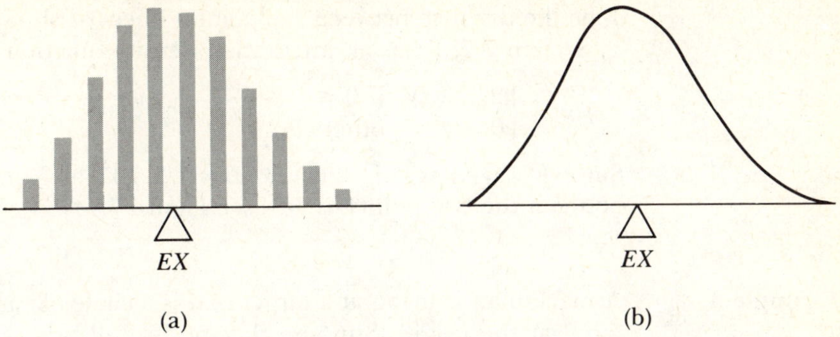

FIGURE 7.12 **(a) Center of gravity of discrete density. (b) Center of gravity of integrating density.**

crete case (Section 2.11) is also appropriate when there is an integrating density (see Figure 7.12).

Example 1 A point is picked at random from [0, 1]. The expected value of X, the numerical value of the point, is

$$\int_0^1 x\,dx = \tfrac{1}{2}$$

It is certainly plausible intuitively that 1/2 is the center of gravity of a uniform mass distribution on [0, 1] (see Figure 7.13).

Example 2 Two points are picked at random independently from [0, 1]. Let

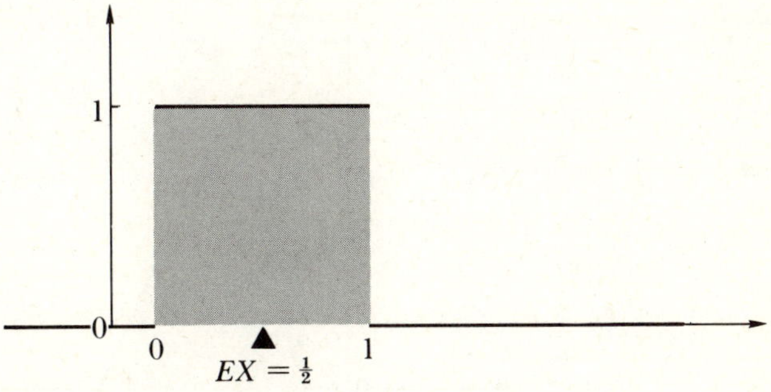

FIGURE 7.13

Y be the distance between the points. Then, as shown in Example 2, Section 7.7, Y has an integrating density function

$$f(x) = \begin{cases} 2(1-x) & \text{if } 0 \leqslant x \leqslant 1 \\ 0 & \text{otherwise} \end{cases}$$

Since $\int_{-\infty}^{\infty} xf(x)\,dx = \int_0^1 x2(1-x)\,dx = 1/3$, EY, the expected distance between the two points, is 1/3 (see Figure 7.14).

Example 3 A marksman is firing at a target that is a circle of radius one, centered at the origin. Suppose that he hits all points of the target randomly—that is, with equal likelihood (this example has already been discussed in Example 2, Section 7.6). Let Z be the distance of the point hit from the origin. Then, for $0 \leqslant t \leqslant 1$,

$$G(t) = P(Z \leqslant t) = \iint_{(u^2+v^2)^{1/2} \leqslant t} f(u, v)\,du\,dv = \frac{\pi t^2}{\pi} = t^2$$

Hence, Z has an integrating density, $g(t) = G'(t)$, which is

$$g(t) = \begin{cases} 2t & \text{if } 0 \leqslant t \leqslant 1 \\ 0 & \text{otherwise} \end{cases}$$

The expected distance of the hit from the origin is

$$EZ = \int_{-\infty}^{\infty} tg(t)\,dt = \int_0^1 t(2t)\,dt = \tfrac{2}{3}$$

Suppose now, more generally, that X has a cdf F, which is a mixture,

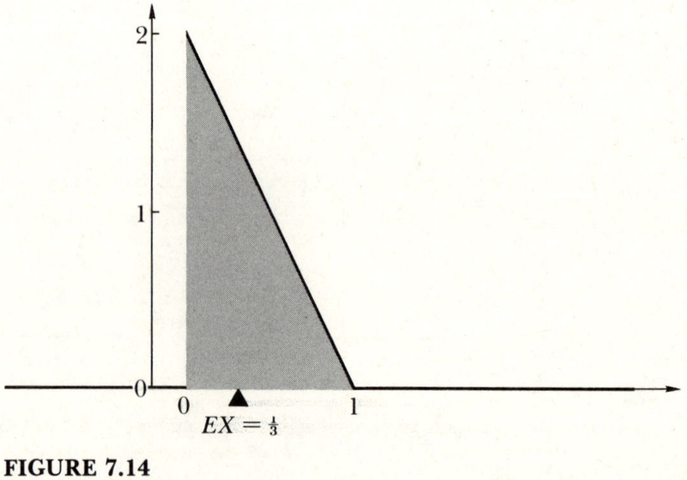

FIGURE 7.14

$$F(t) = pG(t) + (1-p)H(t) = p \sum_{a_i \leq t} g(a_i) + (1-p) \int_{-\infty}^{t} h(x)\, dx$$

The expected value of X is defined as follows.

expected value for a mixture

$$EX = p \sum_i a_i g(a_i) + (1-p) \int_{-\infty}^{\infty} xh(x)\, dx \qquad (2)$$

[The expected value is said to exist only if both

$$\sum_i |a_i| g(a_i) < \infty \quad \text{and} \quad \int_{-\infty}^{\infty} |x| h(x)\, dx < \infty]$$

In other words, if F is a mixture of G and H with weights p and $1-p$, then EX is the same mixture of the two expected values that are associated with G and H. A concise notation for (2) is

$$EX = \int_{-\infty}^{\infty} x\, dF(x) \qquad (3)$$

Denoting

$$\sum_i a_i g(a_i) = \int_{-\infty}^{\infty} x\, dG(x)$$

and

$$\int_{-\infty}^{\infty} xh(x)\, dx = \int_{-\infty}^{\infty} x\, dH(x)$$

then (3) is

$$= \int_{-\infty}^{\infty} x\, dF(x) = p \int_{-\infty}^{\infty} x\, dG(x) + (1-p) \int_{-\infty}^{\infty} x\, dH(x)$$

The reader familiar with the theory of *Stieltjes integration* will recognize that the right side of (3) is a *Stieltjes integral* (Note 9).

7.10 FUNCTIONS OF RANDOM VARIABLES

In Section 7.2, we gave a definition of a random variable that was appropriate for general probability spaces. The purpose of the present section is to point out that "nice" functions of random variables are still random variables. Our attempts to prove these assertions are more suggestive than rigorous. We do want to point out that a problem exists, however. Throughout the remainder of this book you will have to accept that any function of random variables, which we treat as a random variable, is indeed well behaved enough to deserve the name. Hopefully, in your next encounters with probability theory at a more abstract level, you will have these matters straightened out.

Suppose X is a random variable and g is a real-valued function defined over the values of X. Then by $g(X)$ is meant the function over the sample space $\mathfrak{X}$, which assigns to a sample point w the value $g[X(w)]$. For instance, $g(X) = X^2$ assigns to w the value $[X(w)]^2$, and $g(X) = 3X/(4 + X^2)$ assigns to w the value $3X(w)/\{4 + [X(w)]^2\}$. (Compare the discussion in Section 2.3 for discrete random variables.) Now what is really wanted is not only that $g(X)$ be well defined but that it also be a random variable. Recall, according to the definition in Section 7.2, that this statement means that $[g(X)$ in A$]$ is an event whenever A is an interval. In other words, the set of sample points w such that $g[X(w)] = a$, or $g[X(w)] \leq a$, or $a < g[X(w)] < b$, and so forth, must all be in $\mathscr{F}$, the algebra of events.

Example 1 If X is a random variable and c and d are constants, then $g(X) = cX + d$ is a random variable. If $c = 0$, then $g(X)$ is the constant d, which is certainly a random variable. Suppose for simplicity that $0 < c$. Then $[a < g(X) < b] = \{[(a-d)/c] < X < [(b-d)/c]\}$, which is an event. The same argument shows that $[g(X)$ in $A]$ is an event for the most general interval A. The other cases are handled in a similar manner.

Example 2 If X is a random variable then so is X^2. Consider, for example, the set $(X^2 \leq a)$. This set equals $(-\sqrt{a} \leq X \leq \sqrt{a})$ if $0 \leq a$ and is empty if $a < 0$. Hence, $(X^2 \leq a)$ is an event. In a similar way, one shows that $(X^2$ in $A)$ is an event for any interval A, and thus X^2 is a random variable.

Example 3 In the same way, one shows that X^y is a random variable for any real constant y, *as long as* X^y *is well defined.* In particular, if X is a nonnegative random variable, then $\sqrt{X}$ is a random variable also.

Example 4 Suppose X and Y are random variables; then $X + Y$ is a random

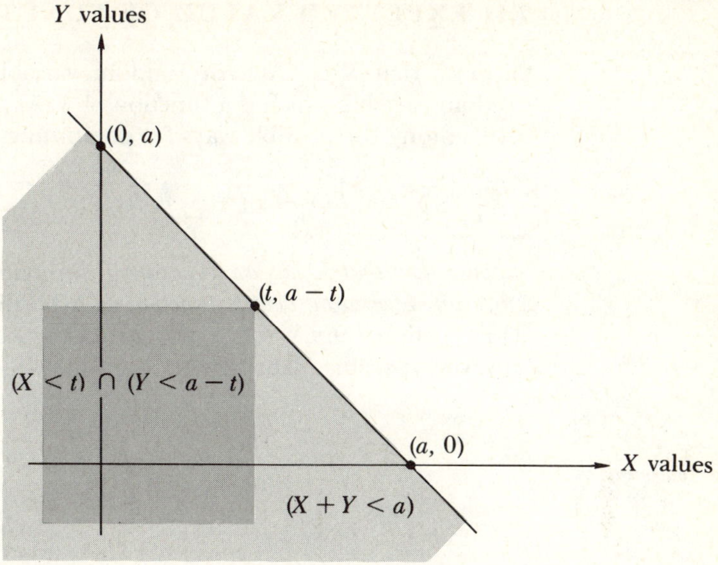

FIGURE 7.15 As t varies over the rational numbers, the union of the sets $(X < t) \cap (Y < a - t)$ equals $(X + Y < a)$.

variable also. Consider, for example, the set $(X + Y < a)$. The argument that this set is actually an event will now be outlined. For any real number t, $(X < t) \cap (Y < a - t)$ is an event, because it is an intersection of two events; this event is contained in the set $(X + Y < a)$ (see Figure 7.15). Now suppose that t varies over *all rational numbers.* The union of the intersecting sets $(X < t) \cap (Y < a - t)$ as t varies over the rationals is an event, because $\mathscr{F}$ is closed under countable unions. This union actually equals the set $(X - Y < a)$. (This fact should be plausible from Figure 7.15, but the details are not given here.) Thus, $(X + Y < a)$ is an event. Variations of this argument will show that $(X + Y \text{ in } A)$ is an event for any interval A.

Example 5 By repeated application of the fact that a sum of two random variables is a random variable, it follows that a sum of n random variables is also a random variable.

Example 6 A product of random variables, XY, is a random variable. More generally, $X_1X_2 \cdots X_n$, a product of n random variables, is a random variable.

7.11 EXPECTED VALUE OF FUNCTIONS

Suppose that X is a *discrete* random variable and $Y = h(X)$ is a random variable which is a function of X. As pointed out in Section 2.10, among the possible ways for computing EY are the following:

$$EY = \sum_i h(u_i)P(X = u_i) = \sum_i v_i P(Y = v_i) \tag{1}$$

Suppose now that X has an integrating density function f and also that Y has an integrating density function g. According to the definition, (1) in Section 7.9, $EY = \int_{-\infty}^{\infty} vg(v)\,dv$. Using Eq. (1) here as a guide, however, it seems plausible that the following is true.

expected value of $h(X)$

$$EY = E[h(X)] = \int_{-\infty}^{\infty} h(u)f(u)\,du = \int_{-\infty}^{\infty} vg(v)\,dv \tag{2}$$

It is a fact that (2) is correct, assuming that EY exists [recall that this means that $\int_{-\infty}^{\infty} |v|g(v)\,dv < \infty$]. That (2) is correct will not be rigorously proved here, but a heuristic argument will be given (Note 10).

Heuristic proof of (2). Express the integral $\int_{-\infty}^{\infty} h(u)f(u)\,du$ as the sum of integrals over the sets where $i/N \leqslant h(u) < (i+1)/N$, i varying over $0, \pm 1, \pm 2, \ldots$. For N large, this sum of integrals is close to $\Sigma_i\,(i/N)P[i/N \leqslant h(X) < (i+1)/N]$. Since $Y = h(X)$ has the density function g, $P[i/N \leqslant h(X) < (i+1)/N]$ is close to $g(i/N)(1/N)$. Hence, $\int_{-\infty}^{\infty} h(u)f(u)$ is close to $\Sigma_i\,(i/N)g(i/N)(1/N)$, which is itself an approximation to $\int_{-\infty}^{\infty} vg(v)\,dv$ (see Figure 7.16). ∎

Example 1 Suppose that X has the integrating density f,

$$f(u) = \begin{cases} 1 & \text{if } 0 \leqslant u \leqslant 1 \\ 0 & \text{otherwise} \end{cases}$$

Let $h(X) = X^3 = Y$. Then Y has the integrating density function

$$g(v) = \begin{cases} \dfrac{1}{3v^{2/3}} & \text{if } 0 \leqslant v \leqslant 1 \\ 0 & \text{otherwise} \end{cases}$$

which follows from the fact that

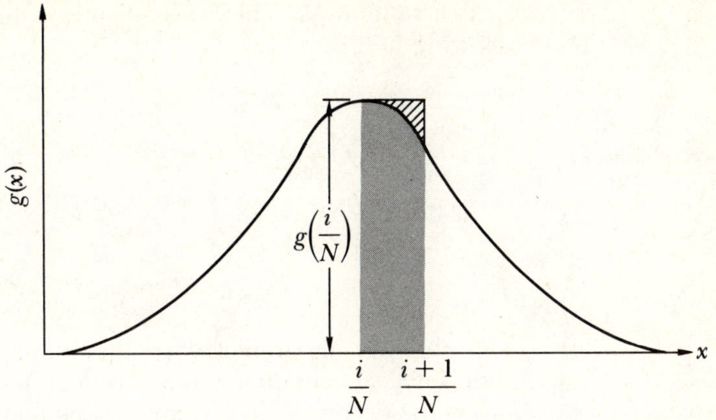

FIGURE 7.16 The value $g(i/N)/N$ is approximately equal to $\int_{i/N}^{(i+1)/N} g(x)\, dx$.

$$G(t) = P(X^3 \leq t) = P(X \leq t^{1/3}) = \begin{cases} 0 & \text{if } t < 0 \\ t^{1/3} & \text{if } 0 \leq t \leq 1 \\ 1 & \text{if } 1 < t \end{cases}$$

Hence,

$$g(t) = G'(t) = \begin{cases} \dfrac{1}{3t^{2/3}} & \text{if } 0 < t < 1 \\ 0 & \text{otherwise} \end{cases}$$

The evaluation of EY can now be done in two different ways. According to the definition, (1), Section 7.9,

$$EY = \int_0^1 v(3v^{2/3})^{-1}\, dv = \tfrac{1}{4}$$

According to (2), an alternative approach is to use the density function of X itself, rather than of X^3, giving

$$EY = \int_0^1 u^3\, du = \tfrac{1}{4}$$

In this example, the advantage of using (2) is that evaluating EY via $\int_0^1 u^3\, du$ does not require the preliminary determination of g, the density of X^3.

There is a version of (2) for several random variables. Suppose that $X_1, \ldots, X_n$ have a joint integrating density f, and $Y = h(X_1, \ldots,$

$X_n)$ is a random variable with an integrating density g. Then the following holds.

expected value of $h(X_1, \ldots, X_n)$

$$EY = \int \cdots \int h(u_1, \ldots, u_n) f(u_1, \ldots, u_n)\, du_1 \cdots du_n = \int_{-\infty}^{\infty} vg(v)\, dv \tag{3}$$

(The absolute integrability of either integral implies that of the other integral and hence is equivalent to the existence of EY.) The heuristic proof of (2) can be adapted word-for-word to (3), but, again, a rigorous proof will not be given here.

An important consequence of (3) is that if $X_1, \ldots, X_n$ have a joint integrating density and $EX_1, \ldots, EX_n$ all exist, then the following is true.

linearity of expected value

$$E(a_1X_1 + \cdots + a_nX_n) = a_1EX_1 + \cdots + a_nEX_n \tag{4}$$

For simplicity, the case $n = 2$ will be considered

$$E(aX + bY) = aEX + bEY \tag{5}$$

Proof of (5). It is a fact that $aX + bY$ has an integrating density (which will be proved in Section 8.4), but the actual form of this density does not enter into the proof. According to (3), taking $h(u_1, u_2) = au_1 + bu_2$,

$$E[h(X, Y)] = E(aX + bY) = \int_{-\infty}^{\infty} \int_{-\infty}^{\infty} (au_1 + bu_2) f(u_1, u_2)\, du_1\, du_2$$

Since by (2), Section 7.6, $\int_{-\infty}^{\infty} f(u_1, u_2)\, du_2$ and $\int_{-\infty}^{\infty} f(u_1, u_2)\, du_1$ are the density functions of X and Y, it follows that

$$\int_{-\infty}^{\infty} \int_{-\infty}^{\infty} (au_1 + bu_2) f(u_1, u_2)\, du_1\, du_2 = aEX + bEY$$

(The absolute integrability of $au_1 + bu_2$ follows from the existence of EX and EY, because $|au_1 + bu_2| \leq |a||u_1| + |b||u_2|$.) Except for replacing sums by integrals, the argument is really the same as in the discrete case, (1), Section 2.5. ▮

Example 2 Suppose that a point is picked at random from the two-dimensional set $0 \leqslant u_1 \leqslant 1$, $0 \leqslant u_2 \leqslant 1$, $0 \leqslant u_1 + u_2 \leqslant 1$ (see Figure 7.17). The mathematical model is that X and Y (the u_1 and u_2 coordinates of the point selected) have a joint integrating density that has a positive constant value over the described set and is 0 outside. That is,

$$f(u_1, u_2) = \begin{cases} 2 & \text{if } 0 \leqslant u_1 \leqslant 1,\ 0 \leqslant u_2 \leqslant 1,\ 0 \leqslant u_1 + u_2 \leqslant 1 \\ 0 & \text{otherwise} \end{cases}$$

(Why is the constant equal to 2?) We shall directly verify that $E(X + Y) = EX + EY$. To do so, we need the integrating density functions of $X + Y$ of X and of Y. It is easy to verify that

$$G(t) = P(X + Y \leqslant t) = \begin{cases} 0 & \text{if } t < 0 \\ t^2 & \text{if } 0 \leqslant t \leqslant 1 \\ 1 & \text{if } 1 < t \end{cases}$$

Hence,

$$g(t) = G'(t) = \begin{cases} 2t & \text{if } 0 \leqslant t \leqslant 1 \\ 0 & \text{otherwise} \end{cases}$$

We compute the marginal density functions of X and Y by (2), Section 7.6, as follows:

$$\text{marginal density of } X = h(x) = \int_{-\infty}^{\infty} f(x, y)\, dy$$

$$= \begin{cases} 2(1 - x) & \text{if } 0 \leqslant x \leqslant 1 \\ 0 & \text{otherwise} \end{cases}$$

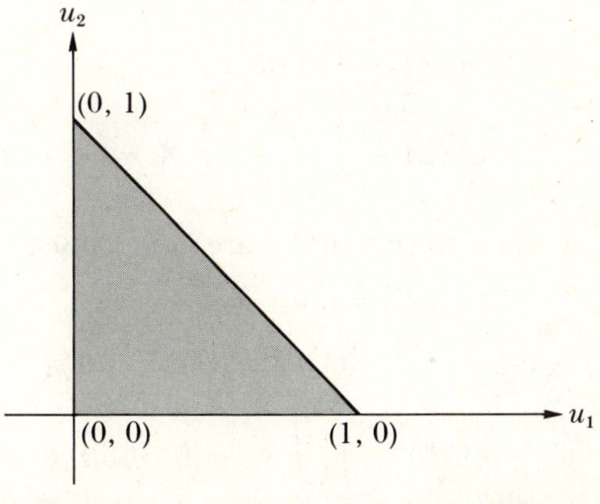

FIGURE 7.17 The set for which $0 \leqslant u_1 \leqslant 1$, $0 \leqslant u_2 \leqslant 1$, and $u_1 + u_2 \leqslant 1$.

$$\text{marginal density of } Y = k(y) = \int_{-\infty}^{\infty} f(x, y)\, dx$$

$$= \begin{cases} 2(1-y) & \text{if } 0 \leq y \leq 1 \\ 0 & \text{otherwise} \end{cases}$$

Using the density function of $X + Y$, we have

$$E(X+Y) = \int_0^1 t(2t)\, dt = \tfrac{2}{3}$$

On the other hand, using the densities of X and Y individually, we have

$$EX = \int_0^1 2x(1-x)\, dx = EY = \tfrac{1}{3}$$

Hence, $E(X+Y) = EX + EY$, as asserted by (5). The advantage of using (5) is that the computation of the density function of $X + Y$ becomes unnecessary for the computation of $E(X+Y)$. The purpose of this example is to verify (5) in a simple, special case.

As another application of (3), it will now be shown that if X and Y are *independent* and have a joint integrating density, and if $E(XY)$, EX, EY all exist, then

$$E(XY) = (EX)(EY) \tag{6}$$

[See (1), Section 4.8, for the earlier version of (6) for discrete random variables.]

Proof of (6). By (3),

$$E(XY) = \int_{-\infty}^{\infty}\int_{-\infty}^{\infty} u_1 u_2 f(u_1, u_2)\, du_1\, du_2$$

But, since X and Y are independent,

$$f(u_1, u_2) = f_1(u_1) f_2(u_2)$$

where f_1 is the density of X and f_2 is the density of Y. Hence,

$$E(XY) = \int_{-\infty}^{\infty}\int_{-\infty}^{\infty} u_1 u_2 f_1(u_1) f_2(u_2)\, du_1\, du_2$$

$$= \left(\int_{-\infty}^{\infty} u f_1(u)\, du\right)\left(\int_{-\infty}^{\infty} u f_2(u)\, du\right) = (EX)(EY) \qquad \blacksquare$$

Similarly, if $X_1, X_2, \ldots, X_n$ are mutually independent, then

$$E(X_1X_2 \cdots X_n) = (EX_1)(EX_2) \cdots (EX_n) \tag{7}$$

assuming that all expected values in (7) exist.

SUMMARY

In a *general probability space,* not every subset of sample points is designated to be an event. The events are a select collection of sets that are closed under countable set operations. This collection is called *the algebra* $\mathscr{F}$. Events are sets to which probability values are assigned, and the probability measure satisfies the same requirements as in discrete sample spaces.

On general probability spaces, a *random variable* X is a function such that (X in A) is an event whenever A is an interval; hence, $P(X \text{ in } A)$ is well defined. The *cumulative distribution function (cdf)* of a random variable X is defined by $F(t) = P(X \leq t)$. A cdf increases from height 0 at $-\infty$ to height 1 at ∞ and is right continuous.

A nonnegative function f is an *integrating density function* for X if $P(X \leq t) = \int_{-\infty}^{t} f(x)\, dx$. Necessarily, $\int_{-\infty}^{\infty} f(x)\, dx = 1$. If there is an integrating density, then the cdf

$$F(t) = P(X \leq t) = \int_{-\infty}^{t} f(x)\, dx$$

is a continuous function of t. If X is discrete, then $F(t)$ is a step function with jumps at the possible values. The jump at a equals $P(X = a)$.

A random variable need not be discrete, nor need it have an integrating density function; it can be a *mixture* of these types.

A nonnegative function f is an *integrating density function of n random variables* $X_1, \ldots, X_n$ if

$$P[(X_1, \ldots, X_n) \text{ in } A] = \int \cdots \int_A f(u_1, \ldots, u_n)\, du_1 \cdots du_n$$

If X has an integrating density f, then

$$EX = \int_{-\infty}^{\infty} uf(u)\, du$$

If $Y = h(X)$ also has an integrating density g, then

$$EY = \int_{-\infty}^{\infty} h(u)f(u)\, du = \int_{-\infty}^{\infty} vg(v)\, dv$$

Similarly, if $Y = h(X_1, \ldots, X_n)$ has an integrating density g, then

$$EY = \int_{-\infty}^{\infty} \cdots \int_{-\infty}^{\infty} h(u_1, \ldots, u_n)\, du_1 \cdots du_n = \int_{-\infty}^{\infty} vg(v)\, dv$$

Just as in the discrete case, if $X_1, \ldots, X_n$ has a joint integrating density function, then

$$E(a_1X_1 + \cdots + a_nX_n) = a_1EX_1 + \cdots + a_nEX_n$$

If X and Y are independent random variables with integrating densities, then

$$E(XY) = (EX)(EY)$$

again, just as in the discrete case, assuming the expected values exist.

EXERCISES

1 A point is picked at random between -1 and 2. If the number selected is negative, it is rounded off to 0; if it is greater than 1, it is rounded off to 1; if it is between 0 and 1, it is not changed.
(a) Show that the cdf of the final number selected is

$$F(x) = \begin{cases} 0 & \text{if } x < 0 \\ \dfrac{1+x}{3} & \text{if } 0 \leq x < 1 \\ 1 & \text{if } 1 \leq x \end{cases}$$

(b) Represent F as a mixture of a discrete cdf and a continuous cdf.
(c) Show that the expected value of the number selected is 1/2. [Use (2), Section 7.9.]

2 Consider the following composite experiment. First a die is thrown. If the die shows 1 or 2, a number is selected at random from the interval [0, 1]. If the die shows 3, 4, 5, or 6, a number is selected at random from the interval [1, 2]. Show that the cdf of the number selected is

$$F(x) = \begin{cases} 0 & \text{if } x < 0 \\ \dfrac{x}{3} & \text{if } 0 \leq x < 1 \\ \dfrac{2x}{3} - \dfrac{1}{3} & \text{if } 1 \leq x < 2 \\ 1 & \text{if } 2 \leq x \end{cases}$$

Notice that F is continuous. What is the integrating density function?

3 The computations in some of the earlier examples should be treated as exercises. Make sure that you can do them, for they will serve as models for some of the exercises that follow. In particular, review the following:
(a) Examples 2 and 3, Section 7.5
(b) Examples 1–5, Section 7.6
(c) Examples 1–4, Section 7.7
(d) Examples 1–3, Section 7.9
(e) Examples 1 and 2, Section 7.11

4 Suppose that X has the integrating density function

$$f(x) = \begin{cases} 1 & \text{if } 0 \leq x \leq 1 \\ 0 & \text{otherwise} \end{cases}$$

(a) If c is a positive constant, show that cX has the integrating density function

$$g(x) = \begin{cases} \dfrac{1}{c} & \text{if } 0 \leq x \leq c \\ 0 & \text{otherwise} \end{cases}$$

(b) Show that X^4 has the integrating density function

$$g(x) = \begin{cases} \dfrac{1}{4x^{3/4}} & \text{if } 0 < x \leq 1 \\ 0 & \text{otherwise} \end{cases}$$

(c) Show that $X^{1/2}$ has the integrating density function

$$g(x) = \begin{cases} 2x & \text{if } 0 \leq x \leq 1 \\ 0 & \text{otherwise} \end{cases}$$

(d) Show that $1/X$ has the integrating density function

$$g(x) = \begin{cases} \dfrac{1}{x^2} & \text{if } 1 < x \\ 0 & \text{otherwise} \end{cases}$$

(e) Show that $-\log X$ (base e) has the integrating density function

$$g(x) = \begin{cases} e^{-x} & \text{if } 0 \leq x \\ 0 & \text{if } x < 0 \end{cases}$$

(f) Show that $X/(1+X)$ has the integrating density function

$$g(x) = \begin{cases} \dfrac{1}{(1-x)^2} & \text{if } 0 \leq x \leq 1/2 \\ 0 & \text{otherwise} \end{cases}$$

[Hint: The style of determining g in every case is about the same. We will outline part (d). First find cdf of $1/X$.

$$P\left(\frac{1}{X} \leq t\right) = \begin{cases} 0 & \text{if } t < 0 \\ P\left(X \geq \frac{1}{t}\right) = 1 - \frac{1}{t} & \text{if } t \geq 1 \\ 0 & \text{if } 0 \leq t < 1 \end{cases}$$

Now differentiate the cdf.]

5 Suppose that X has the integrating density function

$$f(x) = \begin{cases} e^{-x} & \text{if } 0 \leq x \\ 0 & \text{if } x < 0 \end{cases}$$

(a) If c is a positive constant, show that X/c has the integrating density function

$$g(x) = \begin{cases} ce^{-cx} & \text{if } 0 \leq x \\ 0 & \text{if } x < 0 \end{cases}$$

(b) Show that $X/(1+X)$ has the integrating density function

$$g(x) = \begin{cases} \dfrac{e^{-x/(1-x)}}{(1-x)^2} & \text{if } 0 \leq x < 1 \\ 0 & \text{otherwise} \end{cases}$$

(c) Show that $X + c$ has the integrating density function

$$g(x) = \begin{cases} e^{-(x-c)} & \text{if } c \leq x \\ 0 & \text{if } x < c \end{cases}$$

(d) Show that $-X$ has the integrating density function

$$g(x) = \begin{cases} e^{x} & \text{if } x \leq 0 \\ 0 & \text{if } 0 < x \end{cases}$$

[Hint for (b):

$$P\left(\frac{X}{1+X} \leq t\right) = \begin{cases} 0 & \text{if } t \leq 0 \\ P\left(X \leq \frac{t}{1-t}\right) & \text{if } 0 < t < 1 \\ 1 & \text{if } t < 1 \end{cases}$$

Now differentiate this cdf.]

6 Suppose that X has any of the following three integrating density functions:

$$f(x) = \frac{1}{\pi(1+x^2)} \quad \text{if } -\infty < x < \infty$$

$$f(x) = \begin{cases} \dfrac{1}{(1+x)^2} & \text{if } 0 \leq x \\ 0 & \text{if } x < 0 \end{cases}$$

$$f(x) = \begin{cases} \frac{1}{2} & \text{if } 0 \leqslant x \leqslant 1 \\ (1/2x^2) & \text{if } 1 < x \\ 0 & \text{if } x < 0 \end{cases}$$

Show that $1/X$ has exactly the same integrating density function as X does. [Hint for the second case:

$$P\left(\frac{1}{X} \leqslant t\right) = \begin{cases} 0 & t < 0 \\ P\left(X \geqslant \frac{1}{t}\right) = \int_{1/t}^{\infty} \frac{1}{(1+x)^2}\,dx = \frac{t}{1+t}, & t \geqslant 0 \end{cases}$$

Now differentiate this cdf.]

7 Suppose that X is a positive random variable with integrating density function f. Show that X has the same density as does $1/X$ if, and only if,

$$f\left(\frac{1}{x}\right) = x^2 f(x), \qquad x \text{ is positive}$$

Check that this condition is satisfied by the densities in Exercise 6.

8 Suppose X is a positive random variable with integrating density f. Show that $X/(1+X)$ has the integrating density given by

$$g(x) = \begin{cases} f\left(\frac{x}{1-x}\right) \Big/ (1-x)^2 & \text{if } 0 < x < 1 \\ 0 & \text{otherwise} \end{cases}$$

Check that this function agrees with the special cases in Exercises 4f and 5b.

9 Suppose that X is a random variable that lies between 0 and 1 with probability 1, and that it has an integrating density f. Show that $X/(1-X)$ has the integrating density given by

$$g(x) = \begin{cases} f\left(\frac{x}{1+x}\right) \Big/ (1+x)^2 & \text{if } 0 < x \\ 0 & \text{otherwise} \end{cases}$$

10 Suppose that X has the integrating density function f.

(a) If c is a positive constant, show that the density function of cX is

$$g(x) = \left(\frac{1}{c}\right) f\left(\frac{x}{c}\right)$$

(b) Show that the density function of X^2 is

$$g(x) = \begin{cases} \dfrac{f(\sqrt{x}) + f(-\sqrt{x})}{2\sqrt{x}} & \text{if } 0 < x \\ 0 & \text{otherwise} \end{cases}$$

(c) Show that the density function of $|X|^{1/2}$ is

$$g(x) = \begin{cases} 2x[f(x^2) + f(-x^2)] & \text{if } 0 \leqslant x \\ 0 & \text{otherwise} \end{cases}$$

(d) Show that the density function of $1/|X|$ is

$$g(x) = \begin{cases} \dfrac{f(1/x) + f(-1/x)}{x^2} & \text{if } 0 \leqslant x \\ 0 & \text{otherwise} \end{cases}$$

Check that some of the special cases in Exercises 4 and 5 agree with the preceding results.

11 Suppose X and Y are independent random variables, each with the same density function

$$f(x) = \begin{cases} 1 & \text{if } 0 \leqslant x \leqslant 1 \\ 0 & \text{otherwise} \end{cases}$$

(a) Show that XY has the density function

$$g(x) = \begin{cases} -\log x & \text{if } 0 < x \leqslant 1 \\ 0 & \text{otherwise} \end{cases}$$

(b) Show that $X + Y$ has the density function

$$g(x) = \begin{cases} x & \text{if } 0 \leqslant x < 1 \\ 2 - x & \text{if } 1 \leqslant x \leqslant 2 \\ 0 & \text{otherwise} \end{cases}$$

(c) Show that X/Y has the density function

$$g(x) = \begin{cases} \dfrac{1}{2} & \text{if } 0 \leqslant x \leqslant 1 \\ \dfrac{1}{2x^2} & \text{if } 1 < x \\ 0 & \text{otherwise} \end{cases}$$

(d) Show that $\max(X, Y)$ has the density function

$$g(x) = \begin{cases} 2x & \text{if } 0 \leqslant x \leqslant 1 \\ 0 & \text{otherwise} \end{cases}$$

(e) Show that $\min(X, Y)$ has the density function

$$g(x) = \begin{cases} 2(1 - x) & \text{if } 0 \leqslant x \leqslant 1 \\ 0 & \text{otherwise} \end{cases}$$

(f) Show that $[\min(X, Y)]/[\max(X, Y)]$ has the density function

$$g(x) = \begin{cases} 1 & \text{if } 0 \leqslant x \leqslant 1 \\ 0 & \text{otherwise} \end{cases}$$

The calculations in each of the preceding parts are all about the same.

We shall outline the solution to (c) as a typical case. First realize that for this density, $P[(X, Y) \text{ in } A]$ equals the area of A where A is a region in the unit box, because the joint density function of X and Y equals 1 inside the unit box and 0 elsewhere (compare Example 1, Section 7.6 and Figure 7.9). Now let us compute the cdf of X/Y. Inasmuch as $P[(X/Y) \leq t] = 0$ if $t < 0$, we may as well suppose that $t \geq 0$ in what follows. Use Figure 7.18 to verify the following:

$$P\left(\frac{X}{Y} \leq t\right) = \begin{cases} \dfrac{t}{2} & \text{if } t \leq 1 \\ 1 - \dfrac{1}{2t} & \text{if } t > 1 \end{cases}$$

Now differentiate to obtain the density function.

12 Suppose X and Y are independent random variables, each with the same density function

$$f(x) = \begin{cases} e^{-x} & \text{if } 0 \leq x \\ 0 & \text{otherwise} \end{cases}$$

(a) Show that X/Y has the density function

$$g(x) = \begin{cases} \dfrac{1}{(1 + x)^2} & \text{if } 0 \leq x \\ 0 & \text{otherwise} \end{cases}$$

(b) Show that $X + Y$ has the density function

$$g(x) = \begin{cases} xe^{-x} & \text{if } 0 \leq x \\ 0 & \text{otherwise} \end{cases}$$

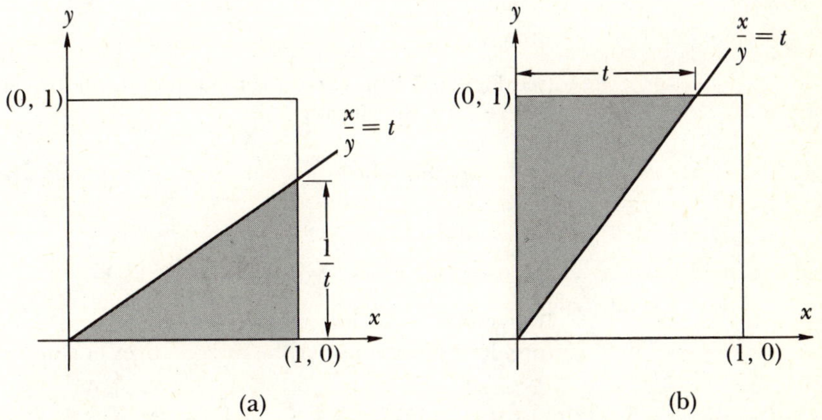

FIGURE 7.18 (a) $t > 1$. The set for which $x/y > t$. Its area is $1/2t$. (b) $t < 1$. The set for which $x/y < t$. Its area is $t/2$.

(c) Show that $X - Y$ has the density function

$$g(x) = \left(\frac{1}{2}\right)e^{-|x|}, \qquad \text{all } x$$

(d) Show that $\max(X, Y)$ has the density function

$$g(x) = \begin{cases} 2(1 - e^{-x})e^{-x} & \text{if } 0 \leqslant x \\ 0 & \text{otherwise} \end{cases}$$

(e) Show that $\min(X, Y)$ has the density function

$$g(x) = \begin{cases} 2e^{-2x} & \text{if } 0 \leqslant x \\ 0 & \text{otherwise} \end{cases}$$

(f) Show that $[\min(X, Y)]/[\max(X, Y)]$ has the density function

$$g(x) = \begin{cases} \dfrac{2}{(1 + x)^2} & \text{if } 0 \leqslant x \leqslant 1 \\ 0 & \text{otherwise} \end{cases}$$

The calculations in each of the preceding parts are all about the same. We shall outline the solution to (f) as a typical special case. The joint density of X and Y is

$$f(x, y) = \begin{cases} e^{-(x+y)} & \text{if } x \geqslant 0 \text{ and } y \geqslant 0 \\ 0 & \text{otherwise} \end{cases}$$

Let $U = \min(X, Y)/\max(X, Y)$. Suppose U is between 0 and 1 with probability 1. Let us suppose that $0 \leqslant t \leqslant 1$. We shall evaluate $P(U \leqslant t)$.

$$P(U \leqslant t) = \iint_A e^{-(x+y)}\, dx\, dy$$

where A is the set of points in the first quadrant, in which

$$\frac{\min(x, y)}{\max(x, y)} < t$$

(see Figure 7.19). Then A is the union of the two shaded regions in Figure 7.19. Specifically,

$$A = (\text{points where } \frac{x}{y} \leqslant t \text{ and } x \leqslant y) \cup (\text{points where } \frac{y}{x} \leqslant t$$

$$\text{and } y \leqslant x)$$

$$= A_1 \cup A_2$$

By symmetry, the integral over each of A_1 and A_2 is the same; therefore, let us just consider the integral over one of these regions.

$$\iint_{A_2} e^{-(x+y)}\, dx\, dy = \int_0^\infty e^{-x}\, dx \int_0^{tx} e^{-y}\, dy = \int_0^\infty (1 - e^{-tx})e^{-x}\, dx = \frac{t}{1 + t}$$

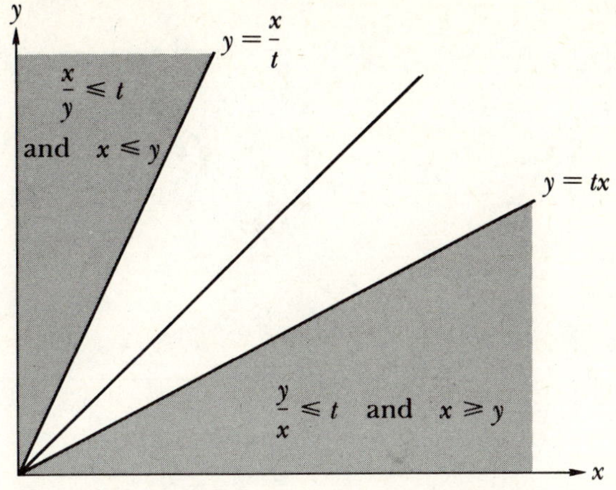

FIGURE 7.19

Hence, $P(U \leqslant t) = 2t/(1 + t)$, for $0 \leqslant t \leqslant 1$. Now differentiate the cdf to obtain the density function.

13 Suppose X and Y are independent random variables, each with the same density function

$$f(x) = \begin{cases} 1 & \text{if } 0 \leqslant x \leqslant 1 \\ 0 & \text{otherwise} \end{cases}$$

(a) Let $U = \max(X, Y)$, $V = \min(X, Y)$. Show that the joint integrating density function of U, V is

$$g(u, v) = \begin{cases} 2 & \text{if } 0 \leqslant v \leqslant u \leqslant 1 \\ 0 & \text{otherwise} \end{cases}$$

(b) Evaluate $\int_{-\infty}^{\infty} g(x, y)\, dy$, $\int_{-\infty}^{\infty} g(x, y)\, dx$ to obtain the densities of U and V individually. [Compare with (d) and (e) of Exercise 11.]

(c) Use (a) to show that the density of $\max(X, Y) - \min(X, Y)$ is

$$g(x) = \begin{cases} 2(1 - x) & \text{if } 0 \leqslant x \leqslant 1 \\ 0 & \text{otherwise} \end{cases}$$

(d) Show that each of $\min(X, Y)$, $\max(X, Y) - \min(X, Y)$, and $1 - \max(X, Y)$ has the same density function.

{Outline for (a): Consider a rectangle A in the u, v plane where

$$a \leqslant u \leqslant a + \Delta, \qquad b \leqslant v \leqslant b + \Delta$$

and

$$a \leqslant a + \Delta \leqslant b \leqslant b + \Delta$$

(See Figure 7.20a.)

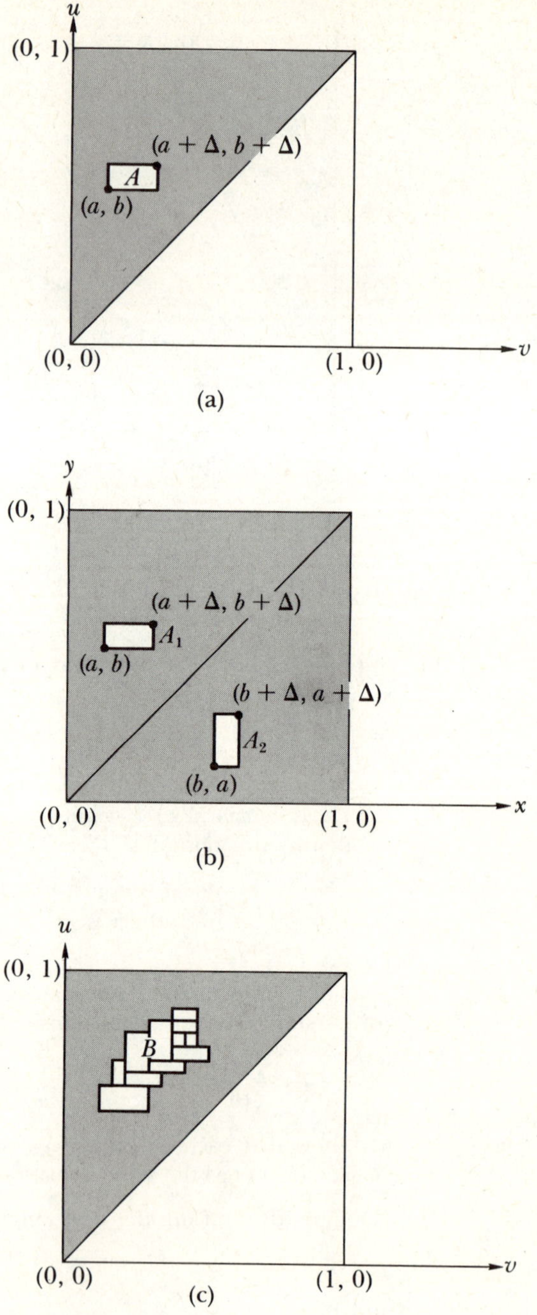
u
(0, 1)
(a + Δ, b + Δ)
A
(a, b)
(0, 0)
(1, 0)
v
(a)
y
(0, 1)
(a + Δ, b + Δ)
A1
(a, b)
(b + Δ, a + Δ)
A2
(b, a)
(0, 0)
(1, 0)
x
(b)
u
(0, 1)
B
(0, 0)
(1, 0)
v
(c)

FIGURE 7.20

$$P[(U,V) \text{ in } A] = P[(a \leq X \leq a+\Delta) \cap (b \leq Y \leq b+\Delta) \cap (X \leq Y)]$$
$$+ P[(a \leq Y \leq a+\Delta) \cap (b \leq X \leq b+\Delta) \cap (Y \leq X)]$$
$$= \iint_{A_1 \cup A_2} 1\, dx\, dy = 2\Delta^2 = 2 \text{ area of } A$$

(See Figure 7.20b.)}

Now suppose that B is a disjoint union of squares like A (see Figure 7.20c). It follows that

$$P[(U, V) \text{ in } B] = 2 \text{ (sum of areas of squares that make up } B)$$
$$= 2 \text{ (area of } B)$$

By continuity, then,

$$P[(U, V) \text{ in } B] = 2 \text{ (area of } B)$$

whenever the area is well defined. Hence, the joint density function of U, V is

$$g(u, v) = \begin{cases} 2 & \text{if } 0 \leq u \leq v \leq 1 \\ 0 & \text{otherwise} \end{cases}$$

[Outline for (c): Let $0 \leq t \leq 1$.

$$P(U - V \leq t) = \iint_A 2\, du\, dv$$

where A is the set shown in Figure 7.21.

$$P(U - V \leq t) = 1 - (1 - t)^2$$

Now differentiate to obtain the density function.]

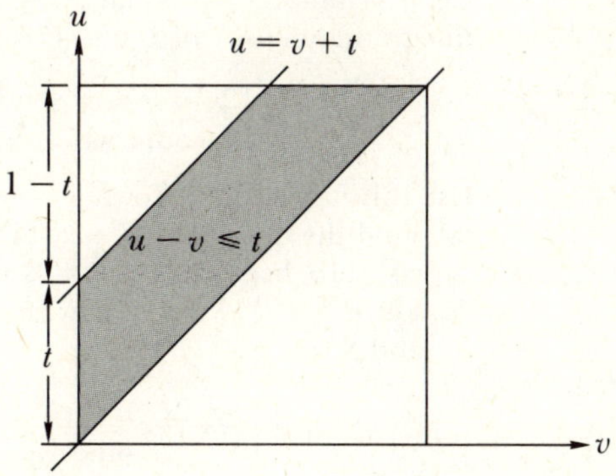

FIGURE 7.21

14 Suppose X and Y are independent random variables each with the exponential density

$$f(x) = \begin{cases} e^{-x} & \text{if } x \geqslant 0 \\ 0 & \text{if } x < 0 \end{cases}$$

(a) Let $U = \min(X, Y)$, $V = \max(X, Y)$. Show that the joint integrating density of U and V is

$$g(u, v) = \begin{cases} 2e^{-(u+v)} & \text{if } 0 \leqslant u \leqslant v \\ 0 & \text{otherwise} \end{cases}$$

(Hint: Follow the outline of the solution for Exercise 13a.)

(b) Show that

$$\int_{-\infty}^{\infty} g(u, v)\, dv \quad \text{and} \quad \int_{-\infty}^{\infty} g(u, v)\, du$$

give the densities already obtained in Exercise 12d and e.

(c) Find the density of $U + V$ from g. (This answer should agree with Exercise 12b since $U + V = X + Y$.)

(d) Let $R = U$, $S = V - U$. Show that the joint integrating density of R and S is

$$h(r, s) = \begin{cases} (2e^{-2r})(e^{-s}) & \text{if } 0 \leqslant r \text{ and } 0 \leqslant s \\ 0 & \text{otherwise} \end{cases}$$

Hence, R and S are independent.

15 Suppose X and Y are independent random variables, each with the uniform density function

$$f(x) = \begin{cases} 1 & \text{if } 0 \leqslant x \leqslant 1 \\ 0 & \text{otherwise} \end{cases}$$

Let $U = \min(X, Y)$, $V = \max(X, Y)$. According to Exercise 13a, the joint integrating density of U and V is

$$g(u, v) = \begin{cases} 2 & \text{if } 0 \leqslant u \leqslant v \leqslant 1 \\ 0 & \text{otherwise} \end{cases}$$

Use this fact in the following.

(a) Find the density of $U + V$. (This answer should agree with Exercise 11b, because $U + V = X + Y$.)

(b) Let $R = U/V$, $S = V$. Show that the joint integrating density of R and S is

$$h(r, s) = \begin{cases} 1 \times 2s & \text{if } 0 \leqslant r \leqslant 1 \text{ and } 0 \leqslant s \leqslant 1 \\ 0 & \text{otherwise} \end{cases}$$

and thus conclude that R and S are independent.

(c) Suppose $T = XY$. Show that the joint integrating density of X and T is

$$k(x, t) = \begin{cases} \frac{1}{x} & \text{if } 0 \leq t < x \\ 0 & \text{otherwise} \end{cases}$$

Check that $\int_{-\infty}^{\infty} k(x, t)\, dx$ agrees with Exercise 11a.

16 Evaluate the expected values of the random variables in Exercises 4 and 5 in two different ways—first, by evaluating $\int_{-\infty}^{\infty} h(x) f(x)\, dx$; second, by evaluating $\int_{-\infty}^{\infty} xg(x)\, dx$ [see (2), Section 7.11). For example, consider Exercise 4c:

$$\int_{-\infty}^{\infty} h(x) f(x)\, dx = \int_0^1 x^{1/2}\, dx = \tfrac{2}{3}$$

$$\int_{-\infty}^{\infty} xg(x)\, dx = \int_0^1 x(2x)\, dx = \tfrac{2}{3}$$

Omit Exercise 5b, for which the integration is not elementary. [Answers: Exercise 4: $c/2$, 1/5, 2/3, ∞, 1; Exercise 5: $1/c$, c, -1.)

17 Evaluate the expected values of the random variables in Exercises 11, parts (a), (b), and (c), and Exercises 12, parts (a), (b), and (c), in two different ways: first, evaluate

$$\int_{-\infty}^{\infty}\int_{-\infty}^{\infty} h(x, y) f(x, y)\, dx\, dy$$

second, evaluate $\int_{-\infty}^{\infty} xg(x)\, dx$. [Hint: For Exercise 12b,

$$\int_{-\infty}^{\infty}\int_{-\infty}^{\infty} h(x, y) f(x, y)\, dx\, dy = \int_0^{\infty}\int_0^{\infty} (x + y)e^{-(x+y)}\, dx\, dy$$

$$= 2\int_0^{\infty} xe^{-x}\, dx \int_0^{\infty} e^{-y}\, dy = 2\int_{-\infty}^{\infty} xg(x)\, dx = \int_0^{\infty} x(xe^{-x})\, dx = 2$$

(integrate by parts or use a table of integrals)]

18 Suppose that X and Y are positive random variables with a joint density function f (either discrete or integrating) that satisfies

$$f(x, y) = f(y, x)$$

Show that

$$E\left(\frac{X}{X + Y}\right) = \frac{1}{2}$$

{Hint: $E[(X + Y)/(X + Y)] = 1$.}

19 Suppose $X_1, X_2, \ldots, X_n$ are mutually independent random variables with cdf's given by

$$P(X_k \leqslant t) = \begin{cases} 0 & \text{if } t < 0 \\ t^k & \text{if } 0 \leqslant t < 1 \\ 1 & \text{if } 1 \leqslant t \end{cases} \qquad k = 1, 2, \ldots, n$$

(a) Let $Y = \max(X_1, X_2, \ldots, X_n)$. Show that

$$EY = \frac{\binom{n+1}{2}}{1 + \binom{n+1}{2}}$$

[Hint: The density function of Y is

$$g(x) = \begin{cases} mx^{n-1} & \text{if } 0 \leqslant x \leqslant 1 \\ 0 & \text{otherwise} \end{cases}$$

where $m = \binom{n+1}{2}$].

(b) Show that $E(X_1X_2 \cdots X_n) = 1/(n+1)$. [Hint: Use (7), Section 7.11.]

20 Suppose that a point is selected at random from the inside of a circle centered at origin and having radius 1. Let X equal the x coordinate of the point and let Y equal the y coordinate of the point.
(a) Give the joint integrating density of X and Y (see Example 2, Section 7.6).
(b) Are X and Y independent?
(c) Derive the density of $(X^2 + Y^2)^{1/2}$. [Answer: $f(x) = 2x$ if $0 \leqslant x \leqslant 1$, and 0 otherwise.]
(d) Evaluate $E(X^2)$. {Hint: *Method 1.* $E(X^2 + Y^2) = \int_0^1 x^2(2x)\,dx = 1/2$. By symmetry. $E(X^2) = 1/4$. *Method 2.* Use Example 5, Section 7.6, and evaluate

$$E(X^2) = \int_1^1 x^2\left[\frac{2}{\pi}(1 - x^2)^{1/2}\right] dx$$

Which method do you like better?}

21 Suppose that X and Y are independent, each with the same integrating density

$$f(x) = g(x) = \begin{cases} 1 & \text{if } 1 \leqslant x \leqslant 2 \\ 0 & \text{otherwise} \end{cases}$$

Evaluate $E(X/Y)$ without first determining the density function of X/Y. {Hint: By independence, X and $1/Y$ are independent; therefore,

$$E\left(\frac{X}{Y}\right) = [EX]\left[E\left(\frac{1}{Y}\right)\right] = \frac{\log 2}{2}$$

$$E\left(\frac{1}{Y}\right) = \int_1^2 \frac{1}{y}\,dy = \log 2\}$$

22 Consider the physics formula $F = MA$ (force = mass × acceleration). Suppose that the force (F) and the mass (M) of a particle are independent random variables, the density function of F being

$$g(x) = \begin{cases} 1 & \text{if } 1 \leq x \leq 2 \\ 0 & \text{otherwise} \end{cases}$$

and the density function of M being

$$h(x) = \begin{cases} 1 & \text{if } 2 \leq x \leq 3 \\ 0 & \text{otherwise} \end{cases}$$

Show that the density function of the acceleration, $A = F/M$, is the following:

$$k(x) = \begin{cases} 0 & \text{if } x < 1/3 \\ 9/2 - 1/(2x^2) & \text{if } 1/3 \leq x < 1/2 \\ \dfrac{5}{2} & \text{if } 1/2 \leq x < 2/3 \\ -2 + \dfrac{2}{x^2} & \text{if } 2/3 \leq x < 1 \\ 0 & \text{if } 1 \leq x \end{cases}$$

Also show that $EA = (3/2)\log(3/2)$.

23 Consider the following simple-minded model of a market situation. Suppose $S(t)$, the proportion of the available stock of a commodity that will be offered at price t, $(0 \leq t \leq 1)$ is given by

$$S(t) = Xt$$

where X is uniformly distributed on $[0, 1]$ (i.e., the density function for X is 1 on $[0, 1]$ and 0 elsewhere). Also $D(t)$, the proportion of the available stock of the commodity that will be demanded at price t, is given by

$$D(t) = (1 - t)Y$$

where Y is also uniformly distributed on $[0, 1]$. Suppose that X and Y are independent random variables, and U, the equilibrium price, is the t value at which the supply and demand curves intersect. Show that the density function of U is

$$f(u) = \begin{cases} \dfrac{1}{2(1-u)^2} & \text{if } 0 \leq u \leq \dfrac{1}{2} \\ \dfrac{1}{2u^2} & \text{if } \dfrac{1}{2} < u \leq 1 \end{cases}$$

24 Exercises 24–26 of Chapter 2 are about relationships among moments of discrete random variables. Show that the suggested proofs go through without any changes if the random variables have integrating density functions.

25 Prove the continuity properties (a) and (b) of Section 7.1. (Hint: For (a) define $C_1 = A_1$, $C_2 = A_2 \cap A_1'$, $C_3 = A_3 \cap A_2'$, ... The C_i's are mutually disjoint with $C_1 \cup C_2 \cup \ldots = A_1 \cup A_2 \cup \ldots$. Now use III (c) of the axioms for a general probability space.)

NOTES

1 The axiomatic approach to probability theory was developed by the great contemporary Russian mathematician A. N. Kolmogrov (1903–), in his short monograph [26]. Abstract probability theory is related to the branch of mathematics called *measure theory*. An advanced but excellent book on measure theory is [27], with a chapter on probability theory. Also see [28], [29], and [30].

2 What we describe here as an "algebra" should strictly speaking be called a "*sigma algebra*." Unfortunately, there are also many other terms in use, such as sigma ring, sigma field, Borel field, and tribe.

3 A standard treatment of Borel sets appears in [27]. A self-contained and readable account is given in Chapters 1 and 2 of [31]. A short, intuitive description occurs on page 105 of [4]. Borel sets are named after the French mathematician Emile Borel (1871–1956).

4 The measure on [0, 1] described in Section 7.1 is called Lebesgue after the French mathematician Henri Lebesgue (1875–1941), one of the creators of the modern theory of measure and integration.

5 Cumulative distribution functions (cdf's) are also called *distribution functions*. For additional reading on cdf's and density functions, see Chapter 3 of [4] and Chapter 15 of [31].

6 The term *integrating density function* is probably new. Usually, *density function* is used for what we call integrating density function. The term *discrete density function* is used in the literature together with a variety of other terms, such as *mass function* and *probability mass function*. *Frequency function* is also used for density function; it can refer to integrating or discrete types.

7 For a review of Riemann integration, see any standard calculus book. Actually, a proper treatment of integrating density functions requires the Lebesgue integral. See Chapter 5 of [31].

8 For a review of multiple integrals, see Chapter 16 of [2] or Chapter 2 of [4]. As in the one-dimensional case, a proper treatment of integrating density functions of several random variables requires the Lebesgue integral. See Chapter 9 of [31].

9 An elementary treatment of the Stieltjes integral appears in Chapter 5 of [32].

10 For a rigorous approach to (2), Section 7.11, see Sections 7.4 and 15.3 of [31].

Chapter Eight

SUMS AND PRODUCTS OF RANDOM VARIABLES

When two random variables have a joint integrating density, it is possible to obtain explicit formulas for the integrating densities of their sum, product, and ratio, among other functions. The proofs require the techniques of change of variables in multiple integrals, which are reviewed in the first part of this chapter. When X and Y are independent, then the integrating density of $X + Y$ is obtained by means of a *convolution formula*. The generating function that was introduced earlier for nonnegative integer-valued random variables can be extended for more general random variables. This leads to *Laplace* and *Fourier transforms* of random variables, which are particularly useful in studying sums of independent random variables.

8.1 CHANGE OF VARIABLES IN MULTIPLE INTEGRATION

In this section, we review techniques of changing variables in multiple integrals. The subject is covered in most calculus books, so we shall present no proofs. Most of our discussion is restricted to integrals in two-dimensional space. This case is easy to visualize and is sufficient for many of our applications.

The first object that needs to be reviewed is the 2×2 determinant. If a_{ij}, $i = 1, 2, j = 1, 2$ are the four entries in a 2×2 matrix, then by the determinant of this matrix is meant

$$\begin{vmatrix} a_{11} & a_{12} \\ a_{21} & a_{22} \end{vmatrix} = a_{11}a_{22} - a_{21}a_{12} \tag{1}$$

It is easy to see that this determinant equals 0 only if the two rows of the matrix are proportional to each other, or, equivalently, only if the two columns are proportional to each other (Note 1).

Next, we focus our attention on an integral in 2-space.

$$\int_{-\infty}^{\infty}\int_{-\infty}^{\infty} f(x, y)\, dx\, dy \tag{2}$$

or, more generally, an integral over a set A that may be only a part of the whole space,

$$\iint_A f(x, y)\, dx\, dy \tag{3}$$

As before, we assume the integral to be a Riemann integral, and the function f and the set A such that the Riemann integral exists. Often, it is difficult to evaluate (3) directly. However, in a new coordinate system, the new version of the integral may become more manageable. Toward this end, let us consider a change of variables described as follows:

$$u = u(x, y) \qquad v = v(x, y) \tag{4}$$

We suppose that this transformation from (x, y) to (u, v) satisfies the following regularity conditions.

(a) If B is the image of A in the (u, v) plane, then the transformation (4) is a one-to-one transformation of A onto B with a well-defined inverse,

$$x = x(u, v) \qquad y = y(u, v) \tag{5}$$

(b) The determinant

$$\begin{vmatrix} \dfrac{\partial x}{\partial u} & \dfrac{\partial x}{\partial v} \\[2ex] \dfrac{\partial y}{\partial u} & \dfrac{\partial y}{\partial v} \end{vmatrix} = J(x, y; u, v)$$

expressed as a function of u and v is continuous in B.

(c) The determinant

$$\begin{vmatrix} \dfrac{\partial u}{\partial x} & \dfrac{\partial u}{\partial y} \\[2ex] \dfrac{\partial v}{\partial x} & \dfrac{\partial v}{\partial y} \end{vmatrix} = J(u, v; x, y)$$

expressed as a function of x and y is continuous in A and vanishes nowhere there.

Then the evaluation of (3) in terms of the new variables (u, v) is the following:

$$\iint_A f(x, y)\, dx\, dy = \iint_B f[x(u, v), y(u, v)]\,|J(x, y; u, v)|\, du\, dv \qquad (6)$$

In other words, the region A is replaced by B, its image under (4). In $f(x, y)$, x and y are replaced by their values in terms of u and v, as expressed by (5). Finally, the differential element $dx\, dy$ is replaced by $|J(x, y; u, v)|\, du\, dv$.

The determinants described in (b) and (c) are called Jacobians. A fact that often eases their computation is that one is the reciprocal of the other. That is,

$$J(x, y; u, v)J(u, v; x, y) = 1$$

whenever this quantity is evaluated in terms of (x, y) and (u, v), which correspond to each other under (4) (Note 2).

Example 1 Let A be a region that excludes the origin, and let

$$x = u \cos v$$

$$y = u \sin v$$

The region B is contained in that part of the (u, v) plane where $0 \leqslant v < 2\pi$ and $0 \leqslant u$. The explicit form of the inverse transformation is

$$u = (x^2 + y^2)^{1/2}$$

$$v = \arc\tan \frac{y}{x}$$

Since

$$J(x, y; u, r) = \begin{vmatrix} \cos v & -u \sin v \\ \sin v & u \cos v \end{vmatrix} = u$$

it follows that

$$\iint_A f(x, y)\, dx\, dy = \iint_B f(u \cos v, u \sin v)\, u\, du\, dv$$

The preceding equation describes the familiar transformation to polar coordinates.

We shall occasionally need the version of this technique that is appropriate to n-dimensional space. Everything really remains the same, word-for-word. If the general version of (4) is expressed as

$$\begin{aligned} u_1 &= u_1(x_1, \ldots, x_n) \\ &\vdots \\ u_n &= u_n(x_1, \ldots, x_n) \end{aligned}$$

then one of the Jacobians is

$$J(x_1, \ldots, x_n; u_1, \ldots, u_n) = \begin{vmatrix} \dfrac{\partial x_1}{\partial u_1} & \cdots & \dfrac{\partial x_1}{\partial u_n} \\ \vdots & & \\ \dfrac{\partial x_n}{\partial u_n} & & \dfrac{\partial x_n}{\partial u_n} \end{vmatrix}$$

where the determinant is now $n \times n$. The other Jacobian is defined analogously in terms of the inverse transformation, and the two determinants are still reciprocal to each other.

8.2 INTEGRATING DENSITY OF A PRODUCT

In this section, we shall obtain a general expression for the integrating density function of a product of independent random variables. It will be needed for various applications later on.

density of a product of independent, nonnegative random variables

Suppose that X and Y are independent, nonnegative random variables having integrating density functions f and g, respectively. The integrating density function of XY is

$$h(x) = \begin{cases} \displaystyle\int_0^\infty \left(\frac{1}{y}\right) f\left(\frac{x}{y}\right) g(y)\, dy & \text{if } x > 0 \\ 0 & \text{otherwise} \end{cases} \tag{1}$$

Proof of (1). First, evaluate the cdf of XY, which is

$$P(XY \leq t) = \iint\limits_{xy \leq t} f(x)g(y)\, dx\, dy \tag{2}$$

The integration is restricted to the first quadrant, since $f(x)g(y)$, the joint density function of X and Y, is zero elsewhere. Make a transformation to new coordinates (u, v) and apply the technique described in Section 8.1. A useful transformation is

$$xy = u \qquad y = v$$

The inverse of this transformation is

$$x = \frac{u}{v} \qquad y = v$$

The Jacobian is

$$J(x, y; u, v) = \begin{vmatrix} \frac{1}{v} & \frac{-u}{v^2} \\ 0 & 1 \end{vmatrix} = \frac{1}{v}$$

The integral (2) is evaluated over the set of all (x, y) points in the first quadrant, where $xy \leq t$. The image of this set under the preceding transformation from (x, y) to (u, v) is the set of (u, v) points where $u \leq t$ and $0 \leq v < \infty$ (see Figure 8.1). For the regularity conditions described in Section 8.1 to hold, we must integrate over a region where $J(x, y; u, v)$ is continuous. This means we should exclude the line $v = 0$, or, equivalently, $y = 0$, a set of probability zero. Hence, for $t > 0$,

$$P(XY \leq t) = P(0 < XY \leq t) = \int_0^t du \int_0^\infty \left(\frac{1}{v}\right) f\left(\frac{u}{v}\right) g(v)\, dv = \int_0^t h(u)\, du$$

is the transformed version of (2), and h is the integrating density of XY, as asserted. ▮

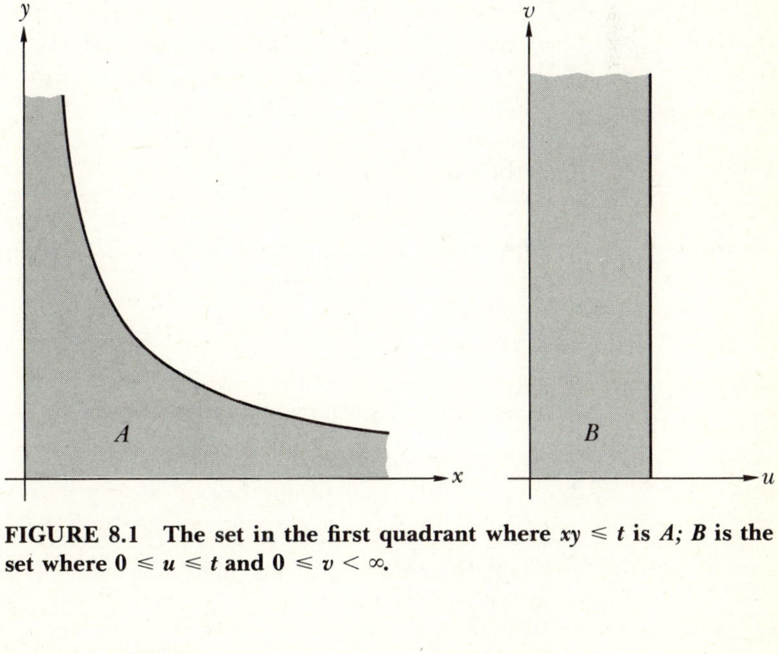

FIGURE 8.1 The set in the first quadrant where $xy \leq t$ is A; B is the set where $0 \leq u \leq t$ and $0 \leq v < \infty$.

Example 1 Suppose X and Y are independent random variables, each with the same integrating density function

$$f(x) = g(x) = \begin{cases} 1 & \text{if } 0 \leqslant x \leqslant 1 \\ 0 & \text{otherwise} \end{cases}$$

We shall compute the density function of XY using (1). Since for $0 \leqslant x \leqslant 1$, $0 \leqslant y \leqslant 1$,

$$f\left(\frac{x}{y}\right) = \begin{cases} 1 & \text{if } y \geqslant x \\ 0 & \text{otherwise} \end{cases}$$

it follows that

$$\frac{1}{y} f\left(\frac{x}{y}\right) g(y) = \begin{cases} \dfrac{1}{y} & \text{if } 0 \leqslant x \leqslant y \leqslant 1 \\ 0 & \text{otherwise} \end{cases}$$

Hence,

$$h(x) = \begin{cases} \displaystyle\int_x^1 \frac{1}{y}\, dy = -\log x & \text{if } 0 \leqslant x \leqslant 1 \\ 0 & \text{otherwise} \end{cases}$$

is the integrating density function of the product XY. (See Exercise 11a, Chapter 7.)

The formula (1) can be iterated to find the density function of the product of n independent, nonnegative random variables by successively finding the density functions of (X_1X_2), $(X_1X_2)(X_3)$, and so forth. This iteration is permissible, because by the mutual independence of $X_1, \ldots, X_n$ it follows that X_1X_2 and X_3 are independent, $X_1X_2X_3$ and X_4 are independent, and so forth. We do this in the next example.

Example 2 Suppose $X_1, X_2, \ldots,$ are mutually independent random variables, each with the same density function

$$f(x) = \begin{cases} 1 & \text{if } 0 \leqslant x \leqslant 1 \\ 0 & \text{otherwise} \end{cases}$$

As shown in Example 1, the density of X_1X_2 is

$$h(x) = \begin{cases} -\log x & \text{if } 0 \leqslant x \leqslant 1 \\ 0 & \text{otherwise} \end{cases}$$

Let us apply (1) again to obtain the density of $(X_1X_2)X_3$ by letting h play the role of g in the right side of (1). It follows that the density of $X_1X_2X_3$ is

$$\int_x^1 \left(\frac{1}{y}\right)(-\log y)\,dy = \frac{(-\log x)^2}{2} \quad \text{if } 0 < x \leq 1$$

$$0 \quad \text{otherwise}$$

Continuing this procedure, we find that, in general, the density of $X_1X_2 \cdots X_n$ is

$$h_n(x) = \begin{cases} \dfrac{(-\log x)^{n-1}}{(n-1)!} & \text{if } 0 < x \leq 1 \\ 0 & \text{otherwise} \end{cases}$$

8.3 INTEGRATING DENSITY OF A RATIO

Following the same method as in Section 8.2, we now obtain a general expression for the density function of the ratio of two independent random variables.

density of a ratio of independent, nonnegative random variables

Suppose that X and Y are independent, nonnegative random variables having integrating density functions f and g, respectively. The integrating density function of X/Y is

$$h(x) = \begin{cases} \displaystyle\int_0^\infty yf(xy)g(y)\,dy & \text{if } x > 0 \\ 0 & \text{otherwise} \end{cases} \tag{1}$$

Proof of (1). We first find the cdf of X/Y, which is

$$P\left(\frac{X}{Y} \leq t\right) = \iint_{(x/y) \leq t} f(x)g(y)\,dx\,dy$$

The integral is restricted to the first quadrant because $f(x)g(y)$, the joint density of X and Y, is zero elsewhere. Make the transformation

$$\frac{x}{y} = u, \qquad y = v$$

whose inverse is

$$uv = x, \qquad v = y$$

The Jacobian is

$$J(x, y; u, v) = \begin{vmatrix} v & u \\ 0 & 1 \end{vmatrix} = v$$

Hence, for $t > 0$,

$$P\left(\frac{X}{Y} \leq t\right) = P\left(0 < \frac{X}{Y} \leq t\right) = \iint\limits_{\substack{0<u<t \\ 0<v<\infty}} vf(uv)g(v)\, du\, dv$$

$$= \int_0^t du \int_0^\infty vf(uv)g(v)\, dv = \int_0^t h(u)\, du$$

Hence, h is the integrating density function of X/Y, as asserted. ∎

Example 1 Suppose that X and Y are independent random variables, each with the same density

$$f(x) = g(x) = \begin{cases} 1 & \text{if } 0 \leq x \leq 1 \\ 0 & \text{otherwise} \end{cases}$$

We apply (1) to find the density of X/Y.

$$yf(xy)g(y) = \begin{cases} y & \text{if } 0 \leq y \leq 1 \text{ and } 0 \leq x \leq 1 \\ y & \text{if } 0 \leq y \leq \dfrac{1}{x} \text{ and } 1 < x \\ 0 & \text{otherwise} \end{cases}$$

Hence,

$$h(x) = \int_0^\infty yf(xy)g(y)\, dy = \begin{cases} \displaystyle\int_0^1 y\, dy = \frac{1}{2} & \text{if } 0 \leq x \leq 1 \\ \displaystyle\int_0^{1/x} y\, dy = \frac{1}{2x^2} & \text{if } 1 < x \end{cases}$$

and $h(x) = 0$ if $x < 0$. (See Exercise 11c, Chapter 7.)

Example 2 Suppose that X and Y are independent random variables, each with the same density

$$f(x) = g(x) = \begin{cases} e^{-x} & \text{if } 0 \leq x \\ 0 & \text{otherwise} \end{cases}$$

For $x \geq 0$, the density of the ratio is given by

$$h(x) = \int_0^\infty ye^{-xy}e^{-y}\, dy = \int_0^\infty ye^{-y(1+x)}\, dy$$

$$= \frac{1}{(1+x)^2} \int_0^\infty ue^{-u}\, du = \frac{1}{(1+x)^2}$$

(after the change of variables $y(1+x)=u$). Hence, the density of the ratio is

$$h(x)=\begin{cases}\dfrac{1}{(1+x)^2} & \text{if } x \geqslant 0\\ 0 & \text{otherwise}\end{cases}$$

(See Exercise 12a, Chapter 7.) Notice that in this example, and in Example 1,

$$\int_0^\infty xh(x)\,dx = E\left(\frac{X}{Y}\right)=\infty$$

8.4 THE CONVOLUTION FORMULA

Having obtained general expressions for the density functions of products and ratios of independent random variables, we now find the density function for a sum of independent random variables. This particular formula is by far the most important one of the three.

density of a sum of independent random variables

Suppose that X and Y are independent random variables with integrating density functions f and g, respectively. Then the integrating density function of $X+Y$ is

$$h(u)=\int_{-\infty}^{\infty} f(u-v)g(v)\,dv=\int_{-\infty}^{\infty} f(v)g(u-v)\,dv \tag{1}$$

The discrete version of (1) is

$$P(X+Y=u)=\sum_i P(X=u-v_i)P(Y=v_i)$$

the sum ranging over $v_1, v_2, \ldots$, the possible values of Y, or

$$P(X+Y=u)=\sum_i P(X=v_i)P(X=u-v_i)$$

the sum ranging over $v_1, v_2, \ldots$, the possible values of X. [See (1), Section 5.5.] Equation (1) is just like these discrete formulas with the integral replacing the summation.

Proof of (1). As usual, we first determine the cdf of $X+Y$,

$$P(X + Y \leq t) = \iint_{x+y\leq t} f(x)g(y)\,dx\,dy$$

Make the change of variables

$$x + y = u \qquad y = v \tag{2}$$

the inverse of which is $u - v = x$, $v = y$. The Jacobian is

$$J(x, y; u, v) = \begin{vmatrix} 1 & -1 \\ 0 & 1 \end{vmatrix} = 1$$

Hence,

$$\iint_{x+y\leq t} f(x)g(y)\,dx\,dy = \iint_{\substack{u\leq t \\ -\infty<v<\infty}} f(u-v)g(v)\,du\,dv$$

$$= \int_{-\infty}^{t} du \int_{-\infty}^{\infty} f(u-v)g(v)\,dv = \int_{-\infty}^{t} h(u)\,du$$

Hence, $h(u) = \int_{-\infty}^{\infty} f(u-v)g(v)\,dv$ is the integrating density function of $X + Y$. If we interchange the roles of X and Y, as we may because $X + Y = Y + X$, we see that the integrating density must also be given by $h(u) = \int_{-\infty}^{\infty} g(u-v)f(v)\,dv$. ▮

The operation described by (1) is an extremely important one in mathematical analysis. Function h is called the *convolution* of the functions f and g, and (1) is called the convolution formula. An easy way to remember (1) is as follows: The functions f and g are integrated out over the straight line in the (x, y) plane where the coordinates sum to u. See Figure 8.2 (Note 3).

If X and Y are nonnegative random variables, then the convolution formula (1) reduces to the following:

$$h(u) = \int_{0}^{u} f(u-v)g(v)\,dv$$

This is so because each of f and g is equal to 0 for negative arguments.

Example 1 Suppose that X and Y are independent random variables, each with the same integrating density

$$f(x) = g(x) = \begin{cases} 1 & \text{if } 0 \leq x \leq 1 \\ 0 & \text{otherwise} \end{cases}$$

Let us find the integrating density of $X + Y$ using (1). Since

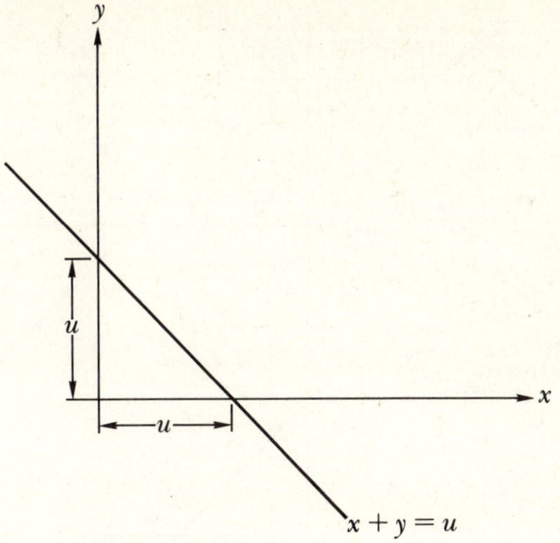

FIGURE 8.2 The line $x + y = u$.

$$f(u-v)g(v) = \begin{cases} 1 & \text{if } 0 \leqslant v \leqslant u \text{ and } 0 \leqslant u \leqslant 1 \\ 1 & \text{if } u-1 \leqslant v \leqslant 1 \text{ and } 1 \leqslant u \leqslant 2 \\ 0 & \text{otherwise} \end{cases}$$

it follows that

$$h(u) = \int_{-\infty}^{\infty} f(u-v)g(v)\,dv = \begin{cases} u & \text{if } 0 \leqslant u \leqslant 1 \\ 2-u & \text{if } 1 \leqslant u \leqslant 2 \\ 0 & \text{otherwise} \end{cases}$$

(See Figure 8.3.)

Example 2 Suppose that X and Y are independent random variables, each with the same integrating density function

$$f(x) = g(x) = \begin{cases} e^{-x} & \text{if } x \geqslant 0 \\ 0 & \text{if } x < 0 \end{cases}$$

The density function of $X + Y$ is the following:

$$h(u) = \int_{-\infty}^{\infty} f(u-v)g(v)\,dv = \begin{cases} \displaystyle\int_0^u e^{-(u-v)}e^{-v}\,dv = ue^{-u} & \text{if } u \geqslant 0 \\ 0 & \text{if } u < 0 \end{cases}$$

The convolution (1) is often referred to as a "smoothing" operation. One feature of what smoothing means can be explained for

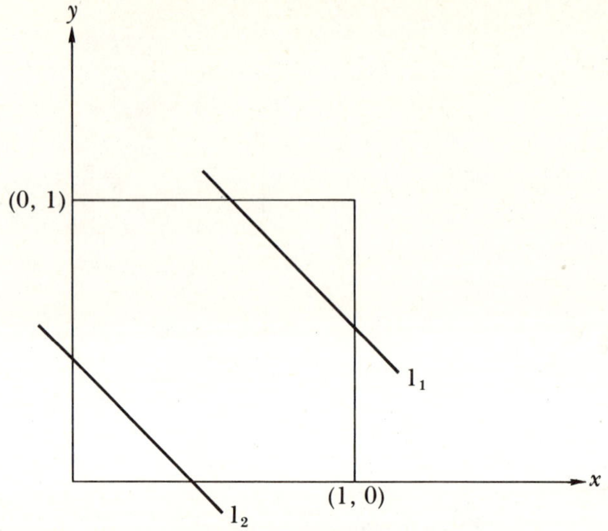

FIGURE 8.3 The convolution integral is over line l_1 if $u > 1$. The convolution integral is over line l_2 if $u < 1$.

Examples 1 and 2 as follows. In each of these examples, the density functions of X and Y have discontinuities (at 0 and 1 in Example 1, and at 0 in Example 2). On the other hand, the density functions of $X+Y$ are continuous. The integrating density function of a sum of two independent random variables is "smoother" than that of either of the components (see Figure 8.4).

The convolution formula (1) can be applied repeatedly to determine the density functions of the sum of more than two random variables. If $X_1, X_2, \ldots$ are mutually independent, then $(X_1 + X_2)$ and X_3 are independent, $(X_1 + X_2 + X_3)$ and X_4 are independent, and so forth. Thus, after finding the density of $X_1 + X_2$ using (1), the same operation can be applied again to find the density of $(X_1 + X_2) + X_3$, and so forth.

Example 3 Suppose that X, Y, and Z are mutually independent random variables, each with the same integrating density, e^{-x} for $x \geqslant 0$ and 0 otherwise. As shown in Example 2, $X+Y$ has the density function that equals xe^{-x} for $x \geqslant 0$ and 0 otherwise. Applying (1) once again to find the density of $X+Y+Z$ shows that density to be equal to

$$k(u) = \begin{cases} \displaystyle\int_0^u (u-v)e^{-(u+v)}e^{-v}\,dv = \frac{u^2}{2}e^{-u} & \text{if } u \geqslant 0 \\ 0 & \text{if } u < 0 \end{cases}$$

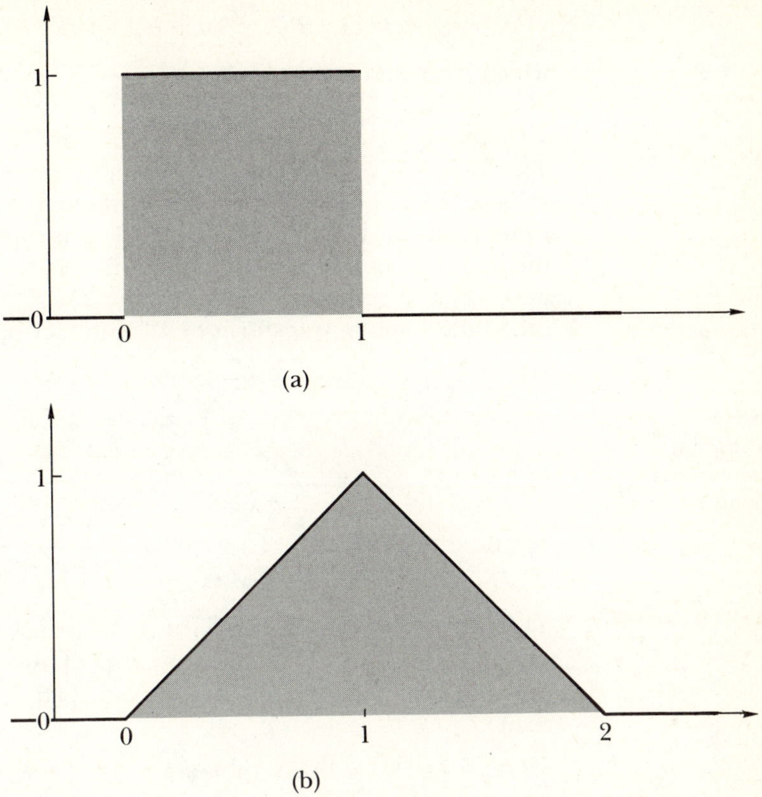

FIGURE 8.4 (a) The integrating density of a number selected at random from [0, 1]. (b) The integrating density of a sum of two numbers selected independently and at random from [0, 1].

These examples are of the *gamma distribution,* which will be investigated in greater detail in Section 9.4.

8.5 GENERATING FUNCTIONS, LAPLACE AND FOURIER TRANSFORMS

We now describe some generalized versions of the generating function, which was introduced in Section 5.1. One of the nice uses of the generating function is to determine the distribution of sums of independent, nonnegative, integer-valued random variables (Section 5.4). One reason for generalizing the generating function is to apply these techniques to more general kinds of random variables. Recall the definition of a generating function of a nonnegative integer-valued random variable X (Section 5.1),

$$\phi(t) = P(X = 0) + P(X = 1)t + P(X = 2)t^2 + \cdots, \qquad |t| \leq 1$$

which is written concisely as

$$\phi(t) = E(t^X) \tag{1}$$

The expression $E(t^X)$ makes sense whether or not X is integer valued; if X is a nonnegative valued random variable, this expected value exists for $0 \leq t \leq 1$, since $t^X \leq 1$. It is conventional to replace the dummy variable t by $e^{-\theta}$, letting θ now be a nonnegative dummy variable. The resulting expected value, $E(e^{-\theta X})$ as a function of θ, is called the Laplace transform of X (Note 3, Chapter 5).

Laplace transform of a nonnegative random variable

Suppose X is a nonnegative random variable. Define a function of a real, nonnegative argument θ by

$$L(\theta) = E(e^{-\theta X}) \tag{2}$$

where L is called the Laplace transform of X (Note 4).

The role of θ is just like that of t in the generating function; θ is a *dummy variable,* which serves no real purpose of its own except in defining (2). If X happens to be nonnegative and integervalued, then

$$L(\theta) = E(e^{-\theta X}) = P(X = 0) + P(X = 1)e^{-\theta} + P(X = 2)e^{-2\theta} + \cdots$$

The relation between the generating function θ and the Laplace transform L in this case is

$$\phi(e^{-\theta}) = L(\theta), \qquad 0 \leq \theta$$

If X is a discrete random variable with nonnegative possible values $u_1, u_2, \ldots,$ then

$$L(\theta) = P(X = u_1)e^{-\theta u_1} + P(X = u_2)e^{-\theta u_2} + \cdots$$

This sum converges when $\theta \geq 0$, because $0 < e^{-\theta u} \leq 1$ when $u \geq 0$. If X has an integrating density function f, then

$$L(\theta) = \int_0^\infty e^{-\theta x} f(x)\, dx$$

This integral converges when $\theta \geq 0$, since $\theta < e^{-\theta u} \leq 1$ when $u \geq 0$. More generally, if F, the cdf of X, is a mixture, as defined in Section 7.8, then

$$L(\theta) = \int_0^\infty e^{-\theta x}\, dF(x) = p \sum_j e^{-\theta u_j} g(u_j) + (1 - p) \int_0^\infty e^{-\theta x} h(x)\, dx$$

where g is the discrete density part of the mixture and h is the integrating density part. In this setting, the Laplace transform is also called a Laplace–Stieltjes transform.

One reason for the importance of the Laplace transform is that it *completely characterizes the distribution of a nonnegative random variable.* This and other properties will be discussed in Section 8.6.

Example 1 Suppose X has the integrating density function

$$f(x) = \begin{cases} 1 & \text{if } 0 \leq x \leq 1 \\ 0 & \text{otherwise} \end{cases}$$

Then $L(\theta) = \int_0^1 e^{-\theta x}\, dx = (1 - e^{-\theta})/\theta$. If one way or another it is determined that the Laplace transform of a random variable is $(1 - e^{-\theta})/\theta$, $0 \leq \theta$, it guarantees that X must have the uniform integrating density f.

Example 2 Suppose X has the integrating density function

$$f(x) = \begin{cases} e^{-x} & \text{if } x \geq 0 \\ 0 & \text{if } x < 0 \end{cases}$$

Then $L(\theta) = \int_0^\infty e^{-\theta x} e^{-x}\, dx = 1/(1 + \theta)$.

A limitation of the Laplace transform is that it is defined only for nonnegative random variables; otherwise, the expected value $E(e^{-\theta X})$ may not exist. Consider, for example, the integrating density function $(1/2)e^{-|x|}$, $-\infty < x < \infty$. This density appeared in Exercise 12c, Chapter 7. For no value of θ except $\theta = 0$ does the integral $\int_{-\infty}^{\infty} e^{-\theta x}(1/2)e^{-|x|}\, dx$ converge. (Check this!) For this reason, another transform is considered which is formally close to the Laplace transform, but which is *always* well defined: the *Fourier transform.*

To introduce the Fourier transform, recall the definition of the complex valued exponential function

$$e^{i\theta} = \cos\theta + i \sin\theta$$

where θ is real and $i = \sqrt{-1}$. The most important properties of this function that will be needed are $|e^{i\theta}| = [(\cos\theta)^2 + (\sin\theta)^2]^{1/2} = 1$ and $(\exp(i\theta_1))(\exp(i\theta_2)) = \exp[i(\theta_1 + \theta_2)]$ (Note 5). One also needs to extend the idea of random variable to allow complex-valued as well as real-valued functions. Thus, if X, Y are real-valued random variables, then $Z = X + iY$ is a complex-valued random variable. [Function Z assigns to the sample point w the complex number $Z(w) = X(w) + iY(w)$.] The definition of expected value of a complex random variable is just what one would expect:

$$EZ = EX + iEY$$

This expected value is said to exist only if each of EX and EY exist. The definition of Fourier transform of a random variable now follows.

Fourier transform of a random variable

The Fourier transform of a random variable X is, by definition,

$$M(\theta) = E(e^{i\theta X}) = E(\cos \theta X) + iE(\sin \theta X) \tag{3}$$

$i = \sqrt{-1}$, θ is real (Note 4).

Since $|\cos \theta x| \leq 1$, $|\sin \theta x| \leq 1$, the expected value $E(e^{i\theta X})$ always exists. In the discrete case, (3) becomes

$$M(\theta) = \sum_j e^{i\theta u_j} P(X = u_j)$$

and when there is an integrating density, (3) becomes

$$M(\theta) = \int_{-\infty}^{\infty} e^{i\theta x} f(x)\, dx$$

In the case of a mixture, (3) becomes

$$M(\theta) = p \sum_j e^{i\theta u_j} g(u_j) + (1-p) \int_{-\infty}^{\infty} e^{i\theta x} h(x)\, dx$$

In this setting, $M(\theta)$ is called a Fourier–Stieltjes transform. As with the generating function and Laplace transform, the Fourier transform of X *completely determines the distribution of* X. This and other aspects of the Fourier transform will be discussed in Section 8.6.

Example 3 Let X have the uniform distribution on $[0, 1]$,

$$f(x) = \begin{cases} 1 & \text{if } 0 \leq x \leq 1 \\ 0 & \text{otherwise.} \end{cases}$$

Then,

$$M(\theta) = \int_0^1 e^{i\theta x}\, dx = \frac{1 - e^{i\theta}}{-i\theta}$$

In a purely formal way, this can be obtained from Example 1 by replacing the θ of that example by $-i\theta$. A more careful check requires the following evaluations, which are left to the reader:

$$\int_0^1 (\cos \theta x)\, dx = \frac{\sin \theta}{\theta}, \qquad \int_0^1 (\sin \theta x)\, dx = \frac{1-\cos \theta}{\theta}$$

$$\frac{\sin \theta}{\theta} + i\frac{(1-\cos \theta)}{\theta} = \frac{1-e^{i\theta}}{-i\theta}$$

8.6 PROPERTIES OF TRANSFORMS

The properties to be described here are standard parts of mathematical analysis. *They will be presented here without proofs* (Note 6).

The various properties will be asserted simultaneously for each of ϕ (the generating function), L (the Laplace transform), and M (the Fourier transform). Recall that M is defined for *any* random variable, L only for *nonnegative* random variables, and ϕ only for *nonnegative,* integer-valued random variables. Each of ϕ, L, M will be called a *transform.*

Property I

Each of the transforms ϕ, L, M is uniquely determined by the distribution.

Suppose, for instance, one knows that L is a Laplace transform of a nonnegative random variable X, and that

$$L(\theta) = \int_0^\infty e^{-\theta x} f(x)\, dx$$

Then one knows that X must have the integrating density f. As a matter of fact, it is not even essential to know that f is an integrating density. If it is simply assumed that $\int_0^\infty |f(x)|\, dx < \infty$, then it must follow that f is an integrating density.

Property II

Each of the transforms ϕ, L, M generates moments. For example,

$$\left.\frac{d\phi(t)}{dt}\right|_{t=1} = EX, \qquad -\left.\frac{dL(\theta)}{d\theta}\right|_{\theta=0} = EX, \qquad -i\left.\frac{dM(\theta)}{d\theta}\right|_{\theta=0} = EX$$

assuming that the transform is appropriately defined and EX exists.

These relationships can be remembered by the heuristic device of "integrating under the expected value sign," which is illustrated for the Laplace transform as follows:

$$\left.\frac{de^{-\theta X}}{d\theta}\right|_{\theta=0} = -Xe^{-\theta X}\Big|_{\theta=0} = -X$$

Hence,

$$\left.\frac{-dEe^{-\theta X}}{d\theta}\right|_{\theta=0} = \left.\frac{-Ede^{-\theta X}}{d\theta}\right|_{\theta=0} = EX$$

The argument is similar for the other transforms. When higher moments exist, the formal relationships obtained in the same way by further differentiations are correct. For ϕ, it has already been pointed out in (2), Section 5.3 that

$$\left.\frac{d^k\phi(t)}{dt^k}\right|_{t=1} = E(X^{(k)}) \qquad \text{(the } k\text{th factorial moment)}$$

For L, this formalism gives

$$(-1)^k \left.\frac{d^kL(\theta)}{d\theta^k}\right|_{\theta=0} = E(X^k) \qquad \text{(the } k\text{th moment)}$$

and for M,

$$(-i)^k \left.\frac{d^kM(\theta)}{d\theta^k}\right|_{\theta=0} = E(X^k) \qquad \text{(the } k\text{th moment)}$$

Example 1 Suppose X has the integrating density

$$f(x) = \begin{cases} e^{-x} & \text{if } x \geq 0 \\ 0 & \text{if } x < 0 \end{cases}$$

Then

$$L(\theta) = \int_0^\infty e^{-\theta x}e^{-x}\,dx = \frac{1}{(1+\theta)}$$

$$\left.\frac{d^k[1/(1+\theta)]}{d\theta^k}\right|_{\theta=0} = \left.\frac{(-1)^k\,k!}{(1+\theta)^{k+1}}\right|_{\theta=0} = (-1)^k\,k!$$

Hence,

$$E(X^k) = \int_0^\infty x^k\,e^{-x}\,dx = k! \tag{1}$$

This is called the *gamma integral,* of great interest in mathematical analysis. An independent verification of (1) is given in Section 9.4.

For simplicity, property III will be first described in a rather special case. Suppose $f_1, f_2, \ldots$ is a sequence of integrating densities concentrated on $(0, \infty)$, and, as $n \to \infty$,

$$\int_0^\infty e^{-\theta x} f_n(x)\, dx \to \int_0^\infty e^{-\theta x} f(x)\, dx \quad \text{for every } \theta \geqslant 0$$

where f is also an integrating density on $(0, \infty)$. Then

$$F_n(t) = \int_0^t f_n(x)\, dx \to \int_0^t f(x)\, dx \quad \text{for every } t \geqslant 0$$

In other words, if a sequence of Laplace transforms converges to the Laplace transform of a random variable X, then the related cdf's also converge to the cdf of X. The general version of III, which applies to the discrete case and to mixtures as well as to the integrating density function case, follows.

Property III (for Laplace transforms)

Suppose that $F, F_1, F_2, \ldots$ are cdf's of *nonnegative* random variables. (These can be discrete, or determined by integrating densities or of mixed type. Moreover, the types can vary from cdf to cdf.) Now suppose that as $n \to \infty$,

$$\int_0^\infty e^{-\theta x}\, dF_n(x) \to \int_0^\infty e^{-\theta x}\, dF(x) \quad \text{for every } \theta \geqslant 0$$

Then

$$F_n(t) \to F(t)$$

for every $t \geqslant 0$ at which F is continuous.

The same property holds for the Fourier transform, as follows.

Property III (for Fourier transforms)

Suppose that $F, F_1, F_2, \ldots$ are cdf's. (Again, they can be discrete, or determined by integrating densities or of mixed type.) Suppose that as $n \to \infty$,

$$\int_{-\infty}^\infty e^{i\theta x}\, dF_n(x) \to \int_{-\infty}^\infty e^{i\theta x}\, dF(x), \quad \text{for every real } \theta$$

Then

$$F_n(t) \to F(t)$$

for every t at which F is continuous.

Example 2 An honest coin is thrown n times. Let X_n be the proportion of heads among the n throws. Since

$$P\left(X_n = \frac{k}{n}\right) = \binom{n}{k}\left(\frac{1}{2}\right)^n, \qquad k = 0, 1, \ldots, n$$

it follows that the Laplace transform of X_n is

$$\sum_{k=0}^{n} e^{-\theta k/n}\binom{n}{k}\left(\frac{1}{2}\right)^n = \left(\frac{1}{2} + \frac{1}{2}e^{-\theta/n}\right)^n$$

Since

$$\frac{1}{2} + \frac{1}{2}e^{-\theta/n} = 1 - \frac{\theta}{2n} + \frac{\theta^2}{4n^2} - \cdots$$

$$\lim_{n\to\infty}\left(1 - \frac{\theta}{2n}\right)^n = e^{-\theta/2}$$

it is plausible that

$$\lim_{n\to\infty}\left(1 - \frac{\theta}{2n} + \frac{\theta^2}{4n^2} - \cdots\right)^n = e^{-\theta/2} \tag{2}$$

by the argument that the smaller terms involving $1/n^2$, $1/n^3$, should not influence the limit. This fact is indeed true, but a rigorous justification will not be given here. Now $e^{-\theta/2}$ is the Laplace transform for the distribution concentrating all its mass on 1/2. The cdf for this mass distribution is

$$F(x) = \begin{cases} 0 & \text{if } x < 1/2 \\ 1 & \text{if } x \geq 1/2 \end{cases}$$

($\int_0^\infty e^{-\theta x}\, dF(x) = e^{-\theta/2}$). According to property III for Laplace transforms, (2) implies that

$$\lim_{n\to\infty} P(X_n \leq t) = \begin{cases} 0 & \text{if } t < 1/2 \\ 1 & \text{if } t > 1/2 \end{cases} \tag{3}$$

[No conclusion can be made from property III for the limit at $t = 1/2$, because that is a point of discontinuity of F. It does happen to be true that $\lim_{n\to\infty} P(X_n \leq 1/2) = 1/2$, however.] Roughly, (3) says that the distribution of the proportion of heads in n tosses, for large n, is highly concentrated about the single point 1/2. Equation (3) is just a careful way of saying the same thing. It is a special case of the law of large numbers, which will be discussed in Section 10.2.

Property I asserts that the Fourier transform M uniquely determines the distribution. There actually exist explicit operational procedures for retrieving the distribution from the Fourier transform. Only a special, but very useful case will be presented here, still without proof.

an inversion formula for Fourier transform

Suppose $M(\theta) = \int_{-\infty}^{\infty} e^{i\theta x} f(x)\, dx$, where f is an integrating density. Suppose also that $\int_{-\infty}^{\infty} |M(\theta)|\, d\theta < \infty$. Then

$$f(x) = \frac{1}{2\pi} \int_{-\infty}^{\infty} e^{-ixy} M(y)\, dy \tag{4}$$

Moreover, the density f is everywhere continuous (Note 7).

Example 3 Consider the integrating density function

$$f(x) = \tfrac{1}{2} e^{-|x|}, \qquad -\infty < x < \infty$$

The Fourier transform is

$$\tfrac{1}{2} \int_{-\infty}^{\infty} e^{i\theta x} e^{-|x|}\, dx = \tfrac{1}{2} \int_{-\infty}^{\infty} (\cos \theta x) e^{-|x|}\, dx = \int_{0}^{\infty} (\cos \theta x) e^{-x}\, dx$$

It is a fact that

$$\int_{0}^{\infty} (\cos \theta x) e^{-x}\, dx = \frac{1}{1+\theta^2} = M(\theta) \qquad \text{(Note 8)} \tag{5}$$

[It is easy to check that

$$\frac{d}{dx}\left[\frac{\theta \sin \theta x - \cos \theta x}{1+\theta^2}\, e^{-x}\right] = (\cos \theta x) e^{-x}$$

from which (5) follows immediately.] According to (4), it follows that

$$\frac{1}{2\pi} \int_{-\infty}^{\infty} e^{-ixy} \left[\frac{1}{1+y^2}\right] dy = \frac{1}{2} e^{-|x|} \tag{6}$$

Since

$$\int_{0}^{t} \frac{1}{1+x^2}\, dx = \arctan t$$

and hence

$$\int_{-\infty}^{\infty} \frac{1}{1+x^2}\, dx = \pi$$

it follows that

$$g(x) = \frac{1}{\pi(1+x^2)}, \qquad -\infty < x < \infty$$

is an integrating density function. [It is called the density of the *Cauchy distribution* (Note 9).] Equation (6) contains, in effect, an evaluation of the Fourier transform for the Cauchy distribution, because it says that

$$\int_{-\infty}^{\infty} e^{i\theta x} \frac{1}{\pi(1+x^2)}\, dx = e^{-|\theta|} \tag{7}$$

The evaluation (7) is a useful fact that will be needed later.

Conversely, once one knows (7), then (4) says that

$$\frac{1}{2\pi} \int_{-\infty}^{\infty} e^{-ixy} e^{-|y|}\, dy = \frac{1}{\pi(1+x^2)}$$

which is what (5) said in the first place.

8.7 SUMS OF INDEPENDENT RANDOM VARIABLES

Recall from Section 5.4 that if X and Y are independent, nonnegative, integer-valued random variables, then the generating function of $X+Y$ is the product of the generating functions of X and of Y. It is worth reviewing the proof. First, $E(t^{X+Y}) = \phi(t)$ is the generating function of $X+Y$. Because X and Y are independent, so are t^X and t^Y. Hence, $E(t^{X+Y}) = E(t^X t^Y) = E(t^X)E(t^Y) = \phi_1(t)\phi_2(t)$, using the fact that the expected value of a product of independent random variables is the product of the expected values. Exactly the same argument works for the Laplace and Fourier transforms.

transform of a sum of independent random variables is the product of the transforms

If X and Y are nonnegative, independent random variables, then

$$E(e^{-\theta(X+Y)}) = E(e^{-\theta X})E(e^{-\theta Y})$$

That is, $L(\theta) = L_1(\theta)L_2(\theta)$, where L, L_1, L_2 are the Laplace transforms of $X+Y, X, Y$, respectively. If X and Y are any independent random variables, then

$$E(e^{i\theta(X+Y)}) = E(e^{i\theta X})E(e^{i\theta Y})$$

That is, $M(\theta) = M_1(\theta)M_2(\theta)$, where M, M_1, M_2 are the Fourier transforms of $X+Y, X, Y$, respectively.

By iterating this fact, one has that the transform of a sum of n independent random variables is the product of the transforms.

Example 1 Suppose that X and Y are independent random variables, each with the exponential density function

$$f(x) = \begin{cases} e^{-x} & \text{if } x \geqslant 0 \\ 0 & \text{if } x < 0 \end{cases}$$

As pointed out in Example 2, Section 8.5, the Laplace transform for this density is

$$\int_0^\infty e^{-\theta x} e^{-x}\, dx = \frac{1}{(1+\theta)} \tag{1}$$

Hence, the Laplace transform of $X + Y$ is $1/(1+\theta)^2$. It is a fact that

$$\int_0^\infty e^{-\theta x}(xe^{-x})\, dx = \frac{1}{(1+\theta)^2} \tag{2}$$

The relation (2) can easily be verified directly. [One quick way of obtaining (2) is to differentiate both sides of (1) with respect to θ.] By property I of Section 8.6,

$$g(x) = \begin{cases} xe^{-x} & \text{if } 0 \leqslant x \\ 0 & \text{if } x < 0 \end{cases}$$

must be the integrating density function of $X + Y$. This fact has already been obtained earlier by means of the convolution formula (Example 2, Section 8.4, and Exercise 12b, Chapter 7). A general approach to sums of exponential random variables appears in Section 9.5.

Example 2 Suppose X and Y are independent and distributed as in Example 1. The Fourier transform of X and of Y is

$$\int_0^\infty e^{i\theta x} e^{-x}\, dx = \frac{1}{1 - i\theta}$$

(In a purely formal way, this can be obtained from Example 2, Section 8.5, by replacing the θ of that example by $-i\theta$. A rigorous justification requires a little more work, however.)

Since $E(e^{i\theta Y}) = 1/(1 - i\theta)$, it follows that $E(e^{i\theta(-Y)}) = 1/(1 + i\theta)$. (Replace θ by $-\theta$.) Hence, the Fourier transform of $X + (-Y) = X - Y$ is $[1/(1+i\theta)][1/(1-i\theta)] = 1/(1+\theta^2)$. As pointed out in Exercise

12c, Chapter 7, the density function of $X - Y$ is $(1/2)e^{-|x|}$, $-\infty < x < \infty$. Hence,

$$\int_{-\infty}^{\infty} e^{i\theta x}\left(\frac{1}{2}\right)e^{-|x|}\,dx = \frac{1}{1+\theta^2}$$

(Compare Example 3, Section 8.6.)

Example 3 Suppose X and Y are independent random variables each with the Cauchy integrating density function

$$f(x) = \frac{1}{\pi(1+x^2)}, \qquad -\infty < x < \infty$$

Then, as pointed out in (6), Section 8.6,

$$\int_{-\infty}^{\infty} e^{i\theta x} f(x)\,dx = e^{-|\theta|}$$

The Fourier transform of $X + Y$ is then

$$Ee^{i\theta(X+Y)} = e^{-2|\theta|} \tag{3}$$

and the Fourier transform of $(X + Y)/2$ is $e^{-|\theta|}$. [Replace θ by $\theta/2$ in (3).] In other words, an average of two independent Cauchy distributed random variables itself has the same Cauchy distribution. Similarly, if $X_1, X_2, \ldots, X_n$ are independent random variables, each with the same Cauchy integrating density function, then the Fourier transform of $(X_1 + \cdots + X_n)/n$ is

$$E \exp i\,\frac{\theta}{n}(X_1 + \cdots + X_n) = (e^{-|\theta|/n})^n = e^{-|\theta|}$$

Thus, an average of independent Cauchy random variables has the same distribution as any one of its components.

SUMMARY

We review the technique for changing variables in multiple integration and use it to determine formulas for the integrating density functions of X/Y, XY, and $X + Y$ when X and Y are independent random variables with integrating densities. (For the ratio and the product, it is assumed that X and Y are positive.) The *integrating density of the sum* is given by

$$h(u) = \int_{-\infty}^{\infty} f(u - v)g(v)\, dv = \int_{-\infty}^{\infty} g(u - v)f(v)\, dv$$

where f and g are the densities of X and Y. This formula is called the *convolution formula.*

The transform of a random variable X is of the form $E(t^X)$, different choices of the "dummy variable" t giving different kinds of transforms. If $|t| \leq 1$ and X has nonnegative integers for its possible values, then $\phi(t) = E(t^X)$ is the *generating function* discussed earlier in Chapter 5. If X is any nonnegative random variable and t is replaced by $e^{-\theta}$, then $L(\theta) = E(e^{-\theta X})$ is the *Laplace transform.* The dummy variable θ ranges over the nonnegative reals. If t is replaced by $e^{i\theta}$, then $M(\theta) = E(e^{i\theta X})$ is the *Fourier transform,* where the dummy variable θ ranges over all real numbers and $i = \sqrt{-1}$. The quantity $M(\theta)$ is well defined for arbitrary random variables, whereas $L(\theta)$ is defined only for nonnegative random variables and $\phi(\theta)$ is defined only for nonnegative integer-valued random variables. Some important properties of transforms are the following:

(a) Two random variables, the transforms of which are identical, must have identical distributions.
(b) If $\{X_n\}$ is a sequence of random variables having transforms that converge to the transform of a random variable X, then the distribution of X_n converges to the distribution of X. (See the text for the exact sense in which this convergence takes place.)
(c) The existing moments of a random variable can be determined from its transform.
(d) The transform of a sum of independent random variables is the product of the transforms.

EXERCISES

In the following exercises, when we refer to the *uniform distribution,* we mean the integration density function

$$f(x) = \begin{cases} 1 & \text{if } 0 \leq x \leq 1 \\ 0 & \text{otherwise} \end{cases}$$

When we refer to the *exponential distribution,* we mean the integrating density function

$$f(x) = \begin{cases} e^{-x} & \text{if } 0 \leq x \\ 0 & \text{otherwise} \end{cases}$$

1 Suppose that X and Y are independent random variables, where X has the exponential distribution and Y has the uniform distribution.
(a) Show that X/Y has the integrating density

$$h(x) = \begin{cases} \int_0^1 (ye^{-xy})\, dy & \text{if } 0 \leq x \\ 0 & \text{otherwise} \end{cases}$$

[Hint: Use (1), Section 8.3.]
(b) Show that XY has the integrating density

$$h(x) = \begin{cases} \int_0^1 \frac{1}{y} e^{-x/y}\, dy = \int_x^\infty \frac{e^{-y}}{y}\, dy & \text{if } x \geq 0 \\ 0 & \text{otherwise} \end{cases}$$

[Hint: Use (1), Section 8.2.]

2 Suppose that X and Y are independent random variables each with the same density function

$$f(x) = g(x) = \begin{cases} ax^{a-1} & 0 < x \leq 1 \\ 0 & \text{otherwise} \end{cases}$$

where $a > 0$.
(a) Show that XY has the density function

$$k(x) = \begin{cases} a^2x^{a-1}(-\log x) & \text{if } 0 < x \leq 1 \\ 0 & \text{otherwise} \end{cases}$$

(b) Suppose that X and Y are independent random variables, X having the density f described above and Y having the density k as described in (a). Show that XY has the following density function:

$$\begin{cases} \frac{a^3}{2} x^{a-1}(\log x)^2 & \text{if } 0 < x \leq 1 \\ 0 & \text{otherwise} \end{cases}$$

3 Suppose that X and Y are independent, nonnegative random variables, with integrating density functions f and g, respectively. Let $h(x)$ be the integrating density of the ratio, which is given by (1), Section 8.3. Show that $X/(X + Y)$ has the integrating density,

$$k(x) = \begin{cases} \frac{1}{(1-x)^2} h\left(\frac{x}{1-x}\right) & \text{if } 0 \leq x < 1 \\ 0 & \text{otherwise} \end{cases}$$

4 Consider h_n, the density of $X_1X_2 \cdots X_n$, obtained in Example 2, Section 8.2. Show, without directly evaluating the integral, that

$$\int_0^1 x \frac{(-\log x)^{n-1}}{(n-1)!}\, dx = \frac{1}{2^n}, \qquad n = 1, 2, \ldots$$

[Hint: By (7), Section 7.11,

$$E(X_1X_2 \cdots X_n) = (EX_1)(EX_2) \cdots (EX_n)$$

which gives the result immediately. Can you also obtain the result by a direct evaluation of the integral?]

5 Show that if $0 \leq x \leq 1$, then

$$\frac{1}{(n-1)!(m-1)!} \int_x^1 \frac{1}{y}\left(-\log \frac{x}{y}\right)^{n-1} (-\log y)^{m-1}\, dy = \frac{(-\log x)^{m+n-1}}{(m+n-1)!}$$

[Hint: Suppose X and Y are independent random variables with densities h_m and h_n, respectively, as described in Example 2, Section 8.2. By that example, XY must have the density h_{m+n}. However, apply (1), Section 8.2, to obtain the desired result.]

6 (a) Suppose that X, Y, and Z are mutually independent random variables, each with the uniform distribution. Show that the integrating density of XY/Z is

$$h(x) = \begin{cases} \dfrac{1}{4} - \dfrac{\log x}{2} & \text{if } 0 < x \leq 1 \\[2ex] \dfrac{1}{4x^2} & \text{if } 1 \leq x \\[2ex] 0 & \text{otherwise} \end{cases}$$

[Hint: Use the density of XY from Example 1, Section 8.2; then use (1), Section 8.3. It will be helpful to know that

$$-\int_0^1 y \log y\, dy = \tfrac{1}{4}$$

as pointed out in Exercise 4.]

(b) Suppose that $X_1, X_2, \ldots, X_n$ and Z are $n+1$ mutually independent random variables, each uniformly distributed. Show that that the integrating density of

$$\frac{X_1X_2 \cdots X_n}{Z}$$

is

$$h(x) = \begin{cases} \dfrac{1}{2^n} \displaystyle\sum_{k=0}^{n-1} \dfrac{(-2 \log x)^k}{k!} & \text{if } 0 \leq x \leq 1 \\[2ex] \dfrac{1}{2^n x^2} & \text{if } 1 \leq x \\[2ex] 0 & \text{otherwise} \end{cases}$$

[Hint: Use the density of $X_1X_2 \cdots X_n$ determined in Example 2, Section 8.2. Then, using (1), Section 8.3,

$$h(x) = \begin{cases} \displaystyle\int_0^1 \frac{y(-\log x - \log y)^{n-1}}{(n-1)!}\,dy & \text{if } 0 \leq x \leq 1 \\ \displaystyle\int_0^{1/x} y\frac{(-\log xy)^{n-1}}{(n-1)!}\,dy = \frac{1}{x^2}\int_0^1 \frac{y(-\log y)^{n-1}}{(n-1)!}\,dy & \text{if } 1 \leq x \end{cases}$$

Expand $(-\log x - \log y)^{n-1}$ by binomial expansion and use the result of Exercise 4.]

7 (a) Suppose that X and Y are independent, nonnegative random variables with integrating density functions f and g, respectively. Function f is arbitrary and g is the uniform distribution. Show that the integrating density of XY is

$$h(x) = \begin{cases} \displaystyle\int_x^\infty \frac{f(y)}{y}\,dy & \text{if } x \geq 0 \\ 0 & \text{otherwise} \end{cases}$$

[Hint:

$$\int_0^1 \frac{1}{y} f\left(\frac{x}{y}\right) dy = \int_x^\infty \frac{f(y)}{y}\,dy \qquad \text{if } x \geq 0$$

Compare Exercise 1b.]

(b) A particle sitting at the origin of the x axis moves to the right to a position X according to an integrating density function f. A second particle then places itself between 0 and X by selecting a position Y "at random" (uniform distribution) between 0 and X. Give an intuitive argument that the second particle is distributed like XY in part (a).

8 Suppose that X and Y are independent random variables with integrating densities f and g, respectively; g is the uniform distribution and f is unspecified, except that

$$\int_0^1 f(x)\,dx = 1$$

(a) Show that X/Y has the density

$$h(x) = \begin{cases} \displaystyle\int_0^1 yf(xy)\,dy & \text{if } 0 \leq x \leq 1 \\ \displaystyle\int_0^{1/x} yf(xy)\,dy = \frac{1}{x^2}\int_0^1 yf(y)\,dy & \text{if } 1 \leq x \\ 0 & \text{otherwise} \end{cases}$$

(Hint: Imitate Example 1, Section 8.3.)

(b) Conclude from (a) that

$$P\left(\frac{X}{Y} \geqslant 1\right) = \int_0^1 yf(y)\, dy$$

and

$$P\left(\frac{X}{Y} \leqslant t \,\middle|\, \frac{X}{Y} \geqslant 1\right) = 1 - \frac{1}{t}, \qquad 1 \leqslant t$$

It is rather remarkable that the conditional distribution of X/Y, given that it exceeds 1, does not depend on the functional form of f.

9 Suppose that X and Y are independent random variables, each exponentially distributed.
Show that the joint integrating density of $U = X/Y$ and $V = X + Y$ is

$$f(u, v) = \begin{cases} (ve^{-v})\left(\dfrac{1}{1+u}\right)^2 & \text{if } u \geqslant 0 \text{ and } v \geqslant 0 \\ 0 & \text{otherwise} \end{cases}$$

and conclude that U and V are independent. (Compare with Example 2, Section 8.3, and Example 2, Section 8.4.) Hint: Evaluate

$$\iint\limits_{\substack{(x/y) \leqslant u \\ x+y \leqslant v}} e^{-(x+y)}\, dx\, dy$$

by the change of variables

$$\frac{x}{y} = r, \qquad x + y = r)$$

10 Suppose that X and Y are independent random variables, each with the same integrating density function

$$f(x) = g(x) = \begin{cases} xe^{-x} & \text{if } x \geqslant 0 \\ 0 & \text{otherwise} \end{cases}$$

Show that $X + Y$ has the integrating density function

$$h(x) = \begin{cases} \frac{1}{6}x^3e^{-x} & \text{if } x \geqslant 0 \\ 0 & \text{otherwise} \end{cases}$$

11 Suppose that X and Y are independent random variables, each with the exponential distribution. Use the convolution formula to show that $X - Y$ has the integrating density function

$$\tfrac{1}{2}e^{-|x|}, \qquad -\infty < x < \infty$$

(Hint: Because $-Y$ has the density that equals e^x, if $x \leqslant 0$ and 0 otherwise,

$$h(u) = \begin{cases} \int_0^\infty e^{-v}e^{u-v}\, dv & \text{if } u \leqslant 0 \\ \int_u^\infty e^{-v}e^{u-v}\, dv & \text{if } u \geqslant 0 \end{cases}$$

The joint density of X and $-Y$ is concentrated in the southeast quadrant of the plane. Draw a picture of the line over which the convolution integration takes place! See Exercise 12c, Chapter 7.)

12 Suppose that X and Y have a joint integrating density function $f(x, y)$, but X and Y are not necessarily independent. Show that $X + Y$ has an integrating density of the following form

$$h(u) = \int_{-\infty}^{\infty} f(u - v, v)\, dv = \int_{-\infty}^{\infty} f(v, u - v)\, dv$$

which generalizes the convolution formula (1), Section 8.4, to non-independent random variables. [Hint: Imitate the proof of (1), Section 8.4.]

13 Suppose that X and Y have the following joint integrating density function:

$$f(x, y) = \begin{cases} 2 & \text{if } x \geqslant 0,\ y \geqslant 0, \text{ and } x + y \leqslant 1 \\ 0 & \text{otherwise} \end{cases}$$

Show that $X + Y$ has the integrating density

$$h(u) = \begin{cases} 2u & \text{if } 0 \leqslant u \leqslant 1 \\ 0 & \text{otherwise} \end{cases}$$

(Hint: Use Exercise 12. Draw a picture of the region in which f is positive.)

14 Suppose that X and Y are independent random variables, each with the same "triangular" density function

$$f(x) = g(x) = \begin{cases} x & \text{if } 0 \leqslant x \leqslant 1 \\ 2 - x & \text{if } 1 \leqslant x \leqslant 2 \\ 0 & \text{otherwise} \end{cases}$$

Show that the Laplace transform of $X + Y$ is

$$L(\theta) = \frac{(1 - e^{-\theta})^4}{\theta^4}$$

(Hint: Use Example 1, Section 8.4, and Example 1, Section 8.5.)

15 Suppose that X has integrating density $f(x)$ and Fourier transform $M(\theta)$.

(a) If a is a constant, show that aX has the Fourier transform $M(a\theta)$.

(b) Show that $-X$ has the Fourier transform $M(-\theta) = \overline{M(\theta)} =$ complex conjugate of $M(\theta)$.

(c) Show that if $f(x) = f(-x)$ for all x, then

$$M(\theta) = \int_{-\infty}^{\infty} (\cos \theta x) f(x)\, dx$$

16 Suppose that X and Y are independent random variables, each with the uniform distribution.

(a) Show that the Fourier transform of $X - Y$ is $2(1 - \cos \theta)/\theta^2$.

(b) Conclude from (a) that if

$$h(x) = \begin{cases} 1 + x & \text{if } -1 \leqslant x \leqslant 0 \\ 1 - x & \text{if } 0 \leqslant x \leqslant 1 \end{cases}$$

then

$$\int_{-\infty}^{1} (\cos \theta x) h(x)\, dx = \frac{2(1 - \cos \theta)}{\theta^2}$$

17 Suppose that $X_1, \ldots, X_n$ are n independent, nonnegative random variables, each with the same distribution. Suppose that $L(\theta)$ is the Laplace transform.

(a) Show that

$$\frac{X_1 + \cdots + X_n}{n}$$

has the Laplace transform $[L(\theta/n)]^n$.

(b) Suppose that

$$L(\theta) = \int_0^1 e^{-\theta x}\, dx = \frac{1 - e^{-\theta}}{\theta}$$

(the Laplace transform for the uniform distribution). Show that

$$\lim_{n \to \infty} \left[\frac{1 - e^{-\theta/n}}{\theta/n}\right]^n = e^{-\theta/2}$$

{Accept the fact that $[1 - (\theta/2n) + \cdots]^n \to e^{-\theta/2}$ as $n \to \infty$.}

(c) Conclude from (b) that if

$$P(X_i \leqslant t) = \begin{cases} 0 & \text{if } t < 0 \\ t & \text{if } 0 \leqslant t \leqslant 1 \\ 1 & \text{if } 1 \leqslant t \end{cases}$$

then

$$\lim_{n \to \infty} P\left(\frac{X_1 + \cdots + X_n}{n} \leqslant t\right) = \begin{cases} 0 & \text{if } t < \frac{1}{2} \\ 1 & \text{if } t > \frac{1}{2} \end{cases}$$

(Hint: Follow Example 2, Section 8.6.)

18 Show that

$$\int_0^\infty e^{i\theta x}(xe^{-x})\,dx = \frac{1}{(1-i\theta)^2}$$

(Hint: See Examples 1 and 2, Section 8.7.)

19 Suppose that X and Y are independent random variables, each with the same integrating density function

$$f(x) = g(x) = \tfrac{1}{2}e^{-|x|}, \qquad -\infty < x < \infty$$

Show that $X + Y$ has the integrating density

$$h(x) = \tfrac{1}{4}e^{-|x|}(1 + |x|), \qquad -\infty < x < \infty$$

{Hint: *Method 1.*

$$\int_{-\infty}^{\infty} e^{i\theta x}h(x)\,dx = \int_{-\infty}^{0} + \int_{0}^{\infty} = \frac{1/4}{1+i\theta} + \frac{1/4}{(1+i\theta)^2} + \frac{1/4}{1-i\theta} + \frac{1/4}{(1-i\theta)^2}$$

$$= \frac{1}{(1+\theta^2)^2}$$

after some algebra.

$$\int_{-\infty}^{\infty} e^{i\theta x}f(x)\,dx = \int_{-\infty}^{0} + \int_{0}^{\infty} = \frac{1/2}{1+i\theta} + \frac{1/2}{1-i\theta} = \frac{1}{1+\theta^2}$$

as in Example 2, Section 8.7.

Method 2. Suppose that X_1, X_2, X_3, and X_4 are mutually independent, each exponentially distributed. Then $X_1 + X_2$ and $X_3 + X_4$ have the density

$$r(x) = xe^{-x}, \qquad x \geqslant 0 \quad \text{and} \quad r(x) = 0, \qquad x < 0$$

Now apply the convolution formula to find s the density of $(X_1+X_2) - (X_3 + X_4)$. For instance, for $y < 0$,

$$s(y) = \int_0^\infty (xe^{-x})[-(y-x)e^{y-x}]\,dx = \tfrac{1}{4}e^{-|y|}(1 + |y|)$$

You will need to know that $\int_0^\infty y^2e^{-y}\,dy = 2$; $\int_0^\infty ye^{-y}\,dy = 1$. Now use the facts that $s(y) = s(-y)$ (why?) and

$$(X_1 + X_2) - (X_3 + X_4) = (X_1 - X_3) + (X_2 - X_4)$$

The density of each of $(X_1 - X_3)$ and of $(X_2 - X_4)$ is $\tfrac{1}{2}e^{-|x|}$, by Exercise 11.}

20 Suppose that X and Y are independent, nonnegative random variables with integrating density functions f and g; g is arbitrary and f is the exponential distribution. Show that the Laplace transform of Y, namely

$$L(\theta) = \int_0^\infty e^{-\theta x} g(x)\, dx$$

has the property that $1 - L(\theta)$ is the cdf of X/Y. That is,

$$P\left(\frac{X}{Y} \le \theta\right) = 1 - L(\theta)$$

[Hint:

$$P\left(\frac{X}{Y} > \theta\right) = \iint_{x/y \to \theta} f(x)g(y)\, dx\, dy = \int_0^\infty g(y)\, dy \int_{\theta y}^\infty e^{-x}\, dx]$$

NOTES

1 For a review of determinants, see Appendix A1 of [2] or Section 13.10 of [3].

2 For a short treatment of Jacobians and change of variables in multiple integrals, see Section 16.4 of [2]. A more elaborate treatment appears in Sections 2.16–2.21 of [4].

3 For additional discussion of the convolution formula, see Section 15.12 of [31].

4 The classic treatment of Laplace and Fourier transforms is [33].

5 If a and b are real, then the absolute value of the complex number $a + ib$ is, by definition, $(a^2 + b^2)^{1/2}$. For a review of basic facts about complex numbers, see Chapter 19 of [2] or Chapter 9 of [3].

6 The propositions in Section 8.6 regarding Fourier transforms are proved in Chapter 10 of [31]. Proofs for Laplace and Fourier transforms appear in [30].

7 For a proof of the inversion formula (4), Section 8.6, see (10.3.2), Section 10.3 of [31].

8 For a proof of (5), Section 8.6, see Section 10.5 of [31].

9 The Cauchy distribution is named after the French mathematician Augustin-Louis Cauchy (1789–1857).

Chapter Nine

SOME SPECIAL DISTRIBUTIONS

A number of important special distributions are given by integrating density functions. Among these are the *uniform distribution,* the *gamma distribution,* the *beta distribution,* the *normal distribution,* and the *chi-square distribution.* This chapter considers some ways in which these distributions arise and some of their interrelationships.

9.1 THE UNIFORM DISTRIBUTION

Consider the experiment of selecting a point "at random" from the interval $[a, b]$. The probability model is that the point selected is a random variable with an integrating density function that is *constant over the interval* $[a, b]$. This distribution is called the *uniform distribution.*

uniform distribution

A random variable X is said to be uniformly distributed over $[a, b]$, $a < b$, if its cdf is

$$P(X \leq t) = \begin{cases} 0 & \text{if } t < a \\ \dfrac{t-a}{b-a} & \text{if } a \leq t \leq b \\ 1 & \text{if } t \geq b \end{cases}$$

X then has an integrating density

$$f(x) = \begin{cases} \dfrac{1}{b-a} & \text{if } a \leq x \leq b \\ 0 & \text{otherwise} \end{cases}$$

(See Figure 9.1.) Because of the rectangular shape of the density function, the uniform distribution is also called the *rectangular distribution.*

FIGURE 9.1 The integrating density of the uniform distribution on [*a*, *b*].

The special case of the uniform distribution over [0, 1] has been the basis for many of the problems and examples in Chapters 7 and 8. It is easy to check that if X is uniformly distributed over $[a, b]$, where $a < b$, then $(X - a)/(b - a)$ is uniformly distributed over [0, 1] (see Exercise 1).

The expected value of a random variable uniformly distributed over $[a, b]$ is $(a + b)/2$. This fact is evident from the symmetry of the density function around the midpoint $(a + b)/2$. The formal evaluation is

$$EX = \int_a^b x\left(\frac{1}{b - a}\right) dx = \frac{a + b}{2}$$

Most of the examples of the uniform distribution assume that the interval is [0, 1], since questions about the more general case can easily be reduced to this one.

9.2 FUNCTIONS OF A UNIFORMLY DISTRIBUTED RANDOM VARIABLE

Various earlier examples concerned functions of a uniformly distributed random variable. Several of these are reviewed in the following example.

Example 1 Suppose X is uniformly distributed over [0, 1).

(a) $1/X$ has the density function

$$g(x) = \begin{cases} 0 & \text{if } x < 1 \\ \dfrac{1}{x^2} & \text{if } x \geq 1 \end{cases}$$

(b) X^2 has the density function

$$g(x) = \begin{cases} \dfrac{1}{2\sqrt{x}} & \text{if } 0 < x < 1 \\ 0 & \text{otherwise} \end{cases}$$

(c) $\sqrt{X}$ has the density function

$$g(x) = \begin{cases} 2x & \text{if } 0 \leq x \leq 1 \\ 0 & \text{otherwise} \end{cases}$$

Example 2 Example 1 can be subsumed under the general case of $Y = X^\alpha$. The form of the distribution and the expected value depends on the value of α. Table 9.1 describes the cdf's and the density functions. If $\alpha \leq -1$, then the expected value equals ∞. Otherwise, the expected value equals $1/(\alpha + 1)$. Figure 9.2 shows the typical appearances of the density functions.

Next we point out that if X is uniformly distributed over $[0, 1]$, then, by choosing h properly, $h(X)$ will have any desired distribution. We prove the assertion in the discrete case first.

> Suppose X is uniformly distributed over $[0, 1]$, and g is a discrete density function over the possible values $u_1, u_2, \ldots$. Then there is a function h such that $h(X)$ has the distribution g.

TABLE 9.1

α	*cdf*		*Integrating density function*	
$\alpha < 0$	$G(x) = \begin{cases} 0 \\ 1 - x^{1/\alpha} \end{cases}$	$\begin{matrix} x < 1 \\ 1 \leq x \end{matrix}$	$g(x) = \begin{cases} \left(\dfrac{-1}{\alpha}\right)x^{(1-\alpha)/\alpha} \\ 0 \end{cases}$	$\begin{matrix} 1 \leq x \\ \text{otherwise} \end{matrix}$
$\alpha > 0$	$G(x) = \begin{cases} 0 \\ x^{1/\alpha} \\ 1 \end{cases}$	$\begin{matrix} x < 0 \\ 0 \leq x \leq 1 \\ 1 < x \end{matrix}$	$g(x) = \begin{cases} \left(\dfrac{1}{\alpha}\right)x^{(1-\alpha)/\alpha} \\ 0 \end{cases}$	$\begin{matrix} 0 \leq x \leq 1 \\ \text{otherwise} \end{matrix}$

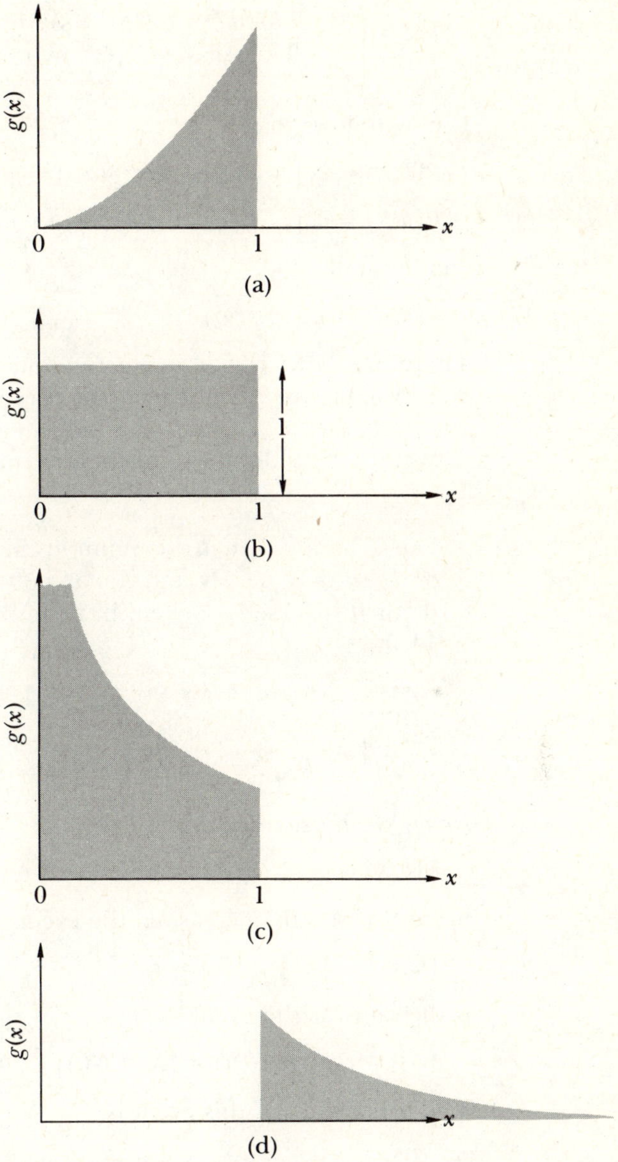

FIGURE 9.2 (a) Typical shape when $0 < \alpha < 1$. (b) Setting $\alpha = 1$ gives the uniform distribution. (c) Typical shape when $\alpha > 1$. (d) Typical shape when $\alpha < 0$.

Proof. Positive numbers $g(u_1)$, $g(u_2)$, . . . sum to 1. Define h as follows:

$$h(x) = \begin{cases} u_1 & \text{if } 0 \leq x < g(u_1) \\ u_2 & \text{if } g(u_1) \leq x < g(u_1) + g(u_2) \\ u_3 & \text{if } g(u_1) + g(u_2) \leq x < g(u_1) + g(u_2) + g(u_3) \\ \text{etc.} \end{cases}$$

It now follows that

$$P[h(X) = u_1] = P[0 \leq X < g(u_1)] = g(u_1)$$

$$P[h(X) = u_2] = P[g(u_1) \leq X < g(u_1) + g(u_2)] = g(u_2)$$

and, in general,

$$P[h(X) = u_n] = g(u_n)$$

Therefore, $h(X)$ is a discrete random variable with the desired density g. Figure 9.3 illustrates the preceding proof graphically. ▮

We now consider the same theorem in a special nondiscrete case.

> Let G be a cdf that is continuous and strictly increasing on (a, b) (a may be $-\infty$ and b may be ∞), so that G has a well-defined inverse G^{-1}; then, if X is uniformly distributed over $[0, 1]$,
>
> $$h(X) = G^{-1}(X) \tag{1}$$
>
> has the cdf G.

Proof. We must show that

$$P[h(X) \leq t] = G(t)$$

Since G is strictly increasing, the event

$$[h(X) \leq t] = [G^{-1}(X) \leq t]$$

is the same as the event

$$\{G[G^{-1}(X)\} \leq G(t)] = [X \leq G(t)]$$

The probability of this event is

$$\int_0^{G(t)} dx = G(t) \quad ▮$$

Actually, it is possible to prove these assertions for the most general cdf, G. The general case is stated here without proof.

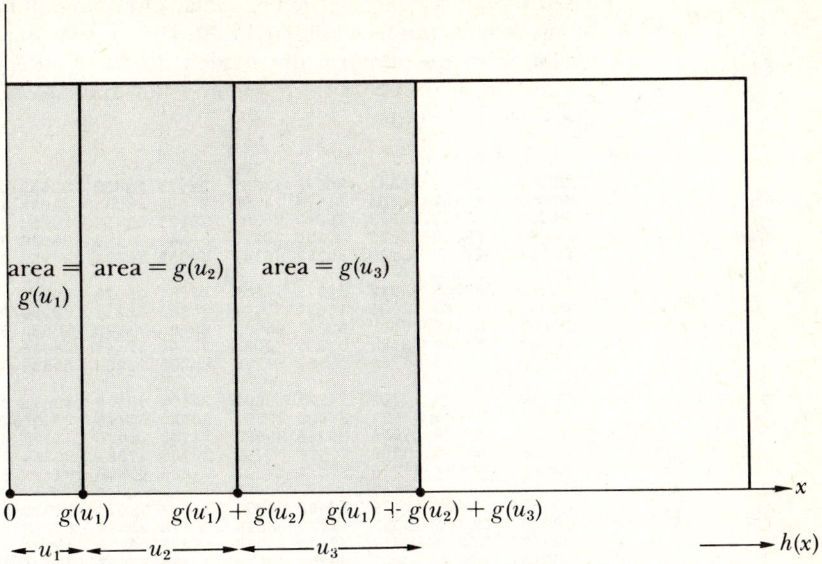

FIGURE 9.3

> If G is *any* cdf, it is possible to choose h so that
>
> $$P[h(X) \leq x] = G(x), \qquad -\infty < x < \infty \tag{2}$$
>
> where X is uniformly distributed over $[0, 1]$.

The importance of (2) is that it can be used to generate random observations having any desired distribution. First, one generates observations uniformly distributed over $[0, 1]$. These can be obtained by various electronic devices and appear in tabulated form as tables of "random numbers" (see Table 9.2) (Note 1). Then the h values of these random numbers can be evaluated, and they are distributed according to cdf G.

Example 3 Suppose that X is uniformly distributed over $[0, 1]$, and

$$h(x) = \begin{cases} 0 & \text{if } 0 \leq x < \frac{1}{2} \\ 1 & \text{if } \frac{1}{2} \leq x < \frac{1}{2}^2 \\ 2 & \text{if } \frac{1}{2}^2 \leq x < \frac{1}{2}^3 \\ & \cdot \\ & \quad\cdot \end{cases}$$

Then $h(x)$ is geometrically distributed over 0, 1, 2, . . . , (Section 6.4).

TABLE 9.2 A typical section of a table of random numbers. A sequence of any k digits can be considered as the first k decimal places of an observation from the uniform distribution on $[0, 1]$. Reproduced by kind permission of the Rand Corporation, from *A Million Random Digits,* Free Press, Glencoe, Illinois, 1955.

00000	10097	32533	76520	13586	34673	54876	80959	09117	39292	74945
00001	37542	04805	64894	74296	24805	24037	20636	10402	00822	91665
00002	08422	68953	19645	09303	23209	02560	15953	34764	35080	33606
00003	99019	02529	09376	70715	38311	31165	88676	74397	04436	27659
00004	12807	99970	80157	36147	64032	36653	98951	16877	12171	76833
00005	66065	74717	34072	76850	36697	36170	65813	39885	11199	29170
00006	31060	10805	45571	82406	35303	42614	86799	07439	23403	09732
00007	85269	77602	02051	65692	68665	74818	73053	85247	18623	88579
00008	63573	32135	05325	47048	90553	57548	28468	28709	83491	25624
00009	73796	45753	03529	64778	35808	34282	60935	20344	35273	88435
00010	98520	17767	14905	68607	22109	40558	60970	93433	50500	73998
00011	11805	05431	39808	27732	50725	68248	29405	24201	52775	67851
00012	83452	99634	06288	98083	13746	70078	18475	40610	68711	77817
00013	88685	40200	86507	58401	36766	67951	90364	76493	29609	11062
00014	99594	67348	87517	64969	91826	08928	93785	61368	23478	34113
00015	65481	17674	17468	50950	58047	76974	73039	57186	40218	16544
00016	80124	35635	17727	08015	45318	22374	21115	78253	14385	53763
00017	74350	99817	77402	77214	43236	00210	45521	64237	96286	02655
00018	69916	26803	66252	29148	36936	87203	76621	13990	94400	56418
00019	09893	20505	14225	68514	46427	56788	96297	78822	54382	14598
00020	91499	14523	68479	27686	46162	83554	94750	89923	37089	20048
00021	80336	94598	26940	36858	70297	34135	53140	33340	42050	82341
00022	44104	81949	85157	47954	32979	26575	57600	40881	22222	06413
00023	12550	73742	11100	02040	12860	74697	96644	89439	28707	25815
00024	63606	49329	16505	34484	40219	52563	43651	77082	07207	31790
00025	61196	90446	26457	47774	51924	33729	65394	59593	42582	60527
00026	15474	45266	95270	79953	59367	83848	82396	10118	33211	59466
00027	94557	28573	67897	54387	54622	44431	91190	42592	92927	45973
00028	42481	16213	97344	08721	16868	48767	03071	12059	25701	46670
00029	23523	78317	73208	89837	68935	91416	26252	29663	05522	82562
00030	04493	52494	75246	33824	45862	51025	61962	79335	65337	12472
00031	00549	97654	64051	88159	96119	63896	54692	82391	23287	29529
00032	35963	15307	26898	09354	33351	35462	77974	50024	90103	39333
00033	59808	08391	45427	26842	83609	49700	13021	24892	78565	20106
00034	46058	85236	01390	92286	77281	44077	93910	83647	70617	42941
00035	32179	00597	87379	25241	05567	07007	86743	17157	85394	11838
00036	69234	61406	20117	45204	15956	60000	18743	92423	97118	96338
00037	19565	41430	01758	75379	40419	21585	66674	36806	84962	85207
00038	45155	14938	19476	07246	43667	94543	59047	90033	20826	69541
00039	94864	31994	36168	10851	34888	81553	01540	35456	05014	51176
00040	98086	24826	45240	28404	44999	08896	39094	73407	35441	31880
00041	33185	16232	41941	50949	89435	48581	88695	41994	37548	73043
00042	80951	00406	96382	70774	20151	23387	25016	25298	94624	61171
00043	79752	49140	71961	28296	69861	02591	74852	20539	00387	59579
00044	18633	32537	98145	06571	31010	24674	05455	61427	77938	91936
00045	74029	43902	77557	32270	97790	17119	52527	58021	80814	51748
00046	54178	45611	80993	37143	05335	12969	56127	19255	36040	90324
00047	11664	49883	52079	84827	59381	71539	09973	33440	88461	23356
00048	48324	77928	31249	64710	02295	36870	32307	57546	15020	09994
00049	69074	94138	87637	91976	35584	04401	10518	21615	01848	76938

Example 4 How should h be chosen so that $h(X)$ has the following integrating density function?

$$g(x) = \begin{cases} 0 & \text{if } x < 1 \\ \dfrac{1}{x^2} & \text{if } 1 \leq x \end{cases}$$

In this case,

$$G(x) = \int_{-\infty}^{x} g(t)\, dt = \begin{cases} 0 & \text{if } x < 1 \\ 1 - \dfrac{1}{x} & \text{if } x \geq 1 \end{cases}$$

The inverse to

$$u = 1 - \frac{1}{x}, \qquad 1 < x < \infty$$

is

$$x = \frac{1}{1-u}, \qquad 0 < u < 1$$
$$= h(u)$$

Hence, $1/(1 - X)$ has the required distribution. Notice that this result is essentially shown in Example 1a. That example shows that $1/X$ has the density g. But X and $1 - X$ are each uniformly distributed over $[0, 1]$; therefore, $1/X$ and $1/(1 - X)$ each have the same distribution.

There is a partial converse to (2) as follows:

Suppose X is a random variable whose cdf is continuous. That is,

$$F(t) = P(X \leq t)$$

is a continuous function of the argument t. Then the random variable $F(X)$ is uniformly distributed on $[0, 1]$.

Proof. It is clear that $F(X)$ must lie in $[0, 1]$ with probability 1. We must show that for any x in $[0, 1]$ we have $P[F(X) \leq x] = x$. But the event $[F(X) \leq x]$ is exactly the same as the event $(X \leq t)$, where $F(t) = x$. (Draw a picture to convince yourself.) Hence, $P[F(X) \leq x] = P(X \leq t) = F(t) = x$. ▮

Example 5 Suppose X has the integrating density function that equals $f(x) = e^{-x}$, for $x \geq 0$, and 0 otherwise. Then

$$F(x) = \begin{cases} 0 & \text{if } x < 0 \\ 1 - e^{-x} & \text{if } x \geq 0 \end{cases}$$

The preceding theorem says that $1 - e^{-X}$ is uniformly distributed on $[0, 1]$. This is easily checked by a direct argument.

9.3 FUNCTIONS OF UNIFORMLY DISTRIBUTED RANDOM VARIABLES

Suppose that $X_1, \ldots, X_n$ are independent random variables, each uniformly distributed over $[0, 1]$. According to (4), Section 7.7, the joint integrating density is

$$f(x_1, \ldots, x_n) = \begin{cases} 1 & \text{if each } x_i \text{ is in } [0, 1] \\ 0 & \text{otherwise} \end{cases} \tag{1}$$

In addition to serving as a model for n points selected independently and at random from $[0, 1]$, $X_1, \ldots, X_n$ can also be interpreted as the coordinates of a single point selected at random from the *unit hypercube* in n-space. (This refers to the set of points $x_1, \ldots, x_n$ satisfying $0 \leq x_i \leq 1, i = 1, \ldots, n$. See Figure 9.4 for $n = 2$ and 3.)

To find the cdf of a function of the X_i's, say, $h(X_1, \ldots, X_n)$, it is necessary to compute

$$P[h(X_1, \ldots, X_n) \leq t] = \int \cdots \int_{h(x_1, \ldots, x_n) \leq t} f(x_1, \ldots, x_n)\, dx_1, \ldots, dx_n$$

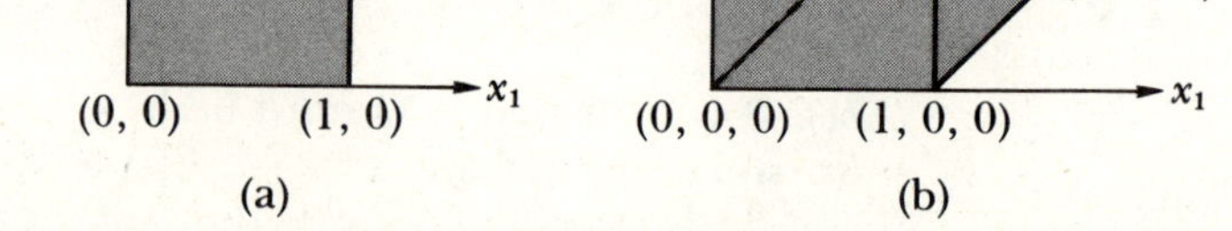

FIGURE 9.4 (a) The two coordinates of a point selected at random from the unit square are independent and uniformly distributed. (b) The three coordinates of a point selected at random from the unit cube are independent and uniformly distributed.

TABLE 9.3

Function	*Integrating density function*
XY	$g(x) = \begin{cases} -\log x & \text{if } 0 < x < 1 \\ 0 & \text{otherwise} \end{cases}$ (Example 1, Section 8.2)
$\dfrac{X}{Y}$	$g(x) = \begin{cases} \dfrac{1}{2} & \text{if } 0 \leqslant x \leqslant 1 \\ \dfrac{1}{2x^2} & \text{if } 1 < x \\ 0 & \text{otherwise} \end{cases}$ (Example 1, Section 8.3)
$X+Y$	$g(x) = \begin{cases} x & \text{if } 0 \leqslant x < 2 \\ 2-x & \text{if } 1 \leqslant x < 2 \\ 0 & \text{otherwise} \end{cases}$ (Example 1, Section 8.4)

which, according to (1), is just the volume of that part of the unit hypercube which intersects the set where $h(x_1, \ldots, x_n) \leqslant t$. The actual computation is often quite indirect, however. When $n = 2$, the distributions of a number of special cases have already been determined in previous examples and exercises. Some of these are summarized in Table 9.3.

The following examples extend the cases of XY and $X + Y$ to the product and sum of n random variables.

Example 1 From Example 2, Section 8.2, the density function of $X_1X_2 \cdots X_n$ is

$$\begin{matrix} \dfrac{(-\log x)^{n-1}}{(n-1)!} & \text{if } 0 < x \leqslant 1 \\ 0 & \text{otherwise} \quad (n = 2, 3, \ldots) \end{matrix} \tag{2}$$

Example 2 The density function of $X_1 + \cdots + X_n$ is

$$k_n(x) = \frac{\binom{n}{0}[c(x)]^{n-1} - \binom{n}{1}[c(x-1)]^{n-1} + \cdots + (-1)^n\binom{n}{n}[c(x-n)]^{n-1}}{(n-1)! \qquad\qquad\qquad\qquad (n-1)!} \tag{3}$$

where

$$c(x) = \begin{cases} 0 & \text{if } x \leqslant 0 \\ x & \text{if } x \geqslant 0 \end{cases}$$

Before proving this assertion, let us consider several special cases. For $n = 2$, (3) equals

$$c(x) - 2c(x-1) + c(x-2) = \begin{cases} 0 & \text{if } x < 0 \\ x & \text{if } 0 \leqslant x < 1 \\ x - 2(x-1) = 2 - x & \text{if } 1 \leqslant x < 2 \\ x - 2(x-1) + x \\ \quad - 2 = 0 & \text{if } 2 \leqslant x \end{cases}$$

Now we will obtain k_3 from k_2 using the convolution formula, (1), Section 8.4. Let f denote the density of the uniform distribution on $[0, 1]$. Then

$$k_3(x) = \int_{\infty}^{\infty} k_2(x-y)f(y)\,dy = \int_0^1 k_2(x-y)\,dy$$

$$k_3(x) = \int_0^1 [c(x-y) - 2c(x-y-1) + c(x-y-2)]\,dy \tag{4}$$

To complete this integration, and for the general proof of (3), we need the following fact:

$$\int_0^1 [c(x-y-k)]^{n-1}\,dy = \frac{[c(x-k)]^n - [c(x-k-1)]^n}{n} \tag{5}$$

The proof of (5) is completely routine and we leave it to the reader. Substitution into (4) gives

$$k_3(x) = \frac{[c(x)]^2 - [c(x-1)]^2 - 2[c(x-1)]^2 + 2[c(x-2)]^2 + [c(x-2)]^2 - [c(x-3)]^2}{2}$$

$$= \frac{\binom{3}{0}[c(x)]^2 - \binom{3}{1}[c(x-1)]^2 + \binom{3}{2}[c(x-2)]^2 - \binom{3}{3}[c(x-3)]^2}{2}$$

which agrees with (3). The general proof is similar and appears as follows. Figure 9.5 shows the graphs of k_1, k_2, and k_3.

Proof of (3). The proof is by induction. We know the assertion is true for $n = 2$. Suppose it is true for all indices $2, \ldots, n$; we will prove that it is also true for $n + 1$.

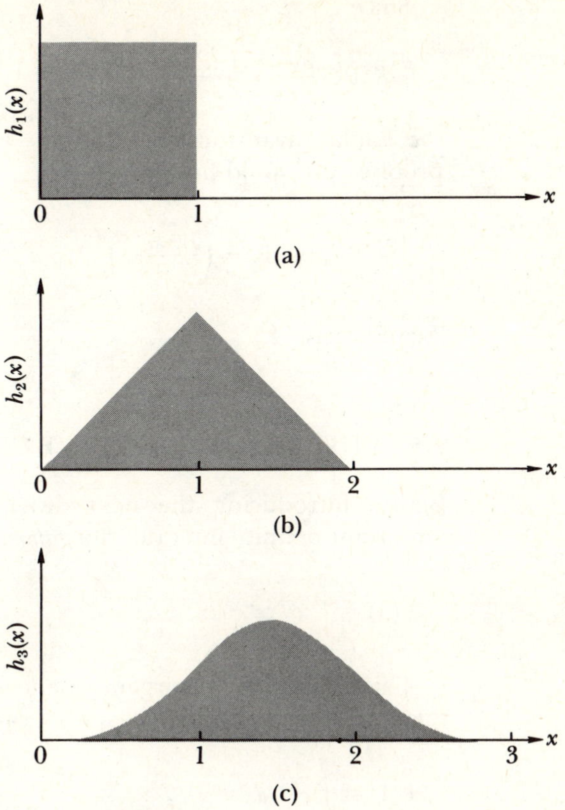

FIGURE 9.5 (a) Density of X_1. (b) Density of $X_1 + X_2$. (c) Density of $X_1 + X_2 + X_3$.

$$k_{n+1}(x) = \int_0^1 k_n(x-y)\,dy, \quad \text{by the convolution formula}$$

$$= \sum_{k=0}^{n} \int_0^1 \binom{n}{k}(-1)^k \frac{[c(x-y-k)]^{n-1}}{(n-1)!}\,dy \quad \text{by the induction hypothesis}$$

$$= \sum_{k=0}^{n} \binom{n}{k}(-1)^k \left\{\frac{[c(x-k)]^n - [c(x-k-1]^n}{n!}\right\} \quad \text{by (5)}$$

$$= \sum_{k=0}^{n} (-1)^k \left[\binom{n}{k} + \binom{n}{k-1}\right] \frac{[c(x-k)]^n + (-1)^n [c(x-n-1)]^n}{n!}$$

$$\left[\text{interpret } \binom{n}{-1} \text{ as } 0\right]$$

This last sum equals (3) since $\binom{n}{k} + \binom{n}{k-1} = \binom{n+1}{k}$ ∎

Since

$$E[\exp(-\theta X_k)] = \frac{1 - e^{-\theta}}{\theta}, \qquad k = 1, \ldots, n$$

the Laplace transform of $X_1 + \cdots + X_n$ is $(1 - e^{-\theta})^n/\theta^n$. Another proof of (3) would be to show that

$$\int_0^\infty e^{-\theta x} k_n(x)\, \mathrm{dx} = \left(\frac{1 - e^{-\theta}}{\theta}\right)^n$$

(See Exercise 4.)

9.4 THE GAMMA DISTRIBUTION

Before introducing the next distribution, we shall review an important definite integral, the *gamma integral,*

$$\Gamma(\alpha) = \int_0^\infty x^{\alpha-1} e^{-x}\, dx \tag{1}$$

This integral is convergent whenever $\alpha > 0$; we shall now check this fact. First, consider the special case $\alpha = 1$,

$$\Gamma(1) = \int_0^\infty e^{-x}\, dx = 1 \tag{2}$$

The convergence for $0 < \alpha < 1$ can be related to (2) as follows. Since

$$\int_0^1 x^{\alpha-1}\, dx = \frac{x^\alpha}{\alpha}\bigg|_0^1 = \frac{1}{\alpha}$$

it follows that

$$0 < \int_0^1 x^{\alpha-1} e^{-x}\, dx \leqslant \int_0^1 x^{\alpha-1}\, dx = \frac{1}{\alpha}$$

We also have that

$$0 < \int_1^\infty x^{\alpha-1} e^{-x}\, dx < \int_1^\infty e^{-x}\, dx < \infty$$

whenever $0 < \alpha < 1$. For $\alpha > 1$, we proceed as follows. By integration by parts,

$$\int_0^\infty x^{\alpha-1}e^{-x}\,dx = -x^{\alpha-1}e^{-x}\Big|_0^\infty + (\alpha-1)\int_0^\infty x^{(\alpha-1)-1}e^{-x}\,dx \tag{3}$$

Since

$$\lim_{x\to\infty} x^{\alpha-1}e^{-x} = 0$$

(3) says that

$$\Gamma(\alpha) = (\alpha-1)\Gamma(\alpha-1) \tag{4}$$

We can iterate (4) to obtain

$$\Gamma(\alpha) = (\alpha-1)(\alpha-2)\Gamma(\alpha-2)$$
$$\vdots$$
$$\Gamma(\alpha) = (\alpha-1)^{(k)}\Gamma(\alpha-k)$$

for any positive integer k, so long as $\alpha - k > 0$. If $\alpha > 1$, then $0 < \alpha - k \leq 1$ for some k. Since the integral for $\Gamma(\alpha - k)$ converges by our earlier remarks, the convergence of (1) for any $\alpha > 1$ follows (Note 2).

Example 1

(a) $\Gamma(\frac{5}{2}) = \frac{3}{2}\frac{1}{2}\Gamma(\frac{1}{2})$

(b) $\Gamma(5) = 4 \cdot 3 \cdot 2 \cdot \Gamma(1) = 4!$

(c) In general, if n is a positive integer, then

$$\Gamma(n) = (n-1)(n-2)\cdots 2\Gamma(1) = (n-1)!$$

A family of important distributions in probability theory are the *gamma distributions,* which have the following integrating density functions:

$$f(x) = \begin{cases} \dfrac{1}{\Gamma(\alpha)}\lambda^\alpha x^{\alpha-1}e^{-\lambda x} & \text{if } x \geq 0 \\ 0 & \text{if } x < 0 \end{cases} \tag{5}$$

In (5), α and λ are *positive* constants and are called the parameters of the distribution.

Since

$$\int_0^\infty \frac{1}{\Gamma(\alpha)}\lambda^\alpha x^{\alpha-1}e^{-\lambda x}\,dx = \int_0^\infty \frac{1}{\Gamma(\alpha)}y^{\alpha-1}e^{-y}\,dy = 1$$

by (1) (make a change of variables $\lambda x = y$), then the functions defined by (5) are, indeed, integrating density functions whenever $\alpha > 0$ and $\lambda > 0$.

The special case of $\alpha = 1$ gives an important distribution called the *exponential distribution.*

exponential distribution

Define an integrating density function f by

$$f(x) = \begin{cases} \lambda e^{-\lambda x} & \text{if } x \geqslant 0 \\ 0 & \text{if } x < 0 \end{cases} \tag{6}$$

It is called the *exponential distribution* with parameter $\lambda > 0$.

Example 2 In quality control engineering, it is often assumed that various components of a piece of apparatus have a lifetime described by the exponential distribution (6). Since $\int_0^\infty x\lambda e^{-\lambda x}\, dx = 1/\lambda$, then λ is the reciprocal of the expected lifetime. To be specific, suppose that the component is a lightbulb. When the lightbulb dies, it is replaced by another, and so forth. Suppose that $X_1, X_2, \ldots$ are the successive lifetimes of the generations of lightbulbs being used in this apparatus. The usual assumption is that each of these X_i's has the same exponential distribution and that they are mutually independent.

Example 3 A geiger counter records the instant of arrival of particles being emitted by some radioactive source. Suppose that the succession of interarrival times between particles is denoted by $X_1, X_2, \ldots$. A usual probability model for this situation is that the X_i's each have the same exponential distribution (6) and they are mutually independent.

If a random variable X is gamma distributed with parameters (α, λ) then λX is gamma distributed with parameters $(\alpha, 1)$; conversely, if X has parameters $(\alpha, 1)$, then X/λ has parameters (α, λ), $(\lambda > 0)$. This fact follows from the simple fact that if X has the integrating density $f(x)$, then X/α has the integrating density $\alpha f(\alpha x)$, if α is a positive constant (see Exercise 10a, Chapter 7). Thus, most computations concerning the gamma distribution can be generalized from the case $\lambda = 1$.

For the $(\alpha, 1)$ gamma distribution, the moments are given by

$$\int_0^\infty x^n \left[\frac{1}{\Gamma(\alpha)}\right] x^{\alpha-1} e^{-x}\, dx = \frac{\Gamma(n+\alpha)}{\Gamma(\alpha)}, \qquad n = 0, 1, 2, \ldots \tag{7}$$

and, for the general case, then the moments are $\Gamma(n+\alpha)/\Gamma(\alpha)\lambda^n$,

$n = 1, 2, \ldots$. In particular, the first moment is

$$\frac{\Gamma(1+\alpha)}{\Gamma(\alpha)\lambda} = \alpha/\lambda \tag{8}$$

The Laplace transform for the $(\alpha, 1)$ gamma distribution is

$$\int_0^\infty e^{-\theta x}\left[\frac{1}{\Gamma(\alpha)}\right]x^{\alpha-1}e^{-x}\,dx = (1+\theta)^{-\alpha}$$

(Make the change of variables $(1+\theta)x = y$.) The Laplace transform for the general (α, λ) distribution is $(1+\theta/\lambda)^{-\alpha}$. The *Fourier transform* for the (α, λ) gamma distribution is $(1 - i\theta/\lambda)^{-\alpha}$. In a formal way, it is made plausible simply by replacing θ in the Laplace transform by $-i\theta$. This "plausible" procedure must be justified, but the details will be omitted here.

Integrals of the form

$$I(\alpha, t) = \int_0^t x^{\alpha-1}e^{-x}\,dx$$

define a function called the *incomplete gamma function.* These are widely tabulated (Note 12, Chapter 6) and can be used to obtain values of the cdf for the gamma distribution.

9.5 FUNCTIONS OF GAMMA-DISTRIBUTED RANDOM VARIABLES

The gamma distribution has the following important *closure property.*

sums of independent gamma random variables are also gamma

> Suppose that X and Y are independent, gamma-distributed random variables, with parameters (α, λ) and (β, λ), respectively (α and β need not be the same, but the λ is the same for both X and Y). Then $X + Y$ is gamma distributed with parameters $(\alpha + \beta, \lambda)$. (1)

First proof of (1). By Section 9.4, the Laplace transforms of X and Y are $(1+\theta/\lambda)^{-\alpha}$ and $(1+\theta/\lambda)^{-\beta}$, respectively. Hence, by independence, the Laplace transform of $X+Y$ is $(1+\theta/\lambda)^{-(\alpha+\beta)}$. Thus, (1) follows from the uniqueness of the Laplace transform. ∎

Second proof of (1). For simplicity, suppose that $\lambda = 1$. By the convolution formula, (1), Section 8.4, the density function of $X + Y$

for positive t is equal to

$$g(t) = \int_0^t \frac{1}{\Gamma(\alpha)\Gamma(\beta)}(t-x)^{\alpha-1}x^{\beta-1}e^{-(t-x)}e^{-x}\,dx$$

$$= \frac{1}{\Gamma(\alpha)\Gamma(\beta)}e^{-t}\int_0^t x^{\beta-1}(t-x)^{\alpha-1}\,dx$$

Make the change of variables $x/t = y$, and then

$$g(t) = \frac{1}{\Gamma(\alpha)\Gamma(\beta)}t^{\alpha+\beta-1}e^{-t}\int_0^1 y^{\beta-1}(1-y)^{\alpha-1}\,dy$$

Since

$$\int_0^\infty g(t)\,dt = 1 = \left[\int_0^1 y^{\beta-1}(1-y)^{\alpha-1}\,dy\right]\left[\frac{1}{\Gamma(\alpha)\Gamma(\beta)}\int_0^\infty t^{\alpha+\beta-1}e^{-t}\,dt\right]$$

and since $\int_0^\infty t^{\alpha+\beta-1}e^{-t}\,dt = \Gamma(\alpha+\beta)$, it follows that

$$\int_0^1 y^{\beta-1}(1-y)^{\alpha-1}\,dy = \frac{\Gamma(\alpha)\Gamma(\beta)}{\Gamma(\alpha+\beta)} \qquad (2)$$

and $g(t) = \dfrac{1}{\Gamma(\alpha+\beta)}t^{\alpha+\beta-1}e^{-t}\,dt \quad \text{if } t \geqslant 0.$

Since $g(t) = 0$ if $t < 0$, Eq. (1) is proved when $\lambda = 1$. The general case follows from the fact that

$$\frac{X}{\lambda}, \frac{Y}{\lambda} \quad \text{and} \quad \frac{X}{\lambda} + \frac{Y}{\lambda}$$

are gamma distributed with parameters (α, λ), (β, λ), and $(\alpha+\beta, \lambda)$, respectively. ∎

We call attention to an important by-product of the proof—the indirect evaluation of the so-called *beta integral*, (2), which will be needed later.

The general version of (1) follows in the same way.

If $X_1, \ldots, X_n$ are mutually independent, gamma-distributed random variables with parameters $(\alpha_1, \lambda), \ldots, (\alpha_n, \lambda)$, respectively, then $X_1 + \cdots + X_n$ is gamma distributed with parameters $(\alpha_1 + \cdots + \alpha_n, \lambda)$.

Example 1 Consider Example 2 in Section 9.4. We can interpret $X_1 + \cdots + X_n$ as the total lifetime of the first n generations of lightbulbs. This total time is gamma distributed with parameters (n, λ). Each lightbulb has expected lifetime $1/\lambda$. The total time that the first n bulbs last has expected value n/λ. [Notice that this agrees with (8), Section 9.4.]

Example 2 Consider Example 3 in Section 9.4. We can interpret $X_1 + \cdots + X_n$ as the total waiting time for the first n particles. This waiting time is gamma distributed with parameters (n, λ). The expected total waiting time for n particles is n/λ.

For various applications to statistics, one needs the distribution of a ratio of independent gamma-distributed random variables. Suppose that X, Y are independent and gamma distributed with parameters (α, λ), (β, λ), respectively. The density function of X/Y is the following.

distribution of a ratio of independent gamma random variables

Suppose that X and Y are independent, gamma-distributed random variables with parameters (α, λ), (β, λ), respectively (λ is the same for both). Then X/Y has the density function

$$h(x) = \begin{cases} \dfrac{\Gamma(\alpha+\beta)}{\Gamma(\alpha)\Gamma(\beta)} \dfrac{x^{\alpha-1}}{(1+x)^{\alpha+\beta}} & \text{if } x \geqslant 0 \\ 0 & \text{otherwise} \end{cases} \tag{3}$$

Proof of (3). Since λX and λY have parameters $(\alpha, 1)$, $(\beta, 1)$, and

$$\frac{X}{Y} = \frac{\lambda X}{\lambda Y}$$

it follows that the distribution does not depend on λ; therefore, we may as well suppose that $\lambda = 1$. According to (1), Section 8.3, for $x \geqslant 0$,

$$h(x) = \int_0^\infty y \left[\frac{1}{\Gamma(\alpha)} (xy)^{\alpha-1} e^{-xy}\right] \left[\frac{1}{\Gamma(\beta)} y^{\beta-1} e^{-y}\right] dy$$

$$= \frac{x^{\alpha-1}}{\Gamma(\alpha)\Gamma(\beta)} \int_0^\infty y^{\alpha+\beta-1} e^{-(1+x)y}\, dy = \frac{\Gamma(\alpha+\beta)}{\Gamma(\alpha)\Gamma(\beta)} \frac{x^{\alpha-1}}{(1+x)^{\alpha+\beta}}$$

after a change of variable $(1+x)y = u$. ∎

For later reference, we also consider the density of $X/(X+Y)$, the values of which are between 0 and 1.

X/(X + Y) is beta distributed

If X and Y are independent, gamma distributed random variables with parameters (α, λ), (β, λ), respectively, then $X/(X+Y)$ has the density function

$$g(x) = \begin{cases} \dfrac{\Gamma(\alpha+\beta)}{\Gamma(\alpha)\Gamma(\beta)} x^{\alpha-1}(1-x)^{\beta-1} & \text{if } 0 \leq x \leq 1 \\ 0 & \text{otherwise} \end{cases} \tag{4}$$

This is an example of a *beta distribution* (Section 9.6).

Proof of (4). The proof follows immediately from (3), using

$$\frac{dP[X/(X+Y) \leq x]}{dx} = \frac{dP[(X/Y) \leq x/(1-x)]}{dx} = \frac{1}{(1-x)^2} h\left(\frac{x}{1-x}\right)$$

(See Exercise 8, Chapter 7.) ▮

It is a curious fact that when the parameters of X and Y are (α, λ) and (β, λ) (λ the same), the pair of random variables $X+Y$ and X/Y are independent. It is not at all obvious, since X and Y each appear in each of the new random variables $X+Y$ and X/Y.

X + Y and X/Y are independent

If X and Y are independent and gamma distributed with parameters (α, λ) and (β, λ), then $X+Y$ and X/Y are independent.

Proof. We suppose $\lambda = 1$, leaving the slightly more general case to the reader. We will evaluate the integral in

$$P\left[(X+Y \leq t) \cap \left(\frac{X}{Y} \leq u\right)\right] = \iint\limits_{\substack{x+y\leq t \\ x/y \leq u}} \frac{1}{\Gamma(\alpha)} x^{\alpha-1} e^{-x} \frac{1}{\Gamma(\beta)} y^{\beta-1} e^{-y}\, dx\, dy$$

This integral is evaluated as follows. Make the transformation $r = x + y$, $s = x/y$. The inverse is $x = sr/(1+s)$, $y = r/(1+s)$, and the Jacobian is

$$J(x, y; r, s) = \begin{vmatrix} \dfrac{\partial x}{\partial s} & \dfrac{\partial x}{\partial r} \\ \dfrac{\partial y}{\partial s} & \dfrac{\partial y}{\partial r} \end{vmatrix} = \begin{vmatrix} \dfrac{r}{(1+s)^2} & \dfrac{s}{(1+s)} \\ \dfrac{-r}{(1+s)^2} & \dfrac{1}{(1+s)} \end{vmatrix} = \frac{r}{(1+s)^2}$$

Hence,

$$P\left[(X+Y \leqslant t) \cap \left(\frac{X}{Y} \leqslant u\right)\right] = \iint\limits_{\substack{r \leqslant t \\ s \leqslant u}} \frac{1}{\Gamma(\alpha)\Gamma(\beta)} \frac{r^{\alpha+\beta-1} s^{\alpha-1} e^{-r}}{(1+s)^{\alpha+\beta}} \, dr \, ds$$

$$= \left[\int_0^u \frac{\Gamma(\alpha+\beta)}{\Gamma(\alpha)\Gamma(\beta)} \frac{s^{\alpha-1}}{(1+s)^{\alpha+\beta}} \, ds\right] \left[\int_0^t \frac{1}{\Gamma(\alpha+\beta)} r^{\alpha+\beta-1} e^{-r} \, dr\right]$$

$$= P(X+Y \leqslant t) P\left(\frac{X}{Y} \leqslant u\right)$$

With no change in argument,

$$P\left[(X+Y \text{ in } A) \cap \left(\frac{X}{Y} \text{ in } B\right)\right] = P(X+Y \text{ in } A) P\left(\frac{X}{Y} \text{ in } B\right)$$

for any sets A and B for which the integral is defined. Hence, $X+Y$ and X/Y are independent. ▮

Notice that as a by-product of the proof, there appear the densities of $X+Y$ and X/Y, which have already been obtained in (1) and (3). Notice also that $X+Y$ and $X/(X+Y)$ are also independent, because

$$\frac{X}{X+Y} = \left[1 + \left(\frac{X}{Y}\right)^{-1}\right]^{-1}$$

An important relationship exists between the gamma distribution and the Poisson distribution. From a purely formalistic point of view, this relationship first appeared as Eq. (5) in Section 6.7. We want to examine this equation from another point of view.

Suppose that $X_1, X_2, \ldots$ is a sequence of mutually independent random variables, each exponentially distributed with the same positive parameter λ. Let

$$S_n = X_1 + \cdots + X_n, \qquad n = 1, 2, \ldots \tag{5}$$

and suppose that $t > 0$. Define Y = number of sums S_n in $[0, t]$. Then Y is Poisson distributed with parameter λt.

Notice that one form of (5) was asserted in (5), Section 6.7.

Proof. That $(Y = 0)$ means $S_1 > t$. Hence,

$$P(Y=0) = \int_t^\infty e^{-\lambda x} \, dx = e^{-\lambda t}$$

Therefore, the assertion is correct for $Y = 0$. If n is a positive integer, then

$$(Y = n) = (S_n \leqslant t) \cap (S_{n+1} > t)$$

(See Figure 9.6.)
Since

$$[(S_n \leqslant t) \cap (S_{n+1} > t)] \cup [(S_n \leqslant t) \cap (S_{n+1} \leqslant t)] = (S_n \leqslant t)$$

and

$$(S_n \leqslant t) \cap (S_{n+1} \leqslant t) = (S_{n+1} \leqslant t)$$

because

$$(S_{n+1} \leqslant t) \subset (S_n \leqslant t)$$

it follows that

$$P(Y = n) + P(S_{n+1} \leqslant t) = P(S_n \leqslant t)$$

Since S_n is gamma distributed with parameters n, λ,

$$P(Y = n) = \int_0^t \frac{1}{\Gamma(n)} \lambda^n x^{n-1} e^{-\lambda x} - \int_0^t \frac{1}{\Gamma(n+1)} \lambda^{n+1} x^n e^{-\lambda x}\, dx$$

Now integrate the second integral by parts, and obtain

$$\int_0^t \frac{1}{\Gamma(n+1)} \lambda^{n+1} x^n e^{-\lambda x}\, dx = \left. \frac{-(\lambda x)^n e^{-\lambda x}}{n!} \right|_0^t + \int_0^t \frac{1}{\Gamma(n)} \lambda^n x^{n-1} e^{-\lambda x}\, dx$$

So, finally,

$$P(Y = n) = \frac{(\lambda t)^n e^{-\lambda t}}{n!} \quad \blacksquare$$

The following example illustrates one typical way in which the preceding relationship arises.

Example 3 A component (such as a lightbulb) is described by the following mathematical model. The first component lives X_1 time units. When it is replaced, its successor lives for a time of X_2, and so forth. Independent random variables $X_1, X_2, \ldots$ are each exponentially distributed with parameter λ. (This was described in Example 2,

FIGURE 9.6

To evaluate the moments of X, it is useful to show first a relation between the normal and gamma distributions.

If X is standard normally distributed, then

X^2 is gamma distributed, $\alpha = \frac{1}{2}$, $\lambda = \frac{1}{2}$

or, equivalently,

$\frac{X^2}{2}$ is gamma distributed, $\alpha = \frac{1}{2}$, $\lambda = 1$

Proof.

$$P\left(\frac{X^2}{2} \leqslant t\right) = P(X^2 \leqslant 2t)$$

$$= \frac{2}{(2\pi)^{1/2}} \int_0^{(2t)^{1/2}} \exp\left(\frac{-x^2}{2}\right) dx$$

$$= \frac{1}{\pi^{1/2}} \int_0^t y^{-1/2} e^{-y}\, dy \quad \left(\text{change of variables, } \frac{x^2}{2} = y\right)$$

Since $\lim_{t\to\infty} P(X^2/2 \leqslant t) = 1$, it follows that

$$\frac{1}{\pi^{1/2}} \int_0^\infty y^{(1/2)-1} e^{-y}\, dy = \frac{\Gamma(1/2)}{\pi^{1/2}} = 1$$

and $X^2/2$ is gamma distributed with parameters (1/2, 1). ∎

We are now in a position to evaluate the moments of X, some of which we shall need for later reference.

moments of standard normal distribution

If X has the standard normal distribution, then

$$E(|X|^n) = \frac{2^{n/2}\Gamma[(n+1)/2]}{\pi^{1/2}}, \qquad n = 1, 2, \ldots \tag{2}$$

$$E(X^{2n+1}) = 0, \qquad n = 0, 1, 2, \ldots \tag{3}$$

According to (2), all moments exist; (3) says that all odd moments are 0.

Proof of (2). The mth moment for the gamma (1/2, 1) distribution is

$$\int_0^\infty x^m \left[\frac{1}{\pi^{1/2}} x^{(1/2)-1} e^{-x}\right] dx = \frac{\Gamma(m+\frac{1}{2})}{\pi^{1/2}} \tag{4}$$

Hence,

$$E(|X|^n) = 2^{n/2} E\left(\frac{X^2}{2}\right)^{n/2} = 2^{n/2} \frac{\Gamma[(n/2)+(1/2)]}{\pi^{1/2}}$$

replacing m by $n/2$ in (4). ▮

Proof of (3). By (2), all moments exist. For an odd moment,

$$\int_{-\infty}^\infty x^{2n+1} f(x)\, dx = 0$$

since $f(x) = f(-x)$. ▮

In particular,

$$EX = 0, \qquad E(X^2) = \frac{2\Gamma(\frac{3}{2})}{\pi^{1/2}} = \frac{2(\frac{1}{2})\Gamma(\frac{1}{2})}{\pi^{1/2}} = 1 \tag{5}$$

We shall now evaluate the Fourier transform.

Fourier transform for the standard normal distribution

If X has the standard normal distribution, then its Fourier transform is $M(\theta) = \exp(-\theta^2/2)$. That is,

$$M(\theta) = E(e^{i\theta X}) = \int_{-\infty}^\infty e^{i\theta x} \frac{1}{(2\pi)^{1/2}} \exp\left(\frac{-x^2}{2}\right) dx = \exp\left(\frac{-\theta^2}{2}\right) \tag{6}$$

Proof of (6). Since $\exp(-x^2/2)$ is an even function of x, it follows that

$$M(\theta) = \int_{-\infty}^\infty (\cos \theta x) \frac{1}{(2\pi)^{1/2}} \exp\left(\frac{-x^2}{2}\right) dx$$

Now differentiate under the integral sign with respect to θ, giving

$$M'(\theta) = -\int_{-\infty}^\infty (\sin \theta x) \frac{1}{(2\pi)^{1/2}} x \exp\left(\frac{-x^2}{2}\right) dx \tag{7}$$

(Note 5.) Now integrate the right side of (7) by parts [let $\sin \theta x = u$, $x \exp(-x^2/2)\, dx = dv$, so $v = -\exp(-x^2/2)$], giving

$$M'(\theta) = (\sin \theta x) \exp\left(\frac{-x^2}{2}\right)\Bigg|_{-\infty}^{\infty} - \theta \int_{\infty}^{\infty} (\cos \theta x) \exp\left(\frac{-x^2}{2}\right) dx = 0 - \theta M(\theta)$$

Since $M'(\theta) = -\theta M(\theta)$, it follows that

$$\int \frac{M'(\theta)}{M(\theta)}\, d\theta = -\int \theta\, d\theta \quad \text{or} \quad \log M(\theta) = \frac{-\theta^2}{2} + \text{constant}$$

Hence, $M(\theta) = \text{constant} \exp(-\theta^2/2)$. Since $M(0) = 1$, it follows that the constant is 1, and therefore $M(\theta) = \exp(-\theta^2/2)$. ▮

An alternative approach to evaluating $M(\theta)$ is to compute

$$A(u) = \int_{-\infty}^{\infty} e^{ux} \frac{1}{(2\pi)^{1/2}} \exp\left(\frac{-x^2}{2}\right) dx$$

$$= \exp\left(\frac{u^2}{2}\right) \int_{-\infty}^{\infty} \frac{1}{(2\pi)^{1/2}} \exp\left[\frac{-(x-u)^2}{2}\right] dx$$

$$= \exp\left(\frac{u^2}{2}\right) \int_{-\infty}^{\infty} \frac{1}{(2\pi)^{1/2}} \exp\left(\frac{-y^2}{2}\right) dy = \exp\left(\frac{u^2}{2}\right)$$

making the change of variables $x - u = y$. Now replace u by $i\theta$ to obtain

$$A(i\theta) = M(\theta) = \exp\left(\frac{-\theta^2}{2}\right)$$

The evaluation of $A(u)$ is certainly correct if u is *real.* It is not at all obvious that replacing u by $i\theta$ is valid without further justification, however.

Extensive tables of the standard normal distribution exist (Note 6). Table 9.4 is a small table of the cumulative distribution, which we shall need later.

In addition to the standard normal distribution, there is the whole family of distributions obtained by linear transformations as follows:

the $N(m, \sigma^2)$ distribution

If X has the standard normal distribution, then the density of

$$\sigma X + m$$

is

$$\frac{1}{(2\pi)^{1/2}\sigma} \exp\left[\frac{-(x-m)^2}{2\sigma^2}\right], \qquad -\infty < x < \infty \tag{8}$$

(m and $\sigma > 0$ are constants). The distribution defined by (8) is called the normal distribution with parameter m and σ. This distribution is usually denoted as $N(m, \sigma^2)$.

TABLE 9.4

t	$(2\pi)^{-1} \int_{-\infty}^{t} \exp(-x^2/2)\,dx$	t	$(2\pi)^{-1} \int_{-\infty}^{t} \exp(-x^2/2)\,dx$
−3.0	0.0013	0	0.5000
−2.9	0.0019	0.1	0.5398
−2.8	0.0026	0.2	0.5793
−2.7	0.0035	0.3	0.6179
−2.6	0.0047	0.4	0.6554
−2.5	0.0062	0.5	0.6915
−2.4	0.0082	0.6	0.7258
−2.3	0.0107	0.7	0.7580
−2.2	0.0139	0.8	0.7881
−2.1	0.0179	0.9	0.8159
−2.0	0.0227	1.0	0.8413
−1.96	0.0250	1.1	0.8643
−1.9	0.0287	1.2	0.8849
−1.8	0.0359	1.3	0.9032
−1.7	0.0446	1.4	0.9192
−1.645	0.0500	1.5	0.9332
−1.6	0.0548	1.6	0.9452
−1.5	0.0668	1.645	0.9500
−1.4	0.0808	1.7	0.9554
−1.3	0.0968	1.8	0.9641
−1.2	0.1151	1.9	0.9713
−1.1	0.1357	1.96	0.9750
−1.0	0.1587	2.0	0.9773
−0.9	0.1841	2.1	0.9821
−0.8	0.2119	2.2	0.9861
−0.7	0.2420	2.3	0.9893
−0.6	0.2742	2.4	0.9918
−0.5	0.3085	2.5	0.9938
−0.4	0.3446	2.6	0.9953
−0.3	0.3821	2.7	0.9965
−0.2	0.4207	2.8	0.9974
−0.1	0.4602	2.9	0.9981
		3.0	0.9987

Proof.

$$P(\sigma X + m \leqslant t) = P\left(X \leqslant \frac{t-m}{\sigma}\right)$$

$$= \int_{-\infty}^{(t-m)/\sigma} f(x)\,dx = \frac{1}{\sigma}\int_{-\infty}^{t} f\left(\frac{x-m}{\sigma}\right) dx$$

Hence, $f[(x-m)/\sigma]/\sigma$ is the density of $\sigma X + m$, which is what (8) asserts. ∎

In terms of the preceding notation, $N(0, 1)$ refers to the standard normal distribution.

Since $E(\sigma X + m) = m$, the center of gravity of the $N(m, \sigma^2)$ distribution is m.

The Fourier transform for the $N(m, \sigma)$ distribution is

$$E[e^{i\theta(\sigma X+m)}] = e^{i\theta m}E(e^{i\theta\sigma x}) = \exp\left(\frac{i\theta m - \theta^2\sigma^2}{2}\right) \tag{9}$$

[The last expected value is evaluated by replacing θ by $\theta\sigma$ in (6).]

It follows from the preceding assertions that if X is $N(m, \sigma^2)$, then $(X - m)/\sigma$ is $N(0, 1)$. It also follows that if X is $N(m, \sigma^2)$, then $aX + b$ is $N(am + b, a^2\sigma^2)$. These facts are easy to verify and are left to the reader.

Figure 9.8 shows the graphs of normal density functions for several values of m and σ.

Example 3 A scientist makes repeated measurements on the temperature of a liquid. Suppose the mathematical model is that the measurements, determined according to the centigrade scale, are normally dis-

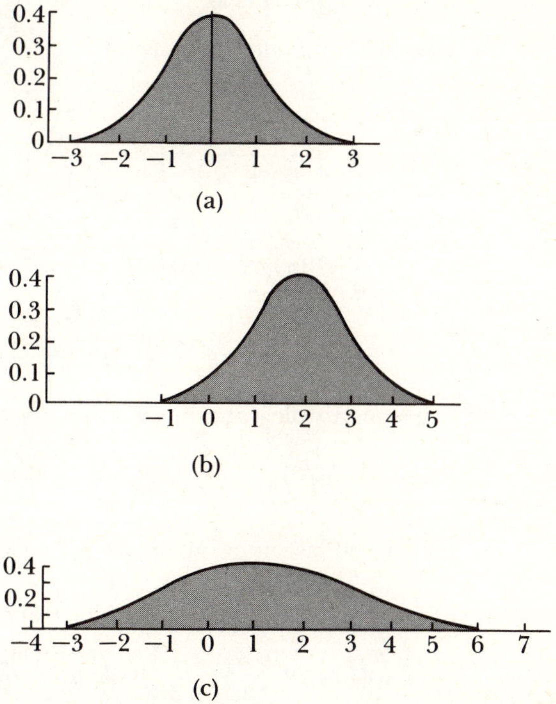

FIGURE 9.8 (a) Normal density, $m = 0$, $\sigma = 1$. (b) Normal density, $m = 2$, $\sigma = 1$. (c) Normal density, $m = 1$, $\sigma = 2$.

tributed with parameters $m = 10$, $\sigma = 1$. If the measurements are referred to a Fahrenheit scale, then, since $F = 1.8C + 32$, the temperature distribution is normal, with parameters

$$m = 1.8(10) + 32 = 50, \qquad \sigma = 1.8$$

9.8 SUMS OF NORMAL RANDOM VARIABLES

An important property of the normal distribution is the *closure* property, that sums of independent normal random variables are still normal.

a sum of independent normal random variables is normal

If X and Y are independent random variables with distributions $N(m, \sigma_1^2)$, $N(m_2, \sigma_2^2)$, respectively, then $X + Y$ has the $N(m_1 + m_2, \sigma_1^2 + \sigma_2^2)$ distribution.

Proof. We shall suppose that $m_1 = m_2 = 0$, since if we establish the result in that case, the general case follows by considering

$$(X + m_1) + (Y + m_2) = (X + Y) + (m_1 + m_2)$$

By the convolution formula—(1), Section 8.4—the density of $X + Y$ is

$$g(x) = \int_{-\infty}^{\infty} \frac{1}{(2\pi)^{1/2}\sigma_1} \exp\left[\frac{-(x-y)^2}{2\sigma_1^2}\right] \frac{1}{(2\pi)^{1/2}\sigma_2} \exp\left(\frac{-y^2}{2\sigma_2^2}\right) dy$$

$$= \frac{1}{[(2\pi)(\sigma_1^2 + \sigma_2^2)]^{1/2}} \exp\left[\frac{-x^2}{2(\sigma_1^2 + \sigma_2^2)}\right]$$

$$\times \int_{-\infty}^{\infty} \frac{1}{a(2\pi)^{1/2}} \exp\left[\frac{-(y-b)^2}{2a^2}\right] dy$$

by some unpleasant, but routine algebra, where

$$a^2 = \frac{(\sigma_1\sigma_2)^2}{\sigma_1^2 + \sigma_2^2}, \qquad b = \frac{xa^2}{\sigma_1^2}$$

Since the last integral equals 1, the result follows. ∎

Second proof. By (9), the Fourier transforms of X and Y are

$$\exp\left(i\theta m_1 - \frac{\theta^2\sigma_1^2}{2}\right), \qquad \exp\left(i\theta m_2 - \frac{\theta^2\sigma_2^2}{2}\right)$$

Hence, the Fourier transform of $X + Y$ is the product of these, and equals

$$\exp\left[i\theta(m_1 + m_2) - \frac{\theta^2}{2}(\sigma_1^2 + \sigma_2^2)\right]$$

This is the Fourier transform of the $N(m_1 + m_2, \sigma_1^2 + \sigma_2^2)$ distribution. ∎

By successive application of this result, we obtain the following.

If $X_1, \ldots, X_n$ are mutually independent random variables having the distributions $N(m_1, \sigma_1^2), \ldots, N(m_n, \sigma_n^2)$, then $X_1 + \cdots + X_n$ is normally distributed with parameters

$$m = m_1 + \cdots + m_n \qquad \sigma^2 = \sigma_1^2 + \cdots + \sigma_n^2$$

An important special case follows.

If $X_1, \ldots, X_n$ are mutually independent, standard normally distributed random variables, then

$$\frac{X_1 + \cdots + X_n}{n^{1/2}}$$

also has the standard normal distribution.

This follows from the fact that each of $X_1/n^{1/2}, \ldots, X_n/n^{1/2}$ is $N(0, n^{-1})$. Hence, the sum is normal with parameters $m = 0$ and $\sigma^2 = (1/n) + \cdots + (1/n) = 1$.

If $X_1, \ldots, X_n$ are mutually independent random variables, each with the same distribution, and

$$\frac{X_1 + \cdots + X_n}{n^{1/2}}, \qquad n = 1, 2, \ldots$$

also has that same distribution, then that distribution *can only be* $N(0, 1)$. We shall indicate later in Section 10.3 why this rather remarkable characterization of the normal distribution works.

A related and even more remarkable fact is the following. Suppose that $X_1, X_2, \ldots$ is any sequence of mutually independent random variables, each with the same distribution (whether normal or not) and

$$EX_i = 0, \qquad EX_i^2 = 1$$

Then the distribution of

$$\frac{X_1 + \cdots + X_n}{n^{1/2}}$$

"approaches" the standard normal distribution as $n \to \infty$. The exact sense in which this "approach to normality" takes place will be explained in Section 10.3. This theorem is called the *central limit theorem,* and it is one of the reasons for the importance of the normal distribution.

9.9 THE CHI-SQUARE DISTRIBUTION

It was shown in Section 9.7 that if X has the $N(0, 1)$ distribution, then X^2 is gamma distributed with parameters $\alpha = 1/2$, $\lambda = 1/2$. It follows that if $X_1, \ldots, X_n$ are mutually independent, each distributed $N(0, 1)$, then $X_1^2 + \cdots + X_n^2$ is gamma distributed. The gamma distribution is called the *chi-square distribution,* with n degrees of freedom when the parameter values are specified to be $\alpha = 1/2$, $\lambda = n/2$.

chi-square distribution

Suppose that $X_1, \ldots, X_n$ are mutually independent, $N(0, 1)$ random variables. Then $X_1^2 + \cdots + X_n^2$ is gamma distributed with parameters $\alpha = n/2$, $\lambda = 1/2$, or, equivalently, it has the *chi-square distribution with n degrees of freedom* (Note 7).

The fact that $X_1^2 + \cdots + X_n^2$ is gamma, $(n/2, 1/2)$, follows from the closure property of the gamma distribution, described in Section 9.5. In particular, $X_1^2 + X_2^2$ is exponentially distributed with parameter $\lambda = 1/2$; equivalently, $(X_1^2 + X_2^2)/2$ is exponentially distributed with $\lambda = 1$.

Example 1 Suppose that a fixed point in the plane is being fired at by projectiles. For convenience, say the target point is the origin (0, 0) of the (x, y) plane. A common model in ballistics is that the hitting point of a projectile is described by (X, Y), a pair of independent, normally distributed random variables, each $N(0, \sigma^2)$. Since $EX = EY = 0$, the "average position" of the projectile is at the target. The distance of the hit from the target is $(X^2 + Y^2)^{1/2}$. Since $(X^2 + Y^2)/2\sigma^2$ is exponentially distributed with parameter $\lambda = 1$ (why?), it follows that

$$P[(X^2 + Y^2)^{1/2} \le t] = P\left(\frac{X^2 + Y^2}{2\sigma^2} \le \frac{t^2}{2\sigma^2}\right) = 1 - \exp\left(\frac{-t^2}{2\sigma^2}\right)$$

Hence, the distance from the target has the density function $(1/\sigma^2)t \exp(-t^2/2\sigma^2)$, $0 \le t$. The expected distance from the target is

$$\int_0^\infty \left(\frac{1}{\sigma^2}\right)t^2 \exp\left(\frac{-t^2}{2\sigma^2}\right) dt = \sqrt{2}\sigma \int_0^\infty u^{(3/2)-1}e^{-u}\, du = \sqrt{2}\sigma\Gamma\left(\frac{3}{2}\right) = \left(\frac{\pi}{2}\right)^{1/2}\sigma$$

Notice the role of the parameter σ. The greater is σ, the more erratic is the behavior of the projectile.

Example 2 [Example (1) continued.] Suppose that n projectiles are fired independently at the target. A model for this situation is that the hitting points $(X_1, Y_1), (X_2, Y_2), \ldots, (X_n, Y_n)$ are $2n$ independent, normally distributed random variables, each $N(0, \sigma^2)$. The closest distance to the target achieved by any of the n targets is

$$\min(X_1^2 + Y_1^2)^{1/2}, \ldots, (X_n^2 + Y_n^2)^{1/2} = D$$

The probability distribution for D can be obtained as follows:

$$\begin{aligned} P(D > t) &= P(D^2 > t^2) \\ &= P\left[\left(\frac{X_1^2 + Y_1^2}{2\sigma^2} > \frac{t^2}{2\sigma^2}\right) \cap \cdots \cap \left(\frac{X_n^2 + Y_n^2}{2\sigma^2} > \frac{t^2}{2\sigma^2}\right)\right] \\ &= \left(\exp\frac{-t^2}{2\sigma^2}\right)^n \end{aligned}$$

by independence. Hence, the density function of D is $(n/\sigma^2)t \cdot \exp(-nt^2/2\sigma^2)$, $0 \leq t$. The expected value of D is

$$\int_0^\infty \left(\frac{n}{\sigma^2}\right)t^2 \exp\left(\frac{-nt^2}{2\sigma^2}\right) dt = \left(\frac{\pi}{2}\right)^{1/2} \frac{\sigma}{\sqrt{n}}$$

(See Exercise 22.)

Example 3 A fact made known by the physicist Maxwell is that the distribution of the velocity of a molecule in a gas is

$$f(v) = \begin{cases} \left(\frac{2}{\pi}\right)^{1/2} \lambda^{3/2} v^2 \exp\left(\frac{-\lambda v^2}{2}\right) & \text{if } 0 < v \\ 0 & \text{if } v \leq 0 \end{cases}$$

where λ is related to the physical nature of the gas (Note 8). If a random variable V has this density, then the density of V^2 is

$$\frac{1}{2\sqrt{v}} f(\sqrt{v}) = \begin{cases} \frac{1}{\sqrt{2\pi}} \lambda^{3/2} v^{(3/2)-1} \exp\left(\frac{-\lambda v}{2}\right) & \text{if } 0 < v \\ 0 & \text{if } v \leq 0 \end{cases}$$

It follows that V^2 is gamma distributed with parameters 3/2, $\lambda/2$. Suppose that V_1, V_2, V_3 are the components of the velocity in each of three orthogonal directions in three-dimensional space. In statistical mechanics, it is usually assumed that V_1, V_2, V_3 are independent random variables the squares of which are gamma distributed with parameters $\lambda/2$ and $\alpha = 1/2$. Hence, $V^2 = V_1^2 + V_2^2 + V_3^2$ is gamma distributed with parameters $\lambda/2$ and $\alpha = 3/2$, as previously asserted. Equivalently, λV^2 is chi-square distributed with three degrees of freedom.

SUMMARY

A random variable is *uniformly distributed* over an interval if it has an integrating density function which is constant over that interval and is 0 elsewhere. Any distribution can be represented as the distribution of $h(X)$ for an appropriately determined h, with X uniformly distributed over $[0, 1]$.

If a random variable X has an integrating density which is $[\Gamma(\alpha)]^{-1}\lambda^\alpha x^{\alpha-1}e^{-\lambda x}$ for $x \geqslant 0$ and is zero otherwise, then X is said to be *gamma distributed.* The parameters α and λ must be positive. In particular, if the parameter α equals 1, the distribution is the *exponential distribution.* If $X_1, \ldots, X_n$ are independent and gamma distributed with parameters $\alpha_1, \lambda; \ldots; \alpha_n, \lambda$, then $X_1 + \cdots + X_n$ is also gamma distributed with parameters $\alpha_1 + \cdots + \alpha_n, \lambda$.

Suppose $X_1, X_2, X_3, \ldots$ are independent and exponentially distributed with the same parameter, λ. The number of sums, X_1, $X_1 + X_2$, $X_1 + X_2 + X_3, \ldots$ that fall in the interval $[0, t]$ is Poisson distributed with parameter λt.

The *beta distribution* is described by the integrating density function, which is $[\Gamma(\alpha+\beta)/\Gamma(\alpha)\Gamma(\beta)]x^{\alpha-1}(1-x)^{\beta-1}$ for x in $[0, 1]$ and 0 elsewhere. Parameters α and β are positive parameters. If X and Y are independent random variables, each gamma distributed with parameters α, λ and β, λ, then $X/(X+Y)$ has the preceding beta distribution. The kth smallest of n points selected at random from the unit interval is beta distributed, the parameters depending on n and k.

A random variable is *normally distributed* if it has an integrating density given by $(2\pi)^{-1/2}\sigma^{-1} \exp -[(x-m)^2/\sigma^2]$. The special case of $m = 0$, $\sigma = 1$, is that of the *standard normal distribution.* A random variable that is normally distributed with parameters m and σ is expressible as $\sigma X + m$ where X has the standard normal distribution. If $X_1, \ldots, X_n$ are independent and normally distributed with parameters $m_1, \sigma_1; \ldots; m_n, \sigma_n$, then $X_1 + \cdots + X_n$ is also normally distributed with $m = m_1 + \cdots + m_n$ and $\sigma^2 = \sigma_1^2 + \cdots + \sigma_n^2$.

The distribution of a sum of squares of n independent, standard normal random variables is called the *chi-square distribution* with n degrees of freedom. This is a special case of a gamma distribution.

EXERCISES

1 Suppose that X is uniformly distributed over $[a, b]$. Show that $cX + d$ is uniformly distributed over $[ca + d, cb + d]$ (suppose $c > 0$). In particular, if X is uniform over $[0, 1]$, then cX is uniform over $[0, c]$.

2 Suppose that $X_1, \ldots, X_n$ are mutually independent random variables, X_k being uniformly distributed over $[k, k+1]$, $k = 1, 2, \ldots, n$. Show that $X_1 + \cdots + X_n - \binom{n+1}{2}$ has the density given by (3), Section 9.3.

3 Show that

$$\sum_{k=0}^{n} \binom{n}{k} (-1)^k [c(x-k)]^{n-1} = 0 \quad \text{if } x > n$$

[Hint: As given by (3), Section 9.3, $k_n(x)$ must be 0 for $x > n$. Why?]

4 (a) Consider the function $c(x)$ defined in Example 2, Section 9.3. Show that

$$\int_0^\infty e^{-\theta x}[c(x-k)]^{n-1}\, dx = \frac{(n-1)!e^{-\theta k}}{\theta^n}$$

[Hint: The preceding integral can be written as

$$\int_k^\infty e^{-\theta(x-k+k)}(x-k)^{n-1}\, dx = e^{-\theta k} \int_0^\infty e^{-\theta y} y^{n-1}\, dy \Bigg]$$

(b) Let $k_n(x)$ be the density defined in (3), Example 2, Section 9.3. Use (a) to show that

$$\int_0^\infty e^{-\theta x} k_n(x)\, dx = \left(\frac{1 - e^{-\theta}}{\theta}\right)^n$$

Why does this give an alternative proof of the assertion in Example 2, Section 9.3? [Hint: The integral equals

$$\sum_{k=0}^{n} \binom{n}{k} (-1)^k \frac{e^{-\theta k}}{\theta^n} \Bigg]$$

5 (a) Suppose that X and Y are independent and each is uniformly distributed over $[0, 1]$. Define

$$U = \begin{cases} X + Y & \text{if } 0 \leq X + Y \leq 1 \\ (X + Y) - 1 & \text{if } 1 \leq X + Y \leq 2 \end{cases}$$

Show that U is uniformly distributed on $[0, 1]$. [Hint: $P(U \leq t) = \int_0^t x\,dx + \int_1^{1+t} (2-x)\,dx = t$.]

(b) Generalize (a). Suppose that $X_1, X_2, \ldots$ are independent random variables, each uniformly distributed in $[0, 1]$. Define, for $t \geq 0$,

$$e(t) = \begin{cases} t & \text{if } 0 \leq t < 1 \\ t-1 & \text{if } 1 \leq t < 2 \\ t-2 & \text{if } 2 \leq t < 3 \\ \vdots & \\ \vdots & \end{cases}$$

(In other words, $e(t) = t - [t]$, the excess of t over $[t]$, the greatest integer contained in t.) Show that $U_n = e(X_1 + \cdots + X_n)$ is uniformly distributed on $[0, 1]$. {Hint: $U_2 = U = e(X_1 + X_2)$ is uniformly distributed over $[0, 1]$ from part (a). Now use the fact that $e[e(X_1 + \cdots + X_{n-1}) + X_n] = e(X_1 + \cdots + X_{n-1} + X_n)$ (draw a picture!) as the basis for an induction argument.}

6 Suppose that $X_1, \ldots, X_n$ are independent random variables, exponentially distributed with parameters $\lambda_1, \ldots, \lambda_n$. Show that $\min(X_1, \ldots, X_n) = Y$ is exponentially distributed with parameters $\lambda = \lambda_1 + \cdots + \lambda_n$. {Hint: $P(Y \leq t) = 1 - P(Y > t) = 1 - P[(X_1 > t) \cap \cdots \cap (X_n > t)] = 1 - P(X_1 > t) \cdots P(X_n > t)$.}

7 Show that an exponentially distributed random variable X has the following property:

$$P(0 \leq X - u \leq t \mid 0 \leq X - u) = P(X \leq t)$$

for any positive constants t and u. In other words, the *conditional* distribution of the amount by which X exceeds u, given that it exceeds u, is the same as the distribution of X itself. [Compare (6), Section 6.4.]

8 Suppose that X and Y are independent random variables, each exponentially distributed with the same parameter λ. Let $U = \min(X, Y)$, $V = \max(X, Y)$.

(a) Show that the joint integrating density of U and V is

$$f(u, v) = \begin{cases} 2e^{-\lambda(u+v)} & \text{if } 0 \leq u \leq v \\ 0 & \text{otherwise} \end{cases}$$

{Hint:

$$\begin{aligned} P[(U, V) \text{ in } A] &= P[(X, Y) \text{ in } A) \cap (X \leq Y)] + P[(Y, X) \text{ in } A) \\ &\qquad \cap (Y < X)] \\ &= 2 \iint_A f(u, v)\,du\,dv\} \end{aligned}$$

(b) Show that U and $V - U$ are independent, exponential, with parameters 2λ and λ, respectively. {Hint: Let f be as in (a).

$$P[(U \leq r) \cap (V - U \leq s)] = \iint\limits_{\substack{0 \leq u \leq r \\ 0 \leq v-u \leq s}} 2e^{-\lambda(u+v)}\, du\, dv$$

Make a change of variables, $u = x$, $v - u = y$.}

9 (a) Suppose that X is a nonnegative random variable with integrating density function f. Show that e^{-X} has the integrating density

$$g(x) = \begin{cases} \dfrac{f(-\log x)}{x} & 0 < x < 1 \\ 0 & \text{otherwise} \end{cases}$$

[Hint: $P(e^{-X} \leq x) = P(X \geq -\log x)$. Now differentiate this cdf.]

(b) If X is gamma distributed with parameters $(\alpha, 1)$, show that e^{-X} has the density

$$g(x) = \begin{cases} \dfrac{(-\log x)^{\alpha-1}}{\Gamma(\alpha)} & 0 < x < 1 \\ 0 & \text{otherwise} \end{cases}$$

(c) Use (b) to explain the density function in Example 1, Section 9.3. (Hint: If Y is uniform on [0, 1], then $-\log Y$ is exponential.)

10 Suppose that X and Y are independent, nonnegative, random variables with integrating density functions. The distribution of X is arbitrary and the distribution of Y is exponential with parameter $\lambda = 1$. Show that the distribution of $Y - X$ has the following property:

$$P(0 \leq Y - X \leq t | Y - X \geq 0) = \lambda \int_0^t e^{-\lambda x}\, dx$$

{Hint: The required conditional probability is

$$\frac{P(0 \leq Y - X \leq t)}{P(0 \leq Y - X)} = \frac{P[1 \leq (e^{-X}/e^{-Y}) \leq e^t]}{P[1 \leq (e^{-X}/e^{-Y})]}$$

e^{-Y} is uniform on [0, 1]. Use Exercise 8b, Chapter 8.}

11 Consider the following process. A point U_1 is picked at random in $[0, 1]$. Then a second point U_2 is picked at random in $[0, U_1]$. Then a third point U_3 is picked at random in $[0, U_2]$, and so forth.

(a) Convince yourself that an appropriate model for the distribution of $U_1, \ldots, U_n, \ldots$ is that

$$U_1 = X_1,\ U_2 = X_1X_2, \ldots, U_n = X_1X_2 \cdots X_n, \ldots$$

where $X_1, \ldots, X_n, \ldots$ are mutually independent, each uniformly distributed on [0, 1].

(b) Let Z be the number of points $U_1, U_2, \ldots$ that fall in $[c, 1]$, $(0 < c < 1)$. Show that Z is Poisson distributed with parameter $\lambda = -\log c$. {Hint:

$$P(Z = k) = P[(X_1 \cdots X_k \geq c) \cap (X_1 \cdots X_{k+1} < c)]$$

$$= P\left[\left(\sum_{i=1}^{k} -\log X_i \leq -\log c\right) \cap \left(\sum_{i=1}^{k+1} -\log X_i > -\log c\right)\right]$$

Now use the fact that $-\log X_1, -\log X_2, \ldots$ are mutually independent, each exponentially distributed with parameter $\lambda = 1$, and consider the relationship between the Poisson and exponential distribution shown in (5), Section 9.5.}

12 Suppose that $X_1, X_2, \ldots, X_n$ are mutually independent random variables, each exponential with the same parameter λ.

(a) Let $M_n = \max(X_1, X_2, \ldots, X_n)$. Show that the integrating density of M_n is

$$f(x) = \begin{cases} n\lambda(1 - e^{-\lambda x})^{n-1}e^{-\lambda x} & \text{if } x \geq 0 \\ 0 & \text{otherwise} \end{cases}$$

(b) Manipulate the integral for expected value,

$$E(M_n) = \int_0^\infty n\lambda x(1 - e^{-\lambda x})^{n-1}e^{-\lambda x}\, dx$$

to show that

$$EM_n - EM_{n-1} = \frac{1}{n\lambda}$$

{Hint: Integrate the following by parts to obtain

$$EM_n = \int_0^\infty [1 - (1 - e^{-\lambda x})^n]\, dx$$

Then write this integral as

$$\int_0^\infty [1 - (1 - e^{-\lambda x})^{n-1}(1 - e^{-\lambda x})]\, dx = EM_{n-1} + \frac{1}{n\lambda}\}$$

(c) Conclude from (b) that

$$EM_n = \frac{[1 + \frac{1}{2} + \cdots + (1/n)]}{\lambda}$$

13 Cars arrive at a tunnel entrance according to the following model. Successive waiting times between car arrivals are independent exponential random variables, each with the same parameter, λ (λ is in units of minutes).

(a) What is the expected number of cars to arrive per hour? (Answer: 60 λ. See Example 3, Section 9.5.)

(b) What is the expected value of the maximum interarrival time among the first 10 cars? (Count the wait for the first car as one of the times.) [Answer: $(1/\lambda)(1 + \frac{1}{2} + \cdots + \frac{1}{10})$ minutes. See Exercise 12c.]

(c) What is the expected value of the minimum interarrival time among the first 10 cars? (Answer: $1/10\lambda$.)

14 Suppose that X and Y are independent random variables, each exponentially distributed with the same parameter λ. Show that

$$\frac{X}{X+Y}$$

is uniformly distributed over $[0, 1]$. (Hint: See Examples 1 and 2, Section 9.6.)

15 Suppose that X is beta distributed with parameters α and β. Show that $1 - X$ is beta distributed with parameters β and α.

16 Suppose X is beta distributed with parameters α, β. Show that

(a) $$E(X^2) = \frac{(\alpha + 1)\alpha}{(\alpha + \beta + 1)(\alpha + \beta)}$$

(b) $$E[X^2(1 - X)^2] = \frac{(\alpha + 1)\alpha(\beta + 1)\beta}{(\alpha + \beta + 3)(\alpha + \beta + 2)(\alpha + \beta + 1)(\alpha + \beta)}$$

17 Suppose that X and Y are independent and normally distributed.

X is $N(5, 4)$, $\quad Y$ is $N(-4, 1)$

(a) What are the values of

$$P(X > 5), \qquad P(Y < -4)$$

$$P(X + Y > 1), \qquad P\left(\frac{3X + 2Y - 7}{12} \geq 0\right)$$

(Answers: $\frac{1}{2}$ in each case.)

(b) Use Table 9.4 for approximate values of the following:

$$P(X + Y > 1 + \sqrt{5}), \qquad P(3X + 2Y - 7 > \sqrt{40})$$

(Answers: 0.1587 in each case.)

18 Suppose X has the $N(m, \sigma^2)$ distribution.
(a) Show that $(X - m)/\sigma$ has the $N(0, 1)$ distribution.
(b) Show that $aX + b$ has the $N(am + b, |a|\sigma)$ distribution.

19 If $X_1, \ldots, X_n$ are mutually independent, each distributed $N(0, 1)$, show that $a_1X_1 + \cdots + a_nX_n$ is

$$N(0, a_1^2 + \cdots + a_n^2), \qquad (a_1, \ldots, a_n \text{ are constants})$$

20 Suppose that X and Y are independent, each distributed $N(0, 1)$.

Show that

$$U = \frac{X+Y}{\sqrt{2}}, \qquad V = \frac{X-Y}{\sqrt{2}}$$

are also independent $N(0, 1)$. {Hint:

$$P[(U \leq u) \cap (V \leq v)] = \frac{1}{2\pi} \iint_{\substack{(x+y)/\sqrt{2} \leq u \\ (x-y)/\sqrt{2} \leq v}} \exp\left[\frac{-(x^2+y^2)}{2}\right] dx\, dy$$

Make a change of variables, $(x+y)/\sqrt{2} = r$, $(x-y)/\sqrt{2} = s$. The Jacobian is 1 and $x^2 + y^2 = r^2 + s^2$. Hence, the required probability is

$$\iint_{\substack{r \leq u \\ s \leq v}} \left[\frac{1}{(2\pi)^{1/2}} \exp\left(\frac{-r^2}{2}\right)\right]\left[\frac{1}{(2\pi)^{1/2}} \exp\left(\frac{-s^2}{2}\right)\right] dr\, ds\}$$

21 Suppose that X and Y are independent, each distributed $N(0, 1)$.
(a) Show that $2XY$ and $X^2 - Y^2$ each have the same distribution. [Hint: $X^2 - Y^2 = 2[(X-Y)/\sqrt{2}][(X+Y)/\sqrt{2}]$. Now use Exercise 20.]
(b) Show that XY has the Fourier transform,

$$M(\theta) = \frac{1}{(1+\theta^2)^{1/2}}$$

[Hint: $X^2/2$ has the transform $(1 - i\theta)^{-1/2}$, by Section 9.9.]

22 Projectiles are fired at the origin of an (x, y) coordinate system. The mathematical model is that the point which is hit, (X, Y), consists of a pair of independent, $N(0, 1)$ random variables. Suppose that two projectiles are fired independently and let D be the distance between the two points that are hit. Show that D^2 is exponentially distributed, $\lambda = 1/4$. [Hint: X_1, Y_1, X_2, Y_2 are independent, $N(0, 1)$. $D^2 = (X_1 - X_2)^2 + (Y_1 - Y_2)^2$ is sum of two gamma variables, $\alpha = 1/2$, $\lambda = ?$. Compare Examples 1 and 2, Section 9.9.]

23 If X has the $N(0, 1)$ distribution, show that the density function of $|X|$ is

$$g(x) = \begin{cases} \left(\dfrac{2}{\pi}\right)^{1/2} \exp\left(\dfrac{-x^2}{2}\right) & \text{if } x \geq 0 \\ 0 & \text{otherwise} \end{cases}$$

24 Suppose that X and Y are independent, each distributed $N(0, 1)$.
(a) Show that $|X|/|Y|$ has the density

$$h(x) = \begin{cases} \dfrac{2}{\pi(1+x^2)} & \text{if } x \geq 0 \\ 0 & \text{otherwise} \end{cases}$$

[Hint: Use (1), Section 8.3, and Exercise 23.]

(b) Show that $X/|Y|$ has the Cauchy distribution, with density,

$$k(x) = \frac{1}{\pi(1 + x^2)}, \qquad -\infty < x < \infty$$

{Hint: First show that k must be symmetric about 0, $[k(x) = k(-x)]$. Then use (a).}

25 Suppose that Y is $N(0, 1)$. Show that the Laplace transform of $1/Y^2$ is

$$L(\theta) = \exp[-(2\theta)^{1/2}]$$

[Hint: Suppose that X and Y are independent, $N(0, 1)$. By Exercise 24b and Eq. (7), Section 8.6, $E(e^{i\theta X/|Y|}) = e^{-|\theta|}$. Evaluate this alternatively as follows. Let f be the $N(0, 1)$ density function

$$E(e^{i\theta X/|Y|}) = \int_{-\infty}^{\infty} f(y)\, dy \int_{-\infty}^{\infty} e^{i\theta(x/|y|)} f(x)\, dx$$

$$= \int_{-\infty}^{\infty} \exp\left(\frac{-\theta^2}{2y^2}\right) f(y)\, dy = L\left(\frac{\theta^2}{2}\right)$$

$\int_{-\infty}^{\infty} e^{i\theta(x/|y|)} f(x)\, dx$ is the Fourier transform of $X/|y|$, which is $\exp(-\theta^2/2y^2)$, by (9), Section 9.7. Since $L(\theta^2/2) = e^{-|\theta|}$, the desired result follows.]

26 Suppose that $n + 1$ points are picked at random from $[0, 1]$. Let $Y_1 \leq Y_2 \leq \cdots \leq Y_{n+1}$ be the values of these points arranged in order. Define

$$X = \begin{cases} Y_{k+1} & \text{with probability } \dfrac{k}{n+1} \\ Y_k & \text{with probability } \dfrac{n-k+1}{n+1} \end{cases}$$

That is, after the $n + 1$ points are selected, either the $(k + 1)$st or kth smallest of these is selected with probabilities $k/(n+1)$, $(n-k+1)/n+1)$, respectively. Show that X has the beta density with parameters $\alpha = k$, $\beta = n - k + 1$. {Hint: If h is the density of Y_k and g the density of Y_{k+1}, then $[k/(n + 1)]g + [(n - k + 1)/(n + 1)]h$ is the density of X.}

NOTES

1 A very large collection of random numbers appears in [34].

2 A treatment of the gamma integral is in Example 6, Section 10.23, [3], and Chapter 11, [32].

3 For additional reading on the beta distribution, see Section 18.4, [31].

4 The normal distribution had been observed as an approximation to

the binomial distribution by DeMoivre. Recognition of the importance of the normal distribution came with the work of Pierre Simon Laplace (1749–1827) and the German mathematician Carl Friedrich Gauss (1777–1855).

5 A discussion of differentiation under the integral sign is in Section 7.3, [31].

6 Tables of the standard normal distribution are in [35]. See [19].

7 See [19] for references to tables of the chi-square distribution.

8 The velocity distribution in Example 3, Section 9.9, is called Maxwell's distribution after the English scientist James Clerk Maxwell (1831–1879).

Chapter Ten

LIMIT LAWS

If $X_1, X_2, \ldots$ is a sequence of independent and identically distributed random variables, then the distribution of the "sample mean" $(X_1 + \cdots + X_n)/n$ becomes more and more concentrated about the single point EX_i as n approaches infinity, assuming that the expected value exists. A particularly easy proof is available when $E(X_i^2)$ exists as well. To deal with the concept of concentration of a distribution, we introduce technical measures called *variance* and *standard deviation.* The previously cited result is called a *law of large numbers.* This chapter also includes the *central limit theorem,* which says that under certain conditions, the distribution of a linear function of $(X_1 + \cdots + X_n)/n$ will approach a standard normal distribution as n approaches ∞.

10.1 VARIANCE AND STANDARD DEVIATION

In this section, we consider a technical device for describing the extent to which a distribution is concentrated about its center of gravity. Consider the two integrating density functions in Figure 10.1. Certainly the density in (b) is more concentrated about the center of gravity than is the density in (a). *How should this degree of concentration be measured*? There is no unique approach to the matter, and there are many competing measures. The conventionally used device is called the *standard deviation.* It has some distinct mathematical advantages, which are the main reasons for its use.

definition of standard deviation and variance

Suppose that X is a random variable such that $E(X^2) < \infty$. The *standard deviation* of X is defined to be

$$\sigma = \{E[(X - EX)^2]\}^{1/2} \qquad \text{(positive square root)}$$

The square of the standard deviation,

$$\sigma^2 = E[(X - EX)^2] \tag{1}$$

is called the *variance of* X. We also denote σ^2 as $\mathrm{Var}(X)$ (Note 1).

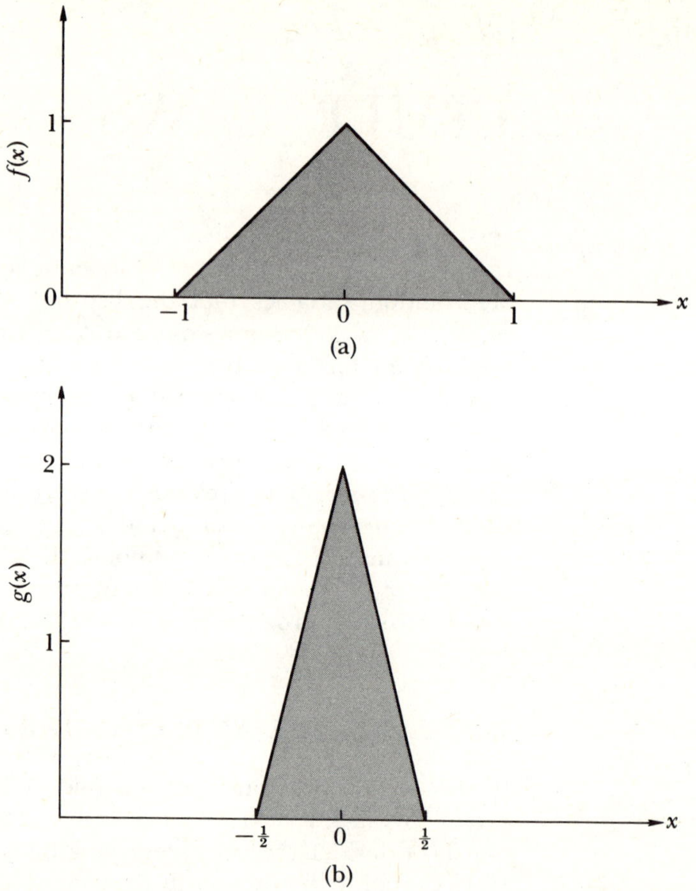

FIGURE 10.1

It is a fact that $E(X^2) < \infty$ implies the existence of EX. (See Exercise 24a, Chapter 2, and Exercise 24, Chapter 7.) If $E(X^2)$ fails to exist, we say that σ is not defined.

When X has an integrating density function f, denoting $E(X)$ by m, we can write (1) as

$$\operatorname{Var}(X) = \sigma^2 = \int_{-\infty}^{\infty} (x - m)^2 f(x)\, dx$$

When X has a discrete density function,

$$\operatorname{Var}(X) = \sigma^2 = \sum_i (u_i - m)^2 P(X = u_i)$$

From a physical point of view, m can be interpreted as the *center of gravity* of the mass distribution described by the density function. Then σ can be interpreted as a measure of how concentrated the mass distribution is around that center of gravity. Roughly, the smaller is σ, the more concentrated is the distribution around the point m. A phrase that is sometimes used in this context is "noisiness of data." The less concentrated a distribution, the more "noisy" are the observations described by such a distribution. If the phenomenon is deterministic and every observation is always the same, then there is no "noise" at all. σ is intended to be a measure of "noise."

Several properties of σ contribute to its plausibility as a measure of concentration. For convenience, these are described in terms of the variance.

properties of σ^2

(a) $\text{Var}(X + c) = \text{Var}(X)$

(b) $\text{Var}(aX) = a^2 \text{Var}(X)$

(c) $\text{Var}(X) = 0$ if, and only if, $P(X = \text{const}) = 1$ (a, c, are constants)

Proof. These properties follow immediately from the definition. For instance,

$$\begin{aligned}\text{Var}(X + c) &= E\{[X + c - E(X + c)]^2\} = E[(X - EX)^2] \\ &= \text{Var}(X)\end{aligned}$$

The remaining parts are similar and are left to the reader. ∎

Property (a) says that shifting the distribution by a constant does not affect σ. This is certainly a desirable property of a measure of concentration. Property (b) says that changing the scale of X values by a multiplicative factor of a multiplies the variance by a^2, and hence multiplies the standard deviation by $|a|$. Properties (a) and (b) combine as

$$\text{Var}(aX + c) = a^2 \text{Var}(X) \tag{2}$$

Example 1 Suppose that X is uniformly distributed on $[0, 1]$. According to (1), $\sigma^2 = \int_0^1 (x - 1/2)^2\,dx = (1/3)(x - 1/2)^3|_0^1 = 1/12$. By properties (a) and (b), the standard deviation for the uniform distribution on $[a, b]$ is $(b - a)/(12)^{1/2}$ (see Figure 10.2).

There is a computational formula for the variance that is often more convenient than the definition (1).

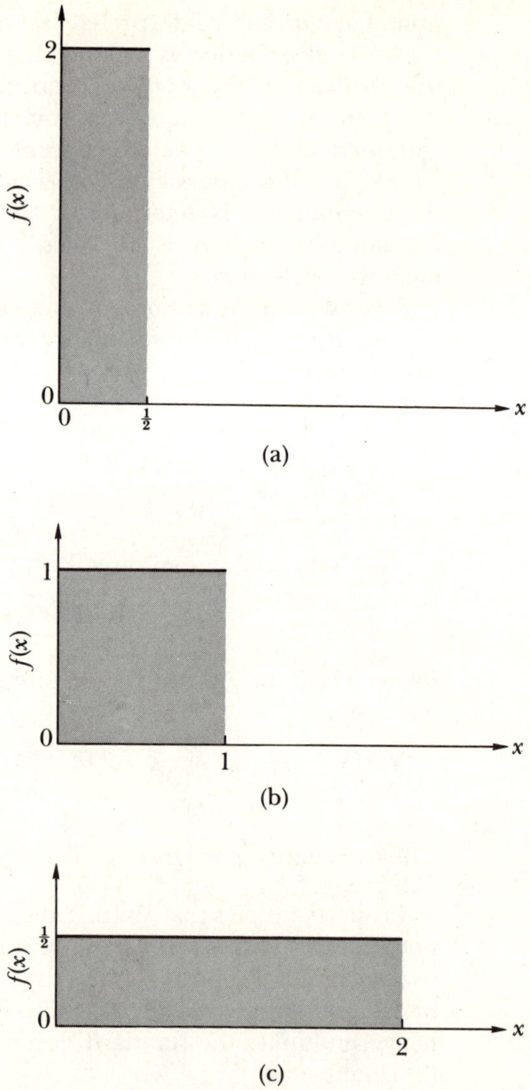

FIGURE 10.2 **(a) $\sigma = (\frac{1}{2})/\sqrt{12} = 0.1444$. (b) $\sigma = 1/\sqrt{12} = 0.2887$. (c) $\sigma = 2/\sqrt{12} = 0.5774$.**

computational formula for variance

$$\sigma^2 = E(X^2) - (EX)^2 \tag{3}$$

Proof of (3). Let $m = EX$. Then $(X - m)^2 = X^2 - 2mX + m^2$. Hence, $E[(X - m)^2] = E(X^2) - 2mEX + m^2 = E(X^2) - m^2$. ∎

Example 2 If X is uniformly distributed on $[0, 1]$, then $E(X^2) = 1/3$ and $EX = 1/2$. According to the computational formula (3), $\sigma^2 = 1/3 - 1/4 = 1/12$, as already seen in Example 1.

Example 3 Suppose X is exponentially distributed with parameter 1. Then $EX = \int_0^\infty xe^{-x}\,dx = 1$, $E(X^2) = \int_0^\infty x^2e^{-x}\,dx = 2$. Hence, $\sigma^2 = 2 - 1 = 1$. If the parameter is λ, then $\sigma^2 = 1/\lambda^2$. Remember that if X has parameter λ, then λX has parameter 1 (Section 9.4). Hence,

$$\text{Var}(\lambda X) = 1 = \lambda^2\,\text{Var}(X), \qquad \text{Var}(X) = \frac{1}{\lambda^2}$$

Example 4 Suppose that X has the $N(0, 1)$ distribution. Then, by (5), Section 9.7, $EX = 0$, $E(X^2) = 1$. Hence, $\sigma^2 = 1$. If X is $N(0, 1)$, then $\sigma X + m$ is $N(m, \sigma^2)$, [(8), Section 9.7]. Thus, the variance for the $N(m, \sigma^2)$ distribution is σ^2, and the standard deviation is σ.

Example 5 Suppose that X has the Cauchy integrating density function $f(x) = 1/\pi(1 + x^2)$, $-\infty < x < \infty$. Since the integral $\int_{-\infty}^{\infty} x^n[1/\pi(1 + x^2)]\,dx$ converges for *no* $n = 1, 2, \ldots$, none of the moments of X exists, and, in particular, Var(X) does not exist.

Probably the main reason for defining variance as in (1) is that the following fact holds true.

variance of a sum of independent random variables is the sum of the variances

Suppose that X and Y are independent random variables with existing variances. Then the variance of $X + Y$ also exists, and

$$\text{Var}(X + Y) = \text{Var}(X) + \text{Var}(Y) \tag{4}$$

Proof of (4). Since $E(X + Y) = EX + EY$,

$$\begin{aligned}\text{Var}(X + Y) &= E[(X - EX + Y - EY)^2]\\ &= E[(X - EX)^2] + E[(Y - EY)^2] + 2E(X - EX)(Y - EY)\end{aligned}$$

$E|(X - EX)(Y - EY)|$ exists by the Schwarz inequality, [Exercise 27, (d) and (e), Chapter 2, and Exercise 24, Chapter 7]. Hence, Var($X + Y$) exists. Since X and Y are independent, $E[(X - EX)(Y - EY)] = E(X - EX)E(Y - EY) = 0$ [(1), Section 4.8, and (6), Section 7.11]. Hence, $\text{Var}(X + Y) = \text{Var}(X) + \text{Var}(Y)$. ∎

By repeated application of (4), we have the following.

If $X_1, \ldots, X_n$ are mutually independent random variables, each with finite variance, then

$$\mathrm{Var}(X_1 + \cdots + X_n) = \mathrm{Var}(X_1) + \cdots + \mathrm{Var}(X_n) \quad (5)$$

Example 6 Suppose that X and Y are independent random variables distributed $N(m_1, \sigma_1^2)$, $N(m_2, \sigma_2^2)$, respectively. Since $\mathrm{Var}(X) = \sigma_1^2$ and $\mathrm{Var}(Y) = \sigma_2^2$, it follows that $\mathrm{Var}(X+Y) = \sigma_1^2 + \sigma_2^2$. This is consistent with the fact that $X + Y$ has the $N(m_1 + m_2, \sigma_1^2 + \sigma_2^2)$ distribution (Section 9.8).

Example 7 Suppose that X is an indicator random variable that equals 1 with probability p and 0 with probability $1-p$. Then $E(X^2) = p$, $EX = p$, and $\mathrm{Var}(X) = p - p^2 = p(1-p)$. Now suppose that $X_1, \ldots, X_n$ are independent indicator random variables, each with the same parameter p. Hence, $\mathrm{Var}(X_1 + \cdots + X_n) = np(1-p)$. But $X_1 + \cdots + X_n$ is binomially distributed with parameters n, p (Section 6.1). Hence, the variance for the binomial distribution is $np(1-p)$. For fixed n, this variance is maximized when $p = 1/2$ (see Figure 10.3). The variance has the smallest possible value, 0, when $p = 0$ or $p = 1$. When a coin with two heads (or two tails) is thrown, there is no question at all as to the number of heads that will appear!

Example 8 Suppose that X is Poisson distributed with parameter λ. Since $EX = \lambda$, $E(X^2) = \lambda^2 + \lambda$, [(4), Section 6.7], it follows that

$$\mathrm{Var}(X) = E(X^2) - (EX)^2 = \lambda$$

If X and Y are not independent random variables, then what can one say about $\mathrm{Var}(X+Y)$? If we examine the proof of (4) again, we see that

$$\mathrm{Var}(X+Y) = \mathrm{Var}(X) + \mathrm{Var}(Y) + 2E[(X - EX)(Y - EY)] \quad (6)$$

In other words, the variance of a sum is the sum of the variances plus the term $2E[(X-EX)(Y-EY)]$. The term $E[(X-EX)(Y-EY)]$ is called the *covariance*. In general, it is not equal to zero if X and Y are not independent.

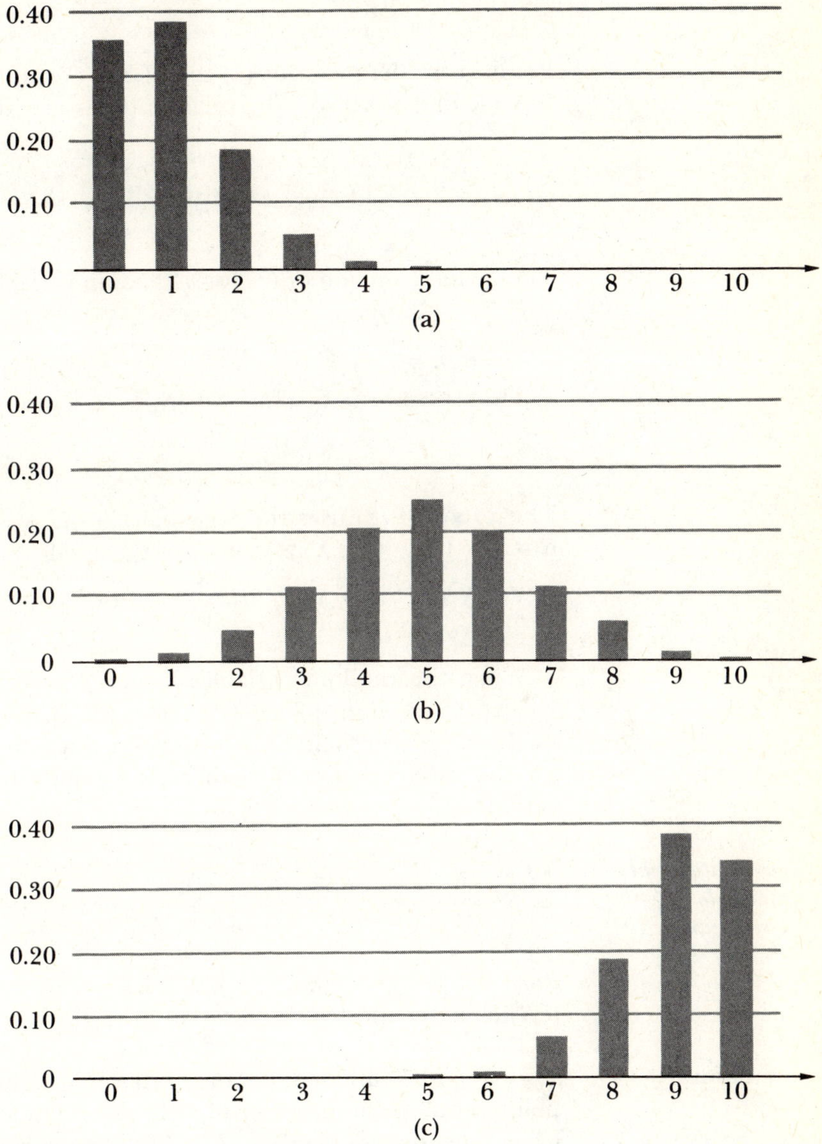

FIGURE 10.3 **(a) Binomial density, $n = 10$, $p = 0.1$, $\sigma^2 = 0.9$. (b) Binomial density, $n = 10$, $p = 0.5$, $\sigma^2 = 2.5$. (c) Binomial density, $n = 10$, $p = 0.9$, $\sigma^2 = 0.9$.**

covariance

$$E[(X - EX)(Y - EY)] = \text{Cov}(X, Y)$$

is called the covariance between the random variables X and Y, assuming this expected value exists.

Recall from the proof of (4) that if Var (X) and Var (Y) exist, then Cov (X, Y) exists also. The relation (6) asserts that

$$\text{Var}(X + Y) = \text{Var}(X) + \text{Var}(Y) + 2\ \text{Cov}(X, Y) \tag{7}$$

The general version of (7) for n random variables is the following

$$\text{Var}(X_1 + \cdot \cdot \cdot + X_n) = \sum_{i=l}^{n} \text{Var}(X_i) + 2 \sum_{i<j} \text{Cov}(X_i X_j) \tag{8}$$

The proof of (8) uses the fact that $\text{Cov}(X_i, X_j) = \text{Cov}(X_j, X_i)$. For $n = 3$, (8) says that $\text{Var}(X_1 + X_2 + X_3)$ equals

$\text{Var}(X_1) + \text{Var}(X_2) + \text{Var}(X_3) + 2\ \text{Cov}(X_1, X_2) + 2\ \text{Cov}(X_1, X_3) + 2\ \text{Cov}(X_2, X_3)$

When the random variables X and Y are identically the same i.e., $X(w) = Y(w)$ for all sample points w], then $\text{Cov}(X, Y) = E[(X - EX)^2] = \text{Var}(X) = \text{Var}(Y)$. Just as there is a computational formula for the variance [Eq. (3)], there is a similar formula for the covariance.

computational formula for covariance

$$\text{Cov}(X, Y) = E(XY) - (EX)(EY) \tag{9}$$

Proof of (9). Since $(X - EX)(Y - EY) = XY - (EX)Y - (EY)X - (EX)(EY)$, its expected value equals the right side of (9). ▮

Example 9

If X and Y are independent, then $\text{Cov}(X, Y) = 0$. [This was pointed out in the proof of (4).] It also follows from (9), because, by independence, $E(XY) = (EX)(EY)$. The converse is, in general, not true; $\text{Cov}(X, Y) = 0$ need not imply that X and Y are independent (see Exercise 6).

Example 10 Suppose that X and Y are indicator random variables. According to (9), $\text{Cov}(X, Y) = E(XY) - (EX)(EY) = P[(X = 1) \cap (Y = 1)] - P(X = 1)P(Y = 1)$. As pointed out earlier, if X and Y are independent, then $\text{Cov}(X, Y) = 0$. This fact can be seen directly in this instance, since $P[(X = 1) \cap (Y = 1] = P(X = 1)P(Y = 1)$. It so happens that for indicator random variables, the converse is also true. Thus $\text{Cov}(X, Y) = 0$ implies that X and Y are independent (see Exercise 5).

Example 11 Suppose that X, Y, and Z are mutually independent random variables. Define

$$U = X + Y, \qquad V = Y + Z$$

Then

$$\begin{aligned}\text{Cov}(U, V) &= E(X - EX + Y - EY)(Y - EY + Z - EZ)\\ &= E[(Y - EY)^2] = \text{Var}(Y)\end{aligned}$$

This fact shows that U and V are not independent if $\text{Var}(Y) > 0$. It is, of course, intuitively plausible that U and V should be dependent, since they both involve the same random variable Y (see Exercise 2).

Example 12 A dichotomous population consists of N objects: M objects are of type 1 and the remaining $N - M$ objects are of type 2. Suppose n drawings are made *without replacement*, $n \leq N$. Define

$$X_i = \begin{cases} 1 & \text{if } i\text{th drawing is of type 1} \\ 0 & \text{otherwise} \end{cases} \qquad i = 1, \ldots, n$$

Then $X = X_1 + \cdots + X_n$ is the total number of type 1 objects drawn (see Section 3.8). If $i \neq j$, then

$$P[(X_i = 1) \cap (X_j = 1)] = \frac{M(M-1)}{N(N-1)} \quad \text{and}$$

$$P(X_i = 1) = P(X_j = 1) = \frac{M}{N}$$

Hence,

$$\text{Cov}(X_i, X_j) = \frac{M(M-1)}{N(N-1)} - \frac{M^2}{N^2} = \frac{-M(N-M)}{N^2(N-1)}$$

$$\text{Var}(X_i) = \text{Var}(X_j) = \frac{M(N-M)}{N^2}$$

It follows from (8) that

$$\mathrm{Var}(X) = n\frac{M(N-M)}{N^2} - 2\binom{n}{2}\frac{M(N-M)}{N^2(N-1)} = \frac{N-n}{N-1}n\frac{M(N-M)}{N^2}$$

Denoting $p = M/N$, the preceding variance can be expressed as

$$\mathrm{Var}(X) = \frac{N-n}{N-1}np(1-p)$$

Recall from Example 7 that if the sampling is *with replacement*, then the X_i's are mutually independent, and $\mathrm{Var}(X) = np(1-p)$. It is interesting to observe that when the sampling is without replacement, the variance is smaller. The random variable X is "noisier" in sampling with replacement because the same objects can be drawn several times. If N is very large compared to n, the variances are practically the same for both types of sampling.

10.2 THE LAW OF LARGE NUMBERS

Suppose that the expected value m and variance σ^2 of a random variable X exist. How much probability information about X can one infer by knowing just m and σ^2? The answer, as one might expect, is *not very much*. One can make some very crude estimates, which turn out to be useful in some cases. The estimates are given by Chebyshev's inequality, as follows:

Chebyshev's inequality

$$P(|X-m| \geq t\sigma) \leq \frac{1}{t^2} \tag{1}$$

or, equivalently,

$$P(|X-m| \geq t) \leq \frac{\sigma^2}{t^2} \tag{2}$$

for any positive number t (Note 2).

Equation (1) says that the probability that a random variable differs from its expected value by t or more standard deviations cannot exceed $1/t^2$ (see Figure 10.4). In (1), if t is less than 1—or in (2), if t is less than σ—then Chebyshev's inequality says nothing of interest; it asserts a probability estimate to be no greater than a number that *exceeds* 1.

Proof of (1). Suppose that X has an integrating density f. By definition,

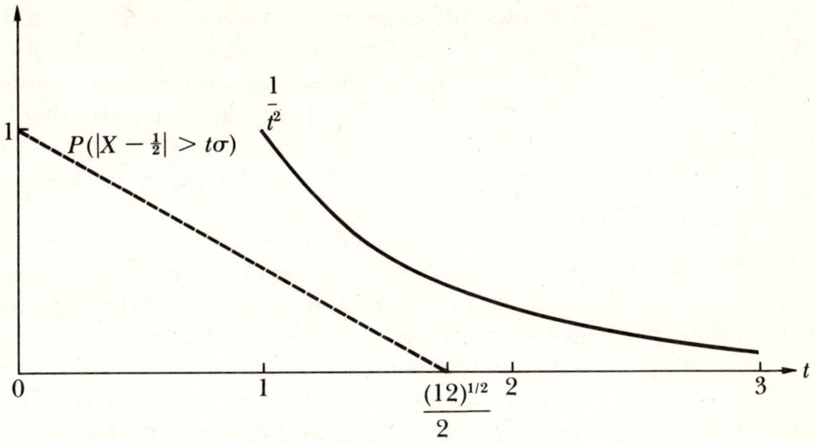

FIGURE 10.4 Comparison of $P(|X - m| > t\sigma)$ (dashed line) with Chebychev approximation (solid line) when X is uniformly distributed on [0, 1].

$$\sigma^2 = \int_{-\infty}^{\infty} (x - m)^2 f(x)\, dx \tag{3}$$

Hence,

$$\sigma^2 \geq \int_{|x-m|\geq t} (x - m)^2 f(x)\, dx \geq t^2 \int_{|x-m|\geq t} f(x)\, dx = t^2 P(|X - m| \geq t)$$

or $P(|X - m| \geq t) \leq \sigma^2/t^2$. If X is discrete, start the argument (3) with $\sigma^2 = \Sigma_i (u_i - m)^2 P(X = u_i)$, but the remaining details are really the same. For a mixture, start the argument (3) with $\sigma^2 = \int_{-\infty}^{\infty} (x - m)^2\, dF(x)$, and the remaining details are still the same. ▮

Proof of (2). Replace t in (2) by $t\sigma$. ▮

Next a theorem called the law of large numbers is described.

law of large numbers

Suppose that $X_1, X_2, \ldots$ is a sequence of mutually independent random variables, each of which has the same expected value m and the same variance σ^2. Then, for any $\epsilon > 0$,

$$\lim_{n>\infty} P\left(\left|\frac{X_1 + \cdots + X_n}{n} - m\right| \geq \epsilon\right) = 0 \tag{4}$$

The quantity $(X_1 + \cdots + X_n)/n$ is usually called the *sample mean* or *sample average.* Notice that $E[(X_1 + \cdots + X_n)/n] = m$. That is, the expected value of the sample average is the same as the expected value of any of its components. Notice also that

$$\text{Var}\left(\frac{X_1 + \cdots + X_n}{n}\right) = \frac{1}{n^2}\text{Var}(X_1 + \cdots + X_n) = \frac{n\sigma^2}{n^2} = \frac{\sigma^2}{n} \quad (5)$$

That is, the variance of the sample average is $1/n$ times the variance of any of the components. If the variance is any kind of decent measure of concentration of the distribution about its center of gravity, then since σ^2/n goes to 0 as n approaches infinity, this fact suggests that the distribution of $(X_1 + \cdots + X_n)/n$ becomes more and more concentrated about the point m as n increases. In the limit, the distribution is the one point mass distribution on m; (4) is just a precise way of saying this.

Proof of (4). By Chebyshev's inequality,

$$P\left(\left|\frac{X_1 + \cdots + X_n}{n} - m\right| \geq \epsilon\right) \leq \frac{\text{Var}[(X_1 + \cdots + X_n)/n]}{\epsilon^2} = \frac{\sigma^2}{n\epsilon^2} \quad (6)$$

Since the right side of (6) goes to 0 as n goes to ∞, the probability on the left side must also go to 0. ▮

There is an alternative way of expressing (4), as follows:

alternative form of law of large numbers

Under the conditions stated for (4),

$$\lim_{n\to\infty} P\left(\frac{X_1 + \cdots + X_n}{n} \leq t\right) = \begin{cases} 0 & \text{if } t < m \\ 1 & \text{if } t > m \end{cases} \quad (7)$$

What (7) asserts is that the cdf of $(X_1 + \cdots + X_n)/n$ "converges" to the cdf of the one-point mass distribution on m (see Figure 10.5). Equation (7) does not say what happens at $t = m$. This matter is more delicate, and it is discussed in Section 10.3.

Proof of (7). Let $t = m - \epsilon$.

$$P\left(\frac{X_1 + \cdots + X_n}{n} \leq m - \epsilon\right) \leq P\left(\left|\frac{X_1 + \cdots + X_n}{n} - m\right| \geq \epsilon\right)$$

Since the right side goes to 0, so does the left side. Let $t = m + \epsilon$.

$$P\left(\frac{X_1 + \cdots + X_n}{n} \leq m + \epsilon\right) \geq 1 - P\left(\left|\frac{X_1 + \cdots + X_n}{n} - m\right| > \epsilon\right)$$

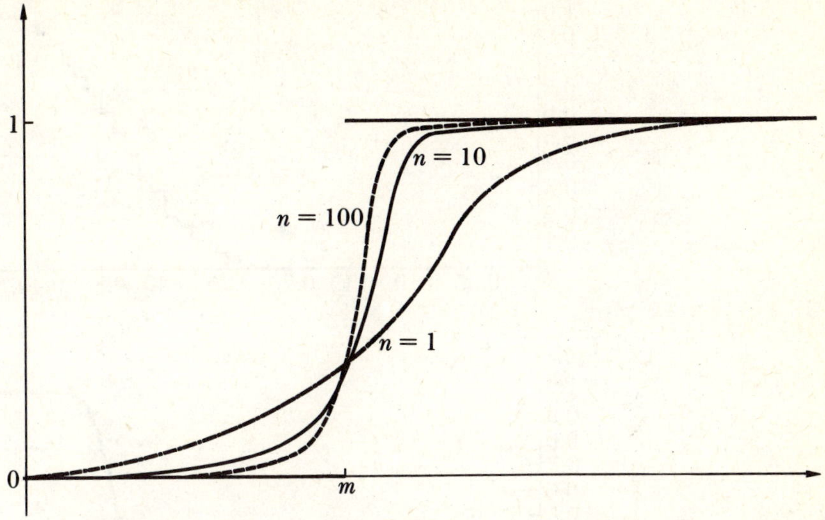

FIGURE 10.5 Schematic description of the cdf's of $(X_1 + \cdots + X_n)/n$ for several increasing values of n.

Since the right side goes to $1 - 0 = 1$, the left side does also. Conversely, (7) implies (4). The proof is left to the reader (Exercise 9). ∎

Example 1 Suppose that $X_1, X_2, \ldots$ are independent indicator random variables,

$$X_i = \begin{cases} 1 & \text{with probability } p \\ 0 & \text{with probability } 1 - p. \end{cases}$$

Then $X_1 + \cdots + X_n$ is binomially distributed with parameters n, p (Section 6.1). And $\text{Var}(X_i) = \sigma^2 = p(1-p)$, $E(X_i) = p$ and $\text{Var}[(X_1 + \cdots + X_n)/n] = p(1-p)/n$. (See Example 7, Section 10.1.) By (7),

$$\lim_{n\to\infty} \sum_{k/n \leq t} \binom{n}{k} p^k (1-p)^{n-k} = \begin{cases} 0 & \text{if } t < p \\ 1 & \text{if } t > p \end{cases}$$

(see Figure 10.6) (Note 3). (See Exercises 19 and 20 for a different approach.)

Example 1 provides somewhat of a mathematical justification for the frequency theory of probability (Chapter 1). Consider a chance experiment and a set of outcomes A, having a probability

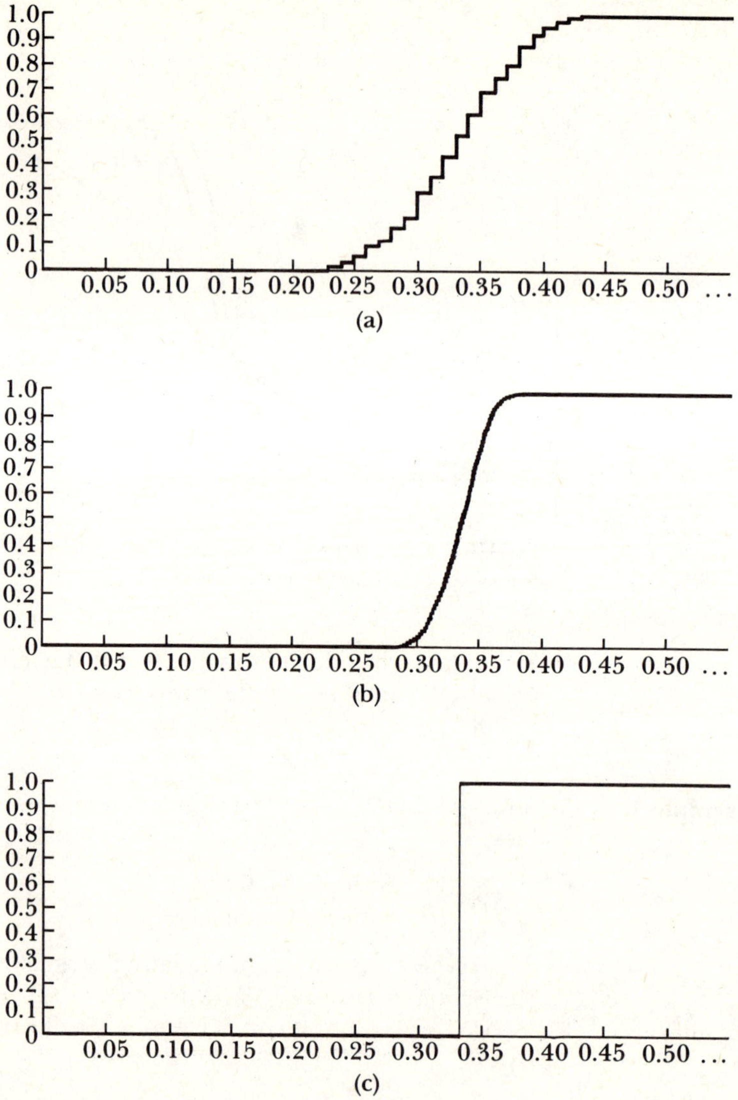

FIGURE 10.6 (a) The cdf of a binomial proportion, $n = 100$, $p = 1/3$. (b) The cdf of a binomial proportion, $n = 1000$, $p = 1/3$. (c) The limiting cdf of a binomial proportion.

of $P(A)$. If the experiment is repeated n times, let $N_n(A)$ denote the number of these repetitions with outcomes falling in A. Then $N_n(A)$ is binomially distributed with parameters n and $p = P(A)$ (each trial's outcome is either in A or not). The frequency theory asserts, somewhat imprecisely, that $N_n(A)/n$ converges to $P(A)$ as $n \to \infty$. In a more precise way, the law of large numbers asserts that

$$\lim_{n\to\infty} P\left(\left|\frac{N_n(A)}{n} - P(A)\right| > \epsilon\right) = 0$$

Example 2 Suppose $X_1, X_2, \ldots$ are independent random variables, each distributed exponentially with $\lambda = 1$. Then, as shown in Example 3, Section 10.1, $EX_i = 1$, $E(X_i^2) = 2$, and $\sigma^2 = \text{Var}(X_i) = 2 - 1 = 1$. Then $X_1 + \cdots + X_n$ has the gamma distribution with density $[1/(n-1)!]x^{n-1}e^{-x}$, $x \geqslant 0$ (Section 9.4), and

$$P\left(\frac{X_1 + \cdots + X_n}{n} \leqslant t\right) = \frac{1}{(n-1)!}\int_0^{nt} x^{n-1}e^{-x}\,dx$$

According to the version of the law of large numbers given by (7),

$$\lim_{n\to\infty} \frac{1}{(n-1)!}\int_0^{nt} x^{n-1}e^{-x}\,dx = \begin{cases} 0 & \text{if } t < 1 \\ 1 & \text{if } t > 1 \end{cases} \tag{8}$$

Example 3 Suppose that $X_1, X_2, \ldots$ are independent, normally distributed random variables, each $N(m, \sigma^2)$. Because $(X_1 + \cdots + X_n)/n$ is itself normally distributed, $N(m, \sigma^2/n)$ (see Section 9.8), it follows that

$$\begin{aligned} P\left(\left|\frac{X_1 + \cdots + X_n}{n} - m\right| < \epsilon\right) &= \int_{m-\epsilon}^{m+\epsilon} \frac{1}{\sqrt{2\pi}\sigma/\sqrt{n}} \exp\left[\frac{-(x-m)^2}{2\sigma^2/n}\right] dx \\ &= \int_{-\sqrt{n}\epsilon/\sigma}^{\sqrt{n}\epsilon/\sigma} \frac{1}{\sqrt{2\pi}} \exp\left(\frac{-y^2}{2}\right) dy \end{aligned} \tag{9}$$

[Make the change of variables $(x - m)/(\sigma/\sqrt{n}) = y$.]

As $n \to \infty$, the right side of (9) approaches

$$\int_{-\infty}^{\infty} \frac{1}{\sqrt{2\pi}} \exp\left(\frac{-y^2}{2}\right) dy = 1$$

Hence,

$$P\left(\left|\frac{X_1 + \cdots + X_n}{n} - m\right| \geq \epsilon\right) \to 0$$

which is just what (4) asserts.

In a typical laboratory situation, an experimenter tries to determine the "true value" of some measurement as follows. He repeats the experiment under the same conditions and each observation is an estimate of the "true value." Presumably, these are not always the same, because there is randomness inherent in the process. A mathematical model for this situation is that the succession of measurements obtained are described by $X_1, X_2, \ldots$, which are independent random variables, where each X_i has the same expected value m, representing the "true value." That the same experiment is being repeated under the same conditions means that, in fact, each X_i has the same distribution. In particular, $\text{Var}(X_i) = \sigma^2$ is the same for each X_i. The experimenter usually averages his several observations as an estimate of the "true value," since such an average is usually thought to be "better" than a single measurement. Thus, for n measurements, $X_1, \ldots, X_n$, the estimate is $(X_1 + \cdots + X_n)/n$. Since $\text{Var}[(X_1 + \cdots + X_n)/n] = \sigma^2/n$, and since variance measures concentration of the distribution about m, the average of n observations provides better and better estimates as n increases. Very roughly, the law of large numbers asserts that one can approach a "perfect" estimate of m by letting the number of observations go to infinity.

It is crucial to point out that (4) or, equivalently, (7) is not valid if the expected value and variance fail to exist. Actually, if the X_i's each have the *same distribution*, it is enough to assume only that $E(X_i)$ exists without the variance necessarily existing (Note 4). The simple proof based on Chebyshev's inequality breaks down, however. The following example shows how bad things can be if $E(X_i)$ does not exist.

Example 4 Suppose that $X_1, X_2, \ldots$ are independent random variables, each with the Cauchy density function $1/[\pi(1 + x^2)]$, $-\infty < x < \infty$. As pointed out in Example 3, Section 8.7, $(X_1 + \cdots + X_n)/n$ has the same Cauchy distribution as any of the X_i's. Although $E(X_i)$ does not exist, certainly 0 should play the role of m, because the density of X_i is symmetric around 0.

Since

$$P\left(\left|\frac{X_1 + \cdots + X_n}{n}\right| \geq \epsilon\right) = P(|X_1| \geq \epsilon)$$

this probability does *not* go to 0 as $n \to \infty$. The reason that (4) breaks down is that the expected value fails to exist. From the experimental point of view discussed earlier, the average of n observations, $(X_1 + \cdots + X_n)/n$, is no better here than a single observation in estimating the "true value."

10.3 CENTRAL LIMIT THEOREM

The law of large numbers says that if $X_1, X_2, \ldots$ is a sequence of independent random variables with common expectations m and variances σ^2, then the distribution of $(X_1 + \cdots + X_n)/n$ becomes "highly concentrated" about the single point m as $n \to \infty$. What makes this work is that

$$\operatorname{Var}\left(\frac{X_1 + \cdots + X_n}{n}\right) = \frac{\sigma^2}{n} \to 0$$

Suppose one makes a linear transformation on $(X_1 + \cdots + X_n)/n$ so that its variance does not go to 0 but remains constant. (It will be seen momentarily that this is possible.) Can one say anything interesting about the distribution, which now presumably does not become concentrated at a single point? The answer is yes—something quite fascinating does happen to the distribution, and this is the content of the central limit theorem that follows. But first we take up a technical preliminary.

Suppose X is a random variable having expected value m and variance σ^2, and suppose $\sigma^2 > 0$.

standardized form of a random variable

$(X - m)/\sigma$ is called the *standardized form* of X.

$$E\left(\frac{X - m}{\sigma}\right) = 0, \qquad \operatorname{Var}\left(\frac{X - m}{\sigma}\right) = 1 \tag{1}$$

The proof of (1) is immediate and is left to the reader.

The standardized form of $(X_1 + \cdots + X_n)/n$, still assuming $\sigma^2 > 0$, is

$$Y_n = \frac{\sqrt{n}}{\sigma}\left(\frac{X_1 + \cdots + X_n}{n} - m\right) = \frac{X_1 + \cdots + X_n}{(n\sigma)^{1/2}} - \frac{\sqrt{n}m}{\sigma} \tag{2}$$

so $E(Y_n) = 0$, $\operatorname{Var}(Y_n) = 1$. The central limit theorem asserts that as $n \to \infty$, the distribution of Y_n converges to the normal $N(0, 1)$ distribution. The hypotheses will require not only that the X_i's have the same expected values and variances but also that they have the same distribution.

central limit theorem

Suppose that $X_1, X_2, \ldots$ is a sequence of independent, identically distributed random variables with

$$E(X_i) = m, \qquad \text{Var}(X_i) = \sigma^2$$

existing and $\sigma^2 > 0$. Let Y_n be defined in (2). Then

$$\lim_{n\to\infty} P(Y_n \leq t) = \frac{1}{(2\pi)^{1/2}} \int_{-\infty}^{t} \exp\left(\frac{-x^2}{2}\right) dx, \qquad -\infty < t < \infty \tag{3}$$

(Note 5).

Proof of (3). The "proof" presented here is really more a plausibility argument than a rigorous proof, since it depends on several points that have not been proved in this book. Consider the Fourier transform of Y_n,

$$\begin{aligned} E(\exp i\theta Y_n) &= E\left\{\exp\left[\frac{i\theta\sqrt{n}}{\sigma}\left(\frac{X_1 + \cdots + X_n}{n} - m\right)\right]\right\} \\ &= E\left\{\exp \frac{i\theta\sqrt{n}}{\sigma}\left[\frac{(X_1 - m) + \cdots + (X_n - m)}{n}\right]\right\} \\ &= \left[E\left(\exp \frac{i\theta}{\sigma}\frac{(X_1 - m)}{\sqrt{n}}\right)\right]^n \end{aligned}$$

This follows from Section 8.7, since $X_1, \ldots, X_n$ are mutually independent and each has the same distribution. Since

$$\exp \frac{i\theta}{\sigma}\left(\frac{X_1 - m}{\sqrt{n}}\right) = 1 + \frac{i\theta}{\sigma}\left(\frac{X_1 - m}{\sqrt{n}}\right) + \left(\frac{i\theta}{\sigma}\right)^2 \frac{[(X_1 - m)/\sqrt{n}]^2}{2} + \cdots$$

and

$$E\left[\exp \frac{i\theta}{\sigma}\frac{(X_1 - m)}{\sqrt{n}}\right] = 1 - \frac{\theta^2}{2n} + \text{terms with higher powers of } \frac{1}{n}$$

it follows that

$$E(\exp i\theta Y_n) = \left(1 - \frac{\theta^2}{2n} + \cdots\right)^n$$

and

$$\lim_{n\to\infty} E(\exp i\theta Y_n) = \lim_{n\to\infty}\left(1 - \frac{\theta^2}{2n}\right)^n = \exp\frac{-\theta^2}{2}$$

Now these steps are not only plausible, but they are also correct. However, the rigorous details to justify them are omitted (Note 6).

Finally, since $\exp -\theta^2/2$ is the Fourier transform of the $N(0, 1)$ distribution [(6), Section 9.7], it follows from Property III, Section 8.6, that the cdf's of the Y_n's converge to the $N(0, 1)$ cdf. This fact is exactly what (3) asserts. ▮

Example 1 Suppose that $X_1, X_2, \ldots$ is a sequence of mutually independent indicator random variables with

$$X_i = \begin{cases} 1 & \text{with probability } p \\ 0 & \text{with probability } 1-p \end{cases}$$

and $0 < p < 1$. Then $X_1 + \cdots + X_n$ is binomially distributed with parameters n, p (Section 6.1). The standardized form of this sum is

$$Y_n = \frac{X_1 + \cdots + X_n - np}{[np(1-p)]^{1/2}}$$

since $E(X_i) = p$, $\operatorname{Var}(X_i) = p(1-p)$. Hence,

$$\lim_{n\to\infty} P(Y_n \leq t) = \frac{1}{(2\pi)^{1/2}} \int_{-\infty}^{t} \exp\frac{-x^2}{2}\, dx$$

Equivalently, in terms of the explicit binomial probabilities,

$$P(Y_n \leq t) = P(X_1 + \cdots + X_n \leq [np(1-p)]^{1/2}t + np)$$

$$= \sum_{0\leq k\leq [np(1-p)]^{1/2}t+np} \binom{n}{k} p^k(1-p)^{n-k} \to \frac{1}{(2\pi)^{1/2}} \int_{-\infty}^{t} \exp\frac{-x^2}{2}\, dx \quad \text{as } n\to\infty \qquad (4)$$

Figure 10.7 shows the cdf of a binomial distribution with the limiting normal cdf superimposed (Note 7).

Example 2 A coin shows heads with probability $p = 0.4$ and tails with probability $1 - p = 0.6$. The coin is thrown 100 times. What is the probability of obtaining 43 or fewer heads? The exact probability is

$$\sum_{k=0}^{43} \binom{100}{k} (0.4)^k(0.6)^{100-k} \qquad (5)$$

which equals 0.76 . . . (Consult a table of the binomial distribution.) To use (4) for an approximation, let

$$43 = [100(0.4)(0.6)]^{1/2}t + 100(0.4) \quad \text{or } t \approx 0.61 \ldots$$

According to (4), the probability (5) is approximately equal to $1/(2\pi)^{1/2} \int_{-\infty}^{0.61} \exp(-x^2/2)\, dx \approx 0.73 \ldots$ (from Table 9.2). Thus, the

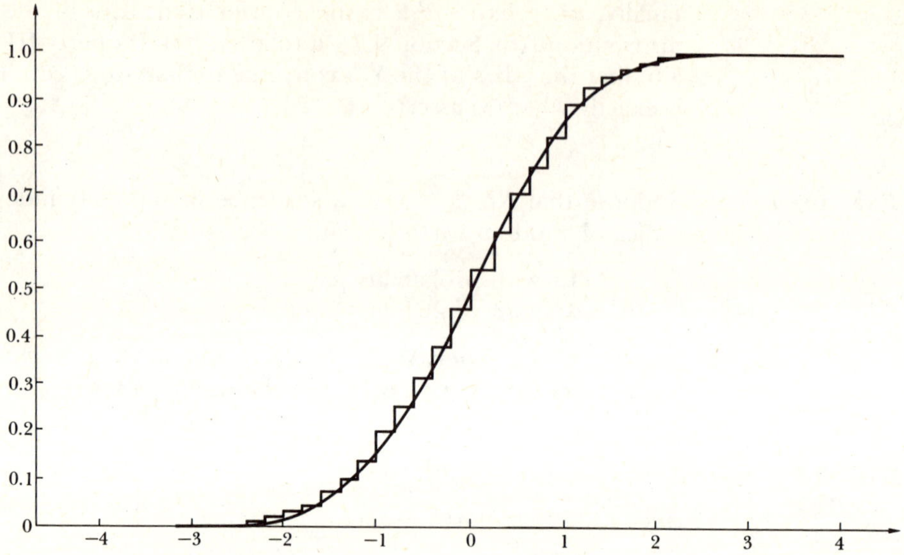

FIGURE 10.7 The stepcurve is the cdf of $(X - np)/[np(1-p)]^{1/2}$ for X binomial, $n = 100$, $p = 1/2$. The smooth curve is the $N(0, 1)$ cdf.

actual probability of 0.76 . . . is approximated by 0.73 . . . . For moderate size n, there are techniques for improving this approximation (Note 8).

Various other distributions can be approximated by the normal distribution. The following example displays this for the Poisson distribution.

Example 3 Suppose that $X_1, X_2, \ldots$ is a sequence of independent random variables, each Poisson distributed with the same parameter, $\lambda > 0$. Also, $E(X_i) = \lambda$, $\text{Var}(X_i) = \lambda$ (Example 8, Section 10.1). Hence,

$$\lim_{n\to\infty} P\left(\frac{X_1 + \cdots + X_n - n\lambda}{(n\lambda)^{1/2}} \le t\right) = \frac{1}{(2\pi)^{1/2}} \int_{-\infty}^{t} \exp\frac{-x^2}{2}\, dx \tag{6}$$

Since $X_1 + \cdots + X_n$ is itself Poisson distributed with parameter $n\lambda$ (Section 6.7), (6) can be written

$$\lim_{n\to\infty} \sum_{0 \le k \le \sqrt{n\lambda}\,t + n\lambda} \frac{(n\lambda)^k e^{-n\lambda}}{k!} = \frac{1}{(2\pi)^{1/2}} \int_{-\infty}^{t} \exp\frac{-x^2}{2}\, dx \tag{7}$$

or, equivalently,

$$\sum_{0 \le k \le u} \frac{(n\lambda)^k e^{-n\lambda}}{k!}$$

is approximated by $1/(2\pi)^{1/2} \int_{\infty}^{a} \exp(-x^2/2)\, dx$, where $a = (u - n\lambda)/(n\lambda)^{1/2}$.

The central limit theorem says that the distribution of a standardized sum of n independent, identically distributed random variables approaches the $N(0, 1)$ normal distribution as $n \to \infty$, assuming the first two moments of the random variables exist. *It is remarkable that the limiting distribution does not depend on the distribution of the random variables being summed.* No matter how much the initial density may differ in appearance from the $N(0, 1)$ density, the density of the sum of even a moderate number of X_i's may begin to assume the bell-shaped appearance of the normal density. For instance, if the X_i's are uniformly distributed over [0, 1], then Figure 10.8 shows the density functions of the stand-

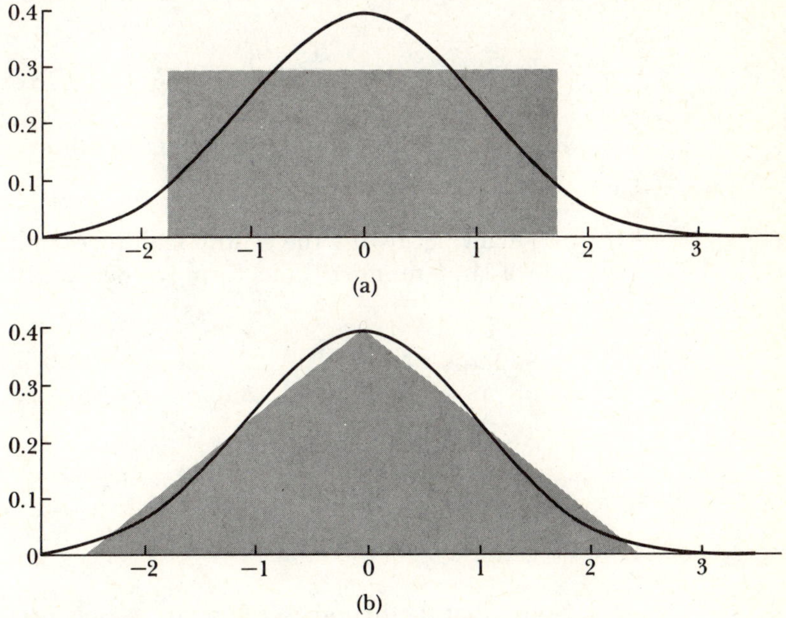

FIGURE 10.8 (a) The density function of $(X_1 - \frac{1}{2})/(1/12)^{1/2}$ with $N(0, 1)$ density superimposed (X_1 is uniformly distributed over [0, 1]). (b) The density function of $(X_1 + X_2 - 1)/(2/12)^{1/2}$ with $N(0, 1)$ density superimposed (X_1, X_2 are independent and uniformly distributed over [0, 1]).

ardized forms of X_1 and of $X_1 + X_2$ with the limiting $N(0, 1)$ density function superimposed. It should be emphasized, however, that the central limit theorem, in its assertion (3), concerns the limit of a sequence of cdf's and not density functions. There are versions of the central limit theorem for density functions but the situation here is more complicated.

Using the central limit theorem, one can show an interesting characterization of the normal distribution.

Suppose that $X_1, X_2, \ldots,$ is a sequence of mutually independent random variables, each with the same distribution, and that

$$Y_n = \frac{X_1 + \cdots + X_n}{n^{1/2}}$$

has the same distribution for every $n = 1, 2, \ldots$. If $E(X_i) = 0$ and $\mathrm{Var}(X_i) = 1$, then the distribution of X_i must be $N(0, 1)$.

Proof. Let $F(t) = P(X_i \leq t)$. By the central limit theorem,

$$\lim_{n\to\infty} P(Y_n \leq t) = \frac{1}{(2\pi)^{1/2}} \int_{-\infty}^{t} \exp\left(\frac{-x^2}{2}\right)$$

But, by the hypothesis, $P(Y_n \leq t) = F(t)$, for every n. Thus, F is the $N(0, 1)$ cdf. ∎

Finally, let us use the central limit theorem to complement the law of large numbers in its form (7), Section 10.2, for $t = m$.

Suppose that $X_1, X_2, \ldots$ is a sequence of independent and identically distributed random variables, with $E(X_i) = m$ and $E(X_i^2) < \infty$. Then

$$\lim_{n\to\infty} P\left(\frac{X_1 + \cdots + X_n}{n} \leq t\right) = \begin{cases} 0 & \text{if } t < m \\ \frac{1}{2} & \text{if } t = m \\ 1 & \text{if } t > m \end{cases} \tag{8}$$

Notice that the hypotheses here are stronger than for (7), Section 10.2. In that case, it was only required that the first and second moments be the same for each X_i. The present hypotheses require the *distributions* to be the same. What is being gained in return is the additional evaluation of the limit for $t = m$.

Proof of (8)

$$P\left(\frac{X_1 + \cdots + X_n}{n} \leq m\right) = P\left[\frac{\sqrt{n}}{\sigma}\left(\frac{X_1 + \cdots + X_n}{n} - m\right) \leq 0\right]$$

By the central limit theorem, the limit as $n \to \infty$ equals

$$\frac{1}{(2\pi)^{1/2}} \int_{-\infty}^{0} \exp\left(\frac{-x^2}{2}\right) dx = \frac{1}{2}$$

The other cases, $t < m$ and $t > m$, were already proved in (7), Section 10.2. ▮

Example 4 Suppose that $X_1, X_2, \ldots$ are independent random variables, each Poisson distributed with parameter 1. By (8), we have

$$\lim_{n \to \infty} \sum_{0 \leq k \leq tn} e^{-n} \frac{n^k}{k!} = \begin{cases} 0 & \text{if } t < 1 \\ \frac{1}{2} & \text{if } t = 1 \\ 1 & \text{if } t > 1 \end{cases}$$

We use the fact that $X_1 + \cdots + X_n$ is Poisson distributed with parameter $\lambda = n$ (Section 6.7).

SUMMARY

The *variance* is a measure of concentration of a random variable X about its mean and is defined by

$$\mathrm{Var}(X) = E[(X - EX)^2]$$

The formula $\mathrm{Var}(X) = E(X^2) - (EX)^2$ is useful; also $\mathrm{Var}(aX + b) = a^2 \mathrm{Var}(X)$ and $\mathrm{Var}(X) = 0$ if, and only if, X is constant with probability 1. The square root of the variance is called the *standard deviation.*

The *covariance* of two random variables X and Y is defined by

$$\mathrm{Cov}(X, Y) = E[(X - EX)(Y - EY)]$$

and can also be computed from the formula $\mathrm{Cov}(X, Y) = E(XY) - (EX)(EY)$.

The variance of a sum of independent random variables equals the sum of the variances. The formula for the variance of a sum of dependent random variables involves the covariances. For instance,

$$\mathrm{Var}(X + Y) = \mathrm{Var}(X) + \mathrm{Var}(Y) + 2\,\mathrm{Cov}(X, Y)$$

If $X_1, \ldots, X_n$ are independent random variables, each with the same expected value m and the same variance σ^2, then the variance of the "sample average" is given by

$$\operatorname{Var}\left(\frac{X_1 + \cdots + X_n}{n}\right) = \frac{\sigma^2}{n}$$

The *law of large numbers* says that under these conditions, the distribution of the sample average becomes highly concentrated about m as n, the sample size, approaches ∞. One version of this assertion is that

$$\lim_{n\to\infty} P\left(\frac{X_1 + \cdots + X_n}{n} \leqslant t\right) = \begin{cases} 0 & \text{if } t < m \\ 1 & \text{if } t > m \end{cases}$$

The main tool used in proving the law of large numbers is *Chebyshev's inequality.*

Suppose that $X_1, X_2, \ldots$ are independent and identically distributed random variables with $E(X_i) = m$ and $\operatorname{Var}(X_i) = \sigma^2$. Define

$$Y_n = \left(\frac{X_1 + \cdots + X_n}{n} - m\right) \Big/ \frac{\sigma}{\sqrt{n}}$$

This is the standardized sample average with $EY_n = 0$, $\operatorname{Var}(Y_n) = 1$. The *central limit theorem* says that for any t,

$$\lim_{n\to\infty} P(Y_n \leqslant t) = \frac{1}{(2\pi)^{1/2}} \int_{-\infty}^{t} \exp\left(\frac{-x^2}{2}\right) dx$$

EXERCISES

1 Suppose that $(X_1, \ldots, X_k)$ has the multinomial distribution with parameters $n; p_1, \ldots, p_k$.
   (a) Show that $\operatorname{Var}(X_1) = np_1(1-p_1)$, $\operatorname{Var}(X_1 + X_2) = n(p_1 + p_2)(1 - p_1 - p_2)$. (Compare Example 7, Section 10.1.)
   (b) Evaluate $\operatorname{Var}(X_1 + X_2)$, using (7), Section 10.1, and conclude that $\operatorname{Cov}(X_1, X_2) = -np_1p_2$. (In general, $\operatorname{Cov}(X_iX_j) = -np_ip_j$, $i \neq j$.)

2 Suppose that $X_1, \ldots, X_n$ have the property that $\operatorname{Cov}(X_i, X_j) = 0$, $i \neq j$, $\operatorname{Var}(X_i) = \sigma^2$, $i = 1, \ldots, n$. Let $U = a_1X_1 + \cdots + a_nX_n$, $V = b_1X_1 + \cdots + b_nX_n$, where the a_i's and b_i's are constants. Show that

$$\operatorname{Cov}(U, V) = \left(\sum_{i=1}^{n} a_ib_i\right)\sigma^2$$

3 Suppose that X and Y have the property that

$$EX = EY, \qquad E(X^2) = E(Y^2)$$

Show that $\mathrm{Cov}(X - Y, X + Y) = 0$. [It is not assumed that X and Y are independent, or even that $\mathrm{Cov}(X, Y) = 0$.]

4 Show that the covariance between $aX + b$ and $cY + d$ is equal to $ac\,\mathrm{Cov}(X, Y)$.

5 (a) Suppose that X and Y are indicator random variables. Show that $\mathrm{Cov}(X, Y) = 0$ implies that X and Y are independent. {Hint: $\mathrm{Cov}(X, Y) = 0$ implies that $P[(X = 1) \cap (Y = 1)] = P(X = 1)P(Y = 1)$. Now use Exercise 23a, Chapter 4.}

(b) Suppose that X and Y each have only two possible values. Show that $\mathrm{Cov}(X, Y) = 0$ implies that X and Y are independent. [Hint: There exist constants a, b, c, d such that $U = aX + b$ and $V = cY + d$ are indicator random variables. Use Exercise 4 to show that $\mathrm{Cov}(U, V) = 0$, and use (a) to conclude that U and V are independent.]

6 Construct a discrete distribution of two random variables X, Y for which $\mathrm{Cov}(X, Y) = 0$ but X and Y are not independent. (Don't try a 2×2 distribution because of Exercise 5!)

7 Show that the assumption of independence can be relaxed in proving the law of large numbers. In particular, show that (4) and (7), Section 10.2, hold under the following conditions: $X_1, X_2, \ldots$ is a sequence of random variables, each of which has the same expected value and the same variance, and $\mathrm{Cov}(X_i, X_j) = 0$, $i \neq j$.

8 Suppose that $X_1, X_2, \ldots$ is a sequence of random variables, each of which has the same expected value, m and $\mathrm{Cov}(X_i, X_j) = 0$, $i \neq j$. $\mathrm{Var}(X_i) = \sigma_i^2$ exists for every i, but these are not necessarily the same. Show that

$$\lim_{n\to\infty} \frac{\sigma_1^2 + \cdots + \sigma_n^2}{n^2} = 0$$

is sufficient for (4) or (7), Section 10.2 to be true.

9 Show that (7), Section 10.2, implies (4), Section 10.2.

10 A point U_1 is picked at random in $[0, 1]$. Then a second point U_2 is picked at random in $[0, U_1]$, and so forth. (See Exercise 11, Chapter 9.) Show that the distribution of $U_n^{1/n}$ becomes concentrated about e^{-1} as $n \to \infty$. That is,

$$\lim_{n\to\infty} P(U_n^{1/n} \geq e^{-t}) = \begin{cases} 0 & \text{if } t < 1 \\ 1 & \text{if } t > 1 \end{cases}$$

(Hint: Use the method described for Exercise 11, Chapter 9, and the law of large numbers.)

11 Suppose n balls are randomly distributed among k cells (n will go to ∞ but k will remain fixed).

(a) Let X_{in} be the number of balls that fall in the ith cell. Show that

$$\lim_{n\to\infty} P\left(\frac{X_{in}}{n} \leqslant t\right) = \begin{cases} 0 & \text{if } t < \dfrac{1}{k} \\ 1 & \text{if } t > \dfrac{1}{k} \end{cases}$$

(b) Let M_n be the maximum number of balls in any cell. That is,

$$M_n = \max(X_{1n}, \ldots, X_{kn})$$

Show that

$$\lim_{n\to\infty} P\left(\frac{M_n}{n} \leqslant t\right) = \begin{cases} 0 & \text{if } t < \dfrac{1}{k} \\ 1 & \text{if } t > \dfrac{1}{k} \end{cases}$$

{Hint:

$$P(M_n \leqslant t) = P[(X_{1n} \leqslant t) \cap \cdots \cap (X_{kn} \leqslant t)]$$

Use the inequalities,

$$1 - \sum_{i=1}^{k} P(X_{in} > t) \leqslant P[(X_{1n} \leqslant t) \cap \cdots \cap (X_{kn} \leqslant t)] \leqslant P(X_{1n} \leqslant t)$$

Use the right inequality when $tn < 1/k$, and use the left one when $tn > 1/k$.}

12 Let T_n be a sequence of random variables such that

$$\lim_{n\to\infty} \text{Var}(T_n) = 0$$

Show that for any $\epsilon > 0$,

$$\lim_{n\to\infty} P(|T_n - ET_n| \leqslant \epsilon) = 0$$

(Hint: Use Chebyshev's inequality.)

13 Consider an infinite sequence of boxes. Box n contains 1 white ball and n black balls, $n = 1, 2, \ldots$. A ball is drawn at random from each box. Let

$$X_n = \begin{cases} 1 & \text{if ball from } n\text{th box is white} \\ 0 & \text{otherwise} \end{cases}$$

The number of white balls drawn from the first n boxes is $S_n = X_1 + \cdots + X_n$. Show that the distribution of $S_n/\log n$ becomes concentrated about 1. That is, given any $\epsilon > 0$,

$$\lim_{n\to\infty} P\left(\left|\frac{S_n}{\log n} - 1\right| \geqslant \epsilon\right) = 0$$

[Hint: Use the fact that $1 + (1/2) + \cdots + (1/n)$ behaves like $\log n$ to show that

$$\lim_{n\to\infty} \frac{E(S_n)}{\log n} = 1$$

(Note 9). Also show that $\mathrm{Var}(S_n/\log n) \to 0$ as $n \to \infty$.]

14 A 10-faced die is thrown 1000 times. Two of the faces are red and the remaining eight are blue. Use the central limit theorem to estimate the probability of obtaining no more than 190 red throws out of the 1000. (The probability from a table of the binomial distribution is 0.227.)

15 A bank accepts rolls of pennies and credits its customers 50¢ per roll without actually counting the contents. Suppose that typically a roll contains 49 pennies 20% of the time, 50 pennies 70% of the time, and 51 pennies 10% of the time. Use the central limit theorem to approximate the following probabilities.
(a) What is the probability that 1000 rolls are worth at least $499.29?
(b) What is the probability that 1000 rolls are worth at least $499.00?
(Answers: (a) 0.045, (b) 1/2.)

16 A six-faced die has its faces marked −2, 2, −4, 4, −14, 14. Let S_n equal the sum of the numbers obtained in n throws of the die. Evaluate the following limits:

(a) $\lim_{n\to\infty} P(S_n \leq 0.041n)$

(b) $\lim_{n\to\infty} P(S_n > 0.0005n)$

(Answers: (a) 1, (b) 0.)

17 Consider the random walk described in Exercise 7, Chapter 6, and suppose that $p = 1/2$. Let Y_n be the position of the particle after n steps.
(a) Show that $\mathrm{Var}(Y_n) = n$.
(b) Use the central limit theorem to show that

$$\lim_{n\to\infty} P\left(\frac{Y_n}{n^{1/2}} \leq t\right) = \frac{1}{(2\pi)^{1/2}} \int_{-\infty}^{t} \exp\left(\frac{-x^2}{2}\right) dx$$

18 Use the central limit theorem to show that

$$\lim_{n\to\infty} \frac{1}{\Gamma(n)} \int_0^{t\sqrt{n}+n} x^{n-1}e^{-x}\, dx = \frac{1}{(2\pi)^{1/2}} \int_{-\infty}^{t} \exp\left(\frac{-x^2}{2}\right) dx$$

(Hint: Follow Example 1, Section 10.3, now supposing the X_i's to be exponential with parameter $\lambda = 1$.)

19 The following is an inequality that does better than Chebyshev's inequality in some cases. If X is a nonnegative random variable and c is a constant, then

$$P(X \leq c) \leq \min_{0\leq t\leq 1} E[t^{(X-c)}]$$

Prove this inequality. Notice that since X is nonnegative and $0 \leq t \leq 1$, then $E(t^X)$ exists, and thus $E[t^{(X-c)}] = t^{-c}E(t^X)$. {Hint:

$$E[t^{(X-c)}] \geq E[t^{(X-c)}; X - c \leq 0]$$

The latter means that the sum or integral that gives $E[t^{(X-c)}]$ is restricted to those values of X where $X - c \leq 0$. Also, $E[t^{(X-c)}; X - c \leq 0] \geq P(X - c \leq 0)$, since $t^{(X-c)} \geq 1$ whenever $X - c \leq 0$. $P(X - c \leq 0) \leq Et^{(X-c)}$ for any t in $[0, 1]$.}

20 (a) If $X_1, \ldots, X_n$ are independent indicator random variables with $P(X_i = 1) = p$, $i = 1, \ldots, n$, then $S_n = X_1 + \cdots + X_n$ has the binomial distribution with parameters n and p. Use Exercise 19 to show that there is a constant ρ, $0 < \rho < 1$, such that

$$P\left(\frac{S_n}{n} \leq c\right) \leq \rho^n, \qquad n = 1, 2, \ldots \quad \text{whenever } c < p$$

(c is fixed and the same constant ρ works for all n. ρ does depend on c, however.) {Hint: By Exercise 19,

$$P(S_n \leq nc) \leq \min_{0 \leq t \leq 1} Et^{(S_n - nc)} = \min_{0 \leq t \leq 1} [A(t)]^n$$

where $A(t) = t^{1-c}p + t^{-c}(1 - p)$. Hence,

$$P(S_n \leq nc) \leq \rho^n$$

where $\rho = \min_{0 \leq t \leq 1} A(t)$. By elementary calculus, $0 < \rho < 1$. For instance, it is sufficient to note that $A'(1) < 1$, using the fact that $c < p$. Draw a picture!}

(b) Show that

$$\lim_{n \to \infty} P\left(\frac{S_n}{n} \leq c\right) = 0$$

using (a). Show that this follows also from the law of large numbers, (7), Section 10.2. Notice that the rate of convergence to 0 given by (a) is much faster than that given by Chebyshev's inequality, which led to (7), Section 10.2.

21 (a) If $E(X^2)$ exists, then show that

$$E[(X - a)^2] = \text{Var}(X) + [E(X) - a]^2$$

(b) Use (a) to show that $E[(X - a)^2]$ is minimized over a by $a = EX$.

(c) The following are some forms of (b) that we will need later. Suppose f is a nonnegative function and

$$\int_{-\infty}^{\infty} f(x)\,dx < \infty, \qquad \int_{-\infty}^{\infty} x^2 f(x)\,dx < \infty$$

Then $\int_{-\infty}^{\infty} (x - a)^2 f(x)\,dx$ is minimized over a by

$$a = \frac{\int_{-\infty}^{\infty} xf(x)\,dx}{\int_{-\infty}^{\infty} f(x)\,dx}$$

More generally, for a given function h, such that $\int_{-\infty}^{\infty} [h(x)]^2 f(x)\,dx < \infty$, $\int_{-\infty}^{\infty} [h(x) - a]^2 f(x)\,dx$ is minimized over a by

$$a = \frac{\int_{-\infty}^{\infty} h(x)f(x)\,dx}{\int_{-\infty}^{\infty} f(x)\,dx}$$

The discrete version of this fact is that if $f \geqslant 0$, $\Sigma_i f(x_i) < \infty$, and $\Sigma_i [h(x_i)]^2 f(x_i) < \infty$, then $\Sigma_i [h(x_i) - a]^2 f(x_i)$ is minimized over a by

$$a = \frac{\sum_i h(x_i)f(x_i)}{\sum_i f(x_i)}$$

The number m is called a *median* of X (or of the distribution of X) if

$$P(X \leqslant m) \geqslant \tfrac{1}{2}, \qquad P(X \geqslant m) \geqslant \tfrac{1}{2}$$

A median is not necessarily unique. (Draw a picture!)

(d) Suppose that X has an integrating density f and that c is any real number. Show that

$$\begin{aligned} E(|X - c|) &= E(|X - m|) + 2\int_m^c (c - x)f(x)\,dx \\ &\quad + \left(\frac{c - m}{2}\right)\left[P(X \leqslant m) - \frac{1}{2}\right] \qquad \text{if } c \geqslant m \\ &= E(|X - m|) + 2\int_c^m (x - c)f(x)\,dx \\ &\quad + \left(\frac{m - c}{2}\right)\left[P(X \geqslant m) - \frac{1}{2}\right] \qquad \text{if } c \leqslant m \end{aligned}$$

[Hint: Suppose that $c \leqslant m$.

$$|X - c| - |X - m| = \begin{cases} c - m & \text{if } X < c \\ 2(X - c) + (c - m) & \text{if } c \leqslant X \leqslant m \\ m - c & \text{if } X > m \end{cases}$$

Now evaluate $E(|X - c| - |X - m|)$. If $c \geqslant m$, the proof is similar.]

(e) Use (d) to show that $E(|X - c|)$ is minimized by choosing $c = m$.

[Hint: The integrands in both integrals of (a) are nonnegative.]

(f) Repeat parts (d) and (e), supposing that X is discrete. There is no change except that $\int$ is replaced by Σ.
[Compare (e) with (b). Part (b) says that $E|X-a|^2)$ is minimized when $a = EX$. Part (f) says that $E(|X-c|)$ is minimized when $c =$ median.]

22 (a) Prove the following inequality. Suppose that Z is a nonnegative random variable the expected value of which exists. Then, for any $t \geqslant 0$,

$$EZ \geqslant tP(Z \geqslant t)$$

[Hint: If Z has an integrating density f, then

$$EZ = \int_0^\infty xf(x)\,dx \geqslant \int_t^\infty xf(x)\,dx \geqslant t\int_t^\infty f(x)\,dx$$

The discrete case is similar.]

(b) Show that Chebyshev's inequality, [(2), Section 10.2] follows from (a) by letting $Z = (X-m)^2$ and replacing t by t^2. {Hint: The result in (a) becomes

$$E[(X-m)^2] \geqslant t^2\,P[(X-m)^2 \geqslant t^2)\,]$$

23 (a) Suppose Z_n is a uniformly bounded sequence of random variables. That is, for finite constants a and b,

$$P(a \leqslant Z_n \leqslant b) = 1, \qquad n = 1, 2, \ldots$$

Suppose that the distribution of Z_n becomes concentrated about m as $n \to \infty$. That is, for any $\epsilon > 0$,

$$\lim_{n\to\infty} P(|Z_n - m| \geqslant \epsilon) = 0$$

(m is necessarily in $[a, b]$). Show that

$$\lim_{n\to\infty} E(Z_n) = m$$

[Hint: $b - Z_n$ and $Z_n - a$ are nonnegative random variables with expected values $b - E(Z_n)$ and $E(Z_n) - a$. By Exercise 22a,

$$b - EZ_n \geqslant (b - m - \epsilon)P(b - Z_n \geqslant b - m - \epsilon)$$

$$EZ_n - a \geqslant (m - a - \epsilon)P(Z_n - a \geqslant m - a - \epsilon)$$

Combining these,

$$a + (m - a - \epsilon)P(Z_n \geqslant m - \epsilon) \leqslant E(Z_n)$$

$$\leqslant b - (b - m - \epsilon)P(Z_n \leqslant m + \epsilon)$$

By hypothesis, $\lim_{n\to\infty} P(Z_n \leqslant m + \epsilon) = \lim_{n\to\infty} P(Z_n \geqslant m - \epsilon) = 1$. Conclude that $\lim_{n\to\infty} EZ_n = m$, since $\epsilon > 0$ is arbitrary.]

(b) Suppose Z_n is a sequence of random variables such that

$$P(Z_n = 0) = 1 - \frac{1}{n}, \qquad P(Z_n = n) = \frac{1}{n}$$

Use this example to show that the result in (a) does not necessarily hold if the condition of uniform boundedness is removed.

NOTES

1 For additional reading on variance and covariance in the discrete case, see Sections 4 and 5 of Chapter IX, [9], and Chapters 6 and 7 of [10].

2 Chebyshev's inequality is named for the eminent Russian mathematician Pafnuti Liwowich Chebyshev (1821–1894).

3 The result in Example 1, Section 10.2, is the earliest version of the law of large numbers to be proved. It was accomplished by elegant but very computational means by James Bernoulli (1654–1705).

4 The law of large numbers with only the first moment existing is the work of Russian mathematician A. Khintchine (1894–1959). For a presentation of his proof, see Section 2, Chapter X, [9].

5 The central limit theorem is one of a broad class of limit theorems for sums of independent random variables which have been developed over the last several decades. For a complete treatment, see [36].

6 For a proof of the central limit theorem, see Section 17.4, [31]. An ingenious proof that does not use Fourier transforms appears in [37].

7 The special case of the central limit theorem in Example 1, Section 10.3, was first proved by DeMoivre. A general proof of the central limit theorem (although unacceptable by modern standards) is due to Laplace. The first rigorous proofs are the work of the Finnish mathematician J. W. Lindeberg (1876–1933) and the contemporary French mathematician Paul Levy (1886–   ).

8 For the improved normal approximation to the binomial referred to in Example 2, Section 10.3, see Equation (3.16), Section VII.3, [9].

9 It is true that

$$\lim_{n\to\infty} \left(1 + \frac{1}{2} + \cdots + \frac{1}{n} - \log n\right)$$

is finite, from which it follows immediately that

$$\lim_{n\to\infty} \frac{1 + \frac{1}{2} + \cdots + \frac{1}{n}}{\log n} = 1$$

See p. 665, [2] and p. 405, [3].

Chapter Eleven

STOCHASTIC PROCESSES

Phenomena that evolve as random functions of time are called *stochastic processes.* Important among these are *counting processes,* which count the accumulating numbers of recurrences of some event where successive waiting times between recurrences are independent and identically distributed. This chapter considers *simple random walk* and the *Polya urn process* as examples of stochastic processes over discrete time. A broad class of important discrete-time stochastic processes are the *Markov chains.* Finally, we look at the *Poisson process,* which is a special but important counting process (Note 1).

11.1 STOCHASTIC PROCESSES

We shall be considering chance phenomena that evolve as functions of time. Some empirical examples follow.

(a) The daily price level of a given commodity is observed over a two-year period.
(b) A record is kept of the numbers of individuals born in each of the first 20 generations of a biological population.
(c) An electrocardiogram record is made, with entries at each of 100 one-second intervals.

Mathematically, each of the preceding examples is described by a sequence of random variables, $Z(t_1)$, $Z(t_2)$, . . . , the indices t_1, t_2, . . . referring to the times at which observations are made. These indices vary over a discrete time set, and in this case the sequence $\{Z(t_i)\}$ is called a *discrete-time stochastic process.* In some cases, a chance phenomenon evolves over a continuum of times, such as the intervals $[a, b]$, $[a, \infty)$, or $(-\infty, \infty)$. Then the family of random variables $\{Z(t)\}$ is called a *continuous-time stochastic process.* Some empirical examples follow.

(a) Two particles are randomly floating about on the surface of a liquid. Let $Z(t)$ be the distance between the two particles for any time t in the interval $[a, b]$.
(b) Consider the earlier example of an electrocardiogram record. Ignoring the practical limitations of the recording equipment, a useful probability model might describe the reading $Z(t)$ with t varying over an interval $[a, b]$.

11.2 WAITING TIME PROCESSES

Many interesting probability problems concern successive waiting times between recurrences of some phenomenon. It is usually assumed that after each recurrence the wait for the next recurrence starts anew, independent of the past history of successive waiting times, and that the distribution of each waiting time is the same. The mathematical model follows.

waiting times

$W_1, W_2, \ldots$ is a sequence of independent and identically distributed, nonnegative random variables. These represent successive waiting times between recurrences.

Example 1

A Geiger counter records the arrivals of radioactive particles. An assumption that seems to fit is that successive waiting times between arrivals are independent and identically distributed random variables, each exponentially distributed with parameter $\lambda > 0$. (The expected waiting time between arrivals is $1/\lambda$.)

Example 2

When a light bulb in a socket burns out, it is replaced by a new bulb. The lifetimes of successive generations of bulbs is represented by $W_1, W_2, \ldots,$ a sequence of independent and identically distributed, nonnegative random variables.

Example 3

Each of two boxes contains a black ball and a white ball. A ball is drawn at random from each box, and these are placed in the opposite boxes. A *recurrence* takes place whenever each box returns to its original state of one black and one white ball. The successive waiting times between recurrences are independent and identically distributed. The waiting time distribution is $P(W_i = 1) = 1/2$, $P(W_i = 2) = 1/2$. (Verify this.)

Associated with the waiting times $W_1, W_2, \ldots$ is a "counting process," $N(t)$, which counts the number of recurrences that have taken place up to time t. $N(t)$ is a stochastic process having the following formal definition:

counting process

$$N(t) = \begin{cases} 0 & 0 \leq t < W_1 \\ 1 & W_1 \leq t < W_1 + W_2 \\ 2 & W_1 + W_2 \leq t < W_1 + W_2 + W_3 \\ \vdots & \end{cases} \tag{1}$$

(See Figure 11.1.) For any $t \geq 0$, $N(t)$ is an integer-valued random variable having the possible values $0, 1, 2, \ldots$.

The relationship between the distribution of $N(t)$ and the distribution of $W_1 + \cdots + W_k$ is the following:

distribution of N(t)

$$\begin{aligned} P[N(t) = 0] &= P(W_1 > t) \\ P[N(t) = 1] &= P(W_1 + W_2 > t) - P(W_1 > t) \\ &\vdots \\ P[N(t) = k] &= P(W_1 + \cdots + W_{k+1} > t) \\ &\quad - P(W_1 + \cdots + W_k > t) \\ &\vdots \qquad 0 \leq t \end{aligned} \tag{2}$$

Proof of (2). From the definition of $N(t)$,

$$[N(t) \leq k] = (W_1 + \cdots + W_{k+1} > t), \qquad k = 0, 1, \ldots \tag{3}$$

{In other words, if the $(k+1)$st recurrence takes place after time t, then, and only then, are there k or fewer recurrences in $[0, t]$.} Hence, from (3),

$$\begin{aligned} P[N(t) = k] &= P[N(t) \leq k] - P[N(t) \leq k-1] \\ &= P(W_1 + \cdots + W_{k+1} > t) - P(W_1 + \cdots + W_k > t) \end{aligned}$$

for $k = 1, 2, \ldots$. The case of $k = 0$ follows from the relationship, $[N(t) = 0] = (W_1 > t)$. (There are no recurrences up to time 0 if, and only if, the first waiting time exceeds t.) ∎

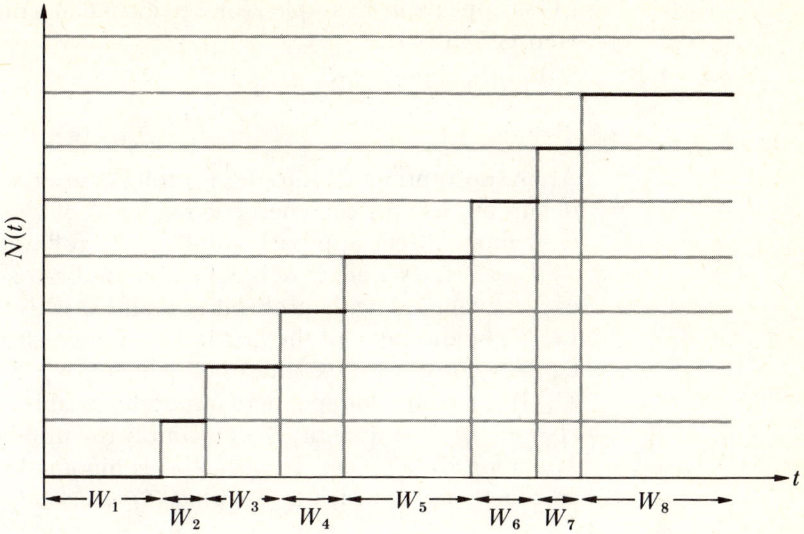

FIGURE 11.1

Example 4 Consider Example 3 again. During the first n exchanges, how many times have the boxes returned to their original state? Since $W_i - 1$ is an indicator random variable, $W_1 + \cdots + W_k - k$ is binomially distributed with parameters k and 1/2. The distribution of $N(4)$, the number of recurrences that have taken place by the fourth step, is computed from (2) to be the following:

i	2	3	4
$P[N(4) = i]$	8/16	7/16	1/16

Example 5 Suppose that W_i has the geometric distribution given by

$$P(W_i = k) = (1-p)^{k-1}p, \qquad k = 1, 2, \ldots$$

Then

$$P[N(n) = k] = \binom{n}{k} p^k (1-p)^{n-k}, \qquad k = 0, 1, \ldots, n; \quad n = 1, 2, \ldots \tag{4}$$

[Since the W_i's are integer valued, $N(t) = N(n)$ for $n \leq t < n + 1$. Only the integer values of t are really relevant.] There are several ways of checking (4). One way is to use first the fact that

$$P(W_1 + \cdots + W_k = n) = \binom{n-1}{k-1} p^k (1-p)^{n-k} \tag{5}$$

[This appears in Example 2(b), Section 6.6 in a different notation.] Hence, by (2),

$$P[N(n) = k] = P(W_1 + \cdots + W_{k+1} > n) - P(W_1 + \cdots + W_k > n) \qquad (6)$$

After substituting (5) into (6), (4) follows after some algebra. The details are left for the energetic reader.

A more direct approach to (4) is the following. Suppose that $X_1, X_2, \ldots$ is a sequence of independent indicator random variables each equaling 1 with probability p and 0 with probability $1 - p$. Let W_1 be the value of the first index i for which $X_i = 1$; let $W_1 + W_2$ be the value of the next index i for which $X_i = 1$; and so forth. Then $W_1, W_2, \ldots$ are independent random variables and $P(W_i = k) = (1 - p)^{k-1}p$. The quantity $N(n)$ is simply the number of X_i's that are 1, among $X_1, \ldots, X_n$. Hence, $N(n)$ is binomially distributed with parameters n, p, which is just what (4) asserts.

11.3 BEHAVIOR OF $N(t)$ FOR LARGE t

We now show that there is a law of large numbers for $N(t)$. Suppose that

$$E(W_i) = m = \text{expected waiting time between recurrences}$$

Then the distribution of $N(t)/t$ is almost all concentrated at $1/m$ for large t. The specific facts follow.

law of large numbers for N(t)

$$\lim_{t\to\infty} P\left[\frac{N(t)}{t} < u\right] = \begin{cases} 0 & \text{if } u < \dfrac{1}{m} \\ 1 & \text{if } u > \dfrac{1}{m} \end{cases} \qquad (1)$$

In fact, (1) holds even if $E(W_i) = \infty$ when $1/m$ is replaced by 0. That is, if $EW_i = \infty$, then

$$\lim_{t\to\infty} P\left[\frac{N(t)}{t} < u\right] = 1 \quad \text{if } u > 0 \qquad (2)$$

The proof uses the law of large numbers proved in Section 10.2.

Proof of (1). First suppose that $E(W_i) = m < \infty$. By (3), Section 11.2,

$$P[N(t) \leq k] = P(W_1 + \cdots + W_{k+1} > t)$$

or, equivalently,

$$P[N(t) < k] = P(W_1 + \cdots + W_k > t)$$

Replace t by kt to obtain

$$\begin{aligned} P[N(kt) < k] &= P\left[\frac{N(kt)}{kt} < \frac{1}{t}\right] \\ &= P(W_1 + \cdots + W_k > kt) \\ &= P\left(\frac{W_1 + \cdots + W_k}{k} > t\right) \end{aligned}$$

By the version of the law of large numbers given by (7), Section 10.2,

$$\lim_{k\to\infty} P\left(\frac{W_1 + \cdots + W_k}{k} > t\right) = \begin{cases} 0 & \text{if } t > m \\ 1 & \text{if } t < m \end{cases}$$

Hence,

$$\lim_{k\to\infty} P\left[\frac{N(kt)}{kt} < \frac{1}{t}\right] = \begin{cases} 0 & \text{if } \dfrac{1}{t} < \dfrac{1}{m} \\ \\ 1 & \text{if } \dfrac{1}{t} > \dfrac{1}{m} \end{cases}$$

which is equivalent to (1). ▮

We leave the proof of (2) to the reader (see Exercise 18).

Example 1 Suppose that the waiting times $W_1, W_2, \ldots$ are geometrically distributed, as in Example 5, Section 11.2. As shown in that example, $N(n)$ is binomially distributed with parameters n and p, for $n = 1, 2, \ldots$. Hence, $E[N(n)/n] = p$. According to the law of large numbers, as it applies to the binomial distribution (Example 1, Section 10.2), the distribution of $N(n)/n$ becomes concentrated about p as n goes to infinity. This fact conforms with the result in (1) since $E(W_i) = 1/p$.

Implicitly, the waiting time random variables W_i have been assumed to satisfy the requirement that $P(W_i < \infty) = 1$; that is, the W_i's assume only *finite* values. In studying waiting time processes, it is very important to extend the concept of random variable to allow the W_i's to equal *infinity* with a positive probability. The understanding must be that if $W_i = \infty$, then any sum $W_1 + \cdots + W_k$ that contains W_i must also equal infinity. The definition of $N(t)$

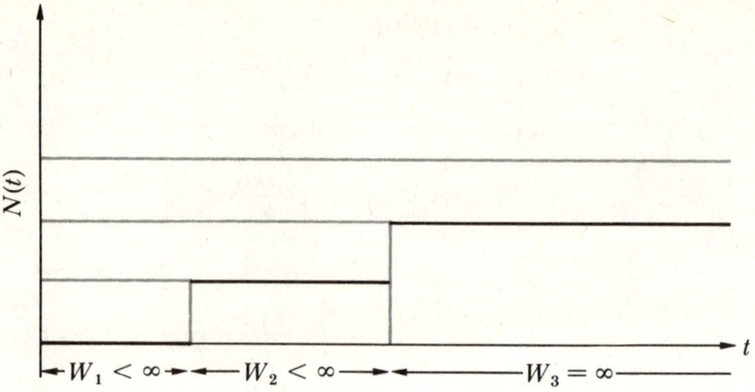

FIGURE 11.2 **If W_1 and W_2 are finite but $W_3 = \infty$, then $N(t)$ is 0 in $[0, W_1)$, $N(t)$ is 1 in $[W_1, W_1 + W_2)$, and $N(t)$ is 2 for all t in $[W_1 + W_2, \infty)$.**

and the relations (1), (2), (3), Section 11.2, are still correct with this extended concept of waiting time (see Figure 11.2).

Example 2 Suppose that

$$W_i = \begin{cases} 1 & \text{with probability } \frac{1}{2} \\ \infty & \text{with probability } \frac{1}{2} \end{cases}$$

Let us compute the distribution of $N(n)$, for $n = 1, 2, \ldots$.

$$\begin{aligned} P[N(n) = 0] &= P(W_1 = \infty) = \tfrac{1}{2} \\ P[N(n) = k] &= P[(W_1 = \cdots = W_k = 1) \cap (W_{k+1} = \infty)] \\ &= (\tfrac{1}{2})^{k+1}, \qquad 1 \leq k < n \\ P[N(n) = n] &= P(W_1 = \cdots = W_n = 1) = (\tfrac{1}{2})^n \\ P[N(n) > n] &= 0 \end{aligned}$$

11.4 THE RENEWAL EQUATION

Consider now the special case of waiting time random variables $W_1, W_2, \ldots$, which are positive and *integer valued.* That is, the possible values of W_i are $1, 2, \ldots$. Denote

$$P(W_i = k) = f_k, \qquad k = 1, 2, \ldots$$

Following the discussion in Section 11.3, we shall allow W_i to assume the value ∞. In other words,

$$f = f_1 + f_2 + \cdots = P(W_i < \infty) \tag{1}$$

may be strictly less than 1. In any case, $f \leq 1$.

Since the waiting times are integer valued, in studying the

counting process one need only consider $N(n), n = 1, 2, \ldots$. Introduce indicator random variables $U(1), U(2), \ldots$ defined as follows:

$$\begin{aligned} U(1) &= N(1) \\ U(2) &= N(2) - N(1) \\ &\vdots \\ U(n) &= N(n) - N(n-1) \\ &\vdots \end{aligned} \tag{2}$$

In other words,

$$U(n) = \begin{cases} 1 & \text{if } W_1 = n \text{ or } W_1 + W_2 = n \text{ or } \cdots W_1 + \cdots + W_n = n \\ 0 & \text{otherwise} \end{cases}$$

$U(n)$ indicates whether one of the partial sums of waiting times equals n. In other words, it indicates whether "a recurrence has taken place" *at* time n. It follows from (2) that

$$U(1) + \cdots + U(n) = N(n)$$

which is the total number of recurrences that have taken place

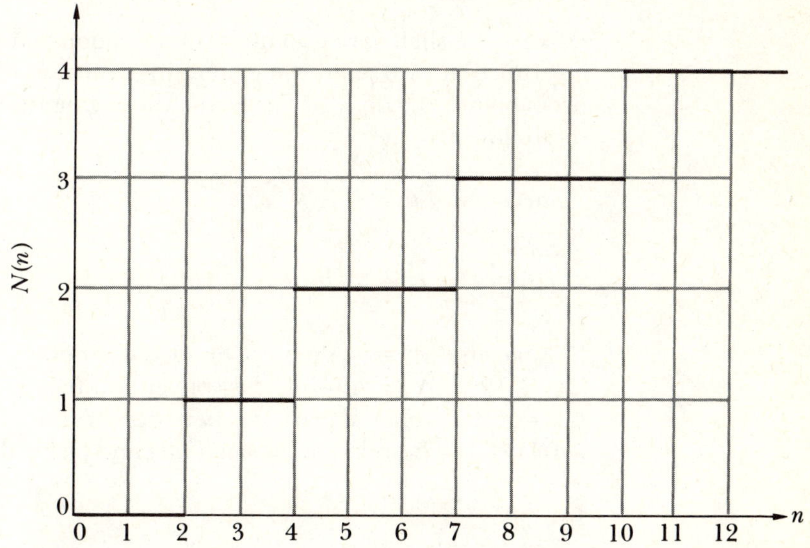

FIGURE 11.3 The values of $U(n)$ and $N(n)$ for the graph are the following:

n	1	2	3	4	5	6	7	8	9	10	11	12
$U(n)$	0	1	0	1	0	0	1	0	0	1	0	0
$N(n)$	0	1	1	2	2	2	3	3	3	4	4	4

through time n (see Figure 11.3). We also denote $P(U(n) = 1)$ as:

recurrence probabilities

$$u_n = P(U(n) = 1) = E[U(n)], \qquad n = 1, 2, \ldots \tag{3}$$

These are called the *recurrence probabilities.*

recurrent event process (renewal process)

The sequence of random variables $U(1)$, $U(2)$, . . . is called a *renewal process* or *recurrent event process.*

In other words, a renewal process is a sequence of random variables indicating the times at which renewals or recurrences take place. The total number of such indications through time n is $N(n)$, the counting process.

We shall use the following terminology.

persistence and transience

If f, as defined by (1), is less than 1, then the renewal process is called *transient.* If $f = 1$, then the process is called *persistent.*

Next, we shall derive an important relation between the generating function of W_i and the generating function of the recurrence probabilities u_n, Eq. (3). Explicitly, these generating functions are the following:

$$u(t) = \sum_{n=0}^{\infty} u_n t^n = 1 + \sum_{n=1}^{\infty} u_n t^n$$

$$f(t) = \sum_{n=1}^{\infty} f_n t^n = \sum_{n=1}^{\infty} P(W_i = n) t^n, \qquad 0 \leq t \leq 1$$

[The term $P(W_i = \infty)$ does not enter the sum for $f(t)$. That is, $f(t) = \Sigma_{i \leq n < \infty} f_n t^n$.] We are using the convention that $u_0 = 1$. Notice that $f(1) = f$. The renewal process is persistent if $f(1) = 1$ and is transient if $f(1) < 1$. The relation between $u(t)$ and $f(t)$ is the following:

the renewal equation

$$u(t) = \frac{1}{1 - f(t)}, \qquad 0 \leq t \leq 1 \tag{4}$$

[If $f(1) = 1$, then $u(1) = \infty$.]

Proof of (4). First consider the following relationships, which follow from the definition of the recurrence probabilities:

$$u_0 = 1$$
$$u_1 = P(W_1 = 1)$$
$$u_2 = P(W_1 = 2) + P(W_1 + W_2 = 2)$$
$$\vdots$$

Hence,

$$u_0 = 1$$
$$u_1 t = P(W_1 = 1)t$$
$$u_2 t^2 = P(W_1 = 2)t^2 + P(W_1 + W_2 = 2)t^2$$
$$\vdots$$

The sum of the left side is $u(t)$. The sum of the right side is

$$1 + \sum_{n=1}^{\infty} P(W_1 = n)t^n + \sum_{n=2}^{\infty} P(W_1 + W_2 = n)t^n + \cdots$$

It should be understood that the terms for $n = \infty$ do not enter any of these sums. This fact can be automatically assured by defining $t^{W_i} = 0$ if $W_i = \infty$. We have the fact that

$$E(t^{W_1+\cdots+W_k}) = \sum_n P(W_1 + \cdots + W_k = n)t^n = [f(t)]^k$$

This fact follows because $t^{W_1}, \ldots, t^{W_k}$ are independent, and, hence, $E(t^{W_1+\cdots+W_k}) = E(t^{W_1}) \cdots E(t^{W_k})$ [(3), Section 5.4]. Hence,

$$u(t) = 1 + f(t) + [f(t)]^2 + \cdots = \frac{1}{1 - f(t)}, \qquad 0 \leq t \leq 1$$

This convergence is assured by the fact that $f(t) < 1$ if $0 \leq t < 1$. If $f(1) = 1$, then (4) still holds, since when $t = 1$,

$$u(1) = 1 + 1 + \cdots = \infty = \frac{1}{1 - f(1)} \qquad ∎$$

Example 1 Consider the experiment described in Example 3, Section 11.2. In this case, $f(t) = (t + t^2)/2$. According to the renewal equation, (4),

$$u(t) = \frac{1}{1 - [(t + t^2)/2]} = \frac{2/3}{1 - t} + \frac{2/3}{2 + t}$$
$$= \sum_{n=0}^{\infty} \left[\frac{2}{3} + \frac{1}{3}\left(\frac{-1}{2}\right)^n\right] t^n$$

Hence, u_n, the probability that the boxes will be in their original state at time n, is

$$u_n = \frac{2}{3} + \frac{1}{3}\left(\frac{-1}{2}\right)^n, \qquad n = 1, 2, \ldots$$

Notice that $\lim_{n\to\infty} u_n = 2/3$. We shall see in (7), which follows, that this limit must coincide with $1/EW_i$. It checks out in this case, since $E(W_i) = (1)(1/2) + (2)(1/2) = 3/2$.

Often, the u_n's are easy to determine directly, whereas an explicit calculation of the waiting time probabilities may not be so easy. The renewal equation can then be used to determine the f_n's.

Example 2 A coin is thrown until the first throw on which a head is obtained that is immediately preceded by a tail. Then the throwing resumes until the next time this combination occurs, and so forth. What is the waiting time distribution? The probability of a recurrence taking place with the nth throw is easily seen to be

$$u_1 = 0, \qquad u_n = \tfrac{1}{4}, \qquad n = 2, 3, \ldots$$

Hence, $u(t) = 1 + t^2/4(1-t)$. Solving for $f(t)$ by (4), we get

$$f(t) = \frac{u(t) - 1}{u(t)} = \frac{t^2/4}{[1 - (t/2)]^2}$$

The coefficients of t^n are

$$f_1 = 0, \qquad f_n = \frac{n-1}{2^n}, \qquad n = 2, 3, \ldots$$

Since $f'(1) = 4$, it follows that the expected number of throws required to obtain first the combination *tail, head* equals 4.

Example 3 (Simple random walk.) A particle performs a random walk over the positions $0, \pm 1, \pm 2, \ldots$ as follows. The particle starts at 0 and makes successive independent steps $+1$ or -1 with probabilities p, $1-p$, respectively. Let $W_1, W_2, \ldots$ be the successive numbers of steps between the repeated returns to 0. Every time the particle is back at the origin, its future evolution is independent of the past. A formal proof of the fact that $W_1, W_2, \ldots$ are independent and identically distributed is not too hard but will be omitted here. $U(n) = 1$ means that at step n the particle is back at the origin, but, of course, not necessarily for the first time. The recurrence probabilities have the following values:

$$u_{2n+1} = P[U(2n+1) = 1] = 0 \quad \text{(a return to 0 in an odd number of steps is not possible)}$$

$$u_{2n} = P[U(2n) = 1] = \binom{2n}{n} p^n (1-p)^n$$

We have the following relationship:

$$\binom{2n}{n} = \frac{(2n)!}{n!n!} = \frac{[(2n)(2n-2)\cdots(4)(2)][(2n-1)(2n-3)\cdots(3)(1)]}{n!n!}$$

$$= \frac{2^{2n}[(n)(n-1)\cdots(2)(1)][(n-\frac{1}{2})(n-\frac{3}{2})\cdots(\frac{3}{2})(\frac{1}{2})]}{n!n!}$$

$$= \binom{-1/2}{n}(-4)^n \tag{5}$$

Hence, using Newton's binomial expansion [(4), Section 6.5],

$$u(t) = \sum_{n=0}^{\infty} \binom{-1/2}{n}[-4p(1-p)t^2]^n = \frac{1}{[1-4p(1-p)t^2]^{1/2}}$$

Solving for $f(t)$ in (4), and simplifying,

$$f(t) = \frac{u(t)-1}{u(t)} = 1 - [1-4p(1-p)t^2]^{1/2}$$

It follows, again using (4), Section 6.5, that the first return probabilities are

$$f_{2n+1} = 0$$

$$f_{2n} = -\binom{1/2}{n}[-4p(1-p)]^n = \frac{2}{n}\binom{2n-2}{n-1}[p(1-p)]^n \tag{6}$$

and that

$$f(1) = f_1 + f_2 + \cdots = 1 - [1-4p(1-p)]^{1/2}$$

$$= 1 - [(2p-1)^2]^{1/2} = 1 - |2p-1|$$

[The relation in (6) follows from the fact that $\binom{1/2}{n} = \frac{1}{2n}\binom{-1/2}{n-1}$, which equals $\binom{2n-2}{n-1}\Big/2n(-4)^{n-1}$ by (5).] Hence, the process is persistent if, and only if, $p = 1/2$. When $p = 1/2$, the probability is 1 that a return to 0 will eventually take place. If $p \neq 1/2$, with probability $|2p-1|$ the walk will "drift off" and never return to 0. It is interesting to compute the expected return time when $p = 1/2$. In this case, $f(t) = 1 - (1-t^2)^{1/2}$, and

$$f'(1) = -\left.\frac{d(1-t^2)^{1/2}}{dt}\right|_{t\uparrow 1} = \infty$$

When $p = 1/2$, even though a return to the origin is certain, the expected return time is infinite (Note 2).

Finally, we want to show that the recurrence probabilities u_n have a Cesaro limit. It means that $\lim_{n\to\infty} (u_1 + \cdots + u_n)/n$ exists.

Cesaro limit of u_n

$$\lim_{n\to\infty} \frac{u_1 + \cdots + u_n}{n} = \frac{1}{E(W_1)} = \frac{1}{m} \tag{7}$$

If $E(W_1) = \infty$, then the limit in (7) equals 0.

Proof of (7). Consider the random variable

$$\frac{N(n)}{n} = \frac{U(1) + \cdots + U(n)}{n}$$

By (1), Section 11.3, the distribution of $N(n)/n$ becomes concentrated around the point $1/m$ as $n \to \infty$. Since

$$E\frac{N(n)}{n} = \frac{u_1 + \cdots + u_n}{n}$$

it is plausible that $(u_1 + \cdots + u_n)/n$ should have the limit $1/m$. This fact is indeed correct, because

$$0 \leqslant \frac{N(n)}{n} \leqslant 1$$

For a proof, see Exercise 23a, Chapter 10. ▮

It is a fact that if $\lim_{n\to\infty} u_n$ exists, then the limit in (7) must exist also and has the same value (Note 3). However, in general, the limit in (7) may not exist, but the limit of the nonzero u_n's must exist (Note 4).

Example 4

(a) In Example 2, $\lim_{n\to\infty} u_n = 1/4$, which agrees with the fact that the expected waiting time for a recurrence equals 4.
(b) In Example 3, when $p = 1/2$, the expected time for return to the origin equals ∞. Hence,

$$\lim_{n\to\infty} u_{2n} = \lim_{n\to\infty} \binom{2n}{n} \frac{1}{4^n} = 0$$

(Can you prove this fact directly?)

11.5 SYMMETRIC RANDOM WALK

We now want to consider, in greater detail, the random walk described in Example 3, Section 11.4. We shall restrict ourselves to the special case that $p = 1/2$. In this case, the random walk is called *symmetric random walk.* The symmetry makes certain aspects of the analysis simpler. A formal way of defining this random walk follows. Let $X_1, X_2, \ldots$ be a sequence of mutually independent random variables, each with the distribution $P(X_i = 1) = P(X_i = -1) = 1/2$. Define

$$S_n = X_1 + \cdots + X_n, \qquad n = 1, 2, \ldots$$

The position of the particle at the nth step is S_n. We denote the initial position by $S_0 = 0$. A useful way of depicting the random walk is shown in Figure 11.4. We refer also to the random variables $U(n)$, W_n, used in Example 3, Section 11.4. In terms of the S_n's,

$$U(n) = 1 \quad \text{if, and only if } S_n = 0$$

The successive waiting times between the zeros of S_n are $W_1, W_2, \ldots$ (see Figure 11.4). As was pointed out in the last section, the generating function of W_i is $f(t) = 1 - (1 - t^2)^{1/2}$, $P(W_i = 2n) = (2/n)\binom{2n-2}{n-1}(1/2^{2n})$ and $P(W_i < \infty) = 1$.

The distribution of S_n is closely related to the binomial and is obtained as follows. First, $(X_i + 1)/2$ assumes the values 0 and 1 with probabilities 1/2, 1/2. Hence,

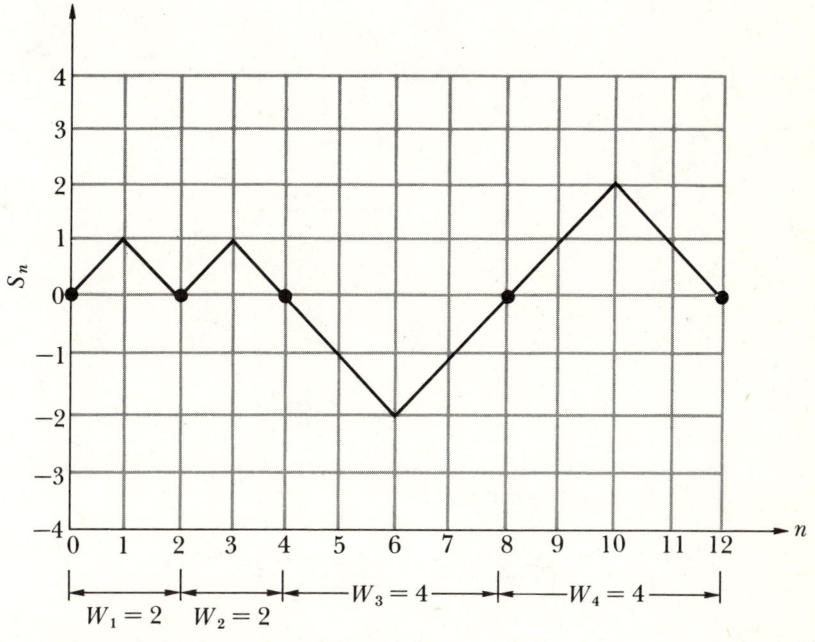

FIGURE 11.4 Graph of random walk the first 12 steps of which are +1, −1, +1, −1, −1, −1, +1, +1, +1, +1, −1, −1.

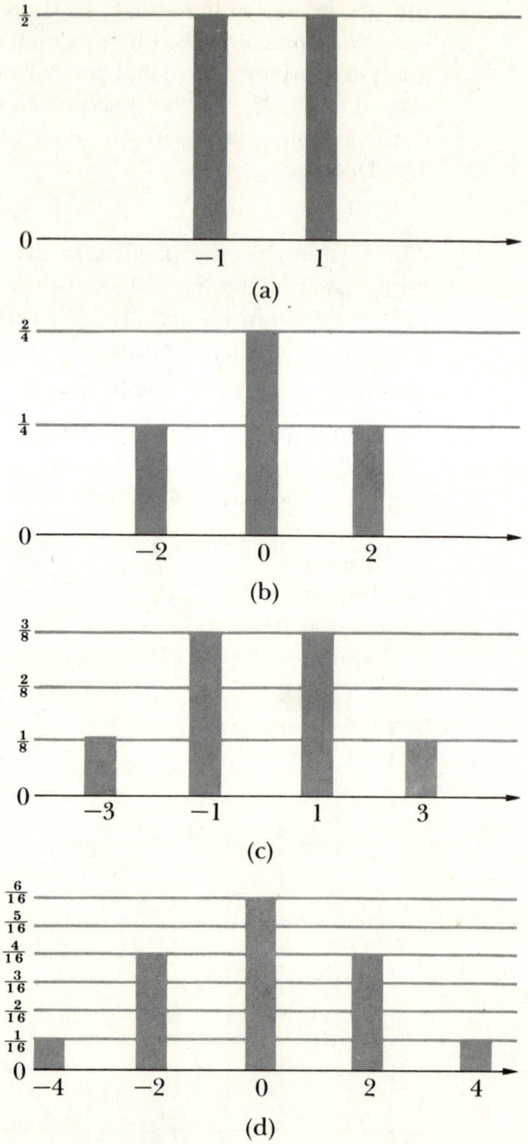

FIGURE 11.5 (a) Density of S_1. (b) Density of S_2. (c) Density of S_3. (d) Density of S_4.

$$\left(\frac{X_1+1}{2}\right)+\cdots+\left(\frac{X_n+1}{2}\right)=\frac{S_n+n}{2}$$

is binomially distributed with parameters n, 1/2, and

$$P\left(\frac{S_n + n}{2} = k\right) = P(S_n = 2k - n) = \binom{n}{k}\frac{1}{2^n}, \qquad k = 0, 1, \ldots, n$$

If n is even, the possible values of S_n are even; otherwise, they are odd (see Figure 11.5).

Since $E(X_i) = 0, E(X_i^2) = 1 = \text{Var}(X_i)$, it follows that $E(S_n) = 0$ and $\text{Var}(S_n) = n$. Hence, by the central limit theorem [(3), Section 10.3], we have the following:

asymptotic normality of particle's position

$$\lim_{n\to\infty} P\left(\frac{S_n}{n^{1/2}} \leq t\right) = \frac{1}{(2\pi)^{1/2}} \int_{-\infty}^{t} \exp\left(\frac{-x^2}{2}\right) dx \qquad (1)$$

11.6 MAXIMUM DISTANCE REACHED BY THE PARTICLE

Since $\text{Var}(S_n) = n$, the distribution of S_n becomes increasingly dispersed as n grows larger. It is interesting to study the maximum distance reached. Specifically, define

$$M_n = \max(0, S_1, S_2, \ldots, S_n), \qquad n = 1, 2, \ldots$$

(see Figure 11.6). The distribution of M_n can be related to that of S_n by means of an interesting device called the *reflection principle.*

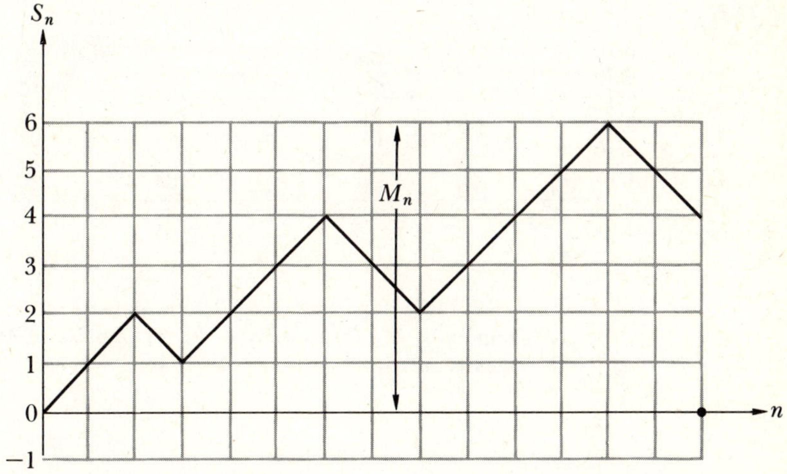

FIGURE 11.6 M_n **is the maximum achieved by random walk up to the** n**th step.**

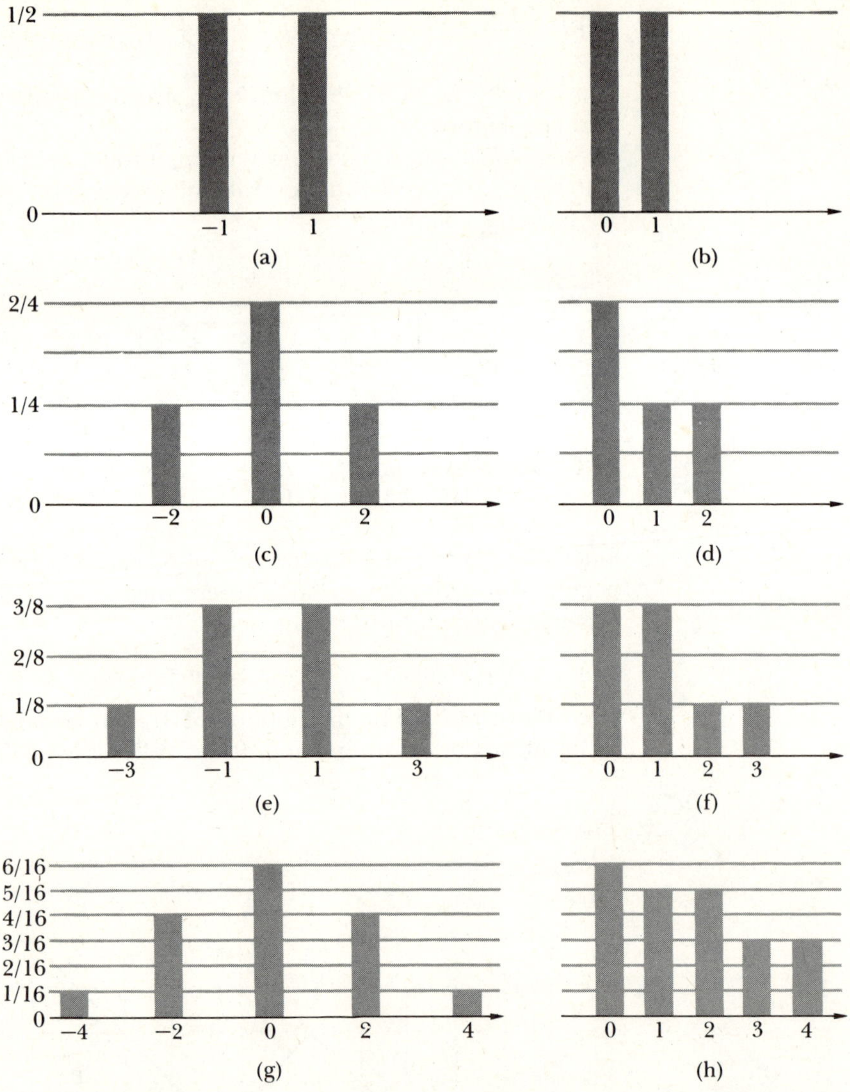

FIGURE 11.7 **(a) Density of S_1. (b) Density of M_1. (c) Density of S_2. (d) Density of M_2. (e) Density of S_3. (f) Density of M_3. (g) Density of S_4. (h) Density of M_4.**

distribution of M_n

(a) $P(M_n \geq k) = 2P(S_n > k) + P(S_n = k) = P(S_n \geq k) + P(S_n > k)$

(b) $P(M_n = k) = P(S_n = k+1) + P(S_n = k), \qquad k = 0, 1, \ldots, n$

(Only one of these terms is not 0. See Figure 11.7.)

(c) $$\lim_{n\to\infty} P\left(\frac{M_n}{n^{1/2}} \leq t\right) = \begin{cases} \left(\frac{2}{\pi}\right)^{1/2} \int_0^t \exp\left(\frac{-x^2}{2}\right) dx & \text{if } t \geq 0 \\ 0 & \text{otherwise} \end{cases}$$

Proof. By breaking up the event $M_n \geq k$, we have

$$P(M_n \geq k) = P[(M_n \geq k) \cap (S_n > k)] + P[(M_n \geq k) \cap (S_n = k)] + P[(M_n \geq k) \cap (S_n < k)] \qquad (1)$$

By the reflection principle,

$$P[(M_n \geq k) \cap (S_n > k)] = P[(M_n \geq k) \cap (S_n < k)] \qquad (2)$$

The proof of (2) is outlined in Figure 11.8. Using the facts that

$$(M_n \geq k) \supset (S_n > k), \qquad (M_n \geq k) \supset (S_n = k)$$

and substituting (2) in (1), we have

$$P(M_n \geq k) = 2P(S_n > k) + P(S_n = k) = P(S_n \geq k) + P(S_n > k)$$

which proves (a). To find the individual probabilities, we take differences.

$$\begin{aligned} P(M_n = k) &= P(M_n \geq k) - P(M_n \geq k+1) \\ &= 2P(S_n > k) - 2P(S_n > k+1) + P(S_n = k) - P(S_n = k+1) \\ &= 2P(S_n = k+1) + P(S_n = k) - P(S_n = k+1) \end{aligned}$$

which proves (b). By the central limit theorem, and part (a), if $t \geq 0$,

$$\begin{aligned} \lim_{n\to\infty} P\left(\frac{M_n}{n^{1/2}} \geq t\right) &= \lim_{n\to\infty} P\left(\frac{S_n}{n^{1/2}} \geq t\right) + \lim_{n\to\infty} P\left(\frac{S_n}{n^{1/2}} > t\right) \\ &= \frac{2}{(2\pi)^{1/2}} \int_t^\infty \exp\left(\frac{-x^2}{2}\right) dx \end{aligned}$$

which proves (c). ∎

Notice that (c) says that $M_n/n^{1/2}$ is asymptotically distributed like $|X|$ where X has the $N(0, 1)$ distribution.

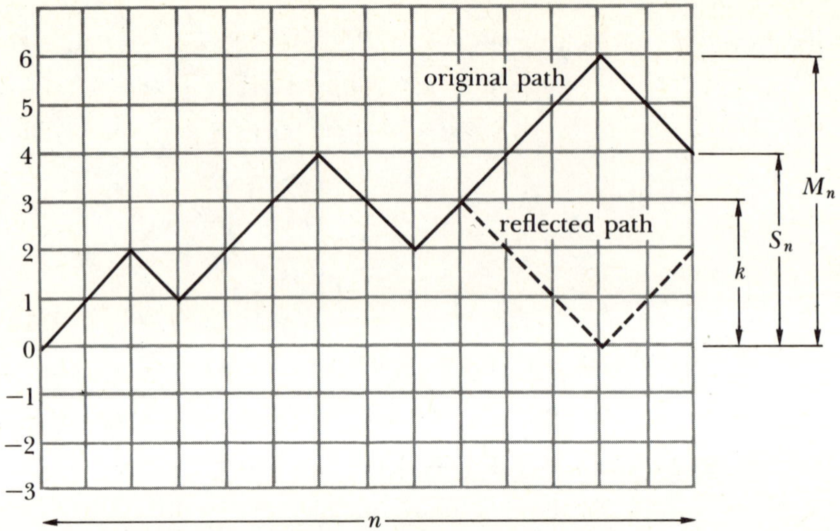

FIGURE 11.8 Consider any path for which $S_n > k \geq 0$. Then $M_n \geq S_n > k$, and, of course, $M_n \geq k$. This path must cross height k at some step between 1 and n. From the last time that the path crosses height k, interchange the +1's and −1's to obtain a reflected path. For the reflected path, $M_n \geq k$ and $S_n < k$. Conversely, consider any path for which $M_n \geq k$ and $S_n < k$. From the last time that the path crosses height k, interchange the +1's and −1's to obtain a reflected path. For the reflected path, $M_n \geq k$ and $S_n > k$. Thus, a one-to-one correspondence exists between the paths in $(M_n \geq k) \cap (S_n > k)$ and the paths in $(M_n \geq k) \cap (S_n < k)$. The probability of any path is $1/2^n$; hence, (2) holds.

11.7 GROWTH OF M_n

In a symmetric random walk, how long does it take the particle to reach the position 1? Let T denote the number of steps required. Explicitly,

$$T = \text{first index } n \text{ such that } S_n = 1$$

The distribution of T is closely related to the distribution of the first return time to the origin. Let W denote the number of steps it takes the particle to first return to the origin, having started from the origin. Then

T and $W - 1$ have the same distribution, which is

$$P(T = 2n) = 0, \qquad P(T = 2n - 1) = \frac{2}{n}\binom{2n-2}{n-1}\frac{1}{4^n},$$

$n = 1, 2, \ldots$ (1)

[See (6), Section 11.4.]

Proof. Since the first step is either right or left, we have

$$P(W = k) = P[(W = k) \cap (X_1 = 1)] + P[(W = k) \cap (X_1 = -1)]$$

By symmetry, each of these probabilities has the same value; thus,

$$P(W = k) = 2P[(W = k) \cap (X_1 = -1)] \tag{2}$$

Let $N(n)$ be the number of n-tuples of 1's and -1's that represent an n-step random walk, starting at the origin, which first hits position 1 at the nth step. Then $P(T = n) = N(n)/2^n$. We now show that $P[(W = k) \cap (X_1 = -1)] = N(k-1)/2^k$. The event $(W = k) \cap (X_1 = -1)$ consists of all k-tuples of $+1$'s and -1's that start with -1 and that describe a random walk which progresses one position to the right in exactly $k - 1$ steps. Of all the 2^k possible paths, exactly $N(k-1)$ have the desired property. Hence, from (2),

$$P(W = k) = \frac{2N(k-1)}{2^k} = P(T = k - 1)$$

which says that $W - 1$ and T have the same distribution. ∎

It now follows from (1) and Example 3, Section 11.4, that $P(T < \infty) = 1$; that is, with probability 1, the random walk will definitely advance one step. It also follows that $ET = \infty$. Advancing one step is certain, but on the average it takes an infinite amount of time!

Now we let T_1 (instead of T) denote the number of steps it takes the random walk to progress to position 1. Let T_2 be the additional number of steps required to progress to position 2, and so forth. Then $T_1, T_2, \ldots$ form a sequence of independent and identically distributed random variables with distributions are given by (1) (see Figure 11.9). The following relationship follows immediately from the definition of the T_i's.

$$P(M_n \geq k) = P(T_1 + \cdots + T_k \leq n) \tag{3}$$

Proof of (3). This follows from the relationship between events,

$$(M_n \geq k) = (T_1 + \cdots + T_k \leq n)$$

It simply says that if the maximum distance achieved in n steps is

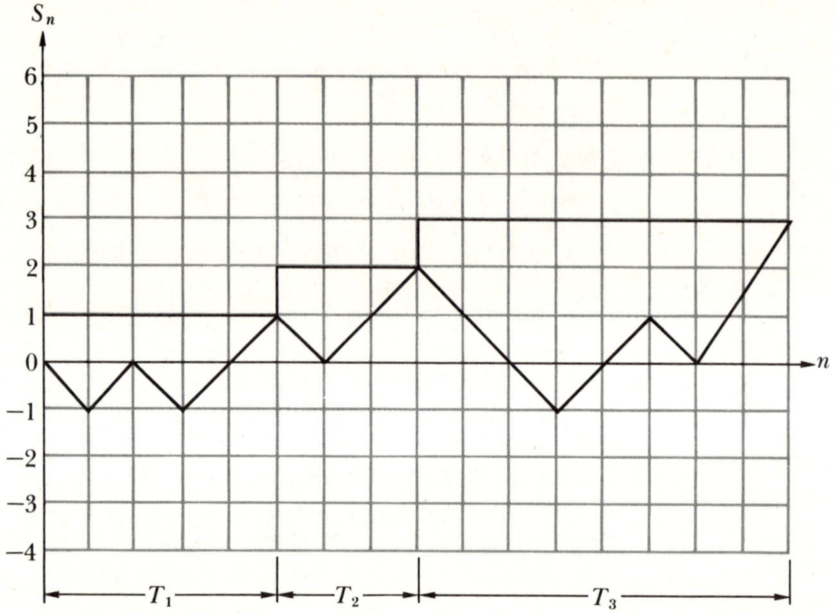

FIGURE 11.9

at least k, then the total time required to make a succession of k positive advances is no greater than n. ▮

From (3), we can now deduce the asymptotic distribution of a sum of T_i's as follows:

$$\lim_{n\to\infty} P\left(\frac{T_1 + \cdots + T_n}{n^2} \leq t\right) = \lim_{n\to\infty} P\left(\frac{M_n}{n^{1/2}} \geq t^{-1/2}\right)$$
$$= \lim_{n\to\infty} P\left[\left(\frac{n^{1/2}}{M_n}\right)^2 \leq t\right] \quad (4)$$

According to the theorem in Section 11.6, $M_n/n^{1/2}$ is asymptotically distributed like $|X|$ where X has the $N(0, 1)$ distribution. By (4), then, $(T_1 + \cdots + T_n)/n^2$ is asymptotically distributed like $1/X^2$, the squared reciprocal of a $N(0, 1)$ random variable. *We have here an interesting example of a limit theorem for sums of random variables, where the central limit theorem does not apply.* It is true that $T_1, T_2, \ldots$ is a sequence of independent and identically distributed random variables. But $E(T_i) = \infty$ and the central limit theorem do not give any information. Even the law of large numbers is not applicable. If it were, the distribution of $(T_1 + \cdots + T_n)/n$ would become concentrated around a single point as n goes to infinity. However, it follows from (4) that

$$\lim_{n\to\infty} P\left(\frac{T_1 + \cdots + T_n}{n} \leqslant t\right) = 0$$

for any positive t, no matter how large. In other words, the distribution of the sample average $(T_1 + \cdots + T_n)/n$ "escapes out to infinity" as n grows large. It occurs because $E(T_i) = \infty$. The limit theorem described by (4), as well as the central limit theorem, are special cases of a broad class of limit theorems for sums of random variables (see Note 5, Chapter 10).

Proof of (4). This proof follows from (3), by replacing k by $(n/t)^{1/2}$. There is a slight technical difficulty in that $(n/t)^{1/2}$ is not necessarily an integer, but it is easily resolved by using the closest integer to it. We leave the details to the reader. ▮

11.8 THE POLYA URN PROCESS

(See Note 2, Chapter 4.) This process evolves as follows. A box contains a white balls and b black balls. A ball is drawn and is replaced *together with an additional ball of the same color.* The procedure is repeated; every time a ball is drawn it is replaced together with an additional ball of that color. (A special case was described in Example 4, Section 4.1.) Let us define a sequence of random variables that indicate the succession of colors drawn. That is,

$$Y_i = \begin{cases} 1 & \text{if } i\text{th drawing is white} \\ 0 & \text{if } i\text{th drawing is black} \end{cases}$$

The sequence of random variables $Y_1, Y_2, \ldots$ is called a Polya urn process. Our main purpose is to study the evolution of the proportion of white balls drawn,

$$\frac{Y_1 + \cdots + Y_n}{n}, \qquad n = 1, 2, \ldots$$

It should be realized that the Y_i's are not mutually independent (why not?), a fact that makes their study more difficult.

The procedure is continued for n steps. Suppose that the balls are individually numbered, $1, \ldots, a, \ldots$ for the whites and $a + 1, \ldots, a + b, \ldots$ for the blacks. Suppose also that we keep track of the numerical designations as well as of the colors. Then, there are

$$(a + b)(a + b + 1) \cdots (a + b + n - 1) = (a + b + n - 1)^{(n)}$$

sample points. Random selection means that each sample point has probability $1/(a+b+n-1)^{(n)}$. Using this probability model, we now evaluate the probability of drawing a specified sequence of colors.

Let $e_1, \ldots, e_n$ be a sequence of 1's and 0's. Then

$$P[(Y_1 = e_1) \cap \cdots \cap (Y_1 = e_n)] = \frac{(a+k-1)^{(k)}(b+n-k-1)^{(n-k)}}{(a+b+n-1)^{(n)}} \tag{1}$$

where $k = e_1 + \cdots + e_n$ is the total number of whites in the sequence.

Proof of (1). The number of sample points in the event under consideration is

$$a(a+1)\cdots(a+k-1)b(b+1)\cdots(b+n-k-1)$$

which is the same as $(a+k-1)^{(k)}(b+n-k-1)^{(n-k)}$. This count does not depend on the particular arrangement of the 1's and 0's, but only on the number of 1's. ∎

It follows from (1) that $(Y_{i_1}, \ldots, Y_{i_n})$ has the same joint distribution as does $(Y_1, \ldots, Y_n)$, where $(i_1, \ldots, i_n)$ is any permutation of $(1, \ldots, n)$. For instance, to show that (Y_3, Y_2, Y_1) has the same joint distribution as (Y_1, Y_2, Y_3), we argue as follows:

$$P[(Y_3 = e_1) \cap (Y_2 = e_2) \cap (Y_1 = e_3)] = P[(Y_1 = e_3) \cap (Y_2 = e_2) \cap (Y_3 = e_1)] = P[(Y_1 = e_1) \cap (Y_2 = e_2) \cap (Y_3 = e_3)]$$

since by (1), the last probability does not depend on the order of e_1, e_2, e_3 but only on the sum $e_1 + e_2 + e_3$. The general proof is similar and we leave it to the reader.

As a consequence, it follows that each of $Y_1, \ldots, Y_n$ has the same marginal distribution. Hence,

$$P(Y_1 = 1) = \cdots = P(Y_n = 1) = \frac{a}{a+b},$$

$$P(Y_1 = 0) = \cdots = P(Y_n = 0) = \frac{b}{a+b}$$

Similarly,

$$P[(Y_i = 1) \cap (Y_j = 1)] = \frac{a(a+1)}{(a+b)(a+b+1)}$$

and

$$P(Y_j = 1 | Y_i = 1) = \frac{a+1}{a+b+1}, \qquad i \neq j$$

Even though the indicator random variables $Y_1, \ldots, Y_n$ are not mutually independent, they can be related to independent random variables in a surprising way, which we shall now describe. Consider the following experiment. First suppose that a number Y is selected from the interval [0, 1] according to some probability distribution. If $Y = p$, then suppose that n independent success-failure experiments are performed, each of which produces success with probability p and failure with probability $1 - p$. Denote the results of the trials by $Y_1, \ldots, Y_n$, where

$$Y_i = \begin{cases} 1 & \text{if } i\text{th trial is a success} \\ 0 & \text{if } i\text{th trial is a failure} \end{cases} \qquad i = 1, \ldots, n$$

What are reasonable probabilities to assign to the 2^n possible sequences of values of $Y_1, \ldots, Y_n$? If Y has an integrating density f, then we claim that a reasonable probability assignment is

$$P[(Y_1 = e_1) \cap \cdots \cap (Y_n = e_n)] = \int_0^1 p^k(1-p)^{n-k} f(p)\, dp \tag{2}$$

where we suppose $k = e_1 + \cdots + e_n$. The intuitive rationale behind (2) is the following. Once Y is known to have the value p, then, under this condition, $Y_1, \ldots, Y_n$ are mutually independent and $Y_1 = e_1, \ldots, Y_n = e_n$ with probability $p^k(1-p)^{n-k}$. To find the overall probability, average these out over all values of p using the density f, which gives the right side of (2). In this way, one assigns probabilities to all 2^n possible sequences $e_1, \ldots, e_n$. It is easy to check that these probabilities sum to 1 as follows. First, sum the right side of (2) over all sequences where $e_1 + \cdots + e_n = k$, for fixed k. This gives $\int_0^1 \binom{n}{k} p^k(1-p)^{n-k} f(p)\, dp$. Sum over k, giving $\int_0^1 [p + (1-p)^n f(p)\, dp = \int_0^1 f(p)\, dp = 1$. Now let us specialize by supposing that f is the beta density (Section 9.6),

$$f(p) = \begin{cases} \dfrac{\Gamma(\alpha+\beta)}{\Gamma(\alpha)\Gamma(\beta)} p^{\alpha-1}(1-p)^{\beta-1} & \text{if } 0 \leq p \leq 1 \\ 0 & \text{otherwise} \end{cases}$$

Substituting this density in (2) gives

$$P[(Y_1=e_1)\cap\cdots\cap(Y_n=e_n)]$$

$$=\int_0^1 \frac{\Gamma(\alpha+\beta)}{\Gamma(\alpha)\Gamma(\beta)}p^{\alpha+k-1}(1-p)^{\beta+n-k-1}\,dp$$

$$=\frac{\Gamma(\alpha+k)\Gamma(\beta+n-k)\Gamma(\alpha+\beta)}{\Gamma(\alpha)\Gamma(\beta)\Gamma(\alpha+\beta+n)} \tag{3}$$

We now observe that the right sides of (3) and (1) are identical if $\alpha=a$ and $\beta=b$. This fact follows from the relations

$$\Gamma(a+k) = (a+k-1)^{(k)}\Gamma(a)$$

$$\Gamma(b+n-k)=(b+n-k-1)^{(n-k)}\Gamma(b)$$

$$\Gamma(a+b+n)=(a+b+n-1)^{(n)}\Gamma(a+b)$$

Let us review, then, an alternate way of describing the evolution of the Polya urn process. "Nature" first selects a value of p according to the beta distribution with parameters $\alpha=a$ and $\beta=b$. Then a sequence of independent success-failure experiments are performed, which produce success or failure with probabilities p and $1-p$. The sequence of indications of successes, $Y_1, Y_2, \ldots$, are distributed in exactly the same way as the originally described Polya urn process. It should be emphasized that the Y_i's are not mutually independent, although they are conditionally independent for fixed p. Despite the fact that this argument is heuristic and not completely rigorous, what is completely correct is the relationship that

$$P[(Y_1=e_1)\cap\cdots\cap(Y_n=e_n)]$$

$$=\frac{(a+k-1)^{(k)}(b+n-k-1)^{(n-k)}}{(a+b+n-1)^{(n)}}=\int_0^1 p^k(1-p)^{n-k}f(p)\,dp \tag{4}$$

where f is the beta density with parameters a and b, and $k = e_1+\cdots+e_n$.

As a consequence of (4), we can now get information about the asymptotic distribution of the proportion of white balls as $n\to\infty$. The fact is asymptotically this proportion has the beta distribution.

$(Y_1+\cdots+Y_n)n$ is asymptotically beta distributed

$$\lim_{n\to\infty} P\left(\frac{Y_1+\cdots+Y_n}{n}\le t\right)=\int_0^t \frac{\Gamma(a+b)}{\Gamma(a)\Gamma(b)}p^{a-1}(1-p)^{b-1}\,dp \tag{5}$$

for any $0<t<1$.

Proof of (5). By (4),

$$P\left(\frac{Y_1 + \cdots + Y_n}{n} \leqslant t\right) = \int_0^1 G_n(p) f(p)\, dp \tag{6}$$

where

$$G_n(p) = \sum_{k/n \leqslant t} \binom{n}{k} p^k (1-p)^{n-k}$$

By the law of large numbers (Example 1, Section 10.2),

$$\lim_{n\to\infty} G_n(p) = G(p) = \begin{cases} 0 & \text{if } p > t \\ 1 & \text{if } p < t \end{cases}$$

This fact makes it plausible that

$$\lim_{n\to\infty} \int_0^1 G_n(p) f(p)\, dp = \int_0^1 G(p) f(p)\, dp = \int_0^t f(p)\, dp \tag{7}$$

which, by virtue of (6), would complete the proof. The limit in (7) is proved as follows. Determine $\Delta > 0$ so that $0 \leqslant t - \Delta \leqslant t + \Delta \leqslant 1$ and

$$\int_{t-\Delta}^{t+\Delta} f(p)\, dp \leqslant \epsilon$$

where $G > 0$. Using the law of large numbers, there is an $N = N(\epsilon)$ such that $G_n(t-\Delta) > 1 - \epsilon$ and $G_n(t+\Delta) < \epsilon$ for all $n \geqslant N$. Since $G_n(p)$ is monotonic in p [see (3), Section 6.1], it follows that in $[0, t-\Delta]$ and in $[t+\Delta, 1]$, $|G_n(p) - G(p)| < \epsilon$ uniformly in p (see Figure 11.10). Hence,

$$\left| \int_0^1 [G_n(p) - G(p)] f(p)\, dp \right|$$

$$\leqslant \int_0^1 |G_n(p) - G(p)| f(p)\, dp \leqslant \int_0^{t-\Delta} + \int_{t-\Delta}^{t+\Delta} + \int_{t+\Delta}^{1} \leqslant 3\epsilon$$

for $n \geqslant N$. ▮

It should be pointed out that for large n, $(Y_1 + \cdots + Y_n)/n$ is practically the same as the proportion of white balls in the box, since this proportion is

$$\frac{a + Y_1 + \cdots + Y_n}{a + b + n}$$

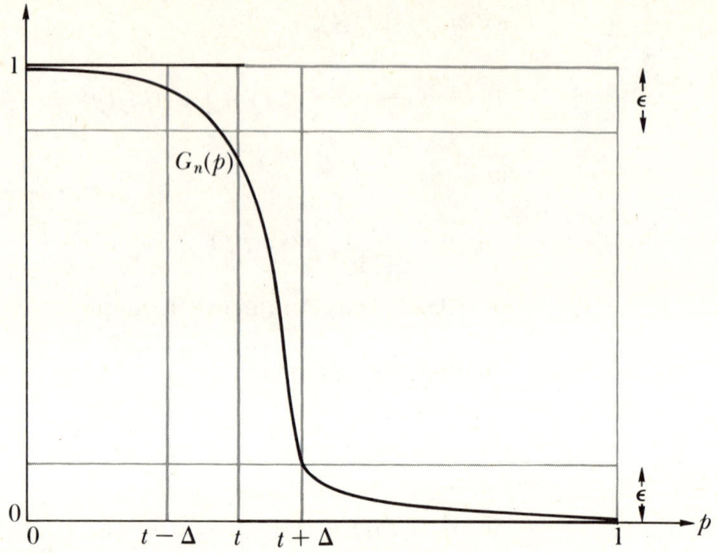

FIGURE 11.10

The difference between this quantity and $(Y_1 + \cdots + Y_n)/n$ goes to 0 as $n \to \infty$. Hence, the proportion of white balls in the box is also asymptotically beta distributed.

Example 1 Suppose that $a = 1$, $b = 2$. Then the limiting distribution of $X_n = (Y_1 + \cdots + Y_n)/n$ is given by the beta density,

$$f(p) = \begin{cases} 2(1-p) & \text{if } 0 < p < 1 \\ 0 & \text{otherwise} \end{cases}$$

The density functions for X_1, X_2, X_3 were given in Figure 4.4 and already show the triangular appearance of the limiting density. These are repeated here in Figure 11.11.

Example 2 Suppose that $a = b = 1$. Then, from (1),

$$P[(Y_1 = e_1) \cap \cdots \cap (Y_n = e_n)] = \frac{k!(n-k)!}{(n+1)!}$$

and

$$P(Y_1 + \cdots + Y_n = k) = \binom{n}{k} \frac{k!(n-k)!}{(n+1)!} = \frac{1}{(n+1)}, \qquad k = 0, 1, \ldots, n$$

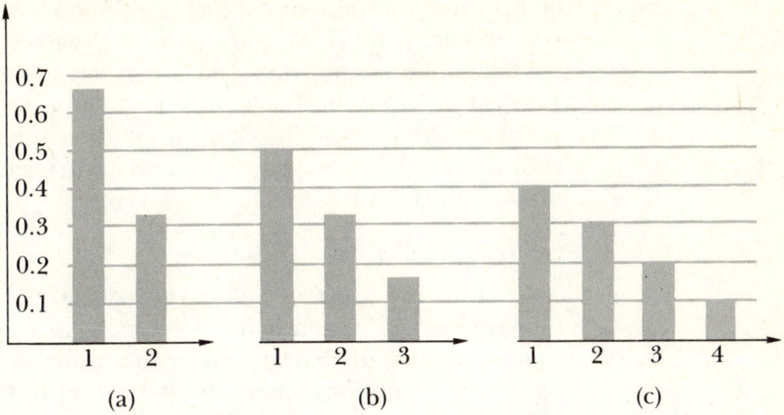

FIGURE 11.11 (a) Density of X_1. (b) Density of X_2. (c) Density of X_3.

Hence, the proportion of white balls drawn has the *discrete* uniform distribution over 0, $1/n, \ldots, n/n$. By (5), the limiting distribution of this proportion has the uniform integrating density over [0, 1] (see Figure 11.12).

11.9 MARKOV CHAINS

We now describe a discrete-time stochastic process called a *Markov chain.* Consider a system that at successive times 0, 1, 2, . . . can be

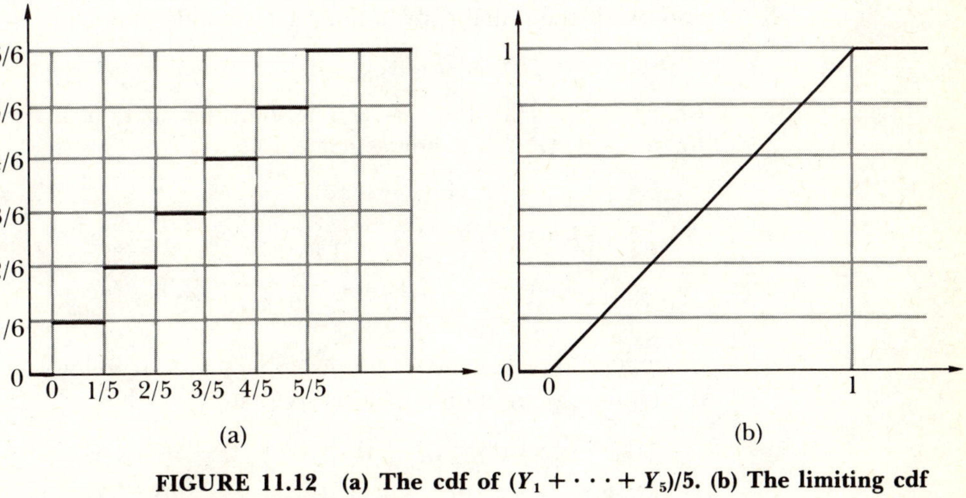

FIGURE 11.12 (a) The cdf of $(Y_1 + \cdots + Y_5)/5$. (b) The limiting cdf of $(Y_1 + \cdots + Y_n)/n$ as $n \to \infty$.

in any one of a finite or countably infinite set of states, $a_1, a_2, \ldots$. For simplicity, we shall denote the states as $1, 2, \ldots$, but for various examples some other labels may be more appropriate. Let $Y_0, Y_1, Y_2, \ldots$ be the states of the system at times $0, 1, 2, \ldots$. If at time n the system is in state i, it will then at time $n+1$ jump to a possibly new state according to a set of probabilities that depend on n and i. That is,

$$P(Y_{n+1}=j\,|Y_n=i)=p_n(i,j)$$

where $p_n(i,1)+p_n(i,2)+\cdots=1$ for each $i=1,2,\ldots$. Moreover, if at time n the system is in state i, then its evolution at the next time is completely determined by the probabilities $p_n(i,1), p_n(i,2), \ldots$, and by nothing else. The preceding history of the process before time n does not influence the transition that takes place from time n to time $n+1$. This fact means that

$$P[Y_{n+1}=j\,|(Y_n=i)\cap(Y_{n-1}=i_{n-1})\cap\cdots\cap(Y_0=i_0)]$$
$$=P(Y_{n+1}=j\,|Y_n=i)=p_n(i,j) \tag{1}$$

The requirement (1) says that where the system goes next depends, probabilistically, on where it is now, and not on the past history of how it got to where it is now.

definition of Markov chain

A sequence of random variables $Y_0, Y_1, Y_2, \ldots$ that satisfies (1) is called a Markov chain.

To describe completely the evolution of a Markov chain, one also needs the initial distribution of Y_0. We denote

$$P(Y_0=i)=p(i), \qquad i=1,2,\ldots$$

where $p(1)+p(2)+\cdots=1$. The joint distribution of the chain for times $0, 1, \ldots$, is now given as follows.

$$P[(Y_0=i_0)\cap(Y_1=i_1)\cap\cdots\cap(Y_n=i_n)]$$
$$=p(i_0)p_0(i_0,i_1)\cdots p_{n-1}(i_{n-1},i_n) \tag{2}$$

Proof of (2). For simplicity, suppose that $n=3$. It follows from the definition of conditional probability that

$$P[(Y_0=i_0)\cap(Y_1=i_1)\cap(Y_2=i_2)\cap(Y_3=i_3)]$$
$$=P(Y_3=i_3|A_2)P(Y_2=i_2|A_1)P(Y_1=i_1|A_0)P(Y_0=i_0)$$

where

$$A_0 = (Y_0 = i_o)$$

$$A_1 = (Y_0 = i_0) \cap (Y_1 = i_1)$$

$$A_2 = (Y_0 = i_0) \cap (Y_1 = i_1) \cap (Y_2 = i_2)$$

However, from (1), the defining property of a Markov chain,

$$P(Y_3 = i_3|A_2) = p_2(i_2, i_3), \quad P(Y_2 = i_2|A_1) = p_1(i_1, i_2),$$

$$P(Y_1 = i_1|A_0) = p_0(i_0, i_1)$$

from which (2) follows. The argument for general n is similar and is left to the reader. ∎

Example 1 Consider again the random walk described in Example 3, Section 11.4. Let S_n denote the position of the particle at the nth step. If the particle is in position i at time n, it will find itself at time $n + 1$ in positions $i + 1$ or $i - 1$ with probabilities p, $1 - p$, respectively. That is,

$$p_n(i, i - 1) = 1 - p, \quad p_n(i, i + 1) = p, \qquad i = 0, \pm 1, \pm 2, \ldots$$

and $p(i, j) = 0$ if j is not $i - 1$ or $i + 1$. Also, S_n is a Markov chain because the particle's evolution out of position i at time n is not influenced by its earlier positions at times $n - 1, n - 2, \ldots, 0$. The states of the system are $\ldots, -2, -1, 0, 1, 2, \ldots$. Notice that in this example the probabilities of transitions, $p_n(i, j)$, do not depend on n. If the particle starts at the origin, then $P(S_0 = 0) = 1$.

Example 2 Consider the Polya urn process $Y_1, Y_2, \ldots$ described in Section 11.8. We have $S_n = a + Y_1 + \cdots + Y_n$ equals the total number of white balls in the box at the nth step with $P(S_0 = a) = 1$. Also, $S_0, S_1, S_2, \ldots$ is a Markov chain with transition probabilities given by

$$p_n(i, i + 1) = \frac{i}{a + b + n}, \qquad p_n(i, i) = \frac{a + b + n - i}{a + b + n} \tag{3}$$

After the nth drawing, there are a total of $a + b + n$ balls in the box. If i of these are white, then the probability of drawing another white ball is $i/(a + b + n)$. It should be intuitively clear that $\{S_n\}$ is a Markov chain; once one knows that $S_n = i$, then the conditional distribution of S_{n+1} is given by the transition probabilities (3). The past history of values of S_m for $m < n$ does not influence this conditional distribution.

Example 3 A mouse is subjected to repeated stimuli. The mouse will respond to a stimulus and be rewarded, or it will fail to respond and then will be punished. Define

$$X_n = \begin{cases} 1 & \text{if mouse responds to } n\text{th stimulus} \\ 0 & \text{otherwise} \end{cases}$$

There are various models of psychological learning theory that describe the evolution of $Y_1 = X_1, Y_2 = X_1 + X_2, Y_3 = X_1 + X_2 + X_3, \ldots$ as a Markov chain. The nonzero transition probabilities are $p_n(i, i), p_n(i, i+1)$. The probability of a response at the nth stage is $p_n(i, i+1)$ if there has been a preceding history of i responses. The psychologist will try to estimate the transition probabilities by running many mice through many sequences of trials (Note 5).

11.10 TRANSITION MATRICES

Denote the matrix of nth step transition probabilities by P_n. That is,

$$\mathbf{P}_n = \begin{pmatrix} P(Y_{n+1} = 1|X_n = 1) & P(Y_{n+1} = 2|X_n = 1) \ldots \\ P(Y_{n+1} = 1|X_n = 2) & P(Y_{n+1} = 2|Y_n = 2) \ldots \\ \vdots & \vdots \end{pmatrix}$$

Also define the row vector

$$\boldsymbol{\pi}_n = [P(Y_n = 1), P(Y_n = 2), \ldots], \qquad n = 0, 1, 2, \ldots$$

In (1), Section 4.5, we showed, in effect, that

$$\boldsymbol{\pi}_{n+1} = \boldsymbol{\pi}_n \mathbf{P}_n \tag{1}$$

and, as a consequence,

$$\boldsymbol{\pi}_1 = \boldsymbol{\pi}_0 \mathbf{P}_0, \ \boldsymbol{\pi}_2 = \boldsymbol{\pi}_0 \mathbf{P}_0 \mathbf{P}_1, \ \ldots, \ \boldsymbol{\pi}_{n+1} = \boldsymbol{\pi}_0 \mathbf{P}_0 \cdots \mathbf{P}_n \tag{2}$$

We emphasize that the relations (1) and (2) hold for any sequence of random variables $Y_0, Y_1, Y_2, \ldots$, whether or not they are a Markov chain. However, most of the examples we shall consider are of Markov chains.

For some sequences of random variables, the transition probabilities $P(Y_{n+1} = j|Y_n = i)$ *do not depend on n*. In such a case, we say that the transitions are *time homogeneous*. If we denote the matrix of transition probabilities simply by **P**, then, in the time-homogeneous case, (2) becomes

$$\boldsymbol{\pi}_n = \boldsymbol{\pi}_0 \mathbf{P}^n, \qquad n = 1, 2, \ldots \tag{3}$$

Example 1 Consider the sequence of random variables $Y_1, Y_2, \ldots$, defined for the Polya urn process in Section 11.8. It follows from the discussion after the proof of (1), Section 11.8, that for every $n = 1, 2, \ldots$,

$$[P(Y_n = 0), P(Y_n = 1)] = \left(\frac{b}{a+b}, \frac{a}{a+b}\right) = \boldsymbol{\pi}$$

and that

$$\begin{pmatrix} P(Y_{n+1} = 0|Y_n = 0) & P(Y_{n+1} = 1|Y_n = 0) \\ P(Y_{n+1} = 0|Y_n = 1) & P(Y_{n+1} = 1|Y_n = 1) \end{pmatrix}$$

$$= \begin{pmatrix} \dfrac{b+1}{a+b+1} & \dfrac{a}{a+b+1} \\ \dfrac{b}{a+b+1} & \dfrac{a+1}{a+b+1} \end{pmatrix} = \mathbf{P}$$

A simple calculation shows that

$$\boldsymbol{\pi}\mathbf{P} = \boldsymbol{\pi}$$

and, hence,

$$\boldsymbol{\pi}\mathbf{P}^n = \boldsymbol{\pi}, \qquad n = 1, 2, \ldots$$

Except for the fact that the indices for Y_n start with $n = 1$ instead of $n = 0$, this relationship agrees with (3). However, notice that $\{Y_n\}$ is not a Markov chain. For instance, consider the following probabilities:

$$P(Y_1 = Y_2 = Y_3 = 1) = \frac{a}{a+b}\frac{a+1}{a+b+1}\frac{a+2}{a+b+2}$$

$$P(Y_1 = Y_2 = 1) = \frac{a}{a+b}\frac{a+1}{a+b+1}$$

Hence,

$$P(Y_3 = 1|Y_1 = Y_2 = 1) = \frac{a+2}{a+b+2}$$

On the other hand, $P(Y_3 = 1|Y_2 = 1) = (a+1)/(a+b+1)$; therefore, $\{Y_n\}$ is not a Markov chain.

Example 2 A box contains two balls. Each ball may be black or white, the colors changing according to the following scheme. At the nth step, a ball is drawn at random; if it is black it is painted white, and if it is white it is painted black. Initially we have 0, 1, or two black balls with probabilities π_0, π_1, π_2, respectively. If Y_0, Y_1,

$Y_2, \ldots$ are the numbers of black balls at the zero step (initially), at the first step, second step, and so forth, then we have the following time-homogeneous transition probabilities:

$$\mathbf{P} = \begin{pmatrix} p(0,0) & p(0,1) & p(0,2) \\ p(1,0) & p(1,1) & p(1,2) \\ p(2,0) & p(2,1) & p(2,2) \end{pmatrix} = \begin{pmatrix} 0 & 1 & 0 \\ 1/2 & 0 & 1/2 \\ 0 & 1 & 0 \end{pmatrix}$$

In this example, $\{Y_n\}$ is a Markov chain since the composition of the box at the $(n + 1)$st step is completely determined by its composition at the nth step. Notice that $\mathbf{P}^3 = \mathbf{P}$ (verify this!). Hence, $\boldsymbol{\pi}_1 = \boldsymbol{\pi}_0\mathbf{P}$, $\boldsymbol{\pi}_2 = \boldsymbol{\pi}_0\mathbf{P}^2$, $\boldsymbol{\pi}_3 = \boldsymbol{\pi}_0\mathbf{P}^3 = \boldsymbol{\pi}_0$. In general, $\boldsymbol{\pi}_0 = \boldsymbol{\pi}_3 = \boldsymbol{\pi}_6 = \ldots$, $\boldsymbol{\pi}_2 = \boldsymbol{\pi}_5 = \boldsymbol{\pi}_8 = \ldots$. By choosing the initial distribution properly, we can insure that each of $Y_0, Y_1, Y_2, \ldots$ has the same distribution. The right choice is

$$\boldsymbol{\pi} = (\pi_0, \pi_1, \pi_2) = [P(Y_0 = 0), P(Y_0 = 1), P(Y_0 = 1)] = (\tfrac{1}{4}, \tfrac{2}{4}, \tfrac{1}{4})$$

For this $\boldsymbol{\pi}$, we have $\boldsymbol{\pi}\mathbf{P} = \mathbf{P}$, which is easily verified, and, hence, $\boldsymbol{\pi}_n = \boldsymbol{\pi}\mathbf{P}^n = \boldsymbol{\pi}$, $n = 1, 2, \ldots$. The intuitive explanation follows. Initially, paint a ball black or white with probabilities 1/2, 1/2, and do so independently for each of the two balls. (This makes the initial distribution the preceding $\boldsymbol{\pi}$.) Then, at each stage, the balls behave as though they were independently painted black or white in the same way. This same example can be developed with any number of balls instead of just two. In general, the Markov chain is called an Ehrenfest urn process and is of interest in statistical mechanics (Note 6).

Example 3 Balls are successively distributed among k cells. Let X_n be the number of empty cells remaining after the nth ball has been distributed. The random variables $X_1, X_2, X_3, \ldots$ form a Markov chain the transitions of which are *not* time homogeneous. The special case of $k = 4$ appears in Example 3, Section 4.5.

11.11 STATIONARY DISTRIBUTIONS AND LONG-RUN BEHAVIOR

For simplicity, the following discussion is restricted to Markov chains with time-homogeneous transitions. Let $Y_0, Y_1, Y_2, \ldots$ be a Markov chain with transition matrix $\mathbf{P}$, the elements of which are $p(i, j) = P(Y_{n+1} = j | Y_n = i)$.

Theorem

Let $p^{(m)}(i, j)$ be the (i, j)th element of $\mathbf{P}^m$, the mth power of $\mathbf{P}$. Then

$$p^{(m)}(i, j) = P(Y_{n+m} = j \mid Y_n = i), \qquad (n, m = 1, 2, \ldots)$$

Proof. Consider the case $m = 2$.

$$P(Y_{n+2} = j \mid Y_n = i) = \frac{P[(Y_{n+2} = j) \cap (Y_n = i)]}{P(Y_n = i)}$$

$$= \sum_k \frac{P[(Y_{n+2} = j) \cap (Y_{n+1} = k) \cap (Y_n = i)]}{P(Y_n = i)}$$

$$= \sum_k P[(Y_{n+2} = j \mid (Y_{n+1} = k) \cap (Y_n = i)] \frac{P[(Y_{n+1} = k) \cap (Y_n = i)]}{P(Y_n = i)}$$

$$= \sum_k P(Y_{n+2} = j \mid Y_{n+1} = k) P(Y_{n+1} = k \mid Y_n = i)$$

$$= \sum_k p(i, k) p(k, j) = p^{(2)}(i, j)$$

This completes the proof for $m = 2$. The case for general m is similar and we leave it to the reader. ▮

We should emphasize that if $\{Y_n\}$ is not a Markov chain, then the preceding theorem does not hold.

Much of the theory of Markov chains with time-homogeneous transitions concerns the behavior of $\lim_{m\to\infty} \mathbf{P}^m$. In general, this limit need not exist as Example 2, Section 11.10 shows. However,

$$\lim_{m\to\infty} \frac{1}{m} (\mathbf{I} + \mathbf{P} + \cdots + \mathbf{P}^m)$$

always exists. [The identity matrix is $\mathbf{I}$. To say that the limit exists means that the limit of the (i, j)th element of the matrix exists for every i and j.] We shall state here, without proofs, some results about $\lim_{m\to\infty} \mathbf{P}^m$ for *finite* transition matrices (Note 7).

Suppose that $\{Y_n\}$ is a Markov chain with time-homogeneous transition matrix $\mathbf{P}$. The possible values of Y_n are $1, 2, \ldots, k$, thus $\mathbf{P}$ is a $k \times k$ matrix. Assume that for some positive integer n *all* the elements of $\mathbf{P}^n$ are *positive*. Then
(a) $\lim_{m\to\infty} \mathbf{P}^m$ exists and is of the form

$$\begin{pmatrix} \pi_1 & \pi_2 & \cdots & \pi_k \\ \vdots & & & \\ \pi_1 & \pi_2 & \cdots & \pi_k \end{pmatrix}$$

(This matrix has k rows that are the same.)
(b) $\boldsymbol{\pi} = (\pi_1, \pi_2, \ldots, \pi_k)$ is a vector of positive elements summing to one, which has the property that

$$\boldsymbol{\pi}\mathbf{P} = \boldsymbol{\pi} \tag{1}$$

(c) $\boldsymbol{\pi}$ is a unique solution of (1) in the following sense: If $\boldsymbol{\lambda} = (\lambda_1, \lambda_2, \ldots, \lambda_k)$ is any vector of positive elements that sum to 1 and $\boldsymbol{\lambda}\mathbf{P} = \boldsymbol{\lambda}$, then $\boldsymbol{\lambda} = \boldsymbol{\pi}$.

What (a) asserts is that

$$\lim_{m\to\infty} p^{(m)}(i, j) = \pi_j \tag{2}$$

In other words, $P(Y_m = j | Y_0 = i)$ has a limit that does not depend on i.

Since $P(Y_m = j) = \Sigma_i\, P(Y_m = j|Y_0 = i)P(Y_0 = i)$, (2) implies that

$$\lim_{m\to\infty} P(Y_m = j) = \sum_i \pi_j P(Y_0 = i) = \pi_j$$

Thus, no matter in what state the system starts, the distribution of Y_m for large m practically does not depend on that starting position.

Suppose now that the initial distribution is given by

$$P(Y_0 = i) = \pi_i$$

where π satisfies (1). Then (1) implies that $\boldsymbol{\pi}\, \boldsymbol{P}^n = \boldsymbol{\pi}$ or

$$P(Y_n = i) = \pi_i$$

In other words, starting the system initially according to distribution $\boldsymbol{\pi}$ insures not only that Y_n has the distribution given by $\boldsymbol{\pi}$ asymptotically, but that Y_n has the same distribution for every n. For this reason, the distribution $\boldsymbol{\pi}$ is called a *stationary distribution.*

Example 1 Suppose the transition matrix is given by

$$\mathbf{P} = \begin{pmatrix} 0.2 & 0.8 \\ 0.6 & 0.4 \end{pmatrix}$$

Since all elements of P are positive, it follows that

$$\lim_{m\to\infty}\begin{pmatrix}0.2 & 0.8\\ 0.6 & 0.4\end{pmatrix}^m = \begin{pmatrix}\pi_1 & \pi_2\\ \pi_1 & \pi_2\end{pmatrix}$$

To find $\boldsymbol{\pi} = (\pi_1, \pi_2)$, we must determine a solution to $\boldsymbol{\pi}\mathbf{P} = \boldsymbol{\pi}$ with $\pi_1 > 0$, $\pi_2 > 0$, and $\pi_1 + \pi_2 = 1$. It is easy to do, and the result is $(\pi_1, \pi_2) = (3/7, 4/7)$. Hence,

$$\lim_{m\to\infty}\begin{pmatrix}0.2 & 0.8\\ 0.6 & 0.4\end{pmatrix}^m = \begin{pmatrix}3/7 & 4/7\\ 3/7 & 4/7\end{pmatrix}$$

You may find it amusing to compute some powers of **P** to determine how fast this approach takes place.

Example 2 Suppose that a particle moves among the three positions 1, 2, 3, which are placed in a circle as in Figure 11.13. At every step, the particle moves one step clockwise with probability p or one step counterclockwise with probability $1 - p$, always proceeding from the previous position. If Y_n is the position at the nth step, then $\{Y_n\}$ is a Markov chain with transition matrix given by

$$\mathbf{P} = \begin{pmatrix}p(1, 1) & p(1, 2) & p(1, 3)\\ p(2, 1) & p(2, 2) & p(2, 3)\\ p(3, 1) & p(3, 2) & p(3, 3)\end{pmatrix} = \begin{pmatrix}0 & p & 1-p\\ 1-p & 0 & p\\ p & 1-p & 0\end{pmatrix}$$

If $0 < p < 1$, it is easy to check that all elements of $\mathbf{P}^2$ are positive, and hence $\lim_{m\to\infty} \mathbf{P}^m$ exists. If $\boldsymbol{\pi} = (1/3, 1/3, 1/3)$, then $\boldsymbol{\pi}\mathbf{P} = \boldsymbol{\pi}$. Hence,

$$\lim\begin{pmatrix}0 & p & 1-p\\ 1-p & 0 & p\\ p & 1-p & 0\end{pmatrix}^m = \begin{pmatrix}1/3 & 1/3 & 1/3\\ 1/3 & 1/3 & 1/3\\ 1/3 & 1/3 & 1/3\end{pmatrix}$$

No matter what the initial distribution of the particle may be,

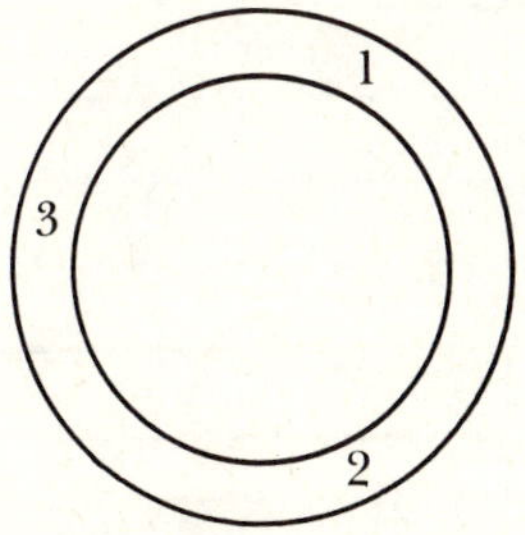

FIGURE 11.13

asymptotically it will be in any of the positions 1, 2, 3 with equal probabilities. Suppose that π is the initial distribution of the particle. That is, suppose initially the particle is in positions 1, 2, 3 with equal probabilities. Then at *any* step thereafter the particle will be in any of the three positions with equal probabilities.

11.12 THE POISSON PROCESS

We now consider a continuous-time stochastic process of great importance, called the Poisson process. It is a counting process, such as the one defined by (1), Section 11.2, with the waiting times being exponentially distributed. Specifically, suppose that W_1, $W_2, \ldots$ is a sequence of mutually independent random variables, each exponentially distributed with the same parameter λ. Let $N(t)$ be the counting process defined as

$$N(t) = \begin{cases} 0 & \text{if } 0 \leq t < W_1 \\ 1 & \text{if } W_1 \leq t < W_1 + W_2 \\ 2 & \text{if } W_1 + W_2 \leq t < W_1 + W_2 + W_3 \\ \vdots & \end{cases} \tag{1}$$

(see Figure 11.14). We have already shown that $N(t)$ is Poisson distributed with parameter λt in (5), Section 9.5. For this reason, $N(t)$ is called the *Poisson process.*

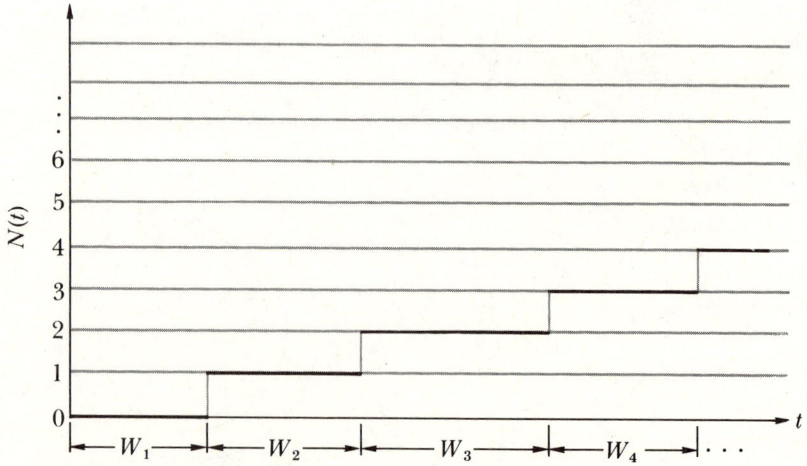

FIGURE 11.14 If the W_i's are exponentially distributed, then $N(t)$ is Poisson distributed.

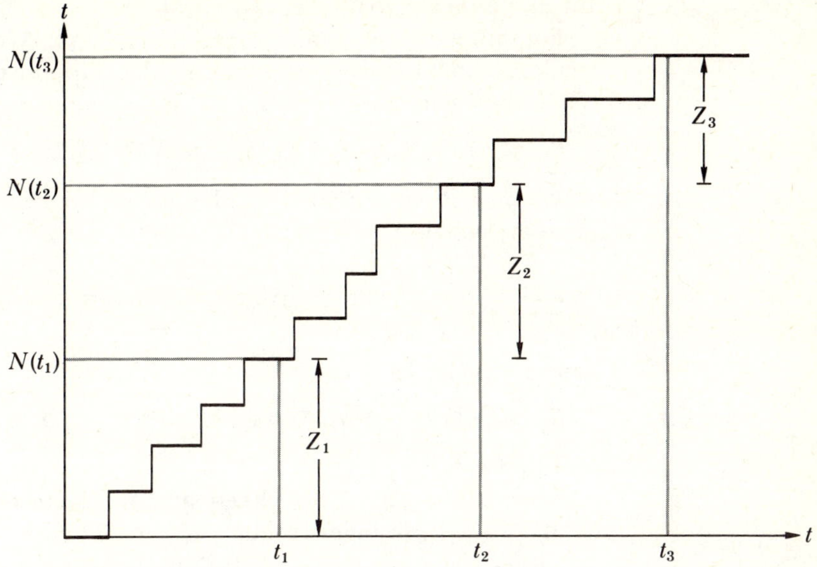

FIGURE 11.15 The increment of $N(t)$ in $[0, t_1)$ is Z_1. The increment in $[t_1, t_2)$ is Z_2, and so forth.

The Poisson process has the remarkable property that its increments are independent random variables, which means the following.

increments in N(t) are independent and Poisson

Suppose that $0 < t_1 < t_2 < \cdots < t_{n-1} < t_n = t$ is a partition of the interval $[0, t]$. Define the increments

$$Z_1 = N(t_1)$$
$$Z_2 = N(t_2) - N(t_1)$$
$$\vdots$$
$$Z_n = N(t_n) - N(t_{n-1})$$

(see Figure 11.15). Then $Z_1, Z_2, \ldots, Z_n$ are mutually independent, Poisson distributed random variables, with parameters $\lambda t_1, \lambda(t_2 - t_1), \ldots, \lambda(t_n - t_{n-1})$, respectively.

Proof. We must show that

$$P[(Z_1 = k_1) \cap \cdots \cap (Z_n = k_n)]$$
$$= \frac{(\lambda t_1)^{k_1}}{k_1!} \exp(-\lambda t_1) \cdots \frac{[\lambda(t_n - t_{n-1})]^{k_n}}{k_n!} \exp[-\lambda(t_n - t_{n-1})] \qquad (2)$$

for all nonnegative integers $k_1, \ldots, k_n$.

For simplicity, we shall prove (2) for $n = 2$. We leave it to the reader to imitate this proof for general n. Suppose that $k > 0, r > 0$. Then the event $(Z_1 = k) \cap (Z_2 = r)$ is the same as the event

$$(S_k \leq t_1) \cap (S_{k+1} > t_1) \cap (S_{k+r} \leq t) \cap (S_{k+r+1} > t)$$

where $S_m = W_1 + \cdots + W_m$ (see Figure 11.16). Hence, $P[(Z_1 = k) \cap (Z_2 = r)]$ equals

$$\lambda^{k+r+1} \int \cdots \int_A \exp\left[-\lambda(x_1 + \cdots + x_{k+r} + x_{k+r+1})\right] dx_1 \cdots dx_{r+1}$$

where A is the set

$$(s_k \leq t_1) \cap (s_{k+1} > t_1) \cap (s_{k+r} \leq t) \cap (s_{k+r+1} > t)$$

and $s_m = x_1 + \cdots + x_m$.

Now hold $x_1, \ldots, x_{k+r}$ fixed and integrate out x_{k+r+1}. Since $s_{k+r+1} > t$, x_{k+r+1} ranges from $t - s_{k+r}$ to ∞, and since

$$\int_{t-s_{k+r}}^{\infty} e^{-\lambda x}\, dx = \frac{1}{\lambda} \exp[-\lambda(t - s_{k+r})]$$

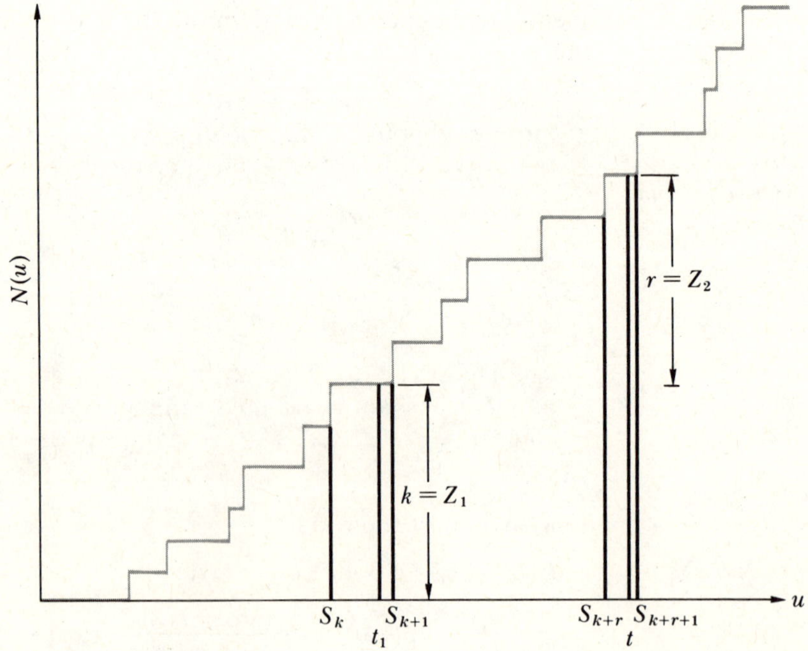

FIGURE 11.16

it follows that

$$P[(Z_1 = k) \cap (Z_2 = r)] = \lambda^{k+r} e^{-\lambda t} \int_B \cdots \int dx_1 \cdots dx_{k+r}$$

where B is the set $(s_k \leq t_1) \cap (s_{k+1} > t_1) \cap (s_{k+r} \leq t)$. We claim that

$$\int_B \cdots \int dx_1 \cdots dx_{k+r} = \frac{1}{k!r!} t_1^k (t - t_1)^r \tag{3}$$

The relation (3) is proved as follows. Suppose $Y_1, \ldots, Y_{r+k}$ are mutually independent random variables, each of which is uniformly distributed over $[0, t]$. The probability that k of the Y_i's fall in $[0, t_1)$ and r fall in $[t_1, t]$ is the binomial probability

$$\binom{r+k}{r} \left(\frac{t_1}{t}\right)^k \left(\frac{t - t_1}{t}\right)^r \tag{4}$$

But this probability is also equal to

$$\frac{1}{t^{k+r}} \int_C \cdots \int dy_1 \cdots dy_{k+r}$$

where C is the set in $r + k$ space where k y_i's are in $[0, t_1)$ and r y_i's are in $[t_1, t]$. Suppose that C is broken up into the $(r + k)!$ parts where

$$y_{i_1} < y_{i_2} < \cdots < y_{i_{r+k}}$$

with $i_1, \ldots, i_{r+k}$ varying over the permutations of $1, \ldots, r + k$. (We can ignore equalities among the y_i's. Why?) Hence, the integral over each part is the same and equals $1/(r + k)!$ times expression (4), and (3) is proved. Hence, with $t = t_2$,

$$P[(Z_1 = k) \cap (Z_2 = r)]$$

$$= \left[\frac{(\lambda t_1)^k \exp(-\lambda t_1)}{k!}\right] \left[\frac{[\lambda(t_2 - t_1)]^r}{r!} \exp[-\lambda(t_2 - t_1)\right]$$

which completes the proof in the special case $n = 2, k > 0, r > 0$. If either or both of k and r are 0, the proof is similar. For instance,

$$P[(Z_1 = 0) \cap (Z_2 = 0)] = P(W_1 > t) = e^{-\lambda t}$$

$$= \exp -\lambda t_1 \exp -\lambda(t_2 - t_1) \quad \blacksquare$$

Notice that as a consequence of the preceding theorem,

$$N(t) = Z_1 + \cdots + Z_n$$

must be Poisson distributed with parameter

$$\lambda t_1 + \lambda(t_2 - t_1) + \cdots + \lambda(t - t_{n-1}) = \lambda t$$

which we already know.

It also follows from (2) that the conditional distribution of $Z_1, \ldots, Z_n$, given that $N(t) = k > 0$, is multinomial with parameters $k; t_1/t, (t_2 - t_1)/t, \ldots, (t - t_{n-1})/t$. In other words,

$$P[(Z_1 = k_1) \cap \cdots \cap (Z_n = k_n) | N(t) = k = k_1 + \cdots + k_n]$$
$$= \frac{k!}{k_1! \cdots k_n!} \left(\frac{t_1}{t}\right)^{k_1} \cdots \left(\frac{t - t_{n-1}}{t}\right)^{k_n} \qquad (5)$$

Suppose that k points are selected at random from $[0, t]$. What is the probability that k_1 of these fall in $[0, t_1]$, k_2 fall in $[t_1, t_2]$, and so forth? This probability is the multinomial probability given by the right side of (5). Hence, (5) can be interpreted as saying the following:

Given that there are k increments in the Poisson process in $[0, t]$, their locations behave as if they are randomly selected in that interval.

Proof of (5). The required conditional probability is

$$\frac{\{[(\lambda t_1)^{k_1} \exp -\lambda t_1]/k_1!\} \cdots}{[(\lambda t)^k/k!]\; e^{-\lambda t}}$$
$$\frac{\{[(\lambda(t - t_{n-1}))^{k_n} \exp -\lambda(t - t_{n-1})]/k_n!\}}{[(\lambda t)^k/k!]\; e^{-\lambda t}}$$

which equals the right side of (5). ∎

Example 1 Customers arrive at a service counter so that successive waiting times between arrivals are independent random variables, each exponentially distributed with the same parameter λ. The number of customers that arrive between times 0 and t, $N(t)$, is Poisson distributed with parameter λt. According to (2), the numbers of customers that arrive in several nonoverlapping time intervals are independent random variables (Note 8).

Example 2 In Example 1, if we replace the word "customers" by the phrase "radioactive emissions," we have another example of a Poisson process that is studied in the theory of radioactive decay (Note 9).

11.13 THE COMPOUND POISSON PROCESS

Another stochastic process of interest in probability theory is the *compound Poisson process.* It evolves much like the Poisson process, except that instead of changing by unit jumps, its jumps are random.

compound Poisson process

Suppose that $W_1, W_2, \ldots, X_1, X_2, \ldots$ are mutually independent random variables. The W_i's are exponentially distributed, each with the same parameter, λ. Each of the X_i's has the same distribution, which is otherwise unspecified. Define a stochastic process $Y(t)$ as follows:

$$Y(t) = \begin{cases} 0 & 0 \leq t < W_1 \\ X_1 & W_1 \leq t < W_2 \\ X_1 + X_2 & W_2 \leq t < W_3 \\ X_1 + X_2 + X_3 & W_3 \leq t < W_4 \\ \vdots & \end{cases} \tag{1}$$

$Y(t)$ is called a *compound Poisson process.*

$Y(t)$ evolves like the Poisson process $N(t)$ defined by (1), Section 11.12, except that instead of the successive jumps being of size one, the jumps are $X_1, X_2, \ldots$ (see Figure 11.17).

Example 1 Suppose that for certain types of insurance policies, waiting times between claims form a sequence of mutually independent and identically distributed random variables, each exponentially distributed with the same parameter. If the successive values of the claims are $X_1, X_2, \ldots$, then $Y(t)$ represents the total claims made on the insurance company up to time t. In the insurance literature, it often is actually assumed that $Y(t)$ is a compound Poisson process (Note 10).

Example 2 Suppose that the mathematical model describing interarrival times between automobiles at a certain traffic point is that these are mutually independent random variables, each exponentially distributed with the same parameter. Hence, $N(t)$, the number of arrivals up to time t, is a Poisson process. Let X_i be the number of persons in the ith car. If the X_i's are mutually independent and independent of the interarrival times, then $Y(t)$, the total number of persons to arrive up to time t, is a compound Poisson process (Note 11).

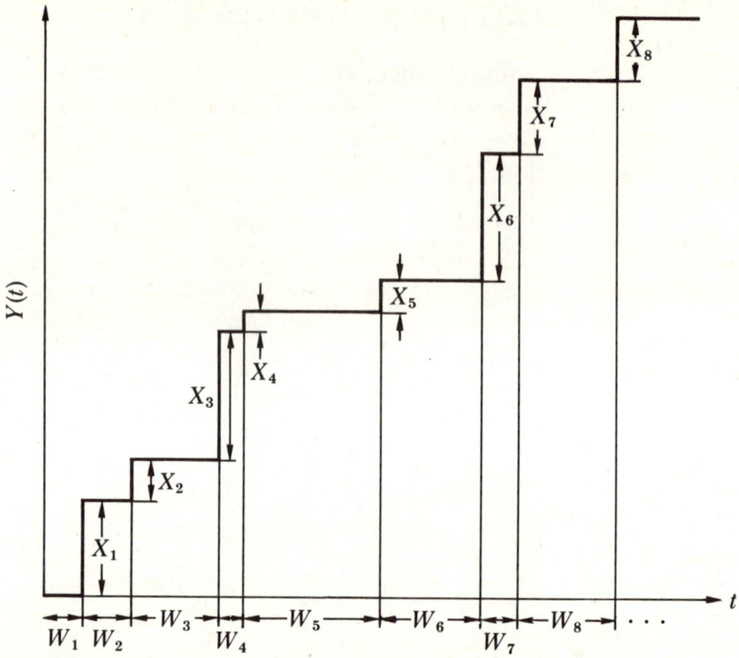

FIGURE 11.17 The compound Poisson process.

In Examples 1 and 2, the jump random variables X_i happen to be positive. In general, definition (1) does not require positive jumps.

The most important property of a compound Poisson process is one that it inherits from the underlying Poisson process—namely, that the process has independent increments.

a compound Poisson process has independent increments

Suppose that $Y(t)$ is a compound Poisson process. If $0 < t_1 < t_2 < \cdots < t_n$, then the increments

$Y(t_1),\ Y(t_2) - Y(t_1),\ \ldots,\ Y(t_n) - Y(t_{n-1})$

are mutually independent random variables.

Proof. For simplicity, consider two increments. Let $N(t)$ be the Poisson process defined in terms of the W_i's, and let

$$Z_1 = N(t_1), \qquad Z_2 = N(t_2) - N(t_1)$$

Then

$$
\begin{aligned}
&P[(Y(t_1) \text{ in } A) \cap (Y(t_2) \text{ in } B)] \\
&= \sum_{\substack{k\geq 0\\ r\geq 0}} P[(Y(t_1) \text{ in } A) \cap (Y(t_2) \text{ in } B) \cap (Z_1 = k) \cap (Z_2 = r)] \\
&= \sum_{\substack{k\geq 0\\ r\geq 0}} P[(X_1 + \cdots + X_k \text{ in } A) \cap (X_{k+1} + \cdots + X_{k+r} \text{ in } B) \\
&\qquad\qquad \cap (Z_1 = k) \cap (Z_2 = r)] \\
&= \sum_{\substack{k\geq 0\\ r\geq 0}} P[(X_1 + \cdots + X_k \text{ in } A) \cap (Z_1 = k)]P[(X_{k+1} + \cdots + X_{k+r} \\
&\qquad\qquad \text{in } B) \cap (Z_2 = r)] \\
&= \sum_{\substack{k\geq 0\\ r\geq 0}} P[(Y(t_1) \text{ in } A) \cap (Z_1 = k)]P[(Y(t_2) - Y(t_1) \text{ in } B) \cap (Z_2 = r)] \\
&= P[Y(t_1) \text{ in } A]P[Y(t_2) - Y(t_1) \text{ in } B]
\end{aligned}
$$

(Interpret $X_1 + \cdots + X_k = 0$ if $k = 0$ and $X_{k+1} + \cdots + X_{k+r} = 0$ if $r = 0$.) We have used the independence of the $N(t)$ process and the X_i's. The proof for more than two increments is similar. ▮

What makes the proof work is that sums of any fixed numbers of X_i's in nonoverlapping intervals are independent, and that the numbers of X_i's being summed are independent.

In general, it is not easy to describe the distribution of $Y(t)$ explicitly, but it is easy to calculate the relevant transform of $Y(t)$, as follows:

transforms of Y(t)

(a) Suppose that the X_i's in (1) are nonnegative, integer valued random variables with generating function $\phi_1(u)$. Then the generating function of $Y(t)$ is

$$\phi(u) = E[u^{Y(t)}] = \exp \lambda t[\phi_1(u) - 1] \tag{2}$$

(b) Suppose that the X_i's in (1) are nonnegative random variables, with Laplace transform $L_1(\theta)$. Then the Laplace transform of $Y(t)$ is

$$L(\theta) = E[e^{-\theta Y(t)}] = \exp \lambda t[L_1(\theta) - 1] \tag{3}$$

(c) Suppose that the X_i's in (1) are random variables with Fourier transform $M_1(\theta)$. Then the Fourier transform of $Y(t)$ is

$$M(\theta) = E[e^{i\theta Y(t)}] = \exp \lambda t[M_1(\theta) - 1] \tag{4}$$

Proof. The proofs of (a), (b), and (c) are practically the same, so we will prove (a) only.

$$\phi(u) = E[u^{Y(t)}] = \sum_{k=0}^{\infty} E\{u^{Y(t)}c_k[N(t)]\}$$

where $N(t)$ is the Poisson process determined by the waiting times W_i in (1) and

$$c_k[N(t)] = \begin{cases} 1 & \text{if } N(t) = k \\ 0 & \text{otherwise} \end{cases}$$

Since $u^{Y(t)}c_k[N(t)] = u^{X_1+\cdots+X_k}c_k[N(t)]$, and since the X_i's and W_i's are independent, it follows that

$$\phi(u) = \sum_{k=0}^{\infty} E[u^{X_1+\cdots+X_k}]E[c_k(N(t))] = \sum_{k=0}^{\infty} [\phi_1(u)]^k P(N(t) = k)$$

$$= \sum_{k=0}^{\infty} \frac{[\phi_1(u)]^k(\lambda t)^k e^{-\lambda t}}{k!} = \exp \lambda t[\phi_1(u) - 1] \quad \blacksquare$$

Example 3 Consider a stochastic process that evolves like a Poisson process, except that whenever it takes a jump, the jump is either 1 or -1 with equal probabilities. In other words, the process is described by (1) with X_i being equal to 1 or -1 with probabilities 1/2, 1/2. Let us compute the Fourier transform of $Y(t)$. Since

$$M_1(\theta) = \frac{e^{i\theta} + e^{-i\theta}}{2} = \cos\theta$$

it follows that the Fourier transform of $Y(t)$ is

$$M(\theta) = e^{\lambda t(\cos\theta - 1)}$$

Since $M'(0) = 0$, $M''(0) = -\lambda t$, it follows that

$$E[Y(t)] = 0, \qquad E[Y^2(t)] = \text{Var}[Y(t)] = \lambda t$$

(See the discussion following Property II, Section 8.6.) If Y_1 and Y_2 are independent Poisson distributed random variables, each with parameter $\lambda t/2$, then the Fourier transform of $Y_1 - Y_2$ is

$$E[\exp i\theta(Y_1 - Y_2)] = e^{\lambda t(\cos\theta - 1)}$$

We leave this to the reader to verify. It shows that $Y(t)$ is distributed like a difference between two independent Poisson random variables. Actually, if $N_1(t)$, $N_2(t)$ are two independent Poisson processes, the parameters of which are both $\lambda/2$, then the processes $N_1(t) - N_2(t)$ and $Y(t)$ are identical, but we shall not prove it (Note 12).

Example 4 A certain item can be produced only one at a time. Production starts at time $t = 0$. After an item is finished, production on the next one starts. The mathematical model is that the succession of production times are $X_1, X_2, \ldots$, a sequence of independent, nonnegative, random variables, each with the same distribution. Customers arrive to purchase these items, and the model prescribes that the interarrival times between customers are mutually independent random variables, each exponentially distributed with the same parameter, λ. As defined by (1), $Y(t)$ can be interpreted as the total time it takes to produce the items needed by the customers that arrive up to time t. If $Y(t) < t$, then at time t the most recently arrived customer has purchased his item and is not waiting in line. If $Y(t) > t$, then $Y(t) - t$ is the amount of time that the most recently arrived customer must wait to be served (see Figure 11.18) (Note 8).

Suppose that $0 < t_1 < t_2$. Then the increment $Y(t_2) - Y(t_1)$ is distributed like $Y(t_2 - t_1)$. This fact can be seen by examining the proof of the independence of $Y(t_1)$ and $Y(t_2) - Y(t_1)$. It also follows from that independence by the following argument with the Fourier transform, (4).

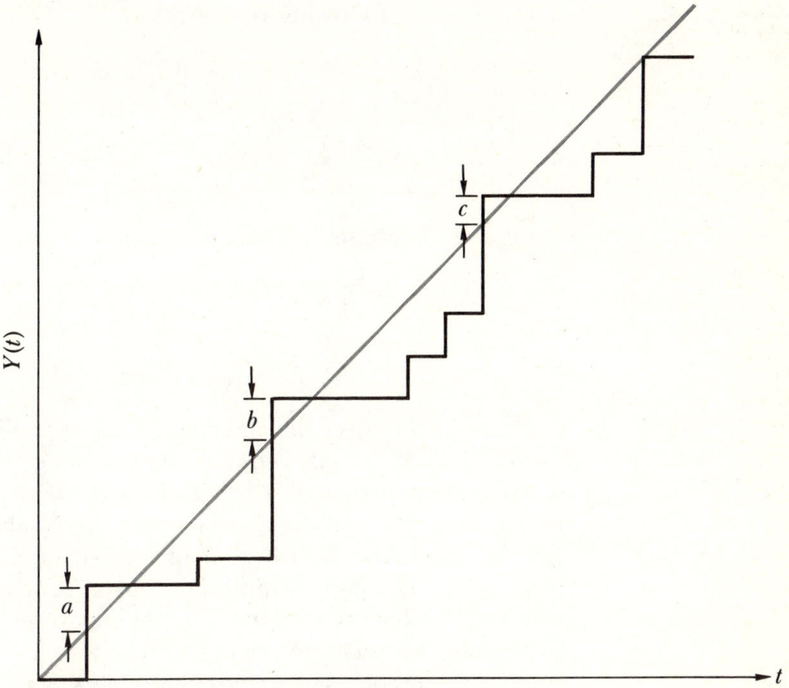

FIGURE 11.18 Time *a* is the waiting time of the first customer, time *b* is the waiting time of the third customer, and time *c* is the waiting time of the sixth customer. The second, fourth, and fifth customers do not have to wait.

$$E[\exp i\theta Y(t_2)] = \exp \lambda t_2[M_1(\theta) - 1]$$
$$= E\{\exp i\theta[Y(t_2) - Y(t_1)]\exp i\theta Y(t_1)\}$$
$$= E\{\exp i\theta[Y(t_2) - Y(t_1)]\} \exp \lambda t_1[M_1(\theta) - 1]$$

The last step uses the independence of $Y(t_2) - Y(t_1)$ and $Y(t_1)$. It follows now that

$$E\{\exp i\theta[Y(t_2) - Y(t_1)]\} = \exp \lambda(t_2 - t_1)[M_1(\theta) - 1]$$

The stochastic process $Y(t)$ is said to have *stationary independent increments.* The word *stationary* refers to the property just described, that the distribution of the increment over $[t_1, t_2]$ is determined by $t_2 - t_1$ only.

11.14 THE EXCESS OVER THE BOUNDARY PROPERTY

It is easy to verify that an exponentially distributed random variable X has the following property

$$P(X \leqslant u + t|X \geqslant u) = P(X \leqslant t) \tag{1}$$

where t and u are positive.

Proof. (See Exercise 7, Chapter 9.)

$$P(X \leqslant u + t|X \geqslant u) = \frac{P(u \leqslant X \leqslant u + t)}{P(X \geqslant u)} = \frac{e^{-\lambda u} - e^{-\lambda(u+t)}}{e^{-\lambda u}}$$
$$= 1 - e^{-\lambda t} = P(X \leqslant t) \quad ▮$$

Property (1) says that given that an exponentially distributed random variable X exceeds u, the excess over u is distributed like X itself. This property is called the *excess over boundary* property, or the *memoryless* property. [Compare it with the similar property for the geometric distribution in (6) Section 6.4.] It is from this basic property that a compound Poisson process inherits its characteristics of having stationary, independent increments. We want to explain this relationship in intuitive terms. Consider the evolution of a compound Poisson process, $Y(t)$. Now, $Y(0)$ equals 0 and remains equal to 0 for an exponentially distributed time, W_1. Then the process jumps to X_1 and remains there for another exponentially distributed amount of time, W_2. At time $W_1 + W_2$, it jumps by the amount X_2 and remains equal to $X_1 + X_2$ for an exponentially distributed time W_2, and so forth. Now consider the process as it evolves *beyond time* $t = t_0$. Whatever the height $Y(t_0)$ may be, the process remains equal to that height for an exponentially distributed time. This fact follows, intuitively at least, from property (1); the additional waiting time is the excess of an exponential random variable beyond a point, *given that it exceeds that point.* Because the counterpart of u in (1) is itself a random variable, it takes some work to make the argument rigorous. Intuitively, we hope that the role of (1) is clear. Taking $(t_0, Y(t_0))$ as a new origin, the process evolves from that point just as it did at the beginning, $t=0$ (see Figure 11.19). Its evolution out of the new origin is not influenced by past history. The sequence of waiting times, including the first excess, are mutually independent and exponentially distributed. The sequence of new jumps X_i are mutually independent and independent of the past history of jumps.

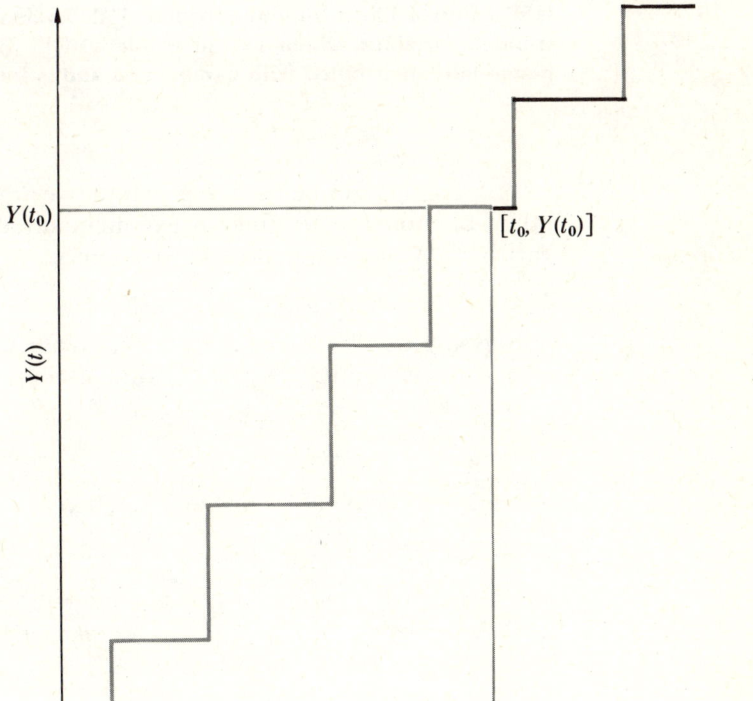

FIGURE 11.19 The evolution of the compound Poisson process out of the origin $(t_0, Y(t_0))$ is a compound Poisson process just like the original process.

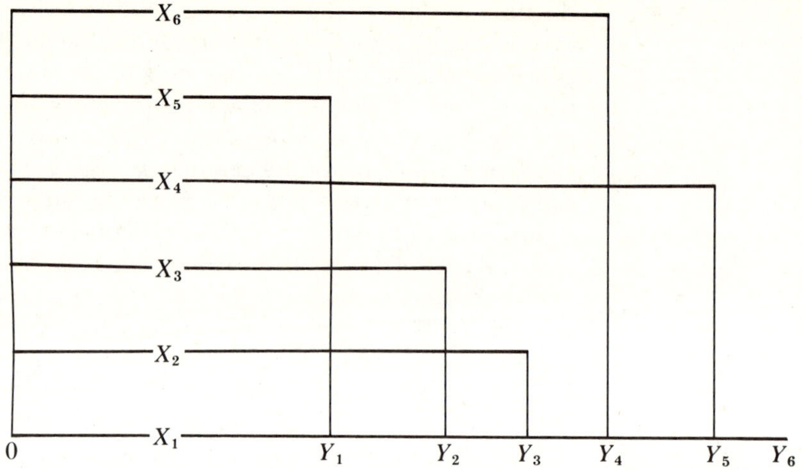

FIGURE 11.20 The minimum of $X_1, \ldots, X_6$ is Y_1, which is exponentially distributed with parameter 6λ. Given that Y_1 is fixed, then $Y_2 - Y_1$ is distributed like a minimum of five, independent, exponentially distributed variables, which are independent of Y_1. Hence, $Y_2 - Y_1$ is exponentially distributed with parameter 5 and is independent of Y_1.

Property (1) can be used to give intuitive arguments for various other phenomena relating to exponentially distributed random variables. We give one more illustration.

Suppose that $X_1, X_2, \ldots, X_n$ are mutually independent random variables, each exponentially distributed with the same parameter, λ. Let

$$Y_1 \leqslant Y_2 \leqslant \cdots \leqslant Y_n$$

be the ordered values of the X_i's. Then

$$Y_1, Y_2 - Y_1, \ldots, Y_n - Y_{n-1}$$

are mutually independent, exponentially distributed with parameters $n\lambda$, $(n-1)\lambda, \ldots, \lambda$, respectively.

(See Exercise 8b, Chapter 9, for a special case.)

Intuitive proof. First observe that Y_1 is exponentially distributed with parameter $n\lambda$. This fact follows from

$$P(Y_1 > t) = P[(X_1 > t) \cap \cdots \cap (X_n > t)]$$
$$= P(X_1 > t) \cdots P(X_n > t) = e^{-\lambda nt}$$

for $t \geq 0$. Now the argument will refer to Figure 11.20. Given that the smallest X_i has a specified value, the excesses of the other $n-1$ X_i's beyond that point behave, by (1), like $n-1$ mutually independent random variables, each exponential with parameter λ. Now $Y_2 - Y_1$, the minimum of these is independent of Y_1 and is exponentially distributed with parameter $(n-1)\lambda$. The argument continues in the same way.

SUMMARY

If $W_1, W_2, W_3, \ldots$ is a sequence of independent and identically distributed, positive random variables, these serve as a model for successive *waiting times* for recurrences of certain phenomena. A *counting process* is a random step function that keeps track of how many recurrences have taken place. Explicitly, $N(t)$ equals 0 in $[0, W_1)$, 1 in $[W_1, W_1 + W_2)$, 2 in $[W_1 + W_2, W_1 + W_2 + W_3)$, and so forth. For large t, the distribution of $N(t)/t$ becomes highly concentrated about the value $1/E(W_i)$, a consequence of the law of large numbers.

In the special case that the preceding waiting times W_i are integer valued, we consider the *recurrent event process* $U(1), U(2), \ldots$. Variable $U(n)$ is an indicator random variable that indicates whether or not a recurrence takes place at time n. In other words, $U(n) = 1$ if some partial sum $W_1, W_1 + W_2, W_1 + W_2 + W_3 + \cdots = n$. The probabilities $u_n = P(U(n) = 1)$ are called *recurrence probabilities.* The *renewal equation* asserts that $1 + u_1 t + u_2 t^2 + \cdots = 1/[1 - E(t^{W_i})]$, $|t| \leq 1$.

The *simple random walk* process describes the random evolution of the positions of a particle that makes independent steps $+1, -1$. The renewal equation is useful in determining information about the random walk, such as the probability that it ever returns to its starting position. Some quite detailed information is obtained about the walk when it is symmetric—that is, when the steps $+1$, -1 occur with equal probabilities.

A *Markov chain* is a sequence of random variables $\{X_n\}$ having the property that the conditional distribution of X_{n+1}, given the values of $X_1, \ldots, X_n$, depends only on the value of X_n. Roughly, it describes the evolution of a system the transition of which into the future depends probabilistically only on where it is now, and not on the past history of how it came to be there. A technical object of importance in studying Markov chains is the *transition matrix* P_n, having an (i, j)th entry of $P(X_{n+1} = j | X_n = i)$. Of special importance is the *time-homogeneous* case, when these *transition probabilities* do not depend on n.

Getting back to the concept of a counting process, suppose that the waiting times are exponentially distributed. Then the counting process $N(t)$ is called a *Poisson process*. For any t, $N(t)$ is Poisson distributed with parameter proportional to t. Moreover, the increments of $N(t)$ over a set of nonoverlapping intervals are mutually independent. This property is intimately related to the fact that the waiting times are exponentially distributed. A *compound Poisson process* is defined like a Poisson process, except that instead of its successive jumps all having the same value 1, these jumps are the random variables $X_1, X_2, \ldots$. The jumps are assumed to be mutually independent as well as independent of the basic counting process that indicates at which times these jumps occur.

EXERCISES

1 Suppose that the recurrence probabilities, (3), Section 11.4, are given by

$$u_1 = u_2 = \cdots = p, \qquad 0 < p < 1$$

Show that $P(W_i = k) = (1-p)^{k-1}p$, $k = 1, 2, \ldots$. In other words, if the recurrence probabilities all have the same value, then the waiting times have a geometric distribution. [Hint: Use the renewal equation, (4), Section 11.4.]

2 A coin is thrown until the combination T, T, H first appears. Then the throwing resumes until the next time this pattern recurs, and so forth.

(a) Show that the recurrence probabilities are

$$u_1 = u_2 = 0, \qquad u_n = \tfrac{1}{8}, \qquad n = 3, 4, \ldots$$

(b) Conclude from (a) that

$$u(t) = 1 + \frac{t^3/8}{1-t}, \qquad f(t) = \frac{t^3}{8 - 8t + t^3}$$

(c) Show, using two different methods, that the expected number of steps between recurrences is eight steps. [Hint: Evaluate $f'(1)$ and also $\lim_{n\to\infty} u_n$. Compare with Example 2, Section 11.4.]

3 Show that if $f_{2n+1} = 0$, then $u_{2n+1} = 0$, $n = 0, 1, \ldots$, and conversely as well. [Hint: Use (4), Section 11.4. Also, f can be expressed as a power series in t^2.]

4 Show that

$$1 + [1 + EN(1)]t + [1 + EN(2)]t^2 + \cdots = \frac{1}{(1-t)(1-f(t))} = \frac{u(t)}{1-t}$$

[Hint: Use (2) and (4), Section 11.4.]

5 Consider the experiment of Example 3, Section 11.2, and Example 1, Section 11.4.

(a) Use Exercise 4 to show that

$$E[N(n)] = \tfrac{2}{3}(n+1) + \tfrac{1}{9}(\tfrac{-1}{2})^n - \tfrac{7}{9}, \qquad n = 1, 2, \ldots$$

(b) The distribution of $N(4)$ appears in Example 4, Section 11.2. Find also the distributions of $N(1)$, $N(2)$, and $N(3)$. Verify directly from these distributions that $E[N(1)], E[N(2)], E[N(3)], E[N(4)]$ agree with (a).

6 Show that for every positive integer n, the first n recurrence probabilities $u_1, \ldots, u_n$ are determined by the first n waiting time probabilities $f_1, \ldots, f_n$, and vice versa. [Hint: Use the renewal equation, (4), Section 11.4.]

7 (a) Show that the joint distribution of $U(1)$, $U(2)$ is the following:

e_1	e_2	$P[(U(1)=e_1) \cap (U(2)=e_2)]$
1	0	$f_1(1-f_1)$
1	1	f_1^2
0	0	$1-f_1-f_2$
0	1	f_2

(b) Show that the joint distribution of $U(1)$, $U(2)$, and $U(3)$ is the following:

e_1	e_2	e_3	$P[(U(1)=e_1) \cap (U(2)=e_2) \cap (U(3)=e_3)]$
1	1	1	f_1^3
1	1	0	$f_1^2(1-f_1)$
1	0	1	f_1f_2
1	0	0	$f_1(1-f_1-f_2)$
0	1	1	f_1f_2
0	1	0	$f_2(1-f_1)$
0	0	1	f_3
0	0	0	$1-f_1-f_2-f_3$

(c) Formulate and prove a general version of (a) and (b). {Hint: For instance, $P[(U(1)=1) \cap (U(2)=1) \cap (U(3)=0) \cap (U(4)=0)] = P[(W_1=1) \cap (W_2=1) \cap (W_3>2)]$. Use the fact that the W_i's are mutually independent.}

8 Suppose that the first return probabilities f_1, f_2, f_3 are positive.

(a) Use Exercise 7a to show that the following three conditions are equivalent:

(1) $U(1)$ and $U(2)$ are independent;

(2) $u_1 = u_2 = f_1$;
(3) $f_2 = (1 - f_1)f_1$.

(b) Use Exercise 7b to show that the following three conditions are equivalent:
(1) $U(1)$, $U(2)$, and $U(3)$ are mutually independent;
(2) $u_1 = u_2 = u_3 = f_1$;
(3) $f_2 = (1 - f_1)f_1$, $f_3 = (1 - f_1)^2 f_1$.

(c) Formulate and try to prove a general version of (a) and (b).

(d) Do you see any connection with Exercise 1? (Hint: Exercise 6 can be useful here.)

9 In a simple, symmetric random walk ($p = 1/2$), let W_1 and W_2 be the waiting times between the first two returns to the origin. Show that

$$P(W_1 + W_2 = 2k) = \begin{cases} 0 & \text{if } k = 1 \\ 2P(W_1 = 2k) & \text{if } k = 2, 3, \ldots \end{cases}$$

{Hint: Compute $[f(t)]^2$, the generating function of $W_1 + W_2$.}

10 (a) Consider a simple, symmetric random walk ($p = 1/2$). If W is the number of steps required for a first return to the origin, show that the generating function of the tail probabilities of W is the following:

$$\sum_{k=0}^{\infty} t^{2k} P(W > 2k) = (1 - t^2)^{-1/2} = u(t)$$

[Hint: Use the generating function $f(t)$ in Example 3, Section 11.4. Imitate the hint for Exercise 8a, Chapter 5.]

(b) Conclude from (a) that

$$P(W > 2k) = u_{2k} = \binom{2k}{k} \frac{1}{4^k}, \qquad k = 0, 1, 2, \ldots$$

11 Let M_n be the maximum distance reached by the particle in a symmetric random walk, as in Section 11.6.

(a) Use the reflection principle (Figure 11.8) to show that

$$\begin{aligned} P[(M_{2n} \geqslant k) \cap (S_{2n} = 0)] &= P[(M_{2n} \geqslant k) \cap (S_{2n} = 2k)] \\ &= P(S_{2n} = 2k) \end{aligned}$$

(b) Deduce from (a) the conditional distribution of M_{2n}, given that the particle is at the origin at the $(2n)$th step. Namely,

$$P(M_{2n} \geqslant k | S_{2n} = 0) = \frac{P(S_{2n} = 2k)}{P(S_{2n} = 0)} = \frac{\binom{2n}{n+k}}{\binom{2n}{n}}$$

(c) From (b), deduce the conditional expected value

$$E(M_{2n} | S_{2n} = 0) = \frac{1}{2P(S_{2n} = 0)} - \frac{1}{2}$$

[Hint: Use Eq. (3), Section 2.10.]

12 Suppose that $V_1, V_2, \ldots$ is a sequence of mutually independent random variables, each geometrically distributed over $0, 1, 2, \ldots$, with parameter p. That is

$$P(V_i = k) = (1-p)^k p, \qquad k = 0, 1, \ldots$$

Let $S_n = V_1 + \cdots + V_n$. Define indicator random variables $U(n)$ by

$$U(n) = \begin{cases} 1 & \text{if } S_n = n \\ 0 & \text{otherwise} \end{cases} \qquad n = 1, 2, \ldots$$

(a) Show that $U(1), U(2), \ldots$ is a recurrent event process with recurrence probabilities

$$u_n = \binom{2n-1}{n} p^n (1-p)^n, \qquad n = 1, 2, \ldots$$

[Hint: The successive waiting times between recurrences of $S_n = n$ are independent and identically distributed. (Draw a picture!) Also, S_n has the negative binomial distribution with parameters n, p (Section 6.6.).]

(b) Show that

$$u_n = (\tfrac{1}{2})\binom{2n}{n} p^n (1-p)^n, \qquad n = 1, 2, \ldots$$

and thus conclude that

$$u(t) = \tfrac{1}{2} + (\tfrac{1}{2})[1 - 4p(1-p)t]^{-1/2}$$

$$f(t) = \frac{1 - [1 - 4p(1-p)t]^{1/2}}{1 + [1 - 4p(1-p)t]^{1/2}}$$

(Hint: The computations are practically the same as those in Example 3, Section 11.4.)

(c) Conclude from (b) that

$$f(1) = \frac{1 - |2p-1|}{1 + |2p-1|} = \begin{cases} \dfrac{p}{1-p} & \text{if } p \leqslant \dfrac{1}{2} \\[2ex] \dfrac{1-p}{p} & \text{if } p \geqslant \dfrac{1}{2} \end{cases}$$

Hence, the process is persistent if, and only if, $p = 1/2$.

(d) Conclude from (b) that when $p = 1/2$, the expected waiting time between recurrences is infinite.

13 Consider the first $2n$ steps of a simple random walk. Let Y_n be the last step at which this random walk was at the origin. The possible values of Y_n are $0, 2, \ldots, 2n$. ($Y_n = 0$ means that the random walk did not return to the origin during its first $2n$ steps.)

(a) Show that

$$P(Y_n = 2k) = u_{2k}(f_{2n-2k+2} + f_{2n-2k+4} + \cdots)$$

(Hint: $Y_n = 2k$ means that at step $2k$, the particle is at the origin, and that the waiting time for the next return exceeds $2n - 2k$.)

(b) Conclude from (a) and from Exercise 10b that

$$P(Y_n = 2k) = u_{2k}u_{2n-2k} = \binom{2k}{k}\binom{2n-2k}{n-k}\frac{1}{4^n}, \qquad k = 0, 1, \ldots, n$$

(c) Verify that

$$P(Y_n = 2k) = \int_0^1 \binom{n}{k} p^k (1-p)^{n-k} f(p)\, dp$$

where f is the beta density with $\alpha = \beta = 1/2$. [Hint: You will need to verify that

$$\frac{\Gamma(m+\frac{1}{2})}{\Gamma(\frac{1}{2})} = \frac{(2m)!}{m!4^m}$$

for any positive integer m. This fact follows from repeated applications of the fact that $\Gamma(b) = (b-1)\Gamma(b-1)$, and some additional algebra.]

(d) Conclude from (c) that $Y_n/2n$ is asymptotically beta distributed with parameters $\alpha = \beta = 1/2$. That is,

$$\lim_{n\to\infty} P\left(\frac{Y_n}{2n} \leq t\right) = \frac{\Gamma(1)}{\Gamma(\frac{1}{2})\Gamma(\frac{1}{2})} \int_0^t p^{(1/2)-1}(1-p)^{(1/2)-1}\, dp, \qquad 0 \leq t \leq 1$$

$$= \int_0^t \frac{1}{\pi p^{1/2}(1-p)^{1/2}}\, dp = \arcsin t$$

The limiting distribution is sometimes called the *arcsin law.* [Hint: The calculations are exactly the same as those in the proof of (5), Section 11.8.]

14 Consider the Polya urn process as defined in Section 11.8. Let W be the number of steps required for the number of black balls drawn to equal the number of white balls drawn for the first time. The possible values of W are 2, 4, ...

(a) Show that

$$P(W = 2n) = \frac{2}{n}\binom{2n-2}{n-1}\frac{\Gamma(n+a)\Gamma(n+b)\Gamma(a+b)}{\Gamma(2n+a+b)\Gamma(a)\Gamma(b)},$$

$$n = 1, 2, \ldots$$

[Hint: Use the representation (4), Section 11.8, and the result about random walk given by (6), Section 11.4.]

(b) In particular, when the box initially consists of one black and one white ball ($a = b = 1$), show that

$$P(W = 2n) = \frac{1}{(2n+1)(2n-1)} = \frac{1}{4n^2 - 1} = \frac{\frac{1}{2}}{2n-1} - \frac{\frac{1}{2}}{2n+1}$$

and, hence,

$$P(W \leq n) = \frac{1}{2} - \frac{1}{6} + \frac{1}{6} - \frac{1}{10} + \cdots + \frac{\frac{1}{2}}{2n-1} - \frac{\frac{1}{2}}{2n+1}$$

$$= \frac{1}{2} - \frac{\frac{1}{2}}{2n+1}$$

It follows that the probability that no equalization ever takes place is

$$P(W = \infty) = \tfrac{1}{2}$$

15 Consider a compound Poisson process as defined by (1), Section 11.13, in which the X_i's are indicator random variables that equal 1 with probability p, $0 < p < 1$. Show that $Y(t)$ is Poisson distributed with parameter λpt.

16 Let $Y(t)$ be a compound Poisson process as defined in Section 11.13. Let $G(u) = P(X_i \leq u)$ be the cdf of the jump variable. Let $Z(t)$ be the maximum jump up to time t. [One possible interpretation, relating to Example 1, Section 11.13, is that $Z(t)$ is the maximum claim presented to the insurance company up to time t.] Show that the cdf of $Z(t)$ is

$$P[Z(t) \leq u] = \exp - \lambda t[1 - G(u)]$$

{Hint: Given that there are n jumps up to time t, the probability that the largest of these is no greater than u equals $[G(u)]^n$. The probability of the condition is $(\lambda t)^n e^{\lambda t}/n!$.}

17 Consider the succession of times at which a Poisson process $Y(t)$ horizontally crosses the 45-degree line out of the origin. (Draw a picture!) Such crossings can occur only at integer-valued times.

(a) Give an intuitive argument to show that $W_1, W_2, \ldots$, the successive waiting times between crossings, are mutually independent and identically distributed random variables. [Hint: Use the *excess over the boundary property*, (1), Section 11.14. The evolution of the process, using the point of crossing as a new origin, is again a Poisson process and is not influenced by the past history up to that point.]

(b) Define

$$U(n) = \begin{cases} 1 & \text{if } Y(n) = n \\ 0 & \text{otherwise} \end{cases} \qquad n = 1, 2, \ldots$$

Show that it follows from (a) that $U(1), U(2), \ldots$ is a recurrent event process (Section 11.4) and the recurrence probabilities are

$$u_n = \frac{(\lambda n)^n e^{-\lambda n}}{n!}, \qquad n = 1, 2, \ldots$$

18 Prove (2), Section 11.3, assuming $E(W_i) = \infty$. {Hint: Define a new sequence of "truncated" waiting times $W_1', W_2', \ldots$, where $W_i' = W_i$ if $W_i \leq T$; otherwise, $W_i' = 0$. Then $E(W_i') \equiv m(T) < \infty$, but

$\lim_{T\to\infty} m(T) = \infty$. Show that $N(t) \leq N'(t)$, where $N'(t)$ is the counting process defined by the W_i', and conclude from (1), Section 11.3, that

$$\lim_{t\to\infty} P\left[\frac{N(t)}{t} < u\right] = 1 \qquad \text{if } u > \frac{1}{m(T)}$$

The desired conclusion follows from the fact that $m(T)$ can be made arbitrarily large.}

19 Consider a sequence of independent success-failure trials, with probability of success p. Let S_n be the number of successes in the first n trials, and define

$$Y_n = \begin{cases} 0 & \text{if } S_n \text{ is even} \\ 1 & \text{if } S_n \text{ is odd} \end{cases} \qquad n = 1, 2, \ldots$$

(a) Show that $\{Y_n\}$ is a Markov chain, with the time-homogeneous transition matrix

$$\mathbf{P} = \begin{pmatrix} p(0, 0) & p(0, 1) \\ p(1, 0) & p(1, 1) \end{pmatrix} = \begin{pmatrix} 1-p & p \\ p & 1-p \end{pmatrix}$$

(b) Use the results of Section 11.11 to show that

$$\lim_{n\to\infty} \mathbf{P}^n = \begin{pmatrix} 1/2 & 1/2 \\ 1/2 & 1/2 \end{pmatrix}$$

whenever $0 < p < 1$.

20 Consider a sequence of independent success-failure experiments, with probability of success p. At time n, let Y_n be the elapsed number of steps since the last success. Thus, if the sequence of outcomes starts out as $S, S, F, F, S, S, F, \ldots$, then $Y_1 = 0, Y_2 = 0, Y_3 = 1, Y_4 = 2, Y_5 = 0, Y_6 = 0, Y_7 = 1, \ldots$. In general, the possible values of Y_n are $0, 1, \ldots, n$, with the understanding that $Y_n = n$ means no successes through the nth step.

(a) Show that $\{Y_n\}$ is a Markov chain with time-homogeneous transition probabilities given by

$$p(i, 0) = p, \qquad p(i, i+1) = 1 - p, \qquad i = 0, 1, 2, \ldots$$

That is,

$$\mathbf{P} = \begin{pmatrix} p & 1-p & 0 & 0 & 0 & 0 & \cdots \\ p & 0 & 1-p & 0 & 0 & 0 & \cdots \\ p & 0 & 0 & 0 & 1-p & 0 & \cdots \\ \vdots & & & & & & \end{pmatrix}$$

(b) Give a direct argument to show that

$$P(Y_n = k) = \begin{cases} (1-p)^k p & \text{if } k = 0, \ldots, n-1 \\ (1-p)^n & \text{if } k = n \end{cases}$$

Hence, for any fixed $k = 0, 1, 2, \ldots$, if $0 < p < 1$,

$$\lim_{n\to\infty} P(Y_n = k) = (1-p)^k p$$

In other words, asymptotically, Y_n is geometrically distributed.

(c) Let $\boldsymbol{\pi} = [p, (1-p)p, (1-p)^2p, \ldots]$, and let P be the transition matrix described in (a). Show that $\boldsymbol{\pi}\mathbf{P} = \boldsymbol{\pi}$.

(d) Interpret (b) and (c) in terms of the analogous results for finite matrices in Section 11.11.

21 Consider a collection of nuclear particles that grows or diminishes in the following way. At time 0, there are Y_0 particles. Each of these, independently, will either die or split in two, with equal probabilities, producing at time 1 a total of Y_1 particles. Again, each of these, independently, will either die or split in two, with equal probabilities, producing at time 2 a total of Y_2 particles, and so forth.

(a) Show that $Y_0, Y_1, Y_2, \ldots$ is a Markov chain with the time-homogeneous transition matrix,

$$\mathbf{P} = \begin{pmatrix} 1 & 0 & 0 & 0 & 0 & 0 & 0 & \cdots \\ 1/2 & 0 & 1/2 & 0 & 0 & 0 & 0 & \cdots \\ 1/4 & 0 & 2/4 & 0 & 1/4 & 0 & 0 & \cdots \\ 1/8 & 0 & 3/8 & 0 & 3/8 & 0 & 1/8 & \cdots \\ \vdots & & & & & & & \end{pmatrix}$$

where, in general,

$$p(i, j) = \begin{cases} \dbinom{i}{j/2}\left(\dfrac{1}{2}\right)^i & \text{if } \dfrac{j}{2} = 0, 1, \ldots, i, \qquad i = 1, 2, \ldots \\ 0 & \text{otherwise} \end{cases}$$

In (b), (c), and (d) that follow, suppose that initially there is one particle. That is, $P(Y_0 = 1) = 1$.

(b) Show that the distributions of Y_1, Y_2, and Y_3 are the following:

i	0	1	2	3	4	5	6	7	8
$P(Y_1 = i)$	1/2	0	1/2						
$P(Y_2 = i)$	5/8	0	2/8	0	1/8				
$P(Y_3 = i)$	89/128	0	20/128	0	14/128	0	4/128	0	1/128

[Hint: Use the iterative relation, Eq. (1), Section 11.10., $\boldsymbol{\pi}_0 = (0, 1, 0, 0, \ldots)$.]

(c) Check computationally in (b) that $E(Y_0) = E(Y_1) = E(Y_2) = E(Y_3) = 1$. Prove the general relation $E(Y_n) = 1$, $n = 0, 1, 2, \ldots$. [Hint: $E(Y_{n+1}|Y_n = i) = i$. Use (5), Section 4.3, to show that $E(Y_{n+1}) = E(Y_n)$.]

(d) Along the same lines as (c), prove that $E(Y_{n+1}^2) = n+1$, and, hence, $\mathrm{Var}(Y_n) = n$, $n = 0, 1, 2, \ldots$. [Hint: Use the fact that the conditional distribution of Y_{n+1}, given that $Y_n = i$, is that of $2X$, where X is binomial with parameters i and 1/2. In fact, this is what (a) asserts. Now use the facts that $\text{Var}(X) = i/4$, $\text{Var}(2X) = i$, and the computational formula for variance to show that

$$E(Y_{n+1}^2 | Y_n = i) = i + i^2$$

Again use (5), Section 4.3, as in (c), to show that $E(Y_{n+1}^2) = E(Y_n) + E(Y_n^2)$.]

Remark: It is true that $\lim_{n\to\infty} P(Y_n = 0) = 1$. This fact suggests that the collection of particles will eventually become extinct, which is also true. The proofs of these assertions are beyond the methods developed here (Note 13). Incidentally, do you find it surprising that $\lim_{n\to\infty} P(Y_n = 0) = 1$, in light of the fact that $\text{Var}(Y_n) \to \infty$?

NOTES

1 For some additional reading on stochastic processes, see [38] and Chapters XIII through XVII of [9].

2 An interesting treatment of random walk in physics is in Chapter 6 of [39].

3 For a proof of the fact that if $\lim u_n$ exists then $\lim (u_1 + \cdots + u_n)/n$ exists also and has the same value, see Chapter 9, Section 7.2, of [32].

4 A proof that (7), Section 11.4, is true, under certain conditions, as an ordinary limit, is in Section XIII.11 of [9].

5 In psychology, a large body of theory relating to learning theory rests heavily on Markov chains. See [40].

6 The general version of Example 2, Section 11.10, was formulated by the physicists Paul and Tatiana Ehrenfest in 1907. For an interesting account of the Ehrenfest model, see [6].

7 The theory of finite Markov chains was first developed by the Russian mathematician A. A. Markov (1856–1922). For a complete treatment of Markov chains with a countable number of states and time-homogeneous transitions, see Chapter XV, [9].

8 A vast literature exists on the theory and application of queues and service systems. For one survey, see [42].

9 For some applications of the Poisson process in physics, see [43] and [44].

10 A journal called *Skandinavisk Aktuaridtidskrift* has many articles on probability theory applied to insurance.

11 For applications of probability theory to auto traffic problems, see [45].

12 Example 3, Section 11.13, is related to a so-called birth and death process, which has applications in biology. For various such applications, see [46], [47], [48], and [49].

13 Exercise 21 describes a special case of a *branching process.* See Chapter XII of [9].

Chapter Twelve

ESTIMATION

In most experimental situations, the experimenter obtains observations the probability distributions of which are not completely known. What may be known is only that the true probability distribution belongs to some class of specified distributions. The experimenter would like to infer information about the true distribution from an analysis of his observed data. In the estimation theory approach to this problem, we suppose there is a real-valued function defined over the various possible distributions that one wants to estimate. An *estimator* is a function of the observations that, in some sense, is close to the function over the distributions to be estimated. A discussion of some intuitive criteria for estimators—*unbiasedness, maximum likelihood* and *consistency*—is presented in this chapter. The notion of a *loss function* is introduced in an attempt to measure the seriousness of disparities between the true value of the number to be estimated and the value of the estimator. An important idea is that the experimenter may be able to appraise the *relative importance* or *likelihood* to him of the various possible true distributions. *Bayes' estimators* take advantage of this prior knowledge and are related to *admissible estimators.* Finally, *confidence interval* procedures of estimation are discussed. (Note 1.)

12.1 ESTIMATION

Suppose that an experimenter makes observations on some numerical phenomenon described by the mathematical model of a random variable, or by a *vector* of random variables,

$$X = X_1, \ldots, X_n$$

Actually, the experimenter does not know the probability distribution of X; we suppose that all he knows is a delimited class to which the distribution belongs. A formal way of describing this situation is as follows. X has a density function (either discrete or integrating) that can be described as $f(x, \theta)$, where θ is an identifying label which varies over a class Ω. The label θ need not be numerical, although it may be; it is simply a device for keeping track of the various possible distributions. If X is discrete, we have

$$P(X \text{ in } A) = \sum_{x \text{ in } A} f(x, \theta)$$

and if X has an integrating density, then

$$P(X \text{ in } A) = \int_{x \text{ in } A} f(x, \theta)\, dx$$

These probabilities depend on θ.

Example 1 A dichotomous population of 1000 objects consists of M objects of type 1 and $1000 - M$ objects of type 2. The experimenter draws three objects without replacement. His observation can be described by $X = (X_1, X_2, X_3)$, where

$$X_i = \begin{cases} 1 & \text{if } i\text{th drawing is of type 1} \\ 0 & \text{otherwise} \end{cases} \qquad i = 1, 2, 3$$

If the experimenter is completely ignorant about M, then he knows that there are 1001 possible distributions of X, determined by the values of the label $M = 0, 1, \ldots, 1000$. If he knows M, then he can completely describe the probability distribution of X. In terms of the earlier discussion, M plays the role of θ and $\Omega = (0, 1, \ldots, 1000)$ (Note 2). Of course, there is nothing special about the size of the population being 1000 or the size of the sample being three. If N is the size of the population and n the size of the sample, and these numbers are known to the experimenter, the example is the same. For some specific versions of this example, consider the following:

(a) A quality control inspector knows that a lot of N produced items consists of an unknown number (M) of defective items. He is required to inspect n items and to determine the defectives among them.

(b) An election pollster intends to interview n persons out of a population of N voters. The interview is simply to determine whether the voter will support a given candidate.

Example 2 Customers arrive at a service counter according to the model that successive waiting times between arrivals are independent random variables, each exponentially distributed with the same parameter, λ. (Now λ will play the role of θ in the previous general discussion.) Suppose first that we keep track of the first n waiting times, which we denote by $W_1, \ldots, W_n$. The integrating density function of this vector is

$$f(x_1, \ldots, x_n, \lambda) = \begin{cases} \lambda^n \exp -\lambda(x_1 + \cdots + x_n) & \text{if } x_i \geq 0, \\ & i = 1, \ldots, n \\ 0 & \text{otherwise} \end{cases}$$

If λ were known, then the distribution of $W_1, \ldots, W_n$ would be completely known. On the other hand, if what is being kept track of is the number of customers to arrive by time t, then this number, $N(t)$, is Poisson distributed with parameter λt (Example 1, Section 11.12). That is,

$$f(x, \lambda) = P[N(t) = x] = \frac{(\lambda t)^x e^{-\lambda tx}}{x!}, \qquad x = 0, 1, 2, \ldots$$

Again, as in the previous case, if λ were known the distribution of $N(t)$ would be completely known. Otherwise, all we can say is that the distribution is determined by the label λ, which varies over the class $\Omega = (0, \infty)$.

Suppose now that the experimenter wants information about the unknown labeling parameter θ, and that the desired information is in the form of $h(\theta)$, a real-valued function of the parameter. For instance, in Example 1a, where $\theta = M$, a function of interest might be $h(M) = M/N$, the proportion of defectives in the lot. Another function of interest might be $M/(N - M)$, the ratio of defectives to nondefectives. In Example 2, where $\theta = \lambda$, the experimenter might be interested in $h(\lambda) = \lambda$, the number of customers arriving per unit time. He might also be interested in $1/\lambda$, the expected waiting time between arrival of customers. A function such as $h(\lambda) = \lambda^2/(1 + \log \lambda)$ may not have any particular experimental significance in Example 2, but nevertheless it will be useful to have the theoretical presentation sufficiently general to allow the study of even such a function.

In general, how should the experimenter go about estimating $h(\theta)$? He has to use his observed data. Mathematically, we will say that he will estimate $h(\theta)$ by means of a function of the observations X. We will call this function $u(X)$. It represents a computation or an analysis done on the observations X, which produces a summarizing number $u(X)$, intended in some sense to "estimate" the desired number $h(\theta)$. Now, what really is meant by "estimating?" The experimenter's observations, X, are random; hence, in general, the result of any analysis, $u(X)$, is also random. It is not likely that $u(X)$ will exactly equal $h(\theta)$. What demands should the experimenter impose on his function $u(X)$? Even though $u(X)$ is random, one may want it to equal $h(\theta)$ on the average; that is, it may be desirable that $E[u(X)]$ be identically equal to $h(\theta)$, no matter what the true value of θ might be. This property is called *unbiasedness,* and it is explained in the next section. It may be that one wants $u(X)$ to be "close" to $h(\theta)$, in terms of some yardstick that measures the undesirability or "loss" associated with various combinations of values of $u(X)$ and $h(\theta)$. Various theoretical approaches to these questions will be discussed.

Let us impose one rather obvious requirement on $u(X)$ at the beginning. We said that $u(X)$ is to represent an analysis of the observed data. It is clear that $u(X)$ cannot depend on the unknown parameter θ, since the experimenter does not know it. Our initial definition of an estimator is that it is a function $u(X)$ *that does not depend on* θ. Any such function, good, bad, or indifferent, will be called an estimator. We then go on to try to decide what it might mean for estimators to be good, bad, or indifferent, and how to recognize the difference.

estimator

Suppose that $X = (X_1, \ldots, X_n)$ has a distribution that depends on θ, where θ varies over a set Ω. Then $h(\theta)$ denotes a real-valued function of θ, which is to be estimated. The random variable $u(X)$ is called an estimator of $h(\theta)$ if its values do not depend on θ.

We should point out that another name commonly used for estimator is *statistic.*

12.2 SOME INTUITIVE CRITERIA

We now want to start discussing criteria according to which one can choose an estimator $u(X)$ of the function $h(\theta)$. One criterion, *unbiasedness,* says that an estimator should on the average be equal to the function it purports to estimate. Remember that since u is an estimator, its values do not depend on θ; θ enters the picture through the distribution of X. Unbiasedness requires that no matter which distribution is used in evaluating the expected value of $u(X)$, this expected value turns out to be $h(\theta)$, where θ is the label of that distribution. For instance, suppose that X has an integrating density $f(x, \theta)$; then the unbiasedness of $u(X)$ means the following:

$$\int u(x) f(x, \theta)\, dx = h(\theta), \quad \text{identically for all } \theta \text{ in } \Omega \tag{1}$$

If X is a single random variable, then the integration is from $-\infty$ to ∞. If $X = (X_1, \ldots, X_n)$, then the integration is over all of n-dimensional space. If X is discrete, then the appropriate version of (1) is

$$\Sigma\, u(x) f(x, \theta) = h(\theta), \quad \text{identically for all } \theta \text{ in } \Omega \tag{2}$$

where the summation is over all possible values of X. To emphasize that the expected values in (1) and (2) are being evaluated with respect to the distribution bearing the label θ, we shall denote them by E_θ. Thus, in terms of this notation, we can rewrite both (1) and (2) as follows:

unbiased estimator

$u(X)$ is called an unbiased estimator of $h(\theta)$ if

$$E_\theta[u(X)] = h(\theta), \quad \text{identically for all } \theta \text{ in } \Omega \tag{3}$$

Example 1

Suppose that $X_1, \ldots, X_n$ are mutually independent, indicator random variables with

$$X_i = \begin{cases} 1 & \text{with probability } p \\ 0 & \text{with probability } 1-p \end{cases} \qquad i = 1, \ldots, n$$

The parameter p is unknown to the experimenter and lies in the set $(0, 1) = \Omega$. The sample average $(X_1 + \cdots + X_n)/n$ is an unbiased estimator of p since

$$E_p\left[\frac{(X_1 + \cdots + X_n)}{n}\right] = p$$

identically in p. There are other unbiased estimators. For instance, X_1 is an unbiased estimator of p also. Intuitively, it seems plausible that $(X_1 + \cdots + X_n)/n$ should be a "better" estimator of p than X_1, if n is greater than 1; the unbiased estimator X_1 does not take advantage of all the data, which does not seem reasonable. Consider also another function of p, namely $h(p) = p^2$. An unbiased estimator of $h(p)$ is given by

$$\frac{(X_1 + \cdots + X_n)^{(2)}}{n(n-1)}$$

This follows from the fact that the second factorial moment of a binomial, n, p random variable is $n^{(2)}p^2$, as pointed out in Section 6.1.

Example 2 Suppose that $X_1, \ldots, X_n$ are mutually independent random variables, each uniformly distributed over $[0, \theta]$. The parameter θ is not known to the experimenter and lies in $(0, \infty) = \Omega$. We consider the estimation of $h(\theta) = \theta$. Since $E(X_i) = \theta/2$, it follows that

$$u_1(X_1, \ldots, X_n) = \frac{2(X_1 + \cdots + X_n)}{n}$$

is an unbiased estimator of θ. We shall also describe another unbiased estimator. Let $Y = \max(X_1, \ldots, X_n)$. The integrating density function of Y is

$$g(y) = \begin{cases} \dfrac{n}{\theta^n} y^{n-1} & \text{if } 0 \leq y \leq \theta \\ 0 & \text{otherwise} \end{cases}$$

since $P(Y \leq y) = P[(X_1 \leq y) \cap \cdots \cap (X_n \leq y)] = (y/\theta)^n$ if $0 \leq y \leq \theta$. Thus $\int_{-\infty}^{\infty} y g(y)\, dy = [n/(n+1)]\theta$. Hence,

$$\frac{n+1}{n} Y = \frac{n+1}{n} \max(X_1, \ldots, X_n) = u_2(X_1, \ldots, X_n)$$

is another unbiased estimator of θ. Which do you think is a better estimator of θ, u_1 or u_2?

Another intuitively plausible criterion for selecting an estimator is called *maximum likelihood.* It estimates $h(\theta)$ by $h(\hat{\theta})$, where $\hat{\theta} = u(X)$ is a value of θ that makes $f(X, \theta)$ as large as possible.

maximum likelihood estimator

Define $u(X) = h(\hat{\theta})$, where $\hat{\theta} = \hat{\theta}(X) =$ a value of θ that maximizes $f(X, \theta)$ with respect to θ in Ω, assuming that such a value exists.

This criterion certainly has intuitive appeal. Having observed X, which θ is likely to be the most correct one? The maximum likelihood approach says choose that θ which makes what has been actually observed as highly probable as possible (Note 3). It should be pointed out that a value of θ that maximizes $f(X, \theta)$ need not be unique.

Example 3 Consider Example 1 again. Since

$$f(X_1, \ldots, X_n, p) = p^{X_1+\cdots+X_n}(1-p)^{n-X_1-\cdots-X_n}$$

the maximizing value of p is, by elementary calculus,

$$\hat{p} = \frac{X_1 + \cdots + X_n}{n}$$

(Check this!) In other words, "the sample proportion" is the maximum likelihood estimator of the "population proportion."

Example 4 Suppose that $X_1, \ldots, X_n$ are mutually independent random variables, each distributed $N(m, \sigma^2)$. Both parameters m and σ are unknown to the experimenter. Thus, $\theta = (m, \sigma)$, m is in $(-\infty, \infty)$, σ is in $(0, \infty)$, and Ω is a half-plane. The function to be maximized is

$$f(X_1, \ldots, X_n, m, \sigma) = \frac{1}{(2\pi\sigma^2)^{n/2}} \exp\left[-\sum_{i=1}^{n} \frac{(X_i - m)^2}{\sigma^2}\right]$$

It is an exercise in elementary calculus to show that the values of m and σ which maximize f for any fixed values $X_1, \ldots, X_n$ are

$$\hat{m} = \frac{X_1 + \cdots + X_n}{n} = \overline{X}, \qquad \hat{\sigma} = \left(\frac{\sum_{i=1}^{n} (X_i - \overline{X})^2}{n}\right)^{1/2} = S$$

(Check this!)

Hence, the maximum likelihood estimator of

$h(\theta) = h(m, \sigma)$ is

$h(\hat{\theta}) = h(\hat{m}, \hat{\sigma}) = h(\overline{X}, S)$

The *sample* mean is $\overline{X}$ and S is called the *sample standard deviation.* Thus, the maximum likelihood estimators of the population mean m and population standard deviation σ are $\overline{X}$ and S, the sample mean and sample standard deviation. {We should point out a slight variation on this terminology. Sometimes $[\Sigma_i (X_i - \overline{X})^2/(n-1)]^{1/2}$ is called the sample standard deviation. See Exercise 2.}

If $u(X)$ is an estimator of $h(\theta)$, it may happen that $u(X)$ takes on values outside the range of values of $h(\theta)$. If this happens with positive probability, it is clearly unreasonable. There is no point in ever estimating $h(\theta)$ by a value it is known never to assume. Maximum likelihood estimators are reasonable in this respect; it follows immediately from their definition that their values never fall outside the range of values of $h(\theta)$.

At first sight, unbiasedness seems like a reasonable criterion. However, there may not exist any unbiased estimators; even if they exist, an insistence on unbiasedness may sometimes lead to estimators with undesirable properties. We elaborate these points in the following examples.

Example 5 Consider the situation of Example 1 again, in which $X_1, \ldots, X_n$ are independent, indicator random variables with $P(X_i = 1) = p$ being the same for each i. Suppose that $h(p) = 1/p$. It is a fact, which we state without proof at this point, that there is no unbiased estimator of $1/p$. That is, there exists no function u such that $E_p[u(X_1, \ldots, X_n)]$ is identically equal to $1/p$ (see Exercise 15). As pointed out in Example 3, $(X_1 + \cdots + X_n)/n$ is the maximum likelihood estimator of p. It follows that its reciprocal, $n/(X_1 + \cdots + X_n)$, is the maximum likelihood estimator of $1/p$. However, $n/(X_1 + \cdots + X_n)$ is not an unbiased estimator of $1/p$. In fact, since this random variable equals ∞ whenever $X_1 + \cdots + X_n$ equals zero, it follows that the expected value is ∞ for all p in $(0, 1)$.

Example 6 We still consider the same situation, except we now suppose that p is a priori known to lie in $(0, 1/2)$. [In other words, $\Omega = (0, 1/2)$.] We state without proof that the only "reasonable" *unbiased* estimator of $h(p) = p$ is the sample proportion $(X_1 + \cdots + X_n)/n$ (see Section 15.4). For any p in $(0, 1/2)$, this estimator will assume values greater than $1/2$. This fact is clearly unreasonable—if p is a priori known to be less than $1/2$, there is no point in an estimator ever asserting it to be greater than $1/2$. Of course, one can define a new estimator as follows:

$$u(X) = \begin{cases} \dfrac{X_1 + \cdots + X_n}{n} & \text{if this value is less than } \dfrac{1}{2} \\[2ex] \dfrac{1}{2} & \text{otherwise} \end{cases} \tag{4}$$

Now the estimator described by (4) is no longer unbiased, although in some respects it is a reasonable estimate. In fact, if, as we are assuming, $\Omega = (0, 1/2)$, then (4) is the maximum likelihood estimator of p. We leave the proof of this assertion to the reader (see Exercise 3).

Finally, we mention another intuitive criterion, that of *consistency*. It concerns an increasing sequence of observations obtained by the experimenter. That is, suppose that X_1 is the first observation,

(X_1, X_2) the first two observations, . . . , $(X_1, \ldots, X_n)$ the first n observations, and so forth. Suppose also that Ω is the same for any number of observations and that there is a single function $h(\theta)$ to be estimated, which does not depend on n. Let $u_1(X_1)$, $u_2(X_1, X_2)$, . . . , $u_n(X_1, \ldots, X_n)$, . . . , be a sequence of estimators of $u(\theta)$. This sequence is said to be *consistent*, if, given any $\epsilon > 0$,

$$\lim_{n\to\infty} P[|u_n(X_1, \ldots, X_n) - h(\theta)| > \epsilon] = 0 \tag{5}$$

Informally, this says that for a very large number of observations, the distribution of $u_n(X_1, \ldots, X_n)$ is highly concentrated about the value $h(\theta)$, no matter what the true value of θ may be.

Example 7 Consider the estimators $2(X_1 + \cdots + X_n)/n$ of Example 2. This sequence of estimators is consistent by virtue of the law of large numbers. We leave the verification to the reader. Similarly, the estimators $\max(X_1, \ldots, X_n)$ are consistent, because

$$\lim_{n\to\infty} P(\max(X_1, \ldots, X_n) \leq y) = \begin{cases} 0 & \text{if } y < \theta \\ 1 & \text{if } y > \theta \end{cases}$$

which is equivalent to (5).

The importance of consistency is somewhat dubious, since a finite number of estimators, for all $n \leq N$, can have very undesirable properties without disturbing the consistency of the infinite sequence.

12.3 EXPECTED LOSS

One approach to choosing an estimator of $h(\theta)$ is to first try to appraise quantitatively the consequences of making mistakes. This approach supposes that there is a real-valued function $L[u(X), \theta]$ that measures the "loss" incurred in estimating $h(\theta)$ by $u(X)$. We call L a *loss function.*

loss and expected loss

Suppose that $L(u(X), \theta)$ is a loss function. Then

$$E_\theta\{L[u(X), \theta]\} = R(u, \theta) \tag{1}$$

is called the *expected loss.* We emphasized that the distribution with respect to which the expected value is taken is the one bearing the label θ. Of course, the definition only makes sense when the expected value exists.

The θ appearing twice in the left side of (1) should be the same. Since (1) has assumed such a forbidding form, we should review exactly what it is. $L[u(X), \theta]$ is the loss incurred in estimating $h(\theta)$ by $u(X)$ when θ is the true parameter. The expected value of this loss is evaluated using the distribution with the same label θ, giving the average loss incurred using the estimator u when the true label of the distribution is θ. The expected loss, $R(u, \theta)$, is a function of two "variables", u and θ; u is an estimator and θ is the label of the true distribution. These two variables combine, via the loss function, into a single number, $R(u, \theta)$, which is a measure of loss suffered when the experimenter uses procedure u and the true distribution label is θ.

Example 1 A commonly used loss function is

$$L[u(X), \theta] = [u(X) - h(\theta)]^2 \tag{2}$$

The use of (2) is motivated more by its mathematical simplicity than anything else. Intuitively, (2) certainly makes sense. When $u(X)$ happens to equal $h(\theta)$, the loss is 0. Also, the further $u(X)$ is from $h(\theta)$, the greater is the loss. But, of course, there are many other loss functions with these same properties, such as

$$|u(X) - h(\theta)|, \qquad [u(X) - h(\theta)]^4, \qquad \left(\frac{|u(X) - h(\theta)|}{|h(\theta)| + 1}\right)^{1/2}$$

If $u(X)$ happens to be an unbiased estimator of $h(\theta)$, then the expected loss using (2) is simply the variance of $u(X)$. In general, the expected loss using (2) equals

$$\text{Var}[h(X)] + \{E[u(X)] - h(\theta)\}^2$$

This follows from the fact that $E(Y^2) = \text{Var}(Y) + (EY)^2$, which is the computational formula for the variance [(3), Section 10.1].

Example 2 Consider Example 2, Section 12.2 again, in which $X_1, \ldots, X_n$ are mutually independent random variables, each uniformly distributed over $[0, \theta]$. We refer to the two unbiased estimators, u_1 and u_2, which were defined in that example. Using the loss function (2), we find the expected loss incurred with the estimator u_1 to be $\theta^2/3n$. It follows from the fact that the variance for the uniform distribution over $[0, \theta]$ is $\theta^2/12$. Using the same loss function, we find the expected loss incurred with the estimator u_2 to be $\theta^2/n(n + 2)$. See Exercise 1d. Notice that when $n \geqslant 2$, then, uniformly in $\theta > 0$,

$$\frac{\theta^2}{n(n+2)} < \frac{\theta^2}{3n}$$

For $n = 1$, u_1 and u_2 are the same. With more than one observation, however, u_2 has a smaller expected loss than u_1. In this sense, u_2 is a better estimator of θ than is u_1, but the use of the word "better" relates to the particular loss function (2).

12.4 ADMISSIBILITY

Consider the problem of estimating $h(\theta)$ using an estimator based on the random observations X, as in the preceding sections. We suppose that there is a given loss function L. How should the experimenter choose among a number of competing estimators? One thing is clear—if an estimator u_2 is *uniformly better* than another estimator u_1, there is no point in considering the use of u_1 at all.

inadmissible and admissible estimators

Suppose that u_1 is an estimator and there exists another estimator u_2 such that

$$R(u_1, \theta) \geq R(u_2, \theta) \quad \text{for all } \theta \text{ in } \Omega$$

with strict inequality holding for at least one θ. Then the estimator u_1 is called *inadmissible.* If an estimator is not inadmissible, it is called *admissible.*

In other words, an estimator is inadmissible if another estimator exists that is uniformly at least as good, and is strictly better for at least one of the possible distributions. If an estimator is admissible, it cannot be uniformly improved upon. It does not mean that it is uniformly better than other estimators. The existence of a uniformly best procedure is most unlikely; life is like that! In general, there may be many admissible procedures. If u_1 and u_2 are both admissible, then, necessarily for some θ's, u_1 is better than u_2, and for some other θ's, u_2 is better than u_1. This situation is described schematically in Figure 12.1.

Example 1 We continue the discussion in Example 2, Section 12.2 and Example 2, Section 12.3. When $n \geq 2$, the estimator u_1 is inadmissible. It does so happen that the estimator u_2 is admissible, but it is not at all obvious.

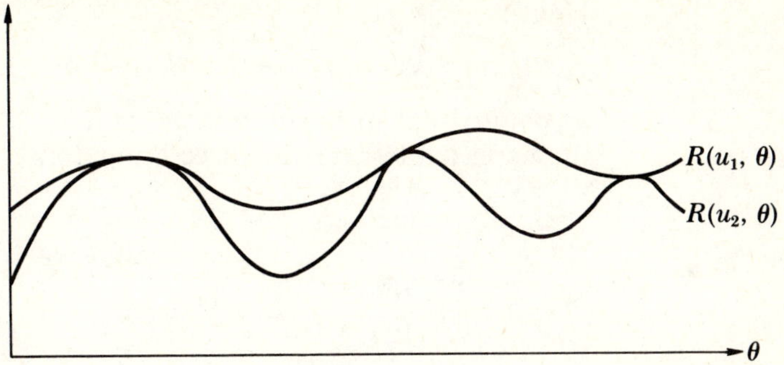

(a)

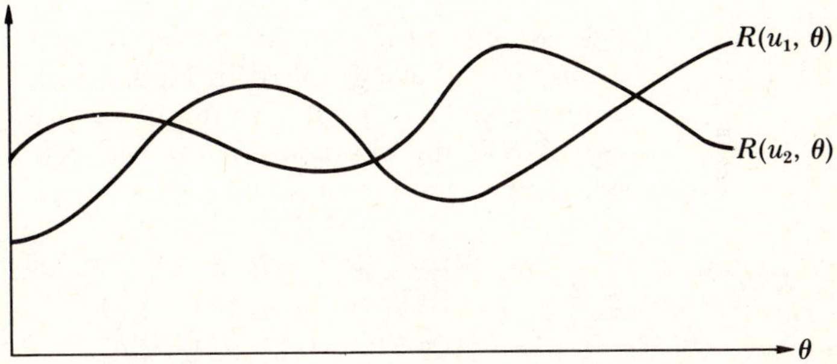

(b)

FIGURE 12.1 (a) The expected loss functions show that u_1 is inadmissible. (b) There is no obvious choice as between u_1 and u_2. For some θ's, u_1 is better than u_2, and for some θ's, u_2 is better than u_1.

Example 2 Suppose that $X_1, \ldots, X_{10}$ are mutually independent, indicator random variables, each equaling 1 with probability p. The parameter p is unknown and lies in $(0, 1) = \Omega$. Denote $X = (X_1, \ldots, X_{10})$ and consider the following estimators of p:

$$u_1(X) = \frac{X_1 + \cdots + X_{10}}{10}$$

$$u_2(X) = \frac{1 + X_1 + \cdots + X_{10}}{11}$$

$$u_3(X) = \frac{X_1 + \cdots + X_5}{5}$$

(It so happens that u_1 and u_3 are both unbiased estimators of p, but u_2 is not.) Suppose that the loss function is the squared error,

$$L[u(X), p] = [u(X) - p]^2$$

Hence,

$$R(u, p) = \text{Var}[u(X)] + \{E[u(X)] - p\}^2$$

as pointed out in Example 1, Section 12.3. The explicit forms of the expected loss for the three estimators u_1, u_2, u_3 are as follows:

$$R(u_1, p) = \frac{p(1-p)}{10}$$

$$R(u_2, p) = \frac{100}{121}\frac{p(1-p)}{10} + \frac{(1-p)^2}{121}$$

$$R(u_3, p) = \frac{p(1-p)}{5}$$

(We leave the calculations to the reader.) The expected losses of u_1, u_2, and u_3 are graphed in Figure 12.2. The estimator u_3 is inadmissible. This fact should not be surprising, since u_3 only makes use of half the observations. As between u_1 and u_2, the choice is not clear. A simple calculation shows that

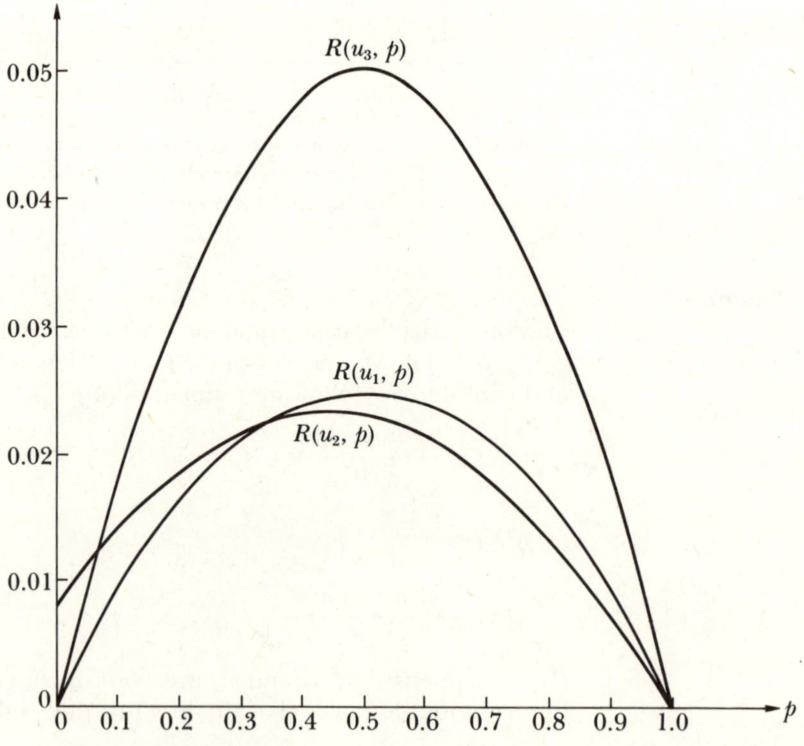

FIGURE 12.2

$R(u_2, p) < R(u_1, p)$ if $\frac{10}{31} < p < 1$

$R(u_2, p) > R(u_1, p)$ if $0 \leq p < \frac{10}{31}$

$R(u_2, p) = R(u_1, p)$ if $p = \frac{10}{31}$ or $p = 1$

12.5 BAYES' ESTIMATORS

The preceding discussion on admissibility should have raised questions in your mind. (a) How do we determine whether an estimator is admissible? (b) How do we go about choosing between several competing admissible estimators? For instance, consider the estimators u_1 and u_2 of Example 2, Section 12.4, with expected loss functions graphed in Figure 12.2. As that figure shows, u_1 has higher expected loss than does u_2 in the interval (10/31, 1). Suppose that the experimenter is more concerned about doing the wrong thing when the true p is in that interval than when it is not. Then he should use the estimator u_2 rather than u_1. What this suggests is that the experimenter might assign weights assessing the relative importance or consequences of errors relating to the various parts of Ω. He could then choose between competing estimators by considering the *expected losses averaged out with respect to this weight function* over all the labels in Ω. We shall now discuss this idea in greater detail. To avoid the technical difficulties having to do with the procedure of averaging out over Ω when Ω is a continuum of values, we first suppose that there are only a finite number of possible distributions.

average expected loss

Suppose that Ω consists of a finite number of distribution labels $\theta_1, \ldots, \theta_m$. Let $(q_1, \ldots, q_m) = q$ be nonnegative *weights* that sum to 1. Let $R(u, \theta)$ be the expected loss determined by an estimator u and distribution θ. Then

$$A(u, q) = R(u, \theta_1)q_1 + \cdots + R(u, \theta_m)q_m$$

is called the *average expected loss.*

Since the quantity A has been built up through several different operations, let us review its meaning. In the background are the possible distributions the labels of which are $\theta_1, \ldots, \theta_m$. There is also a loss function $L(u(X), \theta)$, which tells how well one does in estimating $h(\theta)$ using the estimator $u(X)$ when θ is the true distribution label. The expected value of this loss is $R(u, \theta)$, which depends both on the particular estimator being used, u, and on θ.

The weights $q_1, \ldots, q_m$ assess the relative concern that the experimenter has over the various possible distributions that may prevail. The average, $A(u, q)$, is a single number determined by the estimator u and the weights $q = (q_1, \ldots, q_m)$; it is obtained by averaging out the expected losses $R(u, \theta)$ over the different values of θ. Once one considers the single number $A(u, q)$, the problem of determining a good estimator is considerably simplified. One should choose an estimator u that for fixed q makes $A(u, q)$ as small as possible.

Bayes' estimator

An estimator u that minimizes the average expected loss $A(u, q)$ is called a *Bayes' estimator* with respect to the a priori weights $(q_1, \ldots, q_m) = q$ (Note 4).

In other words, a Bayes' estimator u has the property that for one particular set of weights q,

$$A(u, q) \leqslant A(u_1, q)$$

where u_1 is any other estimator. An estimator that is Bayes with respect to one set of weights may not be Bayes with respect to another set of weights.

Although the following example is quite artificial, it may help to clarify the preceding concepts.

Example 1

A box contains five balls. An unknown number of these balls are black and the remaining ones are white. The object of the experiment is to estimate the number of black balls in the box. The experimenter is restricted to drawing two balls without replacement, without keeping track of the order. Thus, he observes X, the count of the number of black balls he has drawn. To tie in with the previously described notation, $\Omega = (0, 1, 2, 3, 4, 5)$ is the set of possible "states of nature," the six possible values of θ. The function of θ to be estimated is simply $h(\theta) = \theta$, namely the number of black balls in the box. Suppose that the loss function is given by

$$L[u(X), i] = |u(X) - i|, \qquad i = 0, 1, 2, 3, 4, 5$$

That is, if the box really has i balls in it and one estimates the number to be $u(X)$, then the loss suffered is the absolute value of the difference. Finally, suppose that the a priori weights are

$$q_0 = q_1 = \cdots = q_5 = \tfrac{1}{6}$$

which means that the experimenter is equally worried about each of the six possible states of nature.

Let us now compute the expected loss and the average expected loss for the various possible estimators. An estimator is completely determined by the three numbers

$$u(0), \quad u(1), \quad u(2)$$

We also need the six possible probability distributions of X. These are shown in Table 12.1, where $f(k, \theta)$ is the probability of drawing k black balls when the distribution label is θ, the number of black balls in the box.

If we denote the probability of drawing k black balls when the distribution label is θ by $f(k, \theta)$, then the expected loss determined by a procedure u is as follows:

$$|u(0) - \theta| f(0, \theta) + |u(1) - \theta| f(1, \theta) + |u(2) - \theta| f(2, \theta) = R(u, \theta) \tag{1}$$

The average expected loss is

$$A(u, q) = \frac{R(u, 0) + R(u, 1) + \cdots + R(u, 5)}{6} \tag{2}$$

By combining (1) and (2) and using the probabilities in Table 12.1, we have the following expression for $A(u, q)$:

$$A(u, q) = A_0[u(0)] + A_1[u(1)] + A_2[u(2)] \tag{3}$$

where

$$\begin{aligned} A_0[u(0)] &= [|u(0) - 0|10 + |u(0) - 1|6 + |u(0) - 2|3 \\ &\qquad + |u(0) - 3|]/60 \\ A_1[u(1)] &= [|u(1) - 1|4 + |u(1) - 2|6 + |u(1) - 3|6 \\ &\qquad + |u(1) - 4|4]/60 \\ A_2[u(2)] &= [|u(2) - 2| + |u(2) - 3|3 + |u(2) - 4|6 \\ &\qquad + |u(2) - 5|10]/60 \end{aligned} \tag{4}$$

To find a Bayes' estimator, we must determine these numbers—

TABLE 12.1

		θ					
		0	1	2	3	4	5
	0	1	3/5	3/10	1/10		
k	1		2/5	6/10	6/10	2/5	
	2			1/10	3/10	3/5	1

$u(0)$, $u(1)$, $u(2)$—which minimize the right side of (3). We can do this by minimizing each of A_0, A_1, and A_2 separately, since $u(0)$, $u(1)$, and $u(2)$ are at our disposal independently. Figure 12.3 shows the graphs of these three parts. From this graph we can read off the minimizing values and thus describe a Bayes' estimator as follows:

A Bayes' estimator, with respect to a priori weights $q_0 = \cdots = q_5 = 1/6$, is

$u(0) =$ any number in $[0, 1]$

$u(1) =$ any number in $[2, 3]$

$u(2) =$ any number in $[4, 5]$

For instance, $u(0) = 1$, $u(1) = 2$, $u(2) = 5$ is one Bayes' estimator. We should emphasize that there is not a unique Bayes' estimator in this example.

Example 2 Let us repeat Example 1, now supposing that the a priori weights are

$$q_0 = q_1 = q_2 = \tfrac{1}{9}, \qquad q_3 = q_4 = q_5 = \tfrac{2}{9}$$

In this case, the counterparts of Eq. (4) turn out to be

$$A_0[u(0)] = [|u(0) - 0|10 + |u(0) - 1|6 + |u(0) - 2|3 + |u(0) - 3|2]/90$$

$$A_1[u(1)] = [|u(1) - 1|4 + |u(1) - 2|6 + |u(1) - 3|12 + |u(1) - 4|8]/90$$

$$A_2[u(2)] = [|u(2) - 2| + |u(2) - 3|6 + |u(2) - 4|12 + |u(2) - 5|20]/90$$

We leave it to the reader to verify that in this case there is a uniquely determined Bayes' solution; namely,

$$u(0) = 1, \qquad u(1) = 3, \qquad u(2) = 5$$

Notice that in this example the Bayes' estimator tends to estimate "higher" than in Example 1. This tendency is plausible because the a priori weights concentrate more toward the higher parameter values (see Exercise 10b).

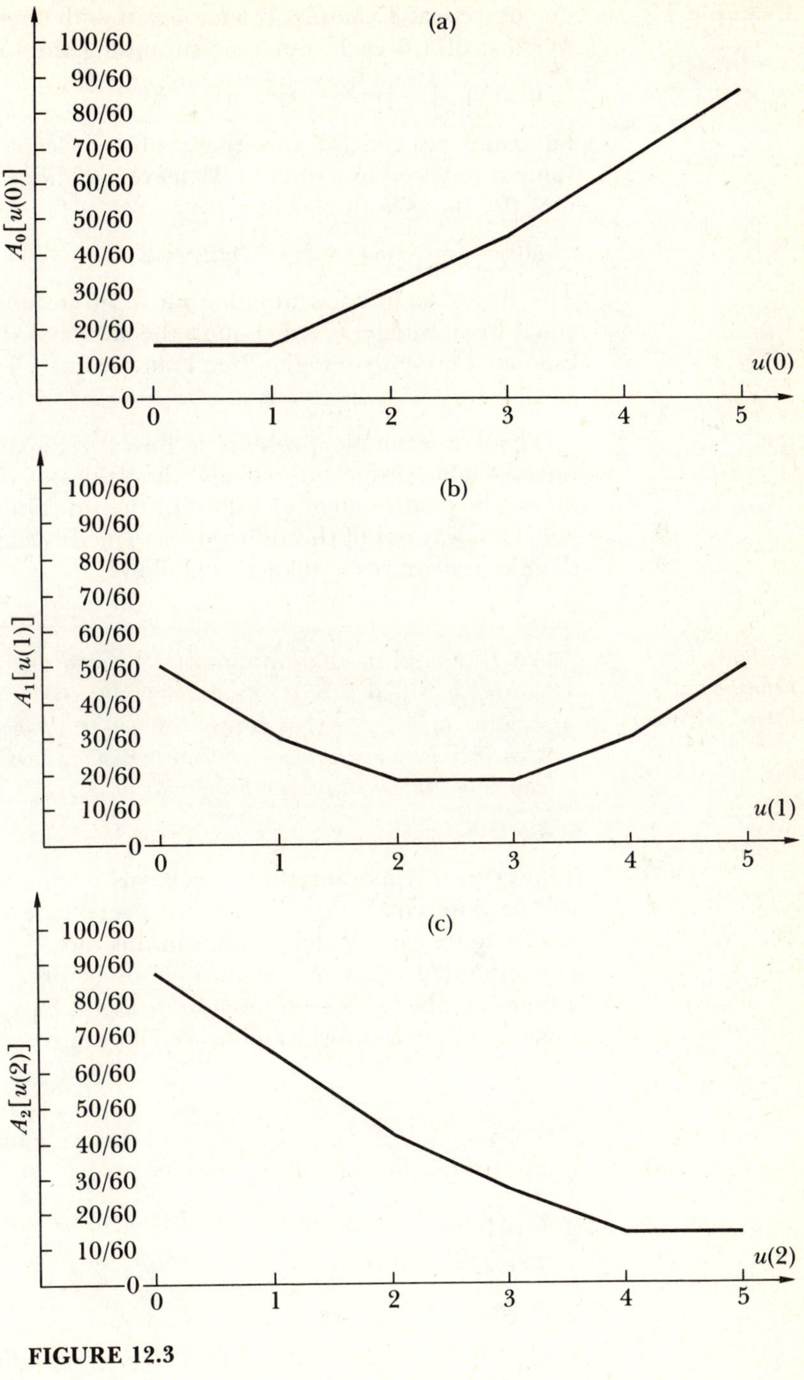
(a)
$A_0[u(0)]$
100/60
90/60
80/60
70/60
60/60
50/60
40/60
30/60
20/60
10/60
0
$u(0)$
0 1 2 3 4 5
(b)
$A_1[u(1)]$
100/60
90/60
80/60
70/60
60/60
50/60
40/60
30/60
20/60
10/60
0
$u(1)$
0 1 2 3 4 5
(c)
$A_2[u(2)]$
100/60
90/60
80/60
70/60
60/60
50/60
40/60
30/60
20/60
10/60
0
$u(2)$
0 1 2 3 4 5

FIGURE 12.3

Example 3 Let us repeat Example 1 once again with the original a priori weights, of 1/6 each, but now supposing the loss function to be

$$L[u(X), i] = [u(X) - i]^2$$

Equations (1) and (3) appear as before, except that the absolute value is replaced by a square. We leave it to the reader to verify that now the Bayes' estimator is

$$u(0) = \tfrac{3}{4}, \qquad u(1) = \tfrac{5}{2}, \qquad u(2) = \tfrac{17}{4}$$

The Bayes' solution is unique, but in no instance is the estimate equal to an integer, even though the values of the function being estimated are only integers (see Exercise 9b).

The last example presents a Bayes' estimator with a clearly unreasonable property—namely, the values of the estimator $u(X)$ do not lie in the range of values of the function being estimated, $h(\theta)$. One way out of this difficulty is to modify slightly the previous definition of Bayes' estimator as follows.

modified definition of Bayes' estimator

An estimator u that minimizes the average expected loss $A(u, q)$ is called a *Bayes' estimator* with respect to the a priori weights $(q_1, \ldots, q_m) = q$. *It is assumed that the only estimators u used in this minimization are those having values in the range of values of h(θ), the function being estimated.*

Subject to this modification, the Bayes' estimator for Example 3 will be different from the one just presented. We leave it to the reader to determine the solution in this case.

Continuing the same assumption as before, that Ω consists of a finite number of θ_i's, we now show that a Bayes' estimator is admissible, if the a priori weights are all positive.

admissibility of Bayes' estimators

Suppose that $\Omega = (\theta_1, \ldots, \theta_m)$ and that u is a Bayes' estimator with respect to the a priori weights

$$q = (q_1, \ldots, q_m), \quad \text{where } q_i > 0, \quad i = 1, \ldots, m$$

Then *u is admissible.*

Proof. We argue by contradiction. If u is not admissible, there is an estimator u_1 such that

$$R(u_1, \theta_i) \leq R(u, \theta_i), \qquad i = 1, \ldots, m \tag{5}$$

with strict inequality for at least one i (Section 12.4). Multiply both sides of (5) by q_i and sum over i. Since the q_i's are strictly positive,

$$A(u_1, q) = \sum_{i=1}^{m} R(u_1, \theta_i)q_i < \sum_{i=1}^{m} R(u, \theta_i)q_i = A(u, q)$$

But $A(u_1, q) < A(u, q)$ is impossible because u is a Bayes' estimator by hypothesis; therefore, u cannot be inadmissible. ∎

The last theorem provides a useful way of recognizing admissible procedures. There are versions of this theorem when Ω consists of more than a finite number of θ_i's. The difficulty in proving a general version is in working with the counterpart of

$$\sum_{i=1}^{m} R(u, \theta_i)q_i$$

In general, this has to be a kind of integration over Ω and demands more mathematical machinery than is supposed at this level. We shall consider one important special case, when Ω is an interval.

Suppose that Ω is the interval (a, b) (a may be $-\infty$ or b may be ∞). Let $q(\theta)$ be an integrating density function. That is,

$$q(\theta) \geq 0, \qquad \int_a^b q(\theta)\, d\theta = 1$$

We call q an *a priori weight function.* Then

$$A(u, q) = \int_a^b R(u, \theta)q(\theta)\, d\theta$$

is called the *average expected loss,* using the estimator u, assuming this integral is well defined. An estimator that minimizes $A(u, q)$ is called a *Bayes' estimator* with respect to the a priori weight function q, provided such a minimizing estimator exists.

In a formal way, these definitions are the same as in the case in which Ω is finite. The theorem proved earlier that a Bayes' estimator is admissible is still true under appropriate conditions. The condition required earlier was that the weight function be strictly positive over all of Ω. We need to assume a little more now.

admissibility of Bayes' estimators

Suppose that Ω is the interval (a, b) and that q is continuous and strictly positive in Ω. Suppose also that for any estimator u, the expected risk, $R(u, \theta)$, is continuous for all θ in Ω. Then a Bayes' estimator is admissible.

Proof. The proof is similar to the previous proof where Ω is finite. If a Bayes' estimator u were not admissible, we would have

$$\int_a^b R(u_1, \theta)q(\theta)\, d\theta < \int_a^b R(u, \theta)q(\theta)\, d\theta$$

for some estimator u_1. {It is true because $[R(u_1, \theta) - R(u, \theta)]q(\theta)$ is continuous, nonnegative in (a, b) and *strictly positive* over some subinterval.} This condition would contradict the fact that u is Bayes. ▮

Example 4

Suppose that $X_1, \ldots, X_n$ are mutually independent, indicator random variables with $P(X_i = 1) = p$, where $\Omega = (0, 1)$. Suppose that $h(p) = p$ is the function to be estimated and $L(u(X), p) = [u(X) - p]^2$ is the loss function, where $X = (X_1, \ldots, X_n)$. Suppose that the *a priori weight function* is $q(p) = 1$, $0 < p < 1$. Thus, the experimenter is *equally* concerned about all parameter values. The average expected risk is

$$A(u, q) = \int_0^1 E_p\{[u(X) - p]^2\}\, dp$$

$$= \int_0^1 \sum_{k_1,\ldots,k_n} [u(k_1, \ldots, k_n) - p]^2 p^{k_1+\cdots+k_n}(1-p)^{n-k_1-\cdots-k_n}\, dp$$

It follows from Exercise 21c, Chapter 10, that $A(u, q)$ is minimized by choosing

$$u(k_1, \ldots, k_n) = \frac{\int_0^1 p[p^k(1-p)^{n-k}]\, dp}{\int_0^1 p^k(1-p)^{n-k}\, dp}$$

$$= \frac{\Gamma(k+2)\Gamma(n-k+1)\Gamma(n+2)}{\Gamma(k+1)\Gamma(n-k+1)\Gamma(n+3)} = \frac{k+1}{n+2}$$

where $k = k_1 + \cdots + k_n$. Hence,

$$u(X) = \frac{X_1 + \cdots + X_n + 1}{n + 2}$$

is the Bayes' estimator with respect to the uniform a priori weight function. It is easy to see that

$$R(u, p) = \sum_{k_1, \ldots, k_n} [u(k_1, \ldots, k_n) - p]^2 p^{k_1 + \cdots + k_n} (1 - p)^{n - k_1 - \cdots - k_n}$$

is a continuous function of p (in fact, a polynomial in p). Hence, the conditions of the above theorem are satisfied, and u is admissible.

Example 5 Consider the same situation as in Example 4, except now the a priori weight function is the beta density,

$$q(p) = \begin{cases} \dfrac{\Gamma(\alpha + \beta)}{\Gamma(\alpha)\Gamma(\beta)} p^{\alpha - 1}(1 - p)^{\beta - 1} & \text{if } 0 < p < 1 \\ 0 & \text{otherwise} \end{cases}$$

($\alpha > 0, \beta > 0$). Following the same argument as in Example 4, we find the Bayes' estimator to be

$$u(X) = \frac{X_1 + \cdots + X_n + \alpha}{n + \alpha + \beta}$$

(See Exercises 6 and 7.)

We should consider now the general evaluation of a Bayes' estimator. For concreteness, suppose that X has an integrating density function and the a priori weight function $q(\theta)$ is also given by an integrating density function. The function to be minimized with respect to estimators u is

$$A(u, q) = \int_a^b R(u, \theta) q(\theta)\, d\theta = \int \left[\int_a^b L(u(x), \theta) f(x, \theta) q(\theta)\, d\theta \right] dx \qquad (6)$$

We have here interchanged the order of integration; the outer integral on the right side of (6) is over the set of values of X. Now consider the right side of (6). If for every fixed x, $u(x)$ is chosen so that the expression in brackets is minimized, then the result of the next integration will also be minimized with respect to u. This fact gives the following criterion:

For every x, let $u(x)$ be a number that minimizes

$$\int_a^b L(y, \theta) f(x, \theta) q(\theta)\, d\theta$$

with respect to the dummy variable $y = u(x)$. The function u thus defined is a Bayes' estimator with respect to q.

If X is discrete, then the outer integral on the right side of (6) becomes a sum over the possible values of X. The preceding criterion remains exactly the same, however. If Ω consists of a finite number of labels, $\theta_1, \ldots, \theta_m$, and if the a priori weight function is given by $q = (q_1, \ldots, q_m)$, then in the preceding criterion the integral is replaced by the sum

$$\sum_{i=1}^{m} L(y, \theta_i) f(x, \theta_i)\, q_i$$

Otherwise, there is no change. We leave the verification to the reader. Notice that the determination of the Bayes' estimator in Example 4 follows exactly along the lines of this criterion. (See Exercises 6, 9, and 10 for some special cases.)

The estimators $u(X)$ are sometimes called *point estimators* of $h(\theta)$, since for each value of the random phenomenon X, $u(X)$ is a single number. We have described some possible rigorous approaches to this estimation. These depend on knowing a loss function L that adequately describes the costs of making mistakes. Usually, the likelihood that the experimenter is able to specify such a loss function is very slight. In well-controlled manufacturing situations, one has a chance of at least roughly appraising the costs of making mistakes. In problems of the social sciences such appraisals may be impossible. How does a medical scientist who is trying to estimate the properties of a vaccine appraise the costs of making a mistake? How does he assign a cost to the mistake of injecting human beings with a vaccine that turns out to have unexpected toxic properties?

In Section 12.7 we shall describe a somewhat more pragmatic approach to estimating, the *method of confidence intervals*.

12.6 BAYES' ESTIMATORS AND BAYESIANS

There is a way of thinking about estimators which is inspired by the theory of games (Note 5). We can suppose that the experimenter and an opponent, whom we call "nature," are playing a game against each other. Each of the players chooses a strategy.

Once the strategies have been chosen, there is a predetermined outcome, as in a Greek tragedy. Nature's strategy consists of presenting a "state of nature" in the form of a distribution θ. More generally, nature presents a weighted combination of strategies, using the a priori weight function q. In other words, nature does a preliminary experiment, according to which she selects a θ out of the available collection Ω according to her probability distribution q. On the other hand, the experimenter's strategy consists of the choice of an estimator u. The result of this pair of choices is the *payoff*, $A(u, q)$. This single number measures the loss to the experimenter as the result of the particular choice of strategies. If we take the uncharitable view that the experimenter's loss is nature's gain, then we can interpret $A(u, q)$ as the gain to nature resulting from the choice of strategies.

If the experimenter knew how nature were going to play—that is, if he knew q—then he should choose his strategy in such a way as to minimize his loss. That is, for fixed q he should choose u that makes $A(u, q)$ as small as possible. In Section 12.5, we called such an estimator *Bayes with respect to q*. An inadmissible estimator is a foolish strategy for the experimenter, because another strategy is at least as good and sometimes better, no matter what nature's choice of θ might be. What we showed in Section 12.5 (at least under certain conditions) is that Bayes' estimators are admissible. In the preceding sections, we have barely alluded to the mathematical theory concerning this approach. The general subject is called *statistical decision theory* (Note 6).

The preceding outlined approach can be looked at as a kind of motivating artifice for understanding the nature of admissible estimators. It becomes especially important in light of the fact that there is a converse to the theorem that Bayes' estimators are admissible. In certain senses, admissible procedures must be Bayes.

Among scientists who use or formulate statistical theory, there are many who take the preceding description seriously to the extent that they insist it is important always to use an a priori weight function, q. Some would insist that by a proper appraisal of the world we live in one can always come up with an appropriate a priori weight function. The proponents of these points of view are called *Bayesians* (Note 7). Whether or not one is a Bayesian, there are situations where a prior distribution arises in a very natural way (Note 8).

12.7 CONFIDENCE INTERVALS

The method of estimation described so far is that of *point estimation*. The number $h(\theta)$ is estimated by a single number, $u(X)$. There is another widely used approach to estimation that uses a pair of numbers $u_1(X)$ and $u_2(X)$. It is the method of *confidence intervals.* The assertion is made by the experimenter that $h(\theta)$ lies between $u_1(X)$ and $u_2(X)$. Associated with this assertion is the probability of its telling the truth, called the *confidence level.*

confidence interval

Suppose that it is possible to find two estimators $u_1(X)$ and $u_2(X)$ having the property that

$$P_\theta[u_1(X) \leq h(\theta) \leq u_2(X)] = 1 - \alpha \quad \text{for every } \theta \text{ in } \Omega \tag{1}$$

Then $[u_1(X), u_2(X)]$ is called a confidence interval estimate of $h(\theta)$, with confidence level $1 - \alpha$.

The notation P_θ means that the probability is evaluated using the distribution with the label θ. Equation (1) asserts that the random interval $[u_1(X), u_s(X)]$ covers the true value $h(\theta)$ with probability $1 - \alpha$ when the true label of the distribution is θ. Equivalently, α is the probability that the random interval fails to tell the truth in its assertion about $h(\theta)$. Typically, α, the probability of being wrong, is close to 0, and $1 - \alpha$ is called the *confidence level,* or *confidence coefficient.*

Example 1

Suppose that $X_1, \ldots, X_n$ are independent random variables, each normally distributed, $N(\theta, 1)$. The experimental situation is that θ, the mean of the population, is not known, whereas the population standard deviation is known and equals 1. Let $\overline{X} = (X_1 + \cdots + X_n)/n$ be the sample mean. It follows from Section 9.8 that the random variable

$$\sqrt{n}(\overline{X} - \theta)$$

has the $N(0, 1)$ distribution. If f is the $N(0, 1)$ density function, suppose that c is a positive number such that

$$\int_{-c}^{c} f(x)\, dx = 1 - \alpha$$

where α is a positive number close to 0 (see Figure 12.4). No matter what the true θ might be, the probability that $n^{1/2}(X - \theta)$ lies between $-c$ and c is $1 - \alpha$. That is,

$$P_\theta[-c \leq \sqrt{n}(\overline{X} - \theta) \leq c] = \int_{-c}^{c} f(x)\, dx = 1 - \alpha \tag{2}$$

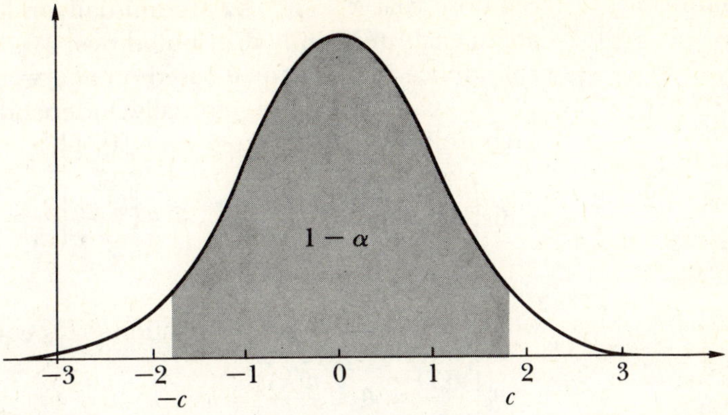

FIGURE 12.4 **$(-c, c)$ contains $1-\alpha$ probability of the standard normal distribution.**

But (2) is equivalent to

$$P_\theta\left(\bar{X} - \frac{c}{\sqrt{n}} \leq \theta \leq \bar{X} + \frac{c}{\sqrt{n}}\right) = 1 - \alpha$$

Thus, to estimate θ, the experimenter uses the confidence interval

$$\left(\bar{X} - \frac{c}{\sqrt{n}}, \bar{X} + \frac{c}{\sqrt{n}}\right) \tag{3}$$

No matter what the true θ might be, the probability that the random interval (3) actually covers θ is $1 - \alpha$.

Example 2 Consider a numerical version of the situation described in Example 1. Five observations on a population described by the $N(\theta, 1)$ distribution are

6.23, 2.94, 5.01, 3.86, 5.78

To construct a 0.95 confidence level estimate, we determine from Table 9.2 that $c = 1.96$ for $\alpha = 0.05$. Here, $\bar{X} = 4.76$, and the confidence interval is

$$\left[4.76 - \frac{1.96}{\sqrt{5}}, 4.76 + \frac{1.96}{\sqrt{5}}\right] = [3.88, 5.64]$$

A confidence interval with a greater probability of telling the truth is necessarily wider. If the probability of telling the truth is smaller, the interval is narrower. What is the confidence interval for $\alpha =$ 0.10?

Example 3 Suppose that $X_1, \ldots, X_n$ are mutually independent random variables, each uniformly distributed over $[0, \theta]$. We shall construct a confidence interval for θ based on $u(X) = \max(X_1, \ldots, X_n)$. Since $(X_1/\theta), \ldots, (X_n/\theta)$ are mutually independent random variables, each uniformly distributed over $[0, 1]$,

$$P_\theta\left[a \leqslant \max\left(\frac{X_1}{\theta}, \cdots, \frac{X_n}{\theta}\right) \leqslant b\right] = P_\theta\left[a \leqslant \frac{u(X)}{\theta} \leqslant b\right] = b^n - a^n \tag{4}$$

whenever $0 \leqslant a \leqslant b \leqslant 1$. Relation (4) is equivalent to

$$P_\theta\left[\frac{u(X)}{b} \leqslant \theta \leqslant \frac{u(X)}{a}\right] = b^n - a^n$$

Hence, if $b^n - a^n = 1 - \alpha$, then

$$\left[\frac{u(X)}{b}, \frac{u(X)}{a}\right]$$

is a $1 - \alpha$ level confidence interval for estimating θ. For fixed α, one should choose a and b to make this interval as narrow as possible. This fact requires minimizing $(1/a) - (1/b)$ subject to $b^n - a^n$ being fixed. It is easy to check that this minimization takes place with $b = 1$. (Verify this!) Hence, finally, a reasonable confidence interval procedure is the following:

1. Determine a so that $1 - a^n = 1 - \alpha$; in other words, $a = \alpha^{1/n}$.
2. Assert that θ lies between $u(X)$ and $u(X)/a$.
3. The probability of telling the truth is $1 - \alpha$, no matter what the true θ might be.

Notice that for fixed α, the width of the confidence interval is $\alpha^{-1/n} - 1$. The larger that n is, the narrower is the confidence interval.

Example 4 A chemist has reason to believe that a certain reaction time is uniformly distributed over some interval $[0, \theta]$ where θ is unknown to him. In repeating the experiment five times, he observes the following reaction times, in seconds:

0.21, 0.35, 0.17, 0.28, 0.30

According to Example 3, a 0.90 confidence interval estimate for θ is that it lies in

$$\left[0.35, \frac{0.35}{(0.10)^{1/5}}\right] = [0.35, 0.55]$$

Some additional examples of confidence intervals are given in Exercises 13 and 14 and in Chapter 15.

SUMMARY

An *estimator* is $u(X)$, a function of the observations, which does not involve the unknown distribution label or parameter, θ. An estimator is intended to be "close" to $h(\theta)$ in some sense for all θ, where $h(\theta)$ is the function of the parameter to be estimated. An unbiased estimator has the property that $E_\theta[u(X)] = h(\theta)$, identically for all parameters θ. If $f(x, \theta)$ is a density function (either discrete or integrating) for the observations X, then a *maximum likelihood* estimator for $h(\theta)$ is defined to be $h(\hat{\theta})$, where $\hat{\theta} = \hat{\theta}(X)$ is a value θ that makes $f(X, \theta)$ as large as possible. A sequence of estimators is called *consistent* if, as the number of observations increases, the distributions of the estimators become highly concentrated about $h(\theta)$, whatever the true θ might be.

If θ is the true value of the parameter, and $u(X)$ the value of the estimator, then one can sometimes assign a *loss* $L[u(X), \theta]$ to this combination. We call L a *loss function.* The expected value $E_\theta\{L[u(X), \theta]\} = R(u, \theta)$ is the *expected loss,* which depends on the true "state of nature" θ, and on the particular estimator u being used. Ideally, the experimenter would like an estimator u that makes $R(u, \theta)$ as small as possible, uniformly for all θ. Unfortunately, this situation is rarely possible; a given estimator may be good for some θ's and bad for others. The first thing one can do is to exclude estimators that can be uniformly improved upon by some other estimator. Such an excluded estimator is called *inadmissible.* If it is not inadmissible, it is called *admissible.*

Sometimes an experimenter may be able to attach a priori weights, $q(\theta)$, to the various possible distribution labels θ. These weights represent his relative concern about the various possible states of nature or appraisal of how likely the various alternatives may be. In such a case, it makes sense to consider the average of the $R(u, \theta)$ values over all θ's using the a priori weights $q(\theta)$. This average is denoted as $A(u, q)$. A *Bayes' estimator* with respect to q is an estimator u that makes $A(u, q)$ as small as possible.

Contrasted with this rather abstract approach to estimating, there is a pragmatic approach called the method of *confidence intervals.* Suppose one can determine two estimators u_1, u_2, such that the random interval $[u_1(X), u_2(X)]$ contains $h(\theta)$, no matter what the true distribution of X might be, with probability $1 - \alpha$. This random interval is called a confidence interval estimate for $h(\theta)$, with *confidence level* $1 - \alpha$.

EXERCISES

1 (a) Suppose that $X_1, \ldots, X_n$ are mutually independent, each uniformly distributed over $[0, \theta]$. Show that $Y = \max(X_1, \ldots, X_n)$ is the maximum likelihood estimator of θ. (Notice that this estimator is not unbiased.)
(b) Show that $[(n+\alpha)/n]\, Y^\alpha$ is an unbiased estimator of θ^α (α is a positive constant).
(c) What is the maximum likelihood estimator of θ^α?
(d) Verify the variance calculation in Example 2, Section 12.3.

2 Suppose that $X_1, \ldots, X_n$ are mutually independent with common expected value, m, and variance σ^2. Show that $\Sigma_{i=1}^n (X_i - \overline{X})^2/(n-1)$ is an unbiased estimator of σ^2 if $n > 1$. [Hint: Use the facts that $\operatorname{Var}(\overline{X}) = \sigma^2/n$ and

$$\sum_{i=1}^{n} (X_i - \overline{X})^2 = \sum_{i=1}^{n} (X_i - m)^2 - n(\overline{X} - m)^2$$

where $\overline{X} = (X_1 + \cdots + X_n)/n$.]

3 Let $X_1, \ldots, X_n$ be mutually independent, indicator random variables with $P(X_i = 1) = p$. The unknown parameter p lies in $[0, 1/2]$. Show that the maximum likelihood estimator of p is given by (4), Example 6, Section 12.2. [Hint: Consider the graph of $p^k(1-p)^{n-k}$ in $[0, 1/2]$ carefully. The maximum may be achieved at an endpoint where the derivative is not zero.]

4 Suppose that $X_1, \ldots, X_n$ are mutually independent random variables with common expected value m and common variants σ^2.
(a) Show that $a_1X_1 + \cdots + a_nX_n$ is an unbiased estimator of m whenever $a_1 + \cdots + a_n = 1$.
(b) Show that in the preceding class of unbiased estimators, the variance is minimized by choosing $a_1 = \cdots = a_n = 1/n$. (Hint: Use the fact that if $a_1 + \cdots + a_n = 1$, $\Sigma_{i=1}^n a_i^2 = \Sigma_{i=1}^n (a_i - 1/n)^2 + 1/n$.)

5 (a) Suppose that $X_1, \ldots, X_n$ are mutually independent random variables, each exponentially distributed with the same parameter λ. Let

$$Y = \min(X_1, \ldots, X_n), \qquad S = X_1 + \cdots + X_n$$

Now, nY and S/n are each unbiased estimators of $1/\lambda$. Using variance as a criterion of goodness, which is the better estimator? [Answer: $\operatorname{Var}(nY) = 1/\lambda^2$, $\operatorname{Var}(S/n) = 1/n\lambda^2$.]
(b) Show that the maximum likelihood estimator of $1/\lambda$ is S/n.

6 Suppose that Ω is an interval (a, b), and that in estimating $h(\theta)$ by $u(X)$, the expected loss is

$$R(u, \theta) = E_\theta[u(X) - h(\theta)]^2$$

Suppose also that the a priori weight function q is an integrating density function over (a, b). Thus, the average expected loss is

$$A(u, q) = \int_a^b R(u, \theta)q(\theta)\, d\theta$$

Show that a Bayes' estimator u is the following:

$$u(X) = \frac{\int_a^b h(\theta) f(X, \theta)q(\theta)\, d\theta}{\int_a^b f(X, \theta)q(\theta)\, d\theta}$$

{Hint: Imitate Example 4, Section 12.5. For instance, if X is discrete, then, interchanging the order of summation and integration,

$$A(u, q) = \sum_x \int_a^b [u(x) - h(\theta)]^2 q(\theta) f(x, \theta)$$

Now use Exercise 21c, Chapter 10. If X is not discrete, the proof is practically the same.}

7 Consider the same situation as in Example 4, Section 12.5, but now determine the Bayes' estimator of p^2. [Answer:

$$\frac{(X + 2)(X + 1)}{(n + 3)(n + 2)} \qquad \text{where } X = X_1 + \cdots + X_n$$

Use Exercise 6.]

8 (a) Suppose that $X_1, \ldots, X_n$ are mutually independent random variables, each Poisson distributed with parameter λ. Suppose that in estimating λ the expected loss is $E_\lambda\{[u(X) - \lambda]^2\}$ and that an a priori weight function over $\Omega = (0, \infty)$ is $q(\lambda) = e^{-\lambda}$. Show that the Bayes' estimator of λ with respect to q is $(X_1 + \cdots + X_n + 1)/(n + 1)$. (Hint: Use Exercise 6.)

(b) Consider the same situation as in (a), except now the function to be estimated is $h(\lambda) = e^{-\lambda}$. Show that the Bayes' estimator of $e^{-\lambda}$ with respect to q is

$$u(X_1, \ldots, X_n) = \left(\frac{n + 1}{n + 2}\right)^{X_1 + \cdots + X_n + 1}$$

(Hint: Use Exercise 6.)

(c) Show that the expected value of the estimator in (b) is

$$\frac{n + 1}{n + 2} \exp -\lambda \frac{n}{n + 2}$$

For large n, this value is close to $e^{-\lambda}$.

9 (a) Repeat Exercise 6 supposing that Ω is a discrete set $(\theta_1, \theta_2, \ldots)$. Show that a Bayes' estimator is the following:

$$u(X) = \frac{\sum_i h(\theta_i) f(X, \theta_i) q(\theta_i)}{\sum_i f(X, \theta_i) q(\theta_i)}$$

(Hint: Use Exercise 21c, Chapter 10.)

(b) Apply (a) to verify the results in Example 3, Section 12.5.

10 (a) Repeat Exercises 6 and 9, now supposing that the expected loss is

$$R(u, \theta) = E_\theta[|u(X) - \theta)|]$$

(We suppose for simplicity that θ is a real number and is the function to be estimated.) If Ω is an interval (a, b) and q is an integrating density over that interval, show that a Bayes' estimator is for every value of X a *median* of the distribution the density of which is

$$h_X(\theta) = \frac{f(X, \theta) q(\theta)}{\int_a^b f(X, \theta) q(\theta)\, d\theta}$$

If Ω is discrete, the result is the same with $\int$ replaced by Σ. (Hint: Use Exercise 21, parts (d), (e), and (f), Chapter 10.)

(b) Apply (a) to verify the results in Example 2, Section 12.5.

11 Suppose that $X_1, \ldots, X_n$ are mutually independent random variables, each exponentially distributed with parameter λ. Suppose that in estimating λ the expected loss is $E_\lambda\{[u(X) - \lambda]^2\}$ and that an a priori weight function over $\Omega = (0, \infty)$ is $q(\lambda) = e^{-\lambda}$. Show that the Bayes' estimator of λ with respect to q is $(n+1)/(X_1 + \cdots + X_n + 1)$. Why is it plausible that the answer here should be the reciprocal of that in Exercise 8a?

12 (a) Verify the Bayes' estimator asserted in Example 3, Section 12.5, using Exercise 6. [For instance, using Table 12.1,

$$u(0) = \frac{0(1) + 1(3/5) + 2(3/10) + 3(1/10)}{1 + 3/5 + 3/10 + 1/10} = 3/4]$$

(b) Verify the Bayes' estimator in Example 1, Section 12.5, using Exercise 10a. {For instance, by Table 12.1, $u(0)$ is a median of the discrete distribution, the probabilities of which are

$$\frac{1}{K}, \frac{3}{5K}, \frac{3}{10K}, \frac{1}{10K}$$

where $K = 1 + 3/5 + 3/10 + 1/10$. Explicitly, these probabilities are 10/20, 6/20, 3/20, 1/20 assigned to θ values of 0, 1, 2, 3, respectively. Median $u(0)$ is any number in $[0, 1]$, since any such number is a median of the above distribution.}

(c) Verify the Bayes' estimator in Example 2, Section 12.5, again using Exercise 10a.

13 Suppose that $X_1, \ldots, X_n$ are mutually independent random variables each $N(0, \sigma^2)$. Show that

$$\left(\sum_{i=1}^{n} X_i^2\right)\Big/a, \quad \left(\sum_{i=1}^{n} X_i^2\right)\Big/b$$

is a confidence interval covering the true value of σ^2 with probability $1-\alpha$ if (a, b) contains $1-\alpha$ probability of the chi-square distribution with n degrees of freedom. [Hint: Use the fact that $(X_1^2+\cdots+X_n^2)/\sigma^2$ is chi-square distributed with n degrees of freedom.]

14 Suppose that $X_1, \ldots, X_n$ are mutually independent random variables, each exponentially distributed with parameter λ. The only measurement available to the experimenter is $Y = \min(X_1, \ldots, X_n)$

(a) Show that

$$\left(\frac{a}{nY}, \frac{b}{nY}\right)$$

is a confidence interval covering the true value of λ with probability $e^{-a}-e^{-b}$ $(0 \leq a < b)$. (Hint: $n\lambda Y$ is exponentially distributed with parameter 1.)

(b) If $e^{-a}-e^{-b}=1-\alpha$ is fixed, how should a and b be chosen to minimize the width of the confidence of the interval?

15 (a) Suppose that $X_1, \ldots, X_n$ are mutually independent, indicator random variables with $P(X_i=1)=p$. For any estimator $u(X_1, \ldots, X_n)$, show that $E_p[u(X_1, \ldots, X_n)]$ is a polynomial in p. [Hint: The required expected value is given by

$$\sum_{k_1,\ldots,k_n} u(k_1, \ldots, k_n)p^k(1-p)^{n-k}$$

where $k = k_1 + \cdots + k_n$. First sum over all combinations of $k_1, \ldots, k_n$, which sum to a given k, and then sum over $k = 0, 1, \ldots, n$. This latter gives $E_p[u(X_1, \ldots, X_n)] = \sum_{k=0}^{n} c_k p^k(1-p)^{n-k}$, where the c_k's are appropriate constants. Clearly, the latter sum is a polynomial in p.]

(b) Use (a) to prove the assertion in Example 5, Section 12.2.

(c) Suppose that $g(p)$ is a polynomial in p of degree no greater than n. Show that there is one and only one unbiased estimator of $g(p)$ based on $X_1+\cdots+X_n$.

NOTES

1 For some additional reading on the theory of estimation, see Chapters 4 and 5 of [50] and Chapters 8 and 11 of [51].

2 Statistical problems relating to dichotomous populations are treated in greater detail in Chapter 15.

3 The method of maximum likelihood is essentially the work of English geneticist and statistician R. A. Fisher (1890–1963). See [52].

4 For additional discussion of Bayes' estimators, see Section 8.13, [51], and Section 5.22, [50].

5 The theory of games was developed by the mathematician John von Neumann (1903–1957) and the economist O. Morgenstern (1902–). See [53] and [54].

6 Statistical decision theory was developed by Abraham Wald (1902–1950), who was one of the all-time outstanding contributors to the theory of statistics. The approach in his book, [55], has since been somewhat simplified. See [56] and [57].

7 For a formulation of the Bayesian point of view, see [58] and [59].

8 See [60] for a description of some circumstances where one does not have to be a Bayesian to consider the real existence of a priori distributions.

9 For some additional reading on confidence intervals, see Chapter 11 of [51].

Chapter Thirteen

TESTING HYPOTHESES

Suppose the collection of distribution labels is partitioned into two parts, Ω_1 and Ω_2. On the basis of observing X, the experimenter wants to decide as to which of these two sets θ actually belongs. We discuss *nonrandomized procedures,* in which the experimenter makes a definite decision on the basis of X. These are extended to *randomized procedures* in which the experimenter may use an additional chance device before he makes his final decision. The *power function* is a tool for examining the effectiveness of a decision procedure; it measures the probability of making one of the decisions as a function of the true parameter θ. *Admissible procedures* are those that cannot be uniformly improved on. As in Chapter 12, an *a priori weight function* for the parameters is a weight function for averaging the errors of a decision procedure over all θ in Ω. A *Bayes' procedure* is a procedure the averaged error of which is as small as possible. As in Chapter 12, Bayes' procedures are admissible under certain conditions. Finally, we consider the rather special case that Ω_1 and Ω_2 each consist of a single θ. An explicit description of the Bayes procedures in this case is given by the *Neyman–Pearson theorem* (Notes 1, 2).

13.1 DECIDING AMONG ALTERNATIVES

In Chapter 12 we considered the estimation approach to statistical decision making. We now want to describe a somewhat different approach to the experimenter's dilemma of making decisions in the face of uncertainty.

Suppose that the experimenter is observing a random variable or random vector X, which, as in Chapter 12, has a density function (either discrete or integrating) $f(x, \theta)$. As before, θ is an identifying label, or parameter, that lies in a known class Ω. The experimenter is faced with the need of making one of a number of possible decisions. The correctness of his decision depends on θ, the true "state of nature." If he knew θ, he would not have to bother experimenting and observing X. He would make that decision which is best in the light of his knowledge that θ prevails in his world. But he does not know θ; he can only obtain clues by observing a random phenomenon X the particular distribution of which is determined by θ.

Example 1

A drug manufacturer is producing a new antibiotic on an experimental basis. He treats 500 persons with the experimental drug. Let X be the number of these persons that are helped by the drug. It is reasonable to suppose that X is binomially distributed with parameters 500 and p; p plays the role of θ and is the probability of a single subject being helped. The manufacturer's object is to obtain information about p. If p were greater than 0.7, he would produce the drug on a commercial basis. If p were less than 0.4, he would abandon the project altogether. If p were between 0.4 and 0.7, he would continue experimentation in the hope of improving the quality of the drug. His decision cannot be based on a sure knowledge of p, but only on inferences about p obtained from X. This is a three-decision problem, since there are three alternatives to choose among.

Throughout the remainder of this chapter, we suppose that there are only two available decisions for the experimenter. The formal description of the situation follows.

two decision problem

A random variable or vector X has a density $f(x, \theta)$; θ varies over Ω, and Ω is partitioned into two disjoint parts, Ω_1 and Ω_2. There are two decisions facing the experimenter. These can be described as follows:

decision 1: decide θ lies in Ω_1

decision 2: decide θ lies in Ω_2

(See Figure 13.1.)

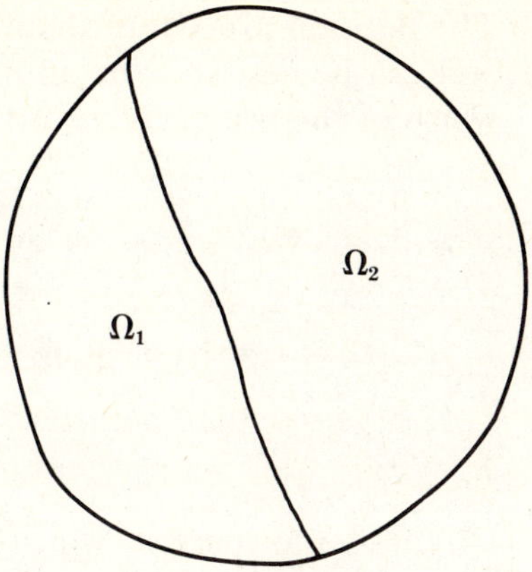

FIGURE 13.1 If θ is in Ω_1, then decision 1 is desired. If θ is in Ω_2, then decision 2 is desired.

Example 2 A space vehicle is scheduled to send a crucial signal from the vicinity of Venus. The message to be transmitted is either a 0 or a 1 according to certain local conditions on the planet. It is known that the signal received is incorrect with probability 0.05. That is, if 0 is transmitted, 1 is received 5% of the time, and if 1 is transmitted, 0 is received 5% of the time. To counteract this difficulty, a sequence of 100 repetitions of the signal is to be sent and it is assumed that errors occur independently within the sequence. The mathematical model is that $X = (X_1, X_2, \ldots, X_{100})$, where the X_i's are independent and identically distributed indicator random variables. Here Ω consists of two parameters, θ_0 and θ_1. If θ_0 is the correct parameter, then

$$P(X_i = 0) = 0.95, \qquad P(X_i = 1) = 0.05$$

If θ_1 is the correct parameter, then

$$P(X_i = 0) = 0.05, \qquad P(X_i = 1) = 0.95$$

The receiving station must decide whether the state observed on Venus is 0 or 1 according to the nature of the observed random vector X.

13.2 DECISION FUNCTIONS

We now want to describe some procedures according to which the experimenter can make a decision in the two-decision problem.

The decision procedure should be based on the X that is observed and can be described by an indicator random variable indicating which of the two decisions is to be made.

decision procedure

A decision procedure is an indicator random variable $\phi(X)$, *where the function ϕ does not depend on θ.*

If $\phi(X) = 1$, then decision 1 is made.

If $\phi(X) = 0$, then decision 2 is made.

In effect, the space in which X assumes its values is partitioned into two sets, A and its complement A' (see Figure 13.2). If the observed value of X lies in A, then the decision is 1; otherwise, it is 2.

To appraise the quality of a decision procedure, we should consider the probabilities of making each of the two decisions as functions of the parameter θ. It is sufficient to consider only the probability of making decision 1, since

$$P_\theta(\text{making decision 1}) + P_\theta(\text{making decision 2}) = 1$$

power function

$$P_\theta[\phi(X) = 1] = E_\theta[\phi(X)] = P_\theta(\text{making decision 1}) = \psi(\theta) \quad (1)$$

is called the power function of the decision procedure ϕ.

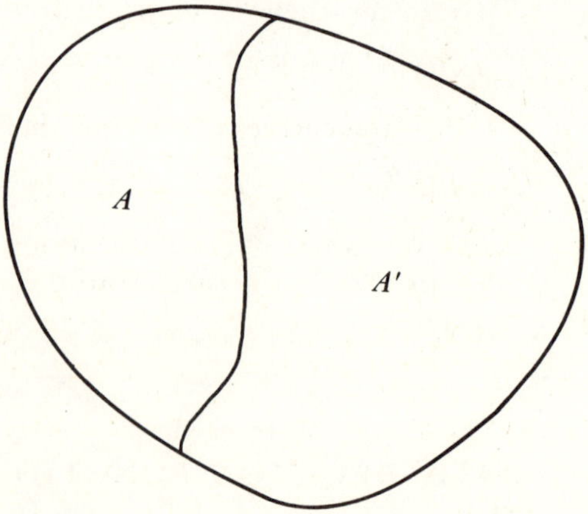

FIGURE 13.2 If the observed value of X is in A, the experimenter will make decision 1; otherwise, he will make decision 2.

Here, P_θ denotes a probability evaluation using the distribution labeled θ, and, just as in Chapter 12, E_θ denotes expected value using that distribution. The ideal power function is shown schematically in Figure 13.3a. Whenever θ is in Ω_1, one would like $\psi(\theta)$ to equal 1, and whenever θ is in Ω_2, one would like $\psi(\theta)$ to be 0. This situation would mean that exactly the right decisions are always being made. Unfortunately, since X is only a clue to the right distribution, some errors are usually inevitable, and the power function is more likely to look like the one shown in Figure 13.3b. The probabilities of making errors are the following:

$$1 - \psi(\theta) = \text{probability of error when } \theta \text{ is in } \Omega_1$$

$$\psi(\theta) = \text{probability of error when } \theta \text{ is in } \Omega_2$$

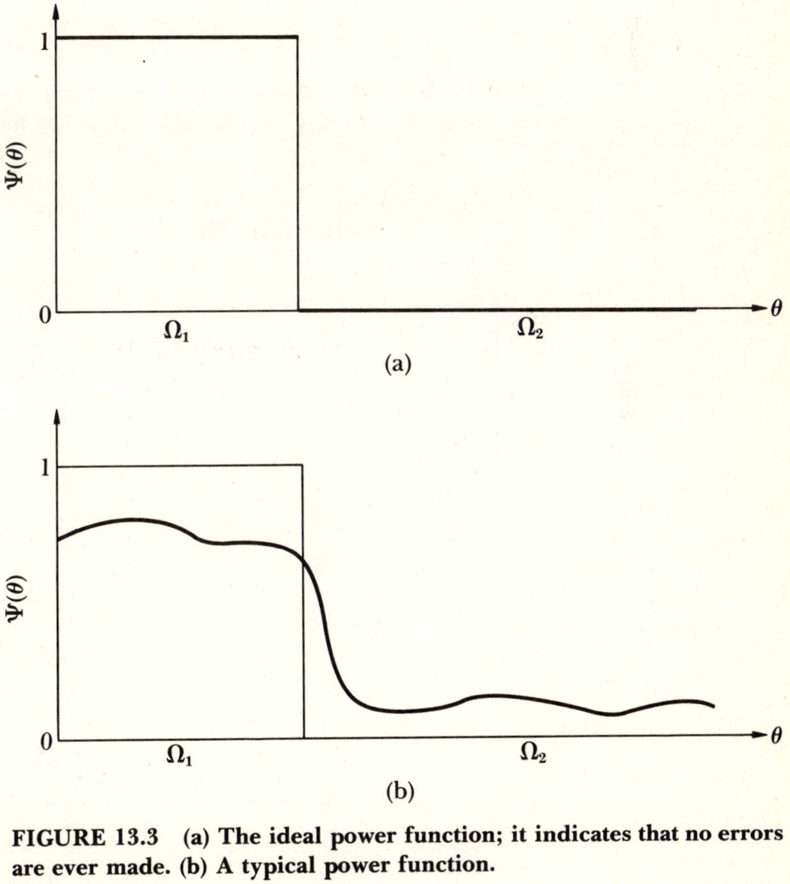

FIGURE 13.3 **(a) The ideal power function; it indicates that no errors are ever made. (b) A typical power function.**

Example 1 Suppose that $X_1, \ldots, X_n$ are mutually independent random variables, each $N(\theta, 1)$. The experimenter wants to decide whether the unknown parameter θ lies in $\Omega_1 = (-\infty, 0)$ or in $\Omega_2 = [0, \infty)$. A reasonable procedure is to base the decision on the sample mean $\overline{X} = (X + \cdots + X_n)/n$. If $\overline{X}$ is in $(-\infty, 0)$, make decision 1; otherwise, make decision 2. The decision function is then

$$\phi(X_1, \ldots, X_n) = \begin{cases} 1 & \text{if } \overline{X} < 0 \\ 0 & \text{if } \overline{X} \geq 0 \end{cases}$$

To compute the power function, we use the fact that $\sqrt{n}(\overline{X} - \theta)$ has the $N(0, 1)$ distribution. Let f denote the density function and F the cdf of the $N(0, 1)$ distribution. Then the power function is

$$\psi(\theta) = P_\theta(\overline{X} < 0) = P_\theta[\sqrt{n}(\overline{X} - \theta) < -\sqrt{n}\theta] = \int_{-\infty}^{-\sqrt{n}\theta} f(x)\, dx$$
$$= F(-\sqrt{n}\theta)$$

Figure 13.4 shows graphs of the power function for several different sample sizes n. Consider the following limits:

$$\lim_{n\to\infty} F(-\sqrt{n}\theta) = \begin{cases} 1 & \text{if } \theta < 0 \\ \frac{1}{2} & \text{if } \theta = 0 \\ 0 & \text{if } \theta > 0 \end{cases}$$

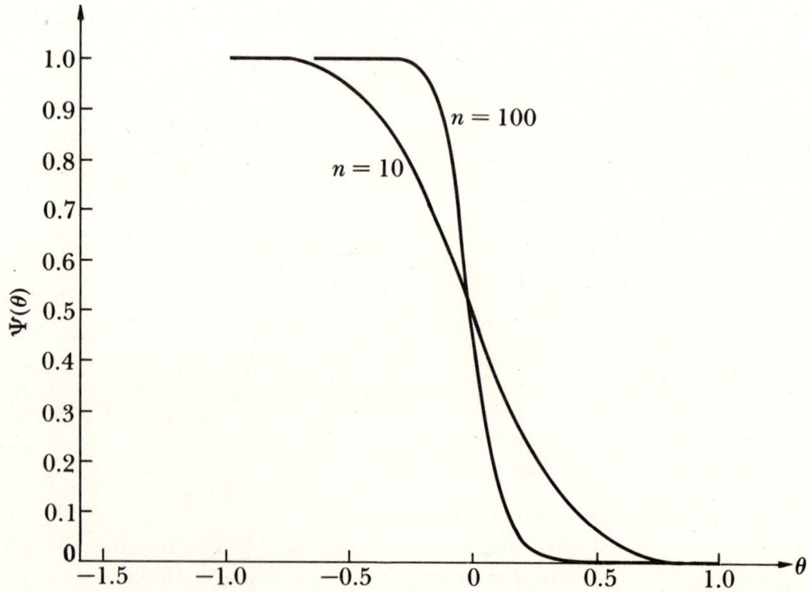

FIGURE 13.4 Power functions for Example 1, Section 13.2, for $n = 10$ and $n = 100$.

Thus, as the sample size grows to infinity, the power function approaches the ideal power function, except at the one difficult point, $\theta = 0$. In determining an adequate sample size, The experimenter should look at the various power functions and decide on one that shows errors he can tolerate. Since additional observations may be costly, he must somehow strike a balance between his desires to minimize both errors and costs of observations.

Example 2 Observations are made on two populations to determine whether their population means are the same. The mathematical model is that $X_1, \ldots, X_n, Y_1, \ldots, Y_n$ are mutually independent, normally distributed random variables, where the X_i's are each $N(\theta_1, 1)$ and the Y_i's are each $N(\theta_2, 1)$. The unknown parameter is $\theta = (\theta_1, \theta_2)$ and Ω is the whole plane. The two possible decisions are the following:

decision 1: θ lies in Ω_1, the subset of the plane where $\theta_1 = \theta_2$

decision 2: θ lies in Ω_2, the subset of the plane where $\theta_1 \neq \theta_2$

(See Figure 13.5.) Suppose that the decision procedure is based on the sample means,

$$\bar{X} = \frac{X_1 + \cdots + X_n}{n}, \qquad \bar{Y} = \frac{Y_1 + \cdots + Y_n}{n}$$

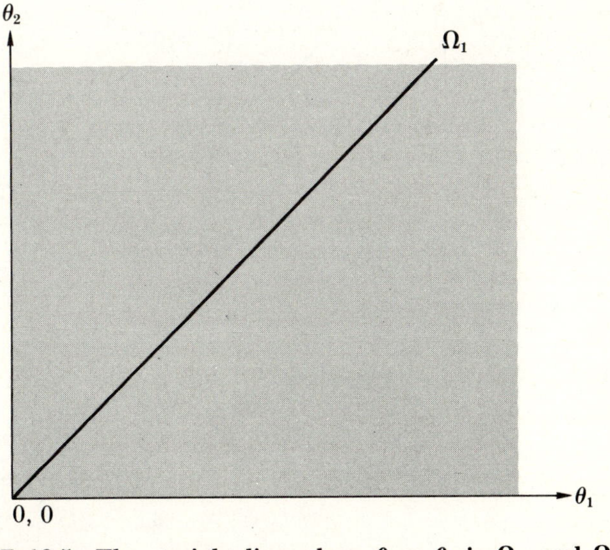

FIGURE 13.5 The straight line where $\theta_1 = \theta_2$ is Ω_1, and Ω_2 is the complement of this set where $\theta_1 \neq \theta_2$.

Specifically, suppose that c is a given constant and we make decision 1 if $|\bar{X}-\bar{Y}| \leq c$. Otherwise—that is, when $|\bar{X}-\bar{Y}|>c$—we make decision 2. Since $(n/2)^{1/2}[\bar{X}-\bar{Y}-(\theta_1-\theta_2)]$ has the $N(0, 1)$ distribution, we can evaluate the power function as follows:

$$\psi(\theta)=\psi(\theta_1, \theta_2)=P_\theta(|\bar{X}-\bar{Y}| \leq c)=\int_{a(n)}^{b(n)} f(x)\, dx$$

$$=F[b(n)]-F[a(n)]$$

where f is the density and F the cdf of the $N(0, 1)$ distribution and

$$b(n)=\left(\frac{n}{2}\right)^{1/2}[c-(\theta_1-\theta_2)], \qquad a(n)=\left(\frac{n}{2}\right)^{1/2}[-c-(\theta_1-\theta_2)]$$

If we use the fact that $F(u)=1-F(-u)$, because of the symmetry of the $N(0, 1)$ density function, it follows that $\psi(\theta_1, \theta_2)=\psi(-\theta_1, -\theta_2)$. We also see that the power function depends on (θ_1, θ_2) only through the value of $|\theta_1-\theta_2|$. Several typical power functions are shown in Figure 13.6.

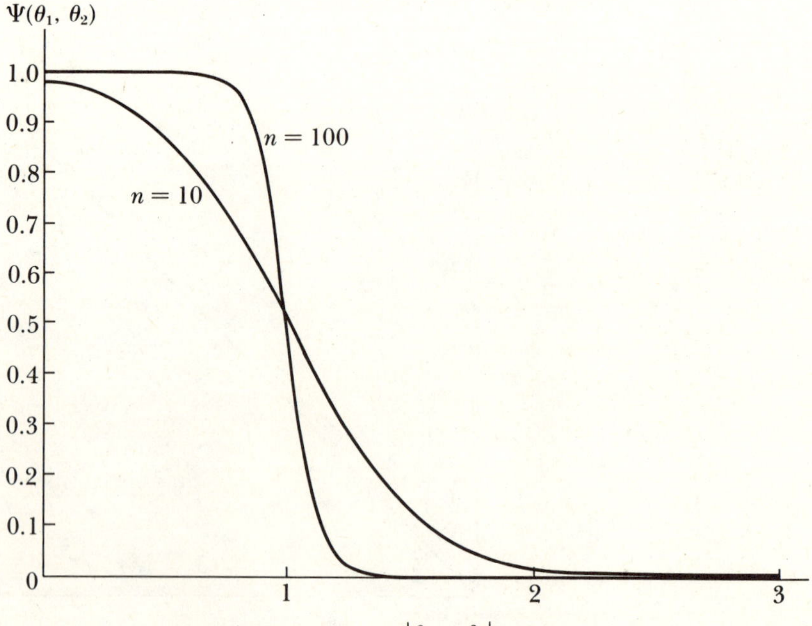

FIGURE 13.6 Power functions for Example 2, Section 13.2, for $n=10$ and $n=100$; $c=1$.

13.3 RANDOMIZED DECISION PROCEDURES

It turns out to be helpful to broaden the scope of what we mean by a decision procedure. Recall from Section 13.2 that a decision procedure is a function based on the observations that equals 1 or 0, indicating that decision 1 or decision 2 is to be taken. We will now admit the possibility of using not only the observations but also an additional chance mechanism, called a randomization device. For instance, suppose that for some values of X one makes decision 1, for some values of X one makes decision 2, and for some values of X one uses a table of random numbers to aid in the decision process.

randomized decision procedure

A randomized decision procedure means a function $\phi(X)$ that does not depend on θ and that satisfies the requirement

$$0 \leq \phi(X) \leq 1$$

We interpret as follows: When X is observed by the experimenter, he then makes decision 1 with probability $\phi(X)$ or decision 2 with probability $1-\phi(X)$.

For instance, if for some X, $\phi(X)=1/2$, then an honest coin can be thrown to determine the decision. If for some X, $\phi(X)=0.312$, then a three-digit drawing can be made from a table of random numbers (see Table 9.2). If any of the numbers 001,002, . . . , 312 are drawn, make decision 1; otherwise, make decision 2. If ϕ assumes only the values 0 or 1, then the decision procedure is a nonrandomized procedure of the type described originally in Section 13.2. A randomized decision procedure may ignore the data altogether. For instance, if $\phi(X)$ is identically equal to 1/2 for all X, the experimenter decides by throwing a coin regardless of the value of X. This procedure may seem irrational, but it is a decision procedure.

As with a nonrandomized decision procedure, the probability of making decision 1 is, as a function of θ, the power function. It is no longer equal to $P_\theta[\phi(X)=1]$, but it still equals $E_\theta[\phi(X)]$ and is the probability of making decision 1. [Compare with (1), Section 13.2.]

power function

The *power function* of a randomized decision procedure ϕ is defined to be

$$\psi(\theta)=E_\theta[\phi(X)]=P_\theta(\text{making decision 1}) \qquad (1)$$

Let us elaborate on the fact that $E_\theta[\phi(X)] = P_\theta$ (making decision 1), in the special case where X is discrete. Denote by A the event that decision 1 is made. Then

$$\phi(a) = P_\theta(A \mid X = a)$$

and

$$P_\theta(A) = \sum_a P_\theta(A \mid X = a) P_\theta(X = a) = E_\theta \phi(X)$$

by (5), Section 4.3. In case X, an integrating density, then Σ is replaced by $\int$, but we omit the details.

Example 1 Four boxes are labeled θ_1, θ_2, θ_3, and θ_4. The box labeled θ_i contains i black balls and $5 - i$ white balls, $i = 1, 2, 3, 4$. Three balls are

TABLE 13.1 Decision procedures

X	$\phi_1(X)$	$\phi_2(X)$	$\phi_3(X)$	$\phi_4(X)$	$\phi_5(X)$
0	1	1	1	0	0.8
1	1	1/2	3/4	0	0.8
2	0	1/2	1/2	1	0.8
3	0	0	1/4	1	0.8

Density functions required to compute power functions

k	$P_{\theta_1}(X = k)$	$P_{\theta_2}(X = k)$	$P_{\theta_3}(X = k)$	$P_{\theta_4}(X = k)$
0	0.4	0.1	0	0
1	0.6	0.6	0.3	0
2	0	0.3	0.6	0.6
3	0	0	0.1	0.4

Power functions

θ	θ_1	θ_2	θ_3	θ_4
$E_\theta[\phi_1(X)]$	1.00	0.70	0.30	0
$E_\theta[\phi_2(X)]$	0.70	0.55	0.45	0.30
$E_\theta[\phi_3(X)]$	0.85	0.70	0.55	0.40
$E_\theta[\phi_4(X)]$	0	0.30	0.70	1.00
$E_\theta[\phi_5(X)]$	0.8	0.8	0.8	0.8

drawn without replacement and X is the number of black balls in the sample. The two decisions are the following:

decision 1: decide that θ is in $\Omega_1 = (\theta_1 \text{ or } \theta_2)$

decision 2: decide that θ is in $\Omega_2 = (\theta_3 \text{ or } \theta_4)$

Table 13.1 describes several decision procedures, together with their power functions and the density functions needed to compute the power functions. Their verification is left to the reader.

One of the advantages of working with randomized decision procedures is that it brings ideas concerning *convexity* into the problem. These will be especially easy to visualize in the special case where Ω_1 and Ω_2 each consist of only one θ apiece; this topic is treated in Section 13.6. Meanwhile, there follow several facts about convexity.

convex combinations of decision procedures and their power functions

Suppose that $\phi_1(X)$ and $\phi_2(X)$ are decision procedures and λ is a constant between 0 and 1. Then

$$\phi(X) = \lambda\phi_1(X) + (1 - \lambda)\phi_2(X)$$

which is called a convex combination of ϕ_1 and ϕ_2, is also a decision procedure. Similarly, a convex combination of power functions $\psi_1(\theta)$, $\psi_2(\theta)$ is defined to be

$$\psi(\theta) = \lambda\psi_1(\theta) + (1 - \lambda)\psi_2(\theta)$$

and is also a power function of some decision procedure. In terms of the above notation, if ψ_1 corresponds to ϕ_1 and ψ_2 corresponds to ϕ_2, then ψ corresponds to ϕ.

Proof. We know that $\phi(X)$ is between 0 and 1 because $\phi_1(X)$ and $\phi_2(X)$ have that property. Moreover, ϕ does not depend on θ because ϕ_1 and ϕ_2 do not depend on θ. Since

$$E_\theta[\psi(X)] = \lambda E_\theta[\phi_1(X)] + (1 - \lambda)E_\theta[\phi_2(X)]$$

the second assertion follows. ▮

13.4 ADMISSIBLE PROCEDURES

The idea of admissibility for decision procedures is similar to that of admissibility for estimators, as discussed in Section 12.4. A decision procedure is admissible if it cannot be uniformly improved upon.

inadmissible and admissible procedures

Let ϕ be a decision procedure and ψ its power function. Suppose that there exists another decision procedure ϕ_1 the power function of which, ψ_1, satisfies the following:

$$\psi_1(\theta) \geq \psi(\theta) \quad \text{for all } \theta \text{ in } \Omega_1 \tag{1}$$

$$\psi_1(\theta) \leq \psi(\theta) \quad \text{for all } \theta \text{ in } \Omega_2 \tag{2}$$

and there is strict inequality for at least one θ in Ω. Then ψ is called *inadmissible.* If a decision procedure is not inadmissible, it is called *admissible.*

Inequality (1) means that ϕ_1 is at least as good as ϕ when decision 1 should be made; (2) means that ϕ_1 is at least as good as ϕ when decision 2 should be made. The fact that strict inequality holds for some θ means that ϕ_1 actually surpasses ϕ somewhere. There is no point in ever using ϕ since ϕ_1 will do at least as well and sometimes better.

Example 1

Consider Example 1, Section 13.3. The decision procedure ϕ_4 is inadmissible. It is thrown out of the competition by each of ϕ_1, ϕ_2, and ϕ_3. Of course, ϕ_4 is a silly procedure because it works just contrary to good sense; when few black balls are drawn, it decides in favor of the boxes with the larger numbers of black balls. Similarly, ϕ_2 and ϕ_3 are inadmissible, since they are each improved on by ϕ_1, and ϕ_5 is a procedure that ignores the experimental data altogether. Although none of ϕ_1, ϕ_2, ϕ_3, or ϕ_4 establishes that ϕ_5 is inadmissible, do you think that there is some other procedure that does?

Although we can exclude inadmissible procedures, in general there is not a uniformly best procedure. The most that we can do is to collect all the admissible procedures; selection from this group must be based on some additional criteria.

13.5 BAYES' PROCEDURES

If one can assign a priori weights to the relative importance of the various states of nature θ in Ω, the problem of choosing a decision procedure becomes easier. The ideas here are practically the same as those concerning estimators, as discussed in Section 12.5. As before, we first consider the special case where Ω consists of only a finite number of θ's.

average error

Suppose that Ω consists of a finite number of distribution labels $\theta_1, \ldots \theta_r, \ldots, \theta_m$, with $\Omega_1 = (\theta_1, \ldots, \theta_r)$, $\Omega_2 = (\theta_{r+1}, \ldots, \theta_m)$. Let $(q_1, \ldots, q_m) = q$ be nonnegative weights that sum to 1. If ϕ is a decision procedure and ψ is its power function, then the average error of ϕ with respect to the a priori weights q is defined to be

$$A(\phi, q) = \sum_{i=1}^{r} [1 - \psi(\theta_i)]q_i + \sum_{i=r+1}^{m} \psi(\theta_i)q_i \tag{1}$$

We purposely are using the same notation A as in Section 12.5 to maintain the analogy between the average error in the two cases of estimation procedures and decision procedures. The first sum in (1) represents the weighted combination of errors when the θ_i's are in Ω_1, and the second sum is the weighted combination of errors when the θ_i's are in Ω_2. If the experimenter is actually able to assign a priori weights q, then the goodness of a procedure ϕ is measured by a single number $A(\phi, q)$. This fact makes the choice of a decision procedure much easier. One must find the procedure that minimizes this average error. As before, such a procedure is called Bayes.

Bayes' procedure

A Bayes' procedure, with respect to the a priori weights $q = q_1, \ldots, q_m$, is a decision procedure ϕ that makes (1) as small as possible. In other words, ϕ is Bayes if

$$A(\phi, q) \leq A(\phi_1, a)$$

whenever ϕ_1 is any other decision procedure.

There is an explicit way of determining a Bayes' procedure, as follows.

explicit form of Bayes' procedure

Let

$$D(x) = \sum_{i=1}^{r} f(x, \theta_i)q_i - \sum_{i=r+1}^{m} f(x, \theta_i)q_i$$

Define

$$\phi(X) = \begin{cases} 1 & \text{if } D(X) > 0 \\ 0 & \text{if } D(X) < 0 \\ \text{any number in } [0, 1] & \text{if } D(X) = 0 \end{cases} \tag{2}$$

Then ϕ is Bayes with respect to the a priori weights $q = (q_1, \ldots, q_m)$.

Proof. The quantity that must be minimized with respect to ϕ is (1), which, expressed in greater detail, equals

$$A(\phi, q) = \sum_{i=1}^{r} \{q_i - E_{\theta_i}[\phi(X)]q_i\} + \sum_{i=r+1}^{m} E_{\theta_i}[\phi(X)]q_i \tag{3}$$

Suppose for simplicity that X is discrete. Then the right side of (3), after interchanging the orders of summation, can be written as

$$\sum_{i=1}^{r} q_i - \sum_{x} \phi(x)\left[\sum_{i=1}^{r} f(x, \theta_i)q_i - \sum_{i=r+1}^{m} f(x, \theta_i)q_i\right]$$

$$= \sum_{i=1}^{r} q_i - \sum_{x} \phi(x)D(x) \tag{4}$$

If X has an integrating density $f(x, \theta)$, then the counterpart of (4) is

$$A(\phi, q) = \sum_{i=1}^{r} q_i - \int \phi(x)\left[\sum_{i=1}^{r} f(x, \theta_i)q_i - \sum_{i=r+1}^{m} f(x, \theta_i)q_i\right] dx$$

$$= \sum_{i=1}^{r} q_i - \int \phi(x)D(x)\, dx \tag{5}$$

The integral in (5) is over one dimension or several, depending on whether X is a random variable or vector, and is over all relevant values of X. Consider the right side of (4). This quantity is minimized by letting $\phi(x)$ be as small as possible when $D(x)$ is negative and as large as possible when $D(x)$ is positive. When $D(x) = 0$, the value of $\phi(x)$ does not influence the value of $A(\phi, q)$. Since $\phi(x)$ must be in $[0, 1]$, these conditions are exactly described by (2). The same considerations apply to (5). ∎

The criterion described by (2) is really quite reasonable. The first part of $D(x)$ is a weighted average of the probabilities of an x value over all the θ's in Ω_1, and the second part is such a weighted average over all the θ's in Ω_2. If the first weighted average is greater than the second, one makes decision 1; otherwise, one makes decision 2.

There is another useful way to interpret (2). Suppose we introduce the fiction that "nature" first randomly selects a parameter θ from among the possible "states of nature," $\theta_1, \ldots, \theta_m$, with probabilities $q_1, \ldots, q_m$, respectively. Then, supposing X to be discrete, for simplicity we can make the interpretation that

$$f(x, \theta_i) = P(X = x | \theta = \theta_i)$$

This probability is of the experimenter observing X, given that "nature" selects θ_i.) Then

$$P(X = x, \theta = \theta_i) = f(x, \theta_i)q_i$$

It follows that

$$P(\theta = \theta_i | X = x) = \frac{f(x, \theta_i)q_i}{\sum_{j=1}^{m} f(x, \theta_j)q_j}$$

and

$$P(\theta \text{ in } \Omega_1 | X = x) = \frac{\sum_{i=1}^{r} f(x, \theta_i)q_i}{\sum_{i=1}^{m} f(x, \theta_i)q_i},$$

$$P(\theta \text{ in } \Omega_2 | X = x) = \frac{\sum_{i=r+1}^{m} f(x, \theta_i)q_i}{\sum_{i=1}^{m} f(x, \theta_i)q_i}$$

It is easy to check that $P(\theta \text{ in } \Omega_1 | X = x) > P(\theta \text{ in } \Omega_2 | X = x)$ if, and only if, $D(x) > 0$. Thus, $D(x) > 0$ means that given the experimenter has observed X to equal x, then the conditional probability that "nature" had selected θ out of Ω_1 is greater than that she selected θ out of Ω_2. This finding provides another rationale for (2).

Example 1 Consider Example 1, Section 13.3 again. Suppose that the a priori weights are 1/4 for each of the θ_i's. Then, using the density functions shown in Table 13.1, we have the following values of D.

$$D(0) = 0.125, \quad D(1) = 0.225, \quad D(2) = -0.225, \quad D(3) = -0.125$$

Hence, the decision procedure denoted as ϕ_1 is the Bayes' procedure. Suppose that the a priori weights are 1/6, 1/6 for θ_1, θ_2 (the θ's in Ω_1) and are 2/6, 2/6 for θ_3, θ_4 (the θ's in Ω_2). Then the values of D are as follows:

$$D(0) = \frac{0.5}{6}, \quad D(1) = 0.1, \quad D(2) = \frac{-(2.1)}{6}, \quad D(3) = \frac{-1}{6}$$

In this case, the decision procedure is also ϕ_1 [see Exercise 1, parts (b) and (c)].

When Ω consists of more than a finite or countable number of distributions, then the ability to define a general version of average error depends on being able to extend the sum in (1) to an integral. We restrict ourselves to the special case that Ω is in the real line or in n-dimensional space. Moreover, we suppose that integrals of the form

$$\int_{\Omega_1} f(x, \theta)q(\theta)\, d\theta, \qquad \int_{\Omega_2} f(x, \theta)q(\theta)\, d\theta$$

make sense where *q is an integrating density function over* Ω. In this case, average error is defined the same as before.

average error

Suppose that q is an integrating density function over Ω. If ϕ is a decision procedure and ψ is its power function, then the average error of ϕ with respect to the a priori weight function q is defined to be

$$A(\phi, q) = \int_{\Omega_1} [1 - \psi(\theta)]q(\theta)\, d\theta + \int_{\Omega_2} \psi(\theta)q(\theta)\, d\theta \tag{6}$$

If X is discrete, then (6), after the order of summation and integration is interchanged, equals

$$A(\phi, q) = \int_{\Omega_1} q(\theta)\, d\theta - \sum_x \phi(x) \int_{\Omega_1} f(x, \theta)q(\theta)\, d\theta$$
$$+ \sum_x \phi(x) \int_{\Omega_2} f(x, \theta)q(\theta)\, d\theta \tag{7}$$

The summation is over the possible values of X. If $f(x, \theta)$ is an integrating density, then (6), after interchanging the order of integrations, equals

$$A(\phi, q) = \int_{\Omega_1} q(\theta)\, d\theta - \int \phi(x) \int_{\Omega_1} f(x, \theta)q(\theta)\, d\theta\, dx$$

$$+ \int \phi(x) \int_{\Omega_2} f(x, \theta)q(\theta)\, d\theta\, dx \tag{8}$$

The outer integrations in (8) are over the relevant values of X. In the special cases described by (7) and (8), the definition of Bayes' procedure is exactly the same as before.

Bayes' procedure

A Bayes' procedure with respect to the a priori weight function q is a decision procedure ϕ that makes (6) [or (7) or (8)] as small as possible.

For average error defined by (7) or (8), the explicit description of a Bayes' procedure given by (2) still holds as follows.

explicit form of Bayes' procedure

Let

$$D(x) = \int_{\Omega_1} f(x, \theta)q(\theta)\, d\theta - \int_{\Omega_2} f(x, \theta)q(\theta)\, d\theta$$

Define

$$\phi(X) = \begin{cases} 1 & \text{if } D(X) > 0 \\ 0 & \text{if } D(X) < 0 \\ \text{any number in } [0, 1] & \text{if } D(X) = 0 \end{cases}$$

Then ϕ is Bayes with respect to the a priori weight function q.

Since the proof is exactly the same as the earlier one, the details are left to the reader.

Example 2

Suppose that $X_1, \ldots, X_n$ are mutually independent indicator random variables, each X_i equaling 1 with probability p and 0 with probability $1 - p$. Suppose that $\Omega_1 = (0, 1/2)$, $\Omega_2 = [1/2, 1)$, $\Omega = (0, 1)$. In other words, the experimenter wants to decide whether p, the probability of success, is in $(0, 1/2)$ (decision 1) or in $[1/2, 1)$ (decision 2). Suppose that all points of Ω are of equal concern to the experimenter. Then the a priori weight function q is the uniform integrating density on $(0, 1)$. To determine a Bayes' procedure, we must calculate $D(k) = \int_0^{1/2} p^k(1-p)^{n-k}\, dp - \int_{1/2}^1 p^k(1-p)^{n-k}\, dp$, which also equals

$$D(k) = \int_0^{1/2} [p^k(1-p)^{n-k} - p^{n-k}(1-p)^k]\, dp \tag{9}$$

As given by (9), D is well defined for all real numbers k between 0 and n (not just the integers). We leave it to the reader to verify that D has the appearance shown in Figure 13.7. That is, D is monotone over $(0, n)$, $D(n/2) = 0$, and $D(k) = -D(n-k)$. [Differentiate (9) with respect to k under the integral sign.] It follows from Figure 13.7 that the Bayes' procedure with respect to the uniform weight function is the following.

If $X_1 + \cdots + X_n < \frac{n}{2}$, make decision 1.

If $X_1 + \cdots + X_n > \frac{n}{2}$, make decision 2.

If $X_1 + \cdots + X_n = \frac{n}{2}$, make either decision or randomize.

Example 3 Suppose that a random variable X has the $N(\theta, 1)$ distribution. The experimenter wants to decide whether θ belongs to $\Omega_1 = (-\infty, 0)$ or to $\Omega_2 = [0, \infty)$. Suppose the a priori weight function is given by a normal density function

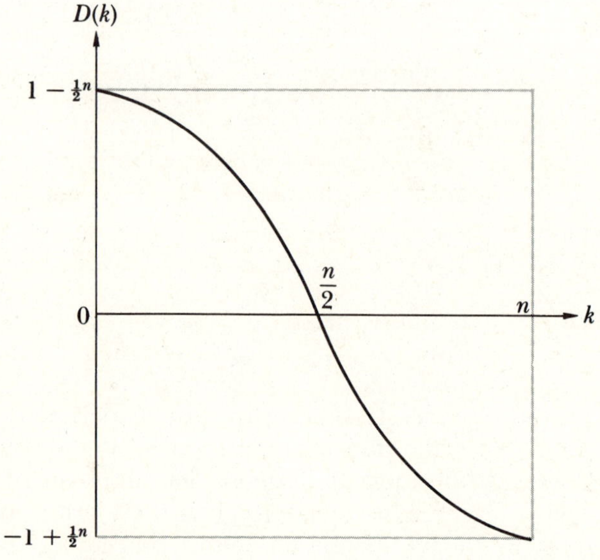

FIGURE 13.7 ***D(k)*, defined by (10), Section 13.5, is monotone decreasing, and *D(n/2)* = 0.**

$$q(\theta) = (2\pi)^{-1/2} \exp \frac{-\theta^2}{2}$$

To determine a Bayes' procedure, we compute

$$D(x) = \frac{1}{2\pi} \int_{-\infty}^{0} \exp\left[-\frac{\theta^2 + (x-\theta)^2}{2}\right] - \exp\left[-\frac{\theta^2 + (x-\theta)^2}{2}\right] d\theta \quad (10)$$

By a simple change of variables in the first integral, (10) becomes,

$$D(x) = \frac{1}{2\pi} \int_{0}^{\infty} \left[\exp - \frac{\theta^2 + (x+\theta)^2}{2} - \exp - \frac{\theta^2 + (x-\theta)^2}{2}\right] d\theta$$

The following facts are not all exactly obvious, but we ask the reader to either verify them or accept them for the moment:

$$D(0) = 0, \quad D(x) = -D(-x), \quad \text{and } D(x) \geqslant 0 \text{ if } x \leqslant 0$$

(See Figure 13.8.) As a consequence, the Bayes' procedure is the following.

If $X < 0$, make decision 1.

If $X > 0$, make decision 2. (11)

If $X = 0$, make either decision.

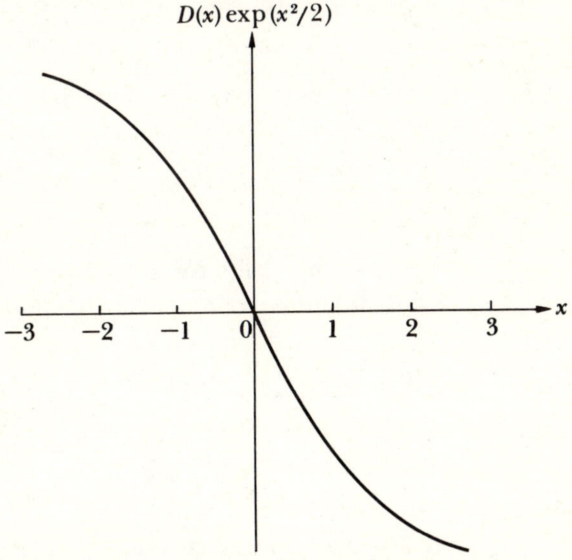

FIGURE 13.8 If $D(x)$ is defined by (11), Section 13.5, then $D(x) \exp(x^2/2)$ monotone decreasing, and $D(0) = 0$.

If instead of a single X we have $X_1, \ldots, X_n$, n mutually independent $N(\theta, 1)$ random variables, but the other details remain the same, then the Bayes procedure is the following.

$$\begin{aligned}
&\text{If } X_1 + \cdots + X_n < 0, \text{ make decision 1.}\\
&\text{If } X_1 + \cdots + X_n > 0, \text{ make decision 2.}\\
&\text{If } X_1 + \cdots + X_n = 0, \text{ make either decision.}
\end{aligned} \tag{12}$$

The details that lead to (12) are practically the same as those that led to (11) and are left to the reader (see Exercise 3a).

Finally, we point out that under appropriate regularity conditions, Bayes' procedures are admissible. For simplicity, we first consider the case where Ω is finite.

admissibility of Bayes' procedures

Suppose that Ω consists of a finite number of distributions and

$$\Omega_1 = (\theta_1, \ldots, \theta_r), \qquad \Omega_2 = (\theta_{r+1}, \ldots, \theta_m)$$

If ϕ is a Bayes' procedure with respect to an a priori weight function $q = (q_1, \ldots, q_m)$, where $q_i > 0$, $i = 1, \ldots, m$, then ϕ is admissible.

Proof. If ϕ were not admissible, then there would be another procedure ϕ_1 such that

$$1 - \psi_1(\theta_i) \leq 1 - \psi(\theta_i), \qquad i = 1, \ldots, r$$

$$\psi_1(\theta_i) \leq \psi(\theta_i), \qquad i = r + 1, \ldots, m$$

with strict inequality for at least one θ_i, where ψ is the power function of ϕ and ψ_1 is the power function of ϕ_1. Hence,

$$\begin{aligned}
A(\phi_1, q) &= \sum_{i=1}^{r} [1 - \psi_1(\theta_i)]q_i + \sum_{i=r+1}^{m} \psi_1(\theta_i)q_i \\
&< \sum_{i=1}^{r} [1 - \psi(\theta_i)]q_i + \sum_{i=r+1}^{m} \psi(\theta_i)q_i = A(\phi, q)
\end{aligned} \tag{13}$$

But this fact contradicts the hypothesis that ϕ is Bayes, so ϕ must be admissible. The strict inequality follows from the fact that the q_i's are strictly positive. ∎

Example 4 The decision procedure referred to as ϕ_1 in Example 1 is admissible.

The same theorem when the a priori weight function is an integrating density requires some regularity conditions for the proof to go through. We shall take the easy way out by supposing that the power function and the weight function are both continuous.

admissibility of Bayes' procedures

Suppose that the a priori weight function is an integrating density q, such that the average error is given by (7) or (8). Suppose also that for any decision procedure the power function is continuous on Ω, and that q is continuous and strictly positive on Ω. Then, if ϕ is Bayes with respect to q, ϕ must be admissible.

Proof. The counterpart of (13) is that if ϕ were not admissible, we would have

$$A(\phi_1, q) - A(\phi, q) = \int_{\Omega_1} \{[1 - \psi_1(\theta)] - [1 - \psi(\theta)]\}q(\theta)\, d\theta + \int_{\Omega_2} [\psi_1(\theta) - \psi(\theta)]q(\theta)\, d\theta < 0 \tag{14}$$

which would contradict the assumption that ϕ is Bayes with respect to q. The strict inequality in (14) follows from the fact that both integrands are nonpositive and continuous and at least one of them is strictly negative somewhere. ∎

Example 5 Consider Example 2. The power function for any procedure ϕ is given by

$$\psi(p) = \sum_{e_1,\ldots,e_n} \phi(e_1, \ldots, e_n)p^{e_1+\cdots+e_n}(1-p)^{n-e_1-\cdots-e_n}$$

Clearly, ψ is a continuous function of p. Hence, the preceding theorem applies in this example, whenever the a priori weight function is continuous.

Example 6 The regularity conditions required in the last theorem are satisfied in Example 3. This fact is not obvious and we ask that it be taken for granted. As a consequence, the decision procedures described by (11) and (12) are admissible.

13.6 THE NEYMAN–PEARSON THEOREM

(See Note 3.) It is instructive to review some of the preceding considerations in the special case that Ω_1 and Ω_2 each consist of one θ apiece. Although this case is really very special, it does allow a nice geometrical interpretation and has some useful applications.

Suppose Ω_1 consists of the single distribution θ_1 and Ω_2 consists of the single distribution θ_2. Letting ψ denote the power function as before, then

$$1 - \psi(\theta_1), \qquad \psi(\theta_2)$$

represent the two errors associated with a decision procedure. The first is the probability of deciding in favor of θ_2 when θ_1 is the correct state of nature, and the second is the probability of deciding in favor of θ_1 when θ_2 is the correct state of nature. We shall denote this pair of errors by (α, β). The errors associated with all possible decision procedures can be plotted in the unit square as in Figure 13.9, and it is called the *error set.* According to Section 13.3, the error set is a convex body. That is, if (α, β) and (α', β') are two error points, then any convex combination

$$[\rho\alpha + (1 - \rho)\alpha', \qquad \rho\beta + (1 - \rho)\beta'] \tag{1}$$

must also be an error point. That is, the error of some decision

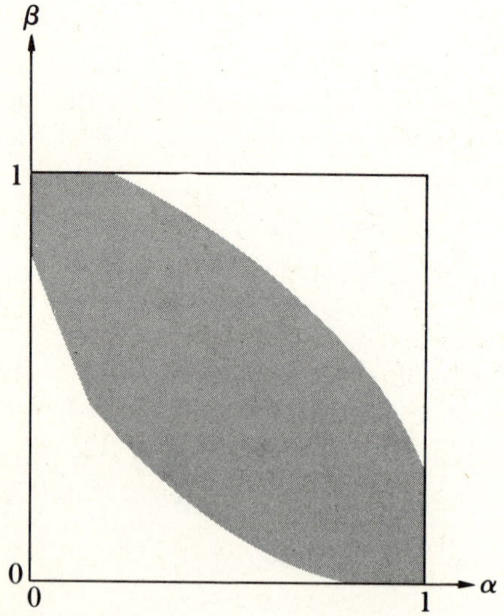

FIGURE 13.9 The set of error points is a convex set.

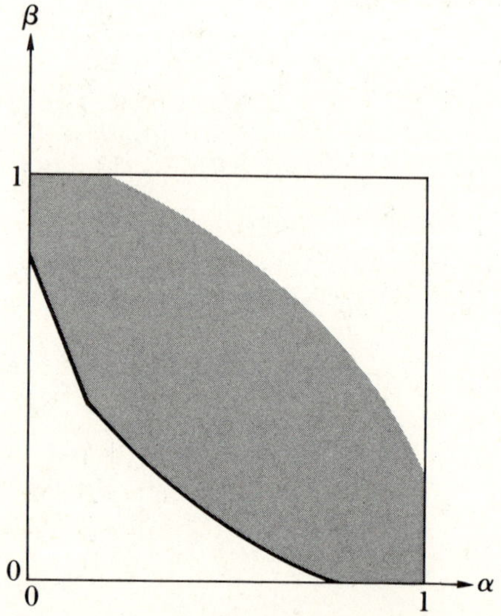

FIGURE 13.10 The heavy line denotes the "lower boundary" of the error set. These are the error points of the *admissible* procedures.

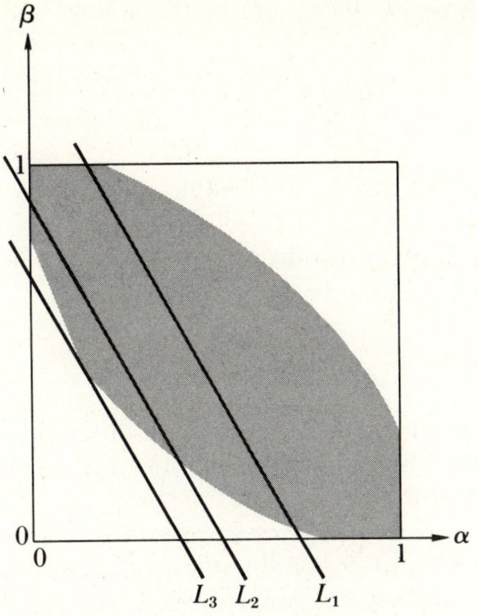

FIGURE 13.11 **Each of the lines L_1, L_2, L_3 is of the form $q_1\alpha + q_2\beta$ = constant. Line L_3 is the line with the *smallest* value of the constant that still intersects the error set. The points in the error set that L_3 intersects are error points of Bayes' procedures with respect to the a priori weights q_1, q_2.**

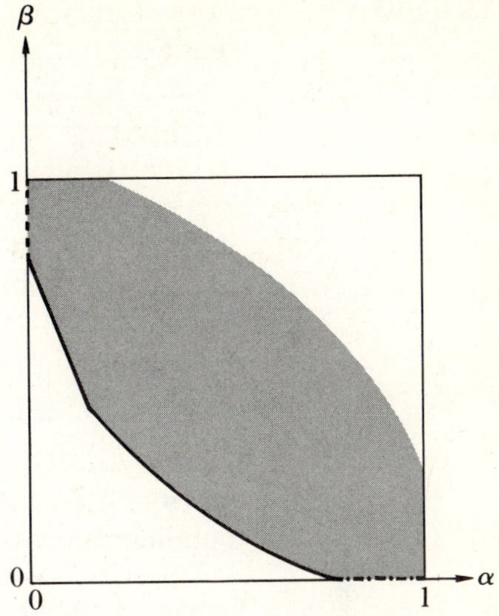

FIGURE 13.12 **The heavy solid line denotes the error points for procedures that are Bayes with respect to a priori weights q_1, q_2, *both of which are positive.* The dashed line denotes the error points for procedures that are Bayes with respect to $q_1 = 1$, $q_2 = 0$. The dot-and-dash line denotes the error points for procedures that are Bayes with respect to $q_1 = 0$, $q_2 = 1$.**

procedure is given by (1) for any ρ between 0 and 1.

One can give a geometrical interpretation to the admissible procedures. These are the procedures having error points lying on the "lower boundary" of the error set, as shown in Figure 13.10. If (α, β) is the error point of an admissible procedure, then the only way we can have both $\alpha' \leq \alpha$ and $\beta' \leq \beta$ is if $\alpha' = \alpha$ and $\beta' = \beta$.

If q_1, q_2 are a priori weights, then a Bayes' procedure with respect to these weights is one that minimizes $q_1\alpha + q_2\beta$. This fact can be interpreted as that procedure lying on the "lower boundary" of the error set and intersecting a straight line of the form $q_1\alpha + q_2\beta$ = constant. Such a straight line with the smallest possible value of the constant intersects all error points of Bayes' procedures with respect to the fixed weights q_1 and q_2 (see Figure 13.11). If $0 < q_1 < 1$, then, according to the theorem in Section 13.5, the Bayes' procedure must be admissible (see Figure 13.12).

Example 1 Suppose that X is known to have one of the following discrete densities:

u	$f(u, \theta_1)$
1	0.3
2	0.2
3	0.5

u	$f(u, \theta_2)$
1	0.4
2	0.5
3	0.1

First, let us consider the nonrandomized decision procedures. Such a procedure is determined by a set A with,

$$\phi(u) = \begin{cases} 1 & \text{if } u \text{ is in } A \\ 0 & \text{if } u \text{ is in } A' \end{cases}$$

In other words, if X is in A, decide in favor of θ_1; otherwise, decide in favor of θ_2. Table 13.2 shows all possible sets A and the corresponding values of α and β.

The error points of the nonrandomized procedures are plotted in Figure 13.13. The set of convex combinations of these points are the error points of all randomized procedures. It is a fact that the error point of any randomized procedure is a convex combination of the error points of nonrandomized procedures. For instance, consider the randomized procedure

$$\phi(1) = 0.2, \quad \phi(2) = 1, \quad \phi(3) = 0.5$$

Then ϕ is a convex combination of ϕ_1, ϕ_2, ϕ_3, where

$$\phi_1(1) = 1, \quad \phi_1(2) = 1, \quad \phi_1(3) = 0$$

$$\phi_2(1) = 0, \quad \phi_2(2) = 1, \quad \phi_2(3) = 0$$

$$\phi_3(1) = 0, \quad \phi_3(2) = 1, \quad \phi_3(3) = 1$$

Explicitly, $\phi = 0.2\phi_1 + 0.3\phi_2 + 0.5\phi_3$. It follows that the error point for ϕ is the same convex combination of the error points of ϕ_1, ϕ_2,

TABLE 13.2

A	α	β
1	0.7	0.4
2	0.8	0.5
3	0.5	0.1
1, 2	0.5	0.9
1, 3	0.2	0.5
2, 3	0.3	0.6
1, 2, 3	0.0	1.0
empty set	1.0	0.0

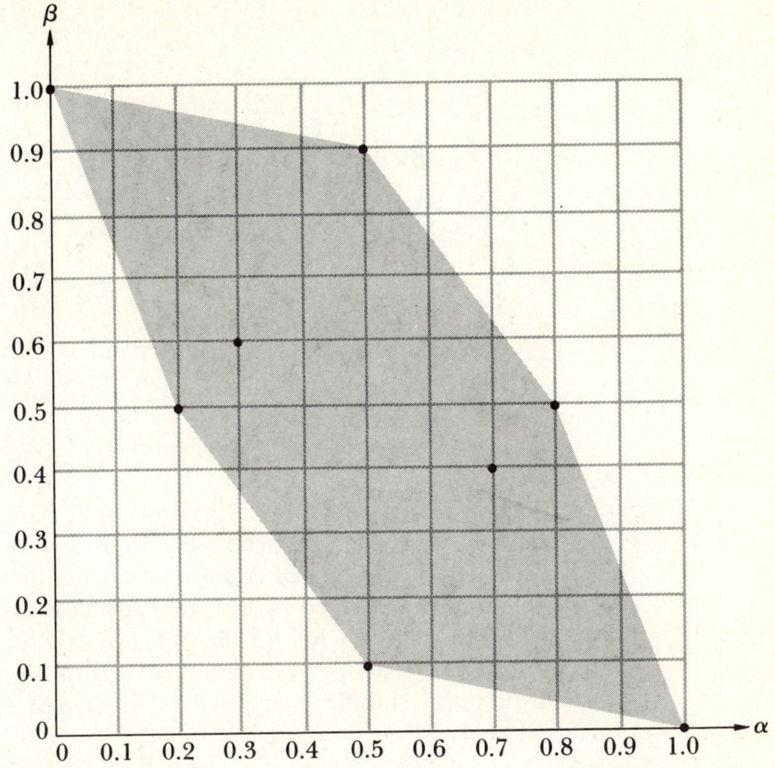

FIGURE 13.13 The error set for Example 1, Section 13.6. Dots indicate error points of nonrandomized procedures.

and ϕ_3. Figure 13.13 shows the error points of all randomized procedures.

When Ω consists of two distributions, the explicit form of the Bayes' procedure shown in Section 13.5, and the fact that for positive weights the Bayes' procedure is admissible, is called the Neyman–Pearson theorem. We now repeat the relevant facts in a slightly different form. We leave it to the reader to check that the form of the decision procedure given as follows is equivalent to the explicit form of the Bayes' procedure presented in Section 13.5.

Neyman–Pearson theorem

Suppose that $f(x, \theta_1)$ and $f(x, \theta_2)$ are two densities (either both discrete or both integrating). Let $0 < q_1 < 1$. Consider the following decision procedure:

$$\phi(X) = \begin{cases} 1 & \text{if } \dfrac{f(X, \theta_1)}{f(X, \theta_2)} > \dfrac{1-q_1}{q_1} \\ & \text{(decide in favor of } \theta_1) \\ 0 & \text{if } \dfrac{f(X, \theta_1)}{f(X, \theta_2)} < \dfrac{1-q_1}{q_1} \\ & \text{(decide in favor of } \theta_2) \\ \text{any number in } [0, 1] & \text{if } \dfrac{f(X, \theta_1)}{f(X, \theta_2)} = \dfrac{1-q_1}{q_1} \\ & \text{(make either decision or randomize)} \end{cases} \qquad (2)$$

This procedure is Bayes with respect to the prior weights q_1, q_2 and is admissible.

The original statement of the theorem by Neyman and Pearson is that if ϕ' is another procedure with an α error smaller than that of ϕ, then the β error of ϕ' must be greater than that of ϕ (Note 4).

The tests described by the Neyman–Pearson theorem are all of the form $\phi(X) = 1$ if $[f(X, \theta_1)/f(X, \theta_2)] > c$, $\phi(X) = 0$ if $[f(X, \theta_1)/f(X, \theta_2)] < c$. Any positive value of c gives a Bayes' procedure, since $(1 - q_1)/q_1 = c$ implies that $q_1 = 1/(1 + c)$ is in $(0, 1)$. The admissibility follows from Section 13.5.

Example 2 Consider the problem of discriminating between the two distributions in Example 1, Section 13.6. If $q_1 = q_2 = 1/2$, then the procedure described by the Neyman–Pearson theorem decides in favor of θ_1 when $X = 3$, and otherwise in favor of θ_2. This is the admissible procedure with an error point of $(\alpha, \beta) = (0.5, 0.1)$. (See Figure 13.13.) Notice that for all q_1 satisfying $1/6 < q_1 < 4/7$, the decision procedure is the same. (Verify this and interpret in terms of Figure 13.13.)

Example 3 Suppose that $X = (X_1, \ldots, X_n)$, where the X_i's are independent random variables, each $N(\theta, 1)$. It is known that θ is either -1 or 1. Suppose that the prior weights are given by $q_1 = q_2 = 1/2$. We have, after a bit of computation,

$$\frac{f(X, 1)}{f(X, -1)} = \frac{\exp - \sum_{i=1}^{n} (X_i - 1)^2/2}{\exp - \sum_{i=1}^{n} (X_i + 1)^2/2} = \exp 2 \sum_{i=1}^{n} X_i$$

Hence, this quantity is greater than or less than 1, accordingly as $\Sigma_{i=1}^{n} X_i$ is positive or negative. Hence, the procedure decides in favor of $\theta=1$ if $\Sigma_{i=1}^{n} X_i > 0$ and in favor of $\theta=-1$ if $\Sigma_{i=1}^{n} X_i < 0$. Since $\Sigma_{i=1}^{n} X_i = 0$ with probability 0, we can ignore this possibility. Notice that the very same procedure would result if the two θ values were $-a$ and a, a being any positive number.

13.7 UNIFORMLY MOST POWERFUL PROCEDURES

The Neyman–Pearson theorem sometimes gives useful information about more general situations. Suppose, as before, that Ω_1 consists of the single distribution θ_1, but that Ω_2 is no longer restricted to contain only one distribution as in Section 13.6. Sometimes it is possible to find a decision procedure that distinguishes between Ω_1 and Ω_2 in a uniformly best way. It is customary to call $1 - E_{\theta_1}[\phi(X)]$ the *significance level* of the test ϕ and denote it by a. This is the probability of making the wrong decision when the true parameter value is θ_1. Let us first consider an example.

Example 1 Let X have the integrating density,

$$f(x, \theta) = \begin{cases} \theta e^{-\theta x} & \text{if } x \geq 0 \\ 0 & \text{if } x < 0 \end{cases}$$

The positive parameter θ is known to be 1, or to be in (0, 1). That is, $\Omega_1 = \{1\}$, $\Omega_2 = (0, 1)$. If θ is any number in (0, 1), then

$$\frac{f(X, 1)}{f(X, \theta)} \begin{cases} > c \\ = c \\ < c \end{cases} \text{ according to whether } X \begin{cases} < d \\ = d \\ > d \end{cases}$$

where d is an appropriate positive number determined by c and θ. [Conversely, given any positive d, there is a corresponding value of c. The explicit relationship is $d = (-\log \theta c)/(1 - \theta)$. Check this!] Hence, for any $d > 0$, the following procedure is admissible.

If $X > d$, decide on Ω_1.

If $X < d$, decide on Ω_2.

(Because $X = d$ with probability 0, this possibility can be ignored.) For fixed d, the significance level is $\alpha = P_1(X < d) = 1 - e^{-d}$. Since $c = (1 - q_1)/q_1$ implies that $q_1 = 1/(1 + c)$, we can make the following summary. For every fixed significance level α, $(0 < \alpha < 1)$, and every fixed θ in (0, 1), there is a $q_1 = q_1(\alpha, \theta)$ such that the preceding decision procedure is the Neyman–Pearson procedure described by (2), Section 13.6. For every fixed α, it is a *uniformly most powerful procedure* the power function of which is $\psi(\theta) = e^{-\theta d}$. This statement means that if ψ_1 is the power function of any other procedure with the same significance level (i.e., $\psi_1(1) = \psi(1) = 1 - \alpha$), then $\psi_1(\theta) \geqslant \psi(\theta) = e^{-\theta d}$, $0 < \theta < 1$. The reason is that the *same* Neyman–Pearson procedure is best against *any* specific θ in (0, 1).

The general situation is described as follows.

uniformly most powerful procedure

Suppose that for given $0 < \alpha < 1$ and given θ in Ω_2 there is a $q_1 = q_1(\alpha, \theta)$ in (0, 1) such that the Neyman–Pearson decision procedure described by (2), Section 13.6, has a fixed significance level $1 - E_{\theta_1}[\phi(X)] = \alpha$. Moreover, suppose that for fixed α, the procedure ϕ does not depend on θ. Then ϕ is called a *uniformly most powerful procedure of significance level* α.

In other words, if the test described by the Neyman–Pearson theorem for distinguishing between θ_1 and θ, and which has a fixed significance level α, is the same for every θ in Ω_2, then the procedure is called uniformly most powerful. The name is appropriate for the following reason.

Suppose that ψ is the power function of a uniformly most powerful procedure of significance level α. If ψ' is the power function of some other procedure, such that

$$1 - \psi'(\theta_1) \leqslant \alpha$$

then it must be true that

$$\psi'(\theta) \geqslant \psi(\theta) \quad \text{for all } \theta \text{ in } \Omega_2 \tag{1}$$

In other words, if there exists some other procedure the significance level of which is not greater than α, then the other errors for all θ in Ω_2 must be uniformly worse, or at least no better. It is in this sense that ϕ is uniformly most powerful (see Figure 13.14).

Proof of (1). If there were some θ in Ω_2 where $\psi'(\theta) < \psi(\theta)$, it would contradict the fact that ϕ is admissible for distinguishing between θ_1 and θ. ∎

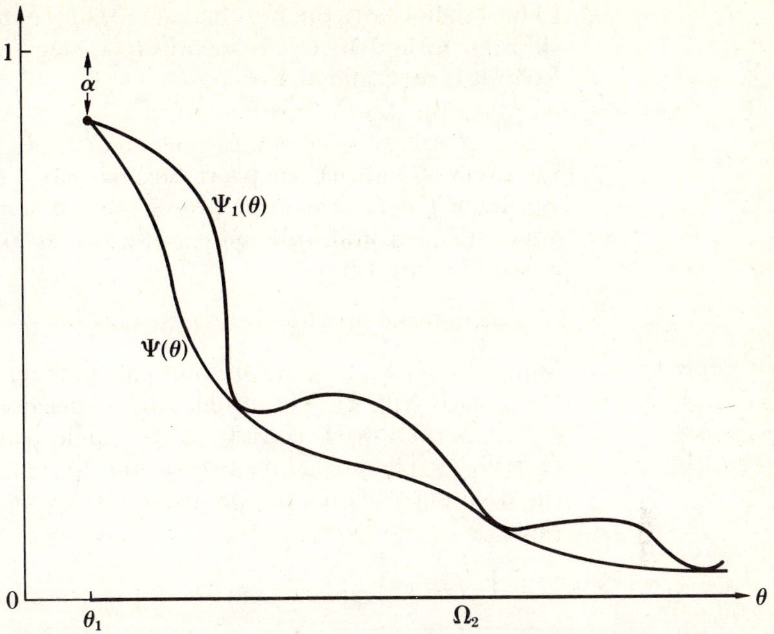

FIGURE 13.14 The power function of a uniformly most powerful procedure is $\psi(\theta)$. If $\psi_1(\theta)$ is the power function of another procedure with $\psi(\theta_1) = \psi_1(\theta_1) = \alpha$, then $\psi(\theta) \leqslant \psi_1(\theta)$ for all θ in Ω_2.

Example 2 Suppose that $X_1, \ldots, X_n$ are mutually independent indicator random variables with $P(X_i = 1) = p$. As usual, p plays the role of θ. Suppose we want to decide whether $p = p_1$, where $0 < p_1 < 1$, or $p > p_1$. In other words, Ω_1 consists of p_1 only and $\Omega_2 = (p_1, 1)$. The procedure given by the Neyman–Pearson theorem for distinguishing between p_1 and fixed p in Ω_2 is of the following form:

Let $X = X_1 + \cdots + X_n$.

If $\dfrac{p_1^X(1-p_1)^{n-X}}{p^X(1-p)^{n-X}} > c$, decide in favor of Ω_1.

If $\dfrac{p_1^X(1-p_1)^{n-X}}{p^X(1-p)^{n-X}} \leqslant c$, decide in favor of Ω_2.

By taking logarithms, we see that the preceding procedure is the same as the following:

$$\begin{aligned}&\text{If } X < d, \text{ decide in favor of } \Omega_1.\\&\text{If } X \geqslant d, \text{ decide in favor of } \Omega_2.\end{aligned} \tag{2}$$

[This relation uses the fact that $p_1(1-p)/p(1-p_1)$ is less than 1 for all $p > p_1$, and hence its logarithm is negative.] By keeping d fixed, α is determined as

$$\alpha = P_{p_1}(X \geqslant d)$$

For every α in (0, 1), the procedure described by (2) is admissible as long as $p > p_1$. The same procedure is best against *any* p in Ω_2; hence, (2) is a uniformly most powerful test of significance level α (see Exercise 12).

Example 3 Suppose that $X_1, \ldots, X_n$ are mutually independent random variables, each $N(\theta, 1)$. The problem is to decide whether $\theta = 0$ or $\theta > 0$. We know Ω_1 consists of the single parameter $\theta = 0$ and $\Omega_2 = (0, \infty)$. The procedure given by the Neyman–Pearson theorem for deciding whether the parameter is 0 or $\theta > 0$ examines the ratio

$$\frac{[1/(2\pi)^{n/2}]\exp\left(-\sum_{i=1}^{n} X_i^2\right)\Big/2}{[1/(2\pi)^{n/2}]\exp\left[-\sum_{i=1}^{n}(X_i-\theta)^2\right]\Big/2} = r(X)$$

The procedure,

$$\begin{aligned} &r(X) > c, \text{ decide in favor of } 0 \\ &r(X) \leqslant c, \text{ decide in favor of } \theta \end{aligned} \tag{3}$$

is admissible as long as $0 < c$. By taking logarithms, we see that (3) is equivalent to

$$\begin{aligned} &\sum_{i=1}^{n} X_i < d, \text{ decide in favor of } 0 \\ &\sum_{i=1}^{n} X_i \geqslant d, \text{ decide in favor of } \theta \end{aligned} \tag{4}$$

Once d is fixed, α is determined by

$$\alpha = P_0\left(\sum_{i=1}^{n} X_i \geqslant d\right)$$

The procedure (4) is uniformly most powerful of significance level α, since (4) does not depend on the value of θ. The calculation showing that (3) implies (4) does depend on the fact that $\theta > 0$. If θ were negative, the inequalities in (4) would be reversed. (Check this!)

Example 4 It is worth pointing out that uniformly most powerful tests need not exist. Consider the same situation as in Example 3, except now $\Omega_1 = (0)$ and Ω_2 is the whole line with the exception of 0. In other words, the experimenter wants to decide whether $\theta = 0$ or $\theta \neq 0$ (as distinct from $\theta > 0$). If we examine the calculations in Example 3, we see that the desired procedure for deciding between 0 and $\theta < 0$ is of the form

$$\sum_{i=1}^{n} X_i > d, \text{ decide in favor of } 0$$

$$\sum_{i=1}^{n} X_i \leq d, \text{ decide in favor of } \theta$$

On the other hand, if $\theta > 0$, the desired procedure with the same significance level is (4). Hence, there is not a uniformly most powerful procedure (see Exercise 14).

SUMMARY

In the two-decision problem, a *decision procedure* is a rule which says that if X is in A, make decision 1 (that θ is in Ω_1); otherwise, make decision 2 (that θ is in Ω_2). This procedure can be described by an indicator random variable $\phi(X)$. More generally, we consider randomized decision procedures in which $\phi(X)$ can assume any values between 0 and 1; $\phi(X)$ is the probability with which the experimenter is to make decision 1 when X is observed.

The probability of making decision 1 is $\psi(\theta) = E_\theta\phi(X)$, considered as a function of the true parameter θ. The *power function* is ψ. Also, $1 - \psi(\theta)$ represents the probability of error if θ is in Ω_1, and $\psi(\theta)$ is the probability of error if θ is in Ω_2.

If one procedure can be improved on in terms of these errors by another procedure, uniformly in θ, then the first procedure is called *inadmissible;* otherwise, it is *admissible.* The experimenter may be able to assign an *a priori weight function* $q(\theta)$ to the space of parameters to express his relative concern or anticipation of likelihood of the various parameters. Averaging out the errors expressed in terms of the power function, using q, gives an overall average error. Explicitly, if Ω is discrete, it is given by,

$$A(\phi, q) = \sum_{\Omega_1} [1 - \psi(\theta)]q(\theta_i) + \sum_{\Omega_2} \psi(\theta_i)q(\theta_i)$$

There is a similar formula where summation can be replaced by integration. For fixed q, a decision procedure ϕ that makes $A(\phi, q)$

as small as possible is called a *Bayes' procedure.* Under certain conditions, *Bayes procedures are admissible.* For instance, this statement is true when Ω consists of finitely many parameters and each $q(\theta)$ is strictly positive.

The *Neyman–Pearson theorem* describes admissible Bayes' procedures in the special case that Ω_1 and Ω_2 each consists of one parameter only. Explicitly, these procedures are of the following form.

$$\text{If } \frac{f(X, \theta_1)}{f(X, \theta_2)} > c, \text{ decide on } \Omega_1.$$

$$\text{If } \frac{f(X, \theta_1)}{f(X, \theta_2)} < c, \text{ decide on } \Omega_2.$$

With equality, make either decision, or randomize.

When Ω_1 consists of one parameter value but Ω_2 may consist of many values, the Neyman–Pearson theorem can sometimes be used to find *uniformly most powerful* procedures.

EXERCISES

1 (a) Verify the power functions in Example 1, Section 13.3.
(b) Verify the assertions in Example 1, Section 13.5.
(c) Again, in the same situation, what is a Bayes' procedure if $q_1 = q_2 = 0.1, \quad q_3 = q_4 = 0.4$?

2 The random variables $X_1, \ldots, X_n$ are mutually independent and uniformly distributed over $[0, \theta]$, $\theta > 0$. The problem is to decide whether $\theta \leq 1$ (decision 1) or $\theta > 1$ (decision 2). Let $Y = \max(X_1, \ldots, X_n)$. The decision procedure follows.

If $Y \leq c$, make decision 1.

If $Y > c$, make decision 2.

(a) Show that the power function of this decision procedure is

$$\psi(\theta) = \begin{cases} 1 & \text{if } \theta < c \\ \left(\dfrac{c}{\theta}\right)^n & \text{if } \theta \geq c \end{cases}$$

(b) Determine the cutoff constant c and the smallest sample size n such that

$$\psi(0.9) = 0.95, \qquad \psi(1.1) = 0.05$$

(Answer: $c = 0.90$; $n = 15$.)

3 (a) Verify the details of Example 3, Section 13.5.

(b) Suppose that $X_1, \ldots, X_n$ are mutually independent random variables, each $N(\theta, 1)$. The problem is to decide whether $|\theta| < 1$ (decision 1) or $|\theta| > 1$ (decision 2). Let $S = X_1 + \cdots + X_n$. The decision procedure follows.

If $|S| \leqslant c$, make decision 1.

If $|S| > c$, make decision 2.

(1) Show that the power function of this decision procedure is

$$\psi(\theta) = F\left(\frac{c - n\theta}{\sqrt{n}}\right) - F\left(\frac{-c - n\theta}{\sqrt{n}}\right)$$

where F is the cdf of the $N(0, 1)$ distribution.

(2) Show that $\psi(\theta) = \psi(-\theta)$ and $\psi(\theta_1) < \psi(\theta_2)$ if $\theta_2 > \theta_1 > 0$.

4 A random variable X is known to have one of the following distributions:

u	$f(u, \theta_1)$	u	$f(u, \theta_2)$	u	$f(u, \theta_3)$	u	$f(u, \theta_4)$
0	0.20	0	0.25	0	0.30	0	0.35
1	0.30	1	0.35	1	0.35	1	0.35
2	0.50	2	0.40	2	0.35	2	0.30

Suppose that $\Omega_1 = (\theta_1, \theta_2)$, $\Omega_2 = (\theta_3, \theta_4)$.

(a) Given a priori weights on $\theta_1, \theta_2, \theta_3, \theta_4$, the values of which are 0.1, 0.4, 0.3, 0.2, respectively, find the Bayes' procedure for deciding between Ω_1 and Ω_2.

(b) Compute the power function of the Bayes' procedure.

5 Consider the same situation as in Exercise 4, but now suppose that two independent observations are made on the same population. That is, $X = (X_1, X_2)$ where X_1 and X_2 are independent, each with the same density $f(u, \theta_i)$. Again, find the Bayes' procedure and compute its power function.

6 Suppose that ϕ is a randomized decision procedure. Define ϕ' by $\phi'(X) = 1 - \phi(X)$.

(a) Show that ϕ' is a randomized decision procedure.

(b) What is the relation between the power functions of ϕ and ϕ'?

7 Consider the same setting as in Exercise 3a and Example 3, Section 13.5, now suppose that the decision procedure is the following:

If $X \geqslant 0$, make decision 1.

If $X < 0$, make decision 2.

Show that this decision procedure is inadmissible. (Hint: Compare the power function for this procedure with that of the procedure in Example 3, Section 13.5.)

8 Suppose that ψ is the power function of a decision procedure ϕ having the properties that

$$\psi(\theta) \geqslant \tfrac{1}{2} \quad \text{for all } \theta \text{ in } \Omega_1$$

$$\psi(\theta) \leqslant \tfrac{1}{2} \quad \text{for all } \theta \text{ in } \Omega_2$$

with strict inequality for at least one θ. Consider the decision procedure ϕ' given by

$$\phi'(X) = 1 - \phi(X)$$

Show that ϕ' is inadmissible. Notice that Exercise 7 is a special case.

9 (a) Suppose that ϕ_1 and ϕ_2 are both inadmissible procedures. Show that $\alpha\phi_1 + (1 - \alpha)\phi_2$ is also an inadmissible procedure for any $0 \leqslant \alpha \leqslant 1$. [Hint: Consider $\alpha\phi_3 + (1 - \alpha)\phi_4$, where ϕ_3 and ϕ_4 improve on ϕ_1 and ϕ_2, respectively.]

(b) Use the graph in Figure 13.13 to convince yourself that a convex combination of admissible procedures is not necessarily admissible.

10 Suppose that a random variable X is known to be exponentially distributed with a parameter that is either λ_1 or λ_2 $(0 < \lambda_1 < \lambda_2)$. Show that the "lower boundary" of the error set consists of the pairs (α, β), which satisfy the relationship $\beta = (1 - \alpha)^{\lambda_2/\lambda_1}$, $0 \leqslant \alpha \leqslant 1$. [Hint: Use the Neyman–Pearson theorem to show that for any positive c, the procedure deciding in favor of λ_1 when $X \geqslant c$ and in favor of λ_2 when $X < c$ is an admissible procedure. Now calculate the error points (α, β) for these procedures.]

11 Let $X_1, \ldots, X_n$ be mutually independent random variables, each exponentially distributed with the same parameter λ. Suppose that

$$\Omega_1 = (0, a] \quad \text{and} \quad \Omega_2 = (a, \infty)$$

Consider the following decision procedure:

If $X_1 + \cdots + X_n > c$, decide λ is in Ω_1.

If $X_1 + \cdots + X_n \leqslant c$, decide λ is in Ω_2.

Let ψ be the power function of this procedure. Show that if ψ_1 is the power function of any other procedure such that $\psi_1(a) \geqslant \psi(a)$, then $\psi_1(\lambda) > \psi(\lambda)$ for all $\lambda > a$. (Hint: Show that the described procedure is uniformly most powerful for deciding whether $\lambda = a$ versus $\lambda > a$.)

12 Repeat Example 2, Section 13.7, now supposing that $\Omega_2 = (0, p_1)$. As before, Ω_1 consists of the single point p_1. Show that for any fixed d, the procedure

$X_1 + \cdots + X_n < d$, decide in favor of Ω_2

$X_1 + \cdots + X_n \geqslant d$, decide in favor of Ω_1

is a uniformly most powerful test.

13 Suppose that $X_1, \ldots, X_n$ are mutually independent random variables with joint integrating density

$$f(x_1, \ldots, x_n, \theta) = \frac{1}{2(2\pi)^{n/2}}\left[\exp -\frac{1}{2}\sum_{i=1}^{n}(x_i - \theta)^2 + \exp -\frac{1}{2}\sum_{i=1}^{n}(x_i + \theta)^2\right]$$

[Notice that when $\theta = 0$, the X_i's are mutually independent and $N(0, 1)$.] Let Ω_1 consist of the single point $\theta = 0$, and let Ω_2 consist of all real numbers with the exception of 0. Consider the following procedure:

If $|X_1 + \cdots + X_n| \leq c$, decide $\theta = 0$.

If $|X_1 + \cdots + X_n| > c$, decide $\theta \neq 0$.

Show that this procedure is uniformly most powerful.

14 Let $X_1, \ldots, X_n$ be mutually independent random variables, each $N(\theta, 1)$. The problem is to decide whether $\theta = 0$ or $\theta \neq 0$. Let ψ be the power function of a procedure and let $\bar{\psi}$ be an "averaged" power function defined as follows:

$$\bar{\psi}(\theta) = \frac{[\psi(\theta) + \psi(-\theta)]}{2}$$

[Notice that $\bar{\psi}(0) = \psi(0)$, $\bar{\psi}(\theta) = \bar{\psi}(-\theta)$.] Consider the following procedure:

If $|X_1 + \cdots + X_n| \leq c$, decide $\theta = 0$.

If $|X_1 + \cdots + X_n| > c$, decide $\theta \neq 0$.

Suppose that ψ_1 is the power function of this procedure and ψ the power function of any other procedure. Show that

$$\psi_1(\theta) \leq \bar{\psi}(\theta), \qquad \text{all } \theta \neq 0$$

whenever

$$\psi_1(0) \leq \bar{\psi}(0)$$

In other words, the described procedure is uniformly most powerful in terms of *averaged power*. However, as Example 4, Section 13.7, shows, there is no uniformly most powerful procedure. (Hint: Use Exercise 13. Show that if ψ is the power function of a procedure, then $\bar{\psi}$ is the power function of the same procedure when the distribution of $X_1, \ldots, X_n$ is as described in Exercise 13.)

NOTES

1 For some additional reading, see Chapter 6, [50] and Chapter 12, [51].

2 The general approach to making decisions is called statistical decision theory. See Note 6, Chapter 12.

3 Jerzy Neyman (1894-   ) and Egon S. Pearson (1895-   ) are contemporary statisticians who have made important contributions to the theory of statistics. Much of their work should be readable by users of this book and can be found in [61].

4 For a slightly different way of expressing the Neyman–Pearson theorem, see Theorem 12.3, p. 292 of [51].

Chapter Fourteen

SUFFICIENT STATISTICS

One of the first problems facing an experimenter who wants to make inferences about the true parameter θ is the following—How should he analyze his data? An important concept is that of *sufficiency.* Statistic S is called a *sufficient statistic* if the conditional distribution of the observations, given S, does not depend on θ. This fact suggests that the first thing the experimenter should do is to compute S, since any other information gleaned from the data, once S is known, cannot possibly shed any light about θ. A *minimal sufficient statistic* represents the "ultimate" or "simplest" computation that the experimenter should do. Any estimation or decision-making procedures can be based on the minimal sufficient statistic without "losing any information." This chapter develops this notion in the special case of discrete sample spaces.

14.1 SUFFICIENT STATISTICS

Suppose that an experimenter does n independent success-failure experiments where the probability of success is p. He does not know the value of p and wants to analyze his experimental outcomes to gain some information about p. The mathematical model is that $X_1, \ldots, X_n$ are mutually independent and each equals 1 with probability p and 0 with probability $1-p$. Whatever analysis of $X_1, \ldots, X_n$ is made to get information about p, only the sum $X_1 + \cdots + X_n$ need be used. This follows because once one knows the sum, the individual values of the X_i's cannot give information about p. The mathematical version of this assertion is the following:

$$P[(X_1=e_1) \cap \cdots \cap (X_n=e_n)|X_1+\cdots+X_n=k],\qquad k=0, 1, \ldots, n \tag{1}$$

does not depend on p $(0<p<1)$.

Proof of (1). If $e_1+\cdots+e_n \neq k$, then the conditional probability is 0. Otherwise, if $e_1+\cdots+e_n=k$, the probability in (1) equals

$$\frac{P[(X_1=e_1) \cap \cdots \cap (X_n=e_n)]}{P(X_1+\cdots+X_n=k)}=\frac{p^{e_1}(1-p)^{1-e_1}\cdots p^{e_n}(1-p)^{1-e_n}}{\binom{n}{k}p^k(1-p)^{n-k}}$$

$$=\frac{1}{\binom{n}{k}} \quad \blacksquare$$

In other words, given that the sum equals k, the conditional probability of any sequence does not involve p. Hence, if one knows the total number of successes, knowing the trials on which these were obtained cannot contribute to a knowledge of p. In this case, we say that $X_1+\cdots+X_n$ is *sufficient for* p, or is a *sufficient statistic for* p.

We want to extend these ideas to a more general setting. Unfortunately, such an extension depends on a rather sophisticated approach to conditional probabilities in the nondiscrete case. *To avoid these difficulties, we shall restrict ourselves to the discrete case* (Notes 1, 2).

The first idea we need is that of a *partition*. Suppose that $\mathfrak{X}$ is a discrete sample space. By a partition of $\mathfrak{X}$ is simply meant a collection of disjoint sets the union of which is all of $\mathfrak{X}$. Since the sample space is discrete, there cannot be more than a countable number of sets making up the partition. Now suppose that X is a random variable defined on $\mathfrak{X}$, with possible values $a_1, a_2, \ldots$. The sets $(X=a_1), (X=a_2), \ldots$, constitute a partition of $\mathfrak{X}$. Of course, there are other random variables that will induce the same partition. For instance, if X is a nonnegative random variable, then X and X^2 induce the same partitions. One partition is called a *subpartition* of a second partition if every set in the second partition is a union of sets of the first (see Figure 14.1). We also say that the first partition is *finer* than the second, or that the second is *rougher* than the first.

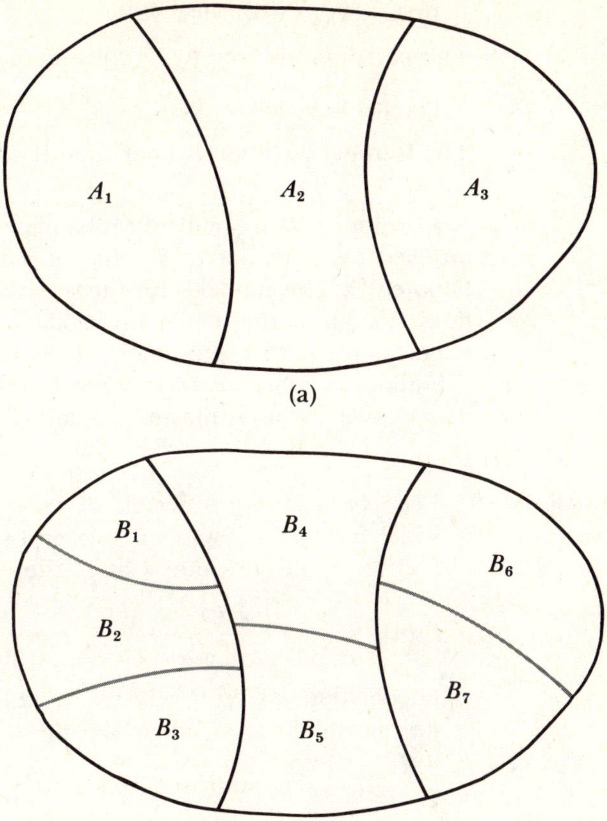

(a)
(b)

FIGURE 14.1 (a) (A_1, A_2, A_3) partitions the sample space. (b) $(B_1, B_2, B_3, B_4, B_5, B_6, B_7)$ is a subpartition of (A_1, A_2, A_3).

Suppose that Y and X are random variables such that $Y = h(X)$. Then the partition induced by X is finer than that induced by Y.

Everything we have said about random variables applies word-for-word if X is a random vector. Thus, if $X = (X_1, \ldots, X_m)$, then the distinct sets of the form $(X_1=a_1) \cap (X_2=a_2) \cap \cdots \cap (X_m=a_m)$, obtained by varying each of the possible values, is a partition of the sample space.

Example 1 Suppose that $\mathfrak{X} = (w_1, w_2, w_3, w_4, w_5)$ and $X(w_1) = -1$, $X(w_2) = -2$, $X(w_3) = 0$, $X(w_4) = 1$, $X(w_5) = 1$. The partition induced by X consists of the four sets,

$(w_1),\quad (w_2),\quad (w_3),\quad (w_4, w_5)$

The partition induced by X^2 consists of the three sets,

$(w_1, w_4, w_5),\quad (w_2),\quad (w_3)$

The four-set partition is finer than the three-set one.

Consider now a family of probability distributions on $\mathfrak{X}$ that are indexed by a parameter θ. The collection of all possible labels is denoted Ω. This model is for the situation where the experimenter does not know the actual probability distribution governing his experiment; but he does know that it is one of a collection of distributions P_θ where θ varies over Ω. Within this setting we shall now describe what is meant by a sufficient statistic.

sufficient statistic

Let S be a random variable or random vector the values of which do not depend on θ. Suppose that for any event A and any possible value a of S, the conditional probabilities

$$P_\theta(A|S=a) = \frac{P_\theta[A \cap (S=a)]}{P_\theta(S=a)} \tag{2}$$

do not depend on θ, whenever these conditional probabilities are well defined. Then S is called a statistic that is sufficient for θ.

In other words, for any A and a, the right-hand side of (2) has the same value for all θ in Ω. Experimentally, S represents one particular way of assigning numbers to the experimental outcomes. If the experimenter wants to get some insight about the correct θ governing his observable outcomes, he cannot improve on his data analysis once he knows S. The reason is the conditional distribution of experimental outcomes given S is not influenced by θ. In this sense, it is sufficient for him to know S.

Example 2

Let us repeat the example described in the initial paragraph of this section. The experiment consists of n independent repetitions of the same success-failure experiment. The sample space has 2^n sample points. A sample point representing k successes and $n-k$ failures is assigned probability $p^k(1-p)^{n-k}$. The unknown parameter p plays the role of θ, and $\Omega = (0, 1)$. Let S be the total number of successes; therefore, S has the binomial distribution with parameters n and p. The conditional distribution of any sample point in

$(S = k)$, given that $S = k$, is $p^k(1-p)^{n-k}/\binom{n}{k}p^k(1-p)^{n-k} = 1/\binom{n}{k}$.
Hence,

$$P_p(A|S=k) = \frac{\text{number of sample points in } A \cap (S=k)}{\binom{n}{k}}$$

The random variable S is a sufficient statistic for p.

Sometimes it is more convenient to talk about the partition induced by S.

sufficient partition

If S is sufficient for θ, then the partition induced by S is called a *partition that is sufficient for* θ.

In other words, if the possible values of S are $a_1, a_2, \ldots$, then the sets $(S = a_1), (S = a_2), \ldots$ make up the sufficient partition. Suppose that $A_1, A_2, \ldots$ are the sets of a partition. For this partition to be sufficient for θ, it must be true that $P_\theta(w|A_i)$ does not depend on θ, for any sample point w and any A_i, whenever the conditional probability is well defined. This fact follows immediately from the definition of sufficiency.

Example 3

Suppose that the sample space consists of the six sample points $w_1, w_2, w_3, w_4, w_5, w_6$, and that there are three possible probability measures:

w_i	$P_{\theta_1}(w_i)$	$P_{\theta_2}(w_i)$	$P_{\theta_3}(w_i)$
w_1	1/4	0	0
w_2	1/4	1/4	0
w_3	1/4	1/4	1/4
w_4	1/4	1/4	1/4
w_5	0	1/4	1/4
w_6	0	0	1/4

We claim that the partition consisting of the five sets

$$\{w_1\}, \quad \{w_2\}, \quad \{w_3, w_4\}, \quad \{w_5\}, \quad \{w_6\}$$

is sufficient for θ. Consider the two-point set, $\{w_3, w_4\}$. We have the following conditional probabilities:

$$P_\theta(w_i|\{w_3, w_4\}) = \begin{cases} \frac{1}{2} & \text{if } i = 3 \text{ or } 4 \\ 0 & \text{otherwise} \end{cases}$$

which holds uniformly for each of the three θ_i's. This fact verifies the required condition for this particular set. For the one-point sets, there is hardly anything to verify. If we define $X(w_1) = 1$, $X(w_2) = 2$, $X(w_3) = X(w_4) = 4$, $X(w_5) = 5$, $X(w_6) = 6$, then this random variable induces the preceding partition and is a sufficient statistic for θ. (Of course, there are infinitely many random variables that induce the same partition.)

Example 4 A box is known to contain marked balls according to one of the following three compositions:

case 1: 1, 2, 3, 4

case 2: 1, 2, 3, 4, 5

case 3: 1, 2, 3, 4, 5, 6

The experiment is to draw two balls at random and to note their numbers. We claim that X = (larger of the two numbers drawn) is sufficient for determining which of the three cases prevails. We leave it to the reader to make a precise formulation of this assertion in terms of the definition of sufficiency. Intuitively, we are asserting that once the experimenter knows the larger of the two numbers drawn, then knowing the value of the other number, or knowing the order in which the two numbers were drawn, cannot possibly add any information as to which of the three cases is the correct one (see Exercise 1).

14.2 MINIMAL SUFFICIENCY

There may be many sufficient statistics. Somehow, one would like to find one that is "least complicated" or most easy to use. This statement is interpreted to mean that the partition induced by a sufficient statistic should be as *rough as possible.* It so happens that there is an extreme partition such that every other sufficient partition is a subpartition of it. We now proceed to describe this partition, which is determined by the so-called *minimal sufficient statistic.*

minimal sufficient partition

Consider the following partition of $\mathfrak{X}$. Two sample points w_1 and w_2 are in the same set of the partition if, and only if,

$$\frac{P_\theta(w_1)}{P_\theta(w_2)} \tag{1}$$

does not depend on θ, whenever this ratio is not of the form 0/0. We can express it as an equivalence relation, $w_1 \sim w_2$. We call the described partition the *minimal sufficient partition.*

It is left to the reader to check that the relation $\sim$ is really an equivalence relation. That is, three properties hold: (1) every sample point is equivalent to itself, $w \sim w$; (2) $w_1 \sim w_2$ implies $w_2 \sim w_1$; and (3) if $w_1 \sim w_2$ and $w_2 \sim w_3$, then $w_1 \sim w_3$. The last property is called *transitivity* (see Exercise 2).

Example 1 Consider Example 2, Section 14.1, once again. The role of θ is played by p. We must examine the ratio of probabilities of two sample points. Suppose $e_1, \ldots, e_n$ is one sample point and $e_1', \ldots, e_n'$ is another. The ratio (1) equals

$$\frac{p^{e_1+\cdots+e_n}(1-p)^{n-(e_1+\cdots+e_n)}}{p^{e_1'+\cdots+e_n'}(1-p)^{n-(e_1'+\cdots+e_n')}}$$

This ratio does not depend on p if, and only if,

$$e_1 + \cdots + e_n = e_1' + \cdots + e_n'$$

In other words, two sample points are in the same set of the partition if, and only if, they both represent the same number of successes. This partition is induced by the sufficient statistic $X_1 + \cdots + X_n$, the total number of successes. This sum will be called a *minimal sufficient statistic.*

We shall now show that the name *minimal sufficient* is justified.

Any sufficient partition must be a subpartition of the minimal sufficient partition. Moreover, the minimal sufficient partition is sufficient.

We must show that if S is a sufficient statistic, then

$$w_1 \sim w_2 \quad \text{whenever} \quad S(w_1) = S(w_2)$$

This fact will imply that every set of the minimal sufficient partition is a union of sets of the form $(S = a)$ (see Figure 14.2). The argument is as follows. Since S is sufficient then whenever w is in $(S = a)$, then

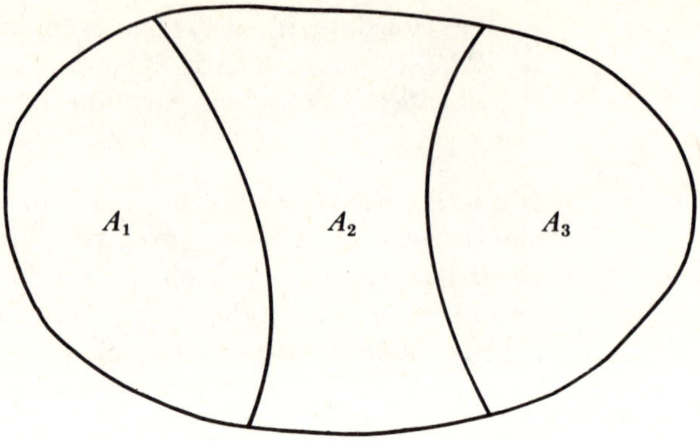

(a)

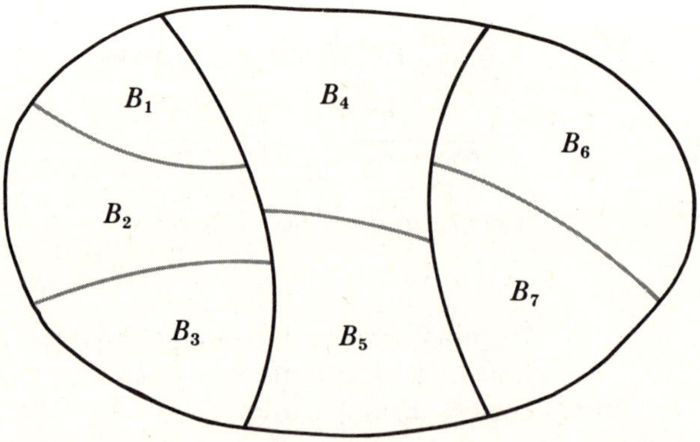

(b)

FIGURE 14.2 (a) The minimal sufficient partition is (A_1, A_2, A_3). (b) If $(B_1, B_2, B_3, B_4, B_5, B_6, B_7)$ is a sufficient partition, it must be a subpartition of (A_1, A_2, A_3).

$$P_\theta(w|S=a) = \frac{P_\theta(w)}{P_\theta(S=a)}$$

does not depend on θ. It follows that if w_1 and w_2 are both in $(S=a)$, then

$$\frac{P_\theta(w_1|S=a)}{P_\theta(w_2|S=a)} = \frac{P_\theta(w_1)}{P_\theta(w_2)}$$

does not depend on θ, and hence $w_1 \sim w_2$.

Now let us check that the minimal sufficient partition is indeed sufficient. Let Y be any random variable that induces the minimal sufficient partition. It is simply a random variable that is constant over any set of the partition and takes on distinct values over distinct sets. If w is in $(Y = a)$, then

$$P_\theta(w|Y = a) = \frac{P_\theta(w)}{\sum_i P_\theta(w_i)} \tag{2}$$

where the summation is over all sample points w_i that satisfy $w_i \sim w$. By (1), $P_\theta(w)/P_\theta(w_i)$ does not depend on θ, and hence (2) also does not depend on θ. It follows from (2) that $P_\theta(A|Y = a)$ does not depend on θ. ∎

minimal sufficient statistic

By a *minimal sufficient statistic* is meant any sufficient statistic Y such that the sets $(Y = a_1)$, $(Y = a_2)$, . . . , are the sets of the minimal sufficient partition. A minimal sufficient statistic must be a function of any other sufficient statistic. That is, if Y is minimal sufficient and S is sufficient, then there is a function h such that

$$Y(w) = h[S(w)] \quad \text{all } w \text{ in } \mathfrak{X} \tag{3}$$

The intuitive idea of a sufficient statistic S is that to get information about θ it suffices to compute S. Since the minimal sufficient statistic Y is a function of any other sufficient statistic, Y represents an "ultimate computation." It is the least complicated of the sufficient statistics. Any two minimal sufficient statistics must be equivalent, in the sense that each must be a one-to-one function of the other.

Proof of (3). By the previous theorem, the partition induced by S is finer than the partition induced by Y. That fact means that any set of the form $(Y = a)$ is made up of sets of the form $(S = b_1)$, $(S = b_2)$, ... Now, simply define $h(b_1) = h(b_2) = \cdots = a$. ∎

We should point out again that throughout all the preceding discussions concerning sufficient statistics, these can be random vectors as well as random variables.

Example 2

In Example 1, the pair of random variables Y_1, Y_2, given by

$$Y_1 = (X_1 + \cdots + X_m), \qquad Y_2 = (X_{m+1} + \cdots + X_n)$$

are jointly sufficient for p. A minimal sufficient statistic is $Y_1 + Y_2$.

Example 3 Suppose that $X_1, \ldots, X_n$ are mutually independent, Poisson-distributed random variables, each with the same parameter λ. Here, λ plays the role of θ and $\Omega = (0, \infty)$. If we consider a sample point to be $(k_1, \ldots, k_n)$, an n-tuple of possible values of the X_i's, then

$$\frac{P_\lambda(k_1, \ldots, k_n)}{P_\lambda(k_1', \ldots, k_n')} = \frac{\lambda^{k_1+\cdots+k_n}e^{-n\lambda}/(k_1! \cdots k_n!)}{\lambda^{k_1'+\cdots+k_n'}e^{-n\lambda}/(k_1'! \cdots k_n'!)}$$

This ratio does not depend on λ if, and only if,

$$k_1 + \cdots + k_n = k_1' + \cdots + k_n'$$

Hence, $X_1 + \cdots + X_n$ is a minimal sufficient statistic. It might be interesting to check the sufficiency of $X_1 + \cdots + X_n$ directly. Since this sum is itself Poisson distributed with parameter $n\lambda$, we have

$$P_\lambda[(X_1 = k_1) \cap \cdots \cap (X_n = k_n) | X_1 + \cdots + X_n = k = k_1 + \cdots + k_n]$$
$$= \frac{\lambda^{k_1+\cdots+k_n}e^{-n\lambda}/(k_1! \cdots k_n!)}{(n\lambda)^k e^{-n\lambda}/k!} = \binom{k}{k_1, \ldots, k_n}\left(\frac{1}{n}\right)^k$$

In other words, the conditional distribution of $X_1, \ldots, X_n$, given that $X_1 + \cdots + X_n = k$, is multinomial with parameters k; $1/n, \ldots, 1/n$. If $k = 0$, the conditional distribution is all concentrated on $(0, \ldots, 0)$.

Example 4 Consider Example 3, Section 14.1. We claim that the partition displayed in that example is the minimal sufficient partition. It is easy to see that $w_3 \sim w_4$ is the only equivalence among two distinct sample points.

14.3 THE FACTORIZATION THEOREM

If S is a sufficient statistic, then it is reflected in the probability measure taking on a special form.

factorization theorem

If S is sufficient for θ, then the probability measure P_θ is of the form

$$P_\theta(w) = h(w)g[S(w), \theta] \tag{1}$$

where h does not depend on θ. As indicated, g does depend on θ, and it depends on the sample point w only through its value $S(w)$. Conversely, if P_θ is of the form (1), then S is sufficient for θ.

Proof. Suppose that S is sufficient for θ and that $S(w) = u$. Then

$$P_\theta\{w|[S = S(w)]\} = \frac{P_\theta(w)}{P_\theta[S = S(w)]} = h(w)$$

does not depend on θ by the definition of sufficiency. Now define

$$g(S(w), \theta) = P_\theta[S = S(w)]$$

to obtain the factorization (1). Conversely, if the probability measure is of the form (1), then, if w is in $(S = u)$,

$$P_\theta(w|S = u) = \frac{P_\theta(w)}{\sum_{S(w)=u} P_\theta(w)} = \frac{h(w)g(S(w), \theta)}{\sum_{S(w)=u} h(w)g(S(w), \theta)} = \frac{h(w)}{\sum_{S(w)=u} h(w)}$$

This conditional probability does not depend on θ; hence, S is sufficient. ∎

We should point out that the factorization (1) is not at all unique. If hg is one factorization, then $(fh)(g/f)$ is another, if this product is well defined, and f is a function of w, which does not depend on θ.

Example 1 Consider n independent repetitions of the same success-failure experiment. The probability of a sample point $w = (e_1, \ldots, e_n)$, (each e_i is 1 or 0), is

$$P_p(w) = p^{e_1+\cdots+e_n}(1 - p)^{n-(e_1+\cdots+e_n)}$$

The role of h in (1) is played by the constant 1 and

$$g(S(w), p) = p^{S(w)}(1 - p)^{n-S(w)}$$

where $S(w) = e_1 + \cdots + e_n$. Hence, as we have seen several times before, the total number of successes is sufficient for p.

Example 2 Consider Example 3, Section 14.2. Since

$$P_\lambda(k_1, \ldots, k_n) = (k_1! \cdots k_n!)^{-1}\lambda^{k_1+\cdots+k_n}e^{-n\lambda}$$

it follows that $S(k_1, \ldots, k_n) = k_1 + \cdots + k_n$ defines a sufficient statistic. We have already seen that the sum is minimal sufficient.

Example 3 A dichotomous population consists of N objects, M of which are of type 1 and $N - M$ of type 2. The population size N is known to the experimenter, but M is not. The role of θ is played by M, and $\Omega = (0, 1, \ldots, N)$. Suppose that n drawings *without replacement* are made, and $X_1, \ldots, X_n$ indicate whether these drawings are of type 1. That is,

$$X_i = \begin{cases} 1 & \text{if } i\text{th drawing is of type 1} \\ 0 & \text{if } i\text{th drawing is of type 2} \end{cases}$$

If $(e_1, \ldots, e_n) = w$ is a sequence of 1's and 0's, then

$$P_M(w) = \frac{M^{(k)}(N-M)^{(n-k)}}{N^{(n)}} = g(k, M)$$

where $k = e_1 + \cdots + e_n$. Again, the role of h in the factorization $P = hg$ is played by $h = 1$. Hence, the total number of type 1 objects drawn is sufficient for M. In fact, it is minimal sufficient; verification is left to the reader (see Exercise 4). It is intuitively plausible that once one knows the total number of 1's drawn the particular order in which they were drawn should not be helpful in getting information about M.

14.4 ONLY SUFFICIENT STATISTICS NEED BE USED

The intuitive idea behind a sufficient statistic is that once it is known then further information about the unknown parameter θ cannot possibly be obtained by further analysis of the experiment. We now want to make this idea somewhat more precise. First, let us consider some problems of estimation.

Recall the intuitive method of maximum likelihood estimation described in Section 12.2.

> The maximum likelihood estimator of θ is a function of the minimal sufficient statistic.

Proof. It follows from the factorization theorem of Section 14.3. If S is sufficient for θ, then the probability measure is given by

$$h(w)g(S(w), \theta) \tag{1}$$

For any given sample point w, the maximum likelihood estimator is a value of θ that makes (1) as large as possible. Such a value is determined by $S(w)$; hence, the maximum likelihood estimator is a function of any sufficient statistic. In particular, it depends on the minimal sufficient statistic. ∎

Example 1 Suppose that $X_1, \ldots, X_n$ are mutually independent random variables, each Poisson distributed with the same parameter λ. The maximum likelihood estimator of λ is obtained by maximizing

$$\lambda^{X_1+\cdots+X_n}e^{-\lambda n}/X_1! \cdots X_n! \tag{2}$$

with respect to λ. By taking the logarithm of (2) and maximizing, we see that the maximum likelihood estimator of λ is

$$\frac{X_1 + \cdots + X_n}{n}$$

which is a minimal sufficient statistic.

Next, we consider estimation procedures using certain special loss functions.

> Suppose that in estimating $h(\theta)$ by $u(X)$, the expected loss is given by
>
> $$E_\theta\{[u(X) - h(\theta)]^2\} = R(u, \theta)$$
>
> There exists an estimator based on a minimal sufficient statistic Y, for which the expected loss is no greater than $R(u, \theta)$, uniformly in θ.

This fact suggests that whatever estimator one uses for $h(\theta)$, it may as well be based on the minimal sufficient statistic, since no other estimator can improve on such a procedure.

Proof. Use formula (2), Section 4.3. According to that formula,

$$E_\theta\{[u(X) - h(\theta)]^2\} = \sum_y E_\theta\{[u(X) - h(\theta)]^2|Y = y\}P_\theta(Y = y) \tag{3}$$

(Even if Y is a random *vector*, this calculation is correct.) Now use the fact that for any random variable Z,

$$E(Z^2) \geqslant (EZ)^2 \tag{4}$$

(See Exercise 25b, Chapter 2. This inequality is true even if the expected value is a conditional expected value. Hence,

$$E_\theta\{[u(X) - h(\theta)]^2|Y = y\} \geqslant \{E_\theta[u(X) - h(\theta)]|Y = y\}^2 \tag{5}$$

Because Y is a sufficient statistic, it follows that the distribution of X, given $Y = y$, does not depend on θ. Hence,

$$E_\theta[u(X)|Y = y] = T(y)$$

defines a function T *that does not depend on* θ. It follows that $T(Y)$ is an estimator, and, from (3) and (5),

$$E_\theta\{[u(X) - h(\theta)]^2\} \geqslant \sum_y [T(y) - h(\theta)]^2 P_\theta(Y = y)$$

$$= E_\theta\{[T(Y) - h(\theta)]^2\} \qquad (6)$$

The inequality (6) says that the estimator $T(Y)$ is at least as good an estimator as is $u(X)$, uniformly in θ. ▮

The preceding theorem also holds true with other loss functions. For instance, suppose that the expected loss is given by

$$E_\theta|u(X) - h(\theta)| = R(u, \theta)$$

Then it is still true that there exists an estimator based on a minimal sufficient statistic Y, for which the expected loss is no greater than $R(u, \theta)$, uniformly in θ. The proof is exactly the same as the previous one, except that now instead of (4) we use the fact that

$$E|Z| \geqslant |EZ| \qquad (7)$$

The proof is left to the reader (see Exercise 10). This will also follow from the proof of (8).

There is a general statement about expected values called Jensen's inequality, of which both (4) and (7) are special cases, which relates to convex functions. There are various ways of defining convex functions. We shall take the following definition: $W(x)$ is called convex if, given any point $(a, W(a))$, there is straight line $cx + d$ having the following properties:

$$ca + d = W(a)$$

$$cx + d \leqslant W(x)$$

In other words, given any point on the graph of $W(x)$, there is a straight line that goes through that point and lies below the graph (see Figure 14.3). Line $cx + d$ is called a *supporting line.* To summarize, we call $W(x)$ convex if, given any point on the graph, there is a supporting line through that point. The general version of (4) and (7) is that if W is a convex function, then

$$E[W(Z)] \geqslant W[E(Z)] \qquad (8)$$

The proof of (8) follows from the fact that

$$E[W(Z)] \geqslant E(cZ + d) = cEZ + d = W[E(Z)]$$

assuming that c and d have been chosen so that the supporting line goes through the point $E(Z), W[E(Z)]$.

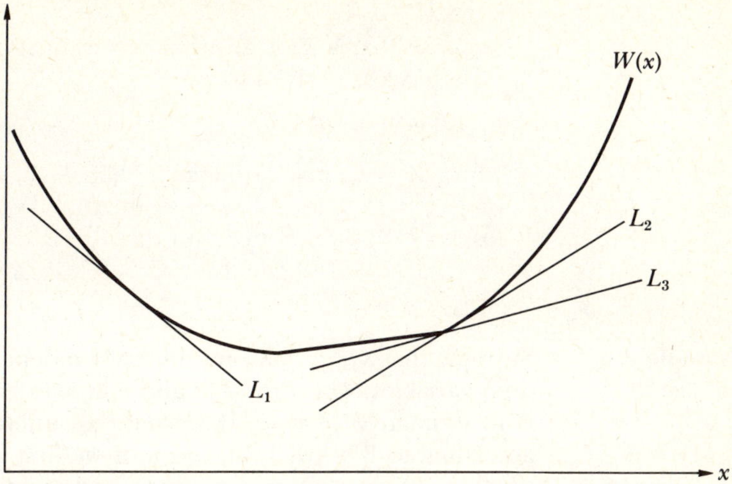

FIGURE 14.3 $W(x)$ **is convex, and** L_1, L_2 **and** L_3 **are each supporting lines.**

Suppose that in estimating $h(\theta)$ by $u(X)$, the expected loss is given by

$$E_\theta\{W[u(X) - h(\theta)]\} = R(u, \theta)$$

where W is a convex function. There exists an estimator based on a minimal sufficient statistic Y, for which the expected loss is no greater than $R(u, \theta)$, uniformly in θ. A specific version of the improved estimator is given by $T(Y)$, where

$$T(y) = E_\theta[u(X)|Y = y] \tag{9}$$

The proof follows through the steps (3) through (6) replacing the square of $u(X) - h(\theta)$ by $W[u(X) - h(\theta)]$ throughout. The details are left to the reader.

Notice that if $u(X)$ happens to be an unbiased estimator of $h(\theta)$, then $T(Y)$ is an unbiased estimator of $h(\theta)$ also. It follows from the fact that

$$E_\theta[T(Y)] = \sum_y T(y)P_\theta(Y = y) = \sum_y E_\theta[u(X)|Y = y]P_\theta(Y = y)$$
$$= E_\theta[u(X)] = h(\theta)$$

again by (2), Section 4.3. The last theorem then becomes the following, which is known as *Blackwell's theorem* (Note 3).

Suppose that $u(X)$ is an unbiased estimator of $h(\theta)$, and the expected loss is given by

$$E_\theta\{W[u(X) - h(\theta)]\}$$

where W is a convex function. Then $T(Y)$, where T is defined by (9), is also an unbiased estimator of $h(\theta)$. The expected loss of $T(Y)$ does not exceed that of $u(X)$, uniformly in θ.

Example 2 Suppose that $X_1, \ldots, X_n$ are mutually independent, indicator random variables, with $P(X_i = p)$ and n at least equal to 2. The object is to estimate $h(p) = p^2$. If we want an unbiased estimator, then, according to Blackwell's theorem, if we first find any unbiased estimator, we can then find an improved estimator based on a minimal sufficient statistic by the construction (9). Recall from Example 1, Section 14.2, that $Y = X_1 + \cdots + X_n$ is a minimal sufficient statistic for p. Let us use the fact that

$$E_p(X_1X_2) = p^2 \quad \text{identically in } p$$

and let X_1X_2 play the role of $u(X)$. To apply (9), we do the following computation:

$$E_p(X_1X_2|Y = k) = \frac{P_p[(X_1 = 1) \cap (X_2 = 1) \cap (X_3 + \cdots + X_n = k-2]}{P_p(Y = k)}$$

$$= \frac{p^2\binom{n-2}{k-2}p^{k-2}(1-p)^{n-k}}{\binom{n}{k}p^k(1-p)^{n-k}} = \frac{k(k-1)}{n(n-1)}$$

Hence,

$$T(Y) = \frac{Y(Y-1)}{n(n-1)}$$

must be an unbiased estimator of p^2. As a matter of fact, $T(Y)$ *is the only unbiased estimator of* p^2 *based on* Y. This follows from the fact that if

$$\sum_{k=0}^{n} a(k)\binom{n}{k}p^k(1-p)^{n-k} = p^2 \quad \text{identically in } p$$

the coefficients $a(k)$ are determined uniquely. (See Exercise 15c, Chapter 12.) Hence, $Y(Y-1)/n(n-1)$ is an unbiased estimator of p^2 which is of minimum variance among all unbiased estimators.

Example 3 Suppose that $X_1, \ldots, X_n$ are mutually independent, Poisson distributed, each with the same parameter λ. By Example 3, Section 14.2, $Y = X_1 + \cdots + X_n$ is a minimal sufficient statistic for λ. Consider the problem of estimating λ^m where m is a positive integer. Since the mth factorial moment of the Poisson distribution with parameter λ equals λ^m [see (4), Section 6.7], it follows that $X_1^{(m)} = u(X)$ is an unbiased estimator of λ^m. The conditional distribution of X_1 given $Y = k$ is binomial with parameters k and $p = 1/n$. [This is easy to check. See (5), Section 11.12, for a similar computation.] Hence, by Section 6.1,

$$E(X_1^{(m)} | Y = k) = \frac{k^{(m)}}{n^m}$$

It follows that

$$T(Y) = \frac{Y^{(m)}}{n^m} = \frac{Y(Y-1) \cdots (Y-m+1)}{n^m}$$

is the improved unbiased estimator of λ^m given by Blackwell's theorem. Just as in the previous example, $T(Y)$ is the only unbiased estimator of λ^m which is a function of Y. This follows from the fact that if

$$\sum_{k=0}^{\infty} \frac{a(k)(n\lambda)^k e^{-n\lambda}}{k!} = \lambda^m \quad \text{identically in } \lambda$$

then the coefficients $a(k)$ are uniquely determined (see Exercise 8). It follows that among all unbiased estimators, $T(Y)$ minimizes the expected loss $E_\lambda\{W[u(X) - \lambda^m]\}$ uniformly in λ if W is convex. In particular, $T(Y)$ is a uniformly minimum variance unbiased estimator.

Example 4 A safety engineer is using the model that the number of accidents per year in a factory is Poisson distributed with parameter λ. He also assumes that accidents in successive years are independent. The model is then that $X_1, \ldots, X_n$ are independent, Poisson-distributed random variables, each with parameter λ. The object is to find a good unbiased estimator of $e^{-\lambda}$, the probability that any given year is free of accidents. Let $Y_1 = 1$ if $X_1 = 0$, and $Y_1 = 0$ otherwise. Then $E(Y_1) = e^{-\lambda}$. We will now use (9) to find an improved unbiased estimator. The required calculation is the following:

$$E(Y_1 | X_1 + \cdots + X_n = k) = \frac{P(X_1 = 0 \text{ and } X_2 + \cdots + X_n = k)}{P(X_1 + \cdots + X_n = k)} = \left(1 - \frac{1}{n}\right)^k$$

Hence,

$$\left(1 - \frac{1}{n}\right)^{X_1+\cdots+X_n} \tag{10}$$

is an improved unbiased estimator based on the minimal sufficient statistic. As in Example 3, there is only one unbiased estimator of $e^{-\lambda}$ based on $X_1 + \cdot\cdot\cdot + X_n$. [If $\Sigma_{k=0}^{\infty} a(k)(\lambda n)^k e^{-\lambda n} = e^{-\lambda}$ identically in λ, it determines the coefficients $a(k)$ uniquely. See Exercise 8.] Hence, (10) must be an unbiased estimator of $e^{-\lambda}$ the variance of which is minimum. Can you show directly that (10) is unbiased? Can you use the law of large numbers to show that the distribution of (10) is highly concentrated about the point $e^{-\lambda}$ when n is large?

As a final elaboration of the idea that only the minimal sufficient statistic need be used in statistical problems, we point out an application to the two-decision problem. Recall the definition of $D(X)$ as given in Section 13.5. It follows from the factorization theorem [(1), Section 14.3] that $D(X)$ depends on the observations only through a minimal sufficient statistic. (We leave the easy verification to the reader. See Exercise 11.) Hence, a Bayes' decision procedure need be based only on the minimal sufficient statistic.

14.5 SUFFICIENCY IN THE NONDISCRETE CASE

In the preceding sections of this chapter, we have purposely restricted ourselves to discrete sample spaces. The ideas about sufficiency go over to a general setting with hardly any change. A general approach requires a definition of conditional probability, which is quite abstract (Note 1, Chapter 7). However, it is a fact that the description of minimal sufficiency in Section 14.2 and the factorization theorem in Section 14.3 are still true in a more general setting. We shall look at several nondiscrete examples, purely from an intuitive point of view (Note 4).

Example 1 Suppose that $X_1, \ldots, X_n$ are mutually independent random variables, each uniformly distributed over $[0, \theta]$, $\Omega = (0, \infty)$. The minimal sufficient statistic for θ is

$$Y = \max(X_1, \ldots, X_n)$$

An intuitive argument that Y is sufficient follows. Suppose that Y has a given value y. From the nature of the uniform distribution, given that the largest X_i equals y, the conditional distribution of the remaining X_i's is that of $n-1$ mutually independent random variables, each uniformly distributed over $[0, y]$. Hence, the conditional distribution does not depend on θ. To show that Y is *minimal* sufficient, we follow the characterization in Section 14.2 as follows. Suppose that a sample point is represented by $(x_1, \ldots, x_n)$. The probability measure is assigned by an integrating density,

$$P(x_1, \ldots, x_n) = \begin{cases} \dfrac{1}{\theta^n} & \text{if each } x_i \text{ is in } [0, \theta] \\ 0 & \text{otherwise} \end{cases}$$

Two sample points $(x_1, \ldots, x_n)$ and $(x_1', \ldots, x_n')$ are in the same equivalence class if, and only if, the ratio

$$\frac{P_\theta(x_1, \ldots, x_n)}{P_\theta(x_1', \ldots, x_n')} \tag{1}$$

does not depend on θ for all θ's in Ω. It is easy to check that this fact is true if, and only if,

$$\max(x_1, \ldots, x_n) = \max(x_1', \ldots, x_n')$$

For instance, suppose $y' = \max(x_1', \ldots, x_n') > y = \max(x_1, \ldots, x_n)$. Then (1) equals one if $\theta > y'$ but equals 0 if $y < \theta < y'$. Hence, $Y = \max(X_1, \ldots, X_n)$ is minimal sufficient. Intuitively, all this is certainly very plausible. The most information that one can possibly find about θ lies in the maximum of the observations. Once it is known that the maximum equals y, the values of the other observations have no relevance at all to learning about θ. It was shown in Example 2, Section 12.2, that $[(n+1)/n]Y$ is an unbiased estimator of θ. Blackwell's theorem, as it relates to this nondiscrete case, says that this unbiased estimator has smallest expected loss, uniformly in θ, among all unbiased estimators.

Example 2 Suppose that $X_1, \ldots, X_n$ are mutually independent random variables, each $N(m, \sigma^2)$. Suppose also that $\theta = (m, \sigma)$ and Ω is the half-plane where $-\infty < m < \infty$ and $\sigma > 0$. The joint integrating density of the X_i's is

$$f(x_1, \ldots, x_n; m, \sigma) = \frac{1}{(2\pi\sigma^2)^{n/2}} \exp - \sum_{i=1}^{n} (x_i - m)^2/2\sigma^2$$

$$= \frac{1}{(2\pi\sigma^2)^{n/2}} \exp -\left[\sum_{i=1}^{n} x_i^2 - 2m \sum_{i=1}^{n} x_i \right.$$

$$\left. + nm^2\right]/2\sigma^2$$

It follows from the factorization theorem, Section 14.3, that

$$Y = \left(\sum_{i=1}^{n} X_i^2, \sum_{i=1}^{n} X_i\right)$$

is sufficient. It is also a fact that Y is minimal sufficient. An equivalent minimal sufficient statistic is given by the pair $\overline{X}$, S, where

$$\overline{X} = \frac{X_1 + \cdots + X_n}{n}$$

is the *sample mean* and

$$S = \left[\frac{\sum_{i=1}^{n} (X_i - \overline{X})^2}{n}\right]^{1/2}$$

is the *sample standard deviation* (see Example 4, Section 12.2).

SUMMARY

We consider the same situation as in Chapters 12 and 13, that the probability distribution for the observed data is labeled by a parameter θ that varies over some parameter set Ω. For simplicity, however, we suppose throughout that the sample space is discrete.

A *partition* of the sample space is called *sufficient* for θ if the conditional probabilities of sample points, given any sets of the partition, do not depend on θ. Any partition is induced by a random variable. If the partition is sufficient, we say that any random variable inducing that partition is a *sufficient statistic.*

It is proved that there exists a *minimal sufficient partition.* That is, there is a sufficient partition such that any other sufficient partition is a subpartition of it. A random variable inducing a minimal sufficient partition is called a *minimal sufficient statistic.*

An important theorem that allows one to recognize sufficient statistics easily is the *factorization theorem.*

If we are looking for unbiased estimators with small variance of the estimator as a criterion of goodness, we prove that such estimators may as well be based on sufficient statistics. Similarly, in the decision-making problem, there is no loss in supposing that Bayes' estimators are based on sufficient statistics.

EXERCISES

1 Verify the assertion in Example 4, Section 14.1.

2 Verify the assertion about $\sim$ following (1), Section 14.2. {Partial hint: The fact that $w_i \sim w_j$ means that $P_\theta(w_i)/P_\theta(w_j)$ does not depend on θ. Hence, $w_1 \sim w_2$ and $w_2 \sim w_3$ imply that $[P_\theta(w_1)/P_\theta(w_2)][P_\theta(w_2)/P_\theta(w_3)]$ does not depend on θ. But this product equals $P_\theta(w_1)/P_\theta(w_3)$. Hence, $w_1 \sim w_3$.}

3 Let $X_1, \ldots, X_n$ be mutually independent random variables, each geometrically distributed with the same parameter p [see (1), section 6.4], where p varies over $(0, 1) = \Omega$. Show that $X_1 + \cdots + X_n$ is minimal sufficient for p.

4 Prove the minimal sufficiency asserted in Example 3, Section 14.3. [Hint: Examine the ratio

$$\frac{M^{(k_1)}(N-M)^{(n-k_1)}}{M^{(k_2)}(N-M)^{(n-k_2)}}$$

If $k_1 > k_2$, find one value of M for which this ratio is well defined and one for which it is of the form $a/0$, $a > 0$. A similar argument works when $k_1 < k_2$.]

5 Suppose that $X = (X_1, \ldots, X_n)$ and $Y = h(X)$ is a random variable that is minimal sufficient for θ.

(a) If $Y \geqslant 0$, show that Y^2 is also minimal sufficient for θ.

(b) If Y is not necessarily nonnegative, is Y^2 still minimal sufficient?

6 Suppose that $\Omega = (\theta_1, \theta_2, \theta_3)$ and the possible densities of X are the following:

u	$f(u, \theta_1)$	u	$f(u, \theta_2)$	u	$f(u, \theta_3)$
0	0.10	0	0.30	0	0.12
1	0.50	1	0.05	1	0.32
2	0.20	2	0.05	2	0.32
3	0.15	3	0.45	3	0.18
4	0.05	4	0.15	4	0.06

(a) Show that the three sets (0, 3, 4), (1), and (2) give a minimal sufficient partition for θ.

(b) Consider the two-decision problem of deciding whether $\theta = \theta_1$ or θ is one of θ_2, θ_3 $[\Omega_1 = (\theta_1), \Omega_2 = (\theta_2, \theta_3)]$. Suppose that a priori weights on $\theta_1, \theta_2, \theta_3$ are 3/6, 2/6, and 1/6, respectively. Determine a Bayes' procedure for this decision problem. [Answer: Decide in favor of Ω_1 if $X = 1$ or $X = 2$; otherwise, decide in favor of Ω_2. Use Section 13.5. Notice that this procedure is constant on the sets of the minimal sufficient partition described in (b). Why?]

(c) Show directly that if ρ_1, ρ_2, ρ_3 are a priori weights on $\theta_1, \theta_2, \theta_3$, then any Bayes' procedure always leads to the same decision when X equals 0, 3, or 4.

7 A box contains N balls. The value N plays the role of the unknown parameter θ, and $\Omega = (1, 2, \ldots)$. A total of n drawings are made with successive replacement between drawings. Let $X_1, \ldots, X_n$ be the succession of n numbers drawn.

(a) Show that $Y = \max(X_1, \ldots, X_n)$ is minimal sufficient for estimating N.

(b) Show that the conditional density of X_1 given Y is the following:

$$P(X_1 = k|Y = r) = \begin{cases} \dfrac{r^{n-1} - (r-1)^{n-1}}{r^n - (r-1)^n} & \text{if } k < r \\[2ex] \dfrac{r^{n-1}}{r^n - (r-1)^n} & \text{if } k = r \end{cases}$$

and, hence,

$$E(X_1|Y = r) = \frac{[r^{n+1} + r^n - (r-1)^{n+1} - (r-1)^n]/2}{r^n - (r-1)^n}$$

(c) Conclude from (b) that

$$E(2X_1 - 1|Y = r) = \frac{r^{n+1} - (r-1)^{n+1}}{r^n - (r-1)^n}$$

(d) Conclude from (c), and from the fact that $E(2X_1 - 1) = N$, that

$$\frac{Y^{n+1} - (Y-1)^{n+1}}{Y^n - (Y-1)^n} = T(Y)$$

is an unbiased estimator of N. [Hint: Use Eq. (9), Section 14.4.]

(e) Verify directly that $T(Y)$ is unbiased. That is, check that

$$\sum_{r=1}^{N} T(r)P(Y = r) = N \quad \text{for } N = 1, 2, \ldots$$

(*f*) Show that there is only one unbiased estimator of N based on Y. Hence, by Blackwell's theorem, $T(Y)$ is the minimum variance unbiased estimator of N. {Hint: Suppose that f is a function that does not depend on N, such that

$$\sum_{r=1}^{N} f(r)P(Y = r) = N, \qquad N = 1, 2, \ldots$$

Observe that $N = 1$ determines $f(1)$, $N = 2$ determines $f(2)$, and so forth. Use the fact that $P(Y = r) = [r^n - (r-1)^n]/N^n$. Check also that f agrees with T, as it must.}

8 Prove the assertion in Example 3, Section 14.4, that $Y^{(m)}/n^m$ is the only unbiased estimator of λ^m based on Y. [Hint: Unbiasedness means that

$$\sum_{k=0}^{\infty} \frac{a(k)(n\lambda)^k}{k!} = \lambda^m e^{n\lambda}$$

The coefficients $a(k)n^k/k!$ of λ^k must be uniquely determined.]

9 A factory produces 100 items daily. These items are, upon inspection, clearly classifiable as *acceptable* or *defective*. Consider one given day's production and let θ denote the total number of defectives produced. The quality control plan selects a sample for inspection as follows. First, a single item is inspected. If it is acceptable, no further inspection is done. If the first item is defective, an additional item is selected and inspected. Denote the sample points for this experiment as follows:

0	first item acceptable
1, 0	first item defective, second item acceptable
1, 1	first item defective, second item defective

(a) Show that the only sufficient partition of the sample space is the partition into three sets, each consisting of one of the sample points.

(b) Show that the only unbiased estimator of the proportion of defectives in the population is X, where

$$X = \begin{cases} 1 & \text{if first item is defective} \\ 0 & \text{otherwise} \end{cases}$$

(Since it seems unreasonable to ignore the information obtained from inspecting a second item, what does this fact suggest about the advisability of unbiased estimators in this case?)

(c) Show that the maximum likelihood estimator of the proportion of defectives is Y, the proportion of defectives among all the items inspected. That is, $Y(0) = 0$, $Y(1, 0) = 1/2$, $Y(1, 1) = 1$.

(d) Compute and compare $E\{[X - (\theta/100)]^2\}$ and $E\{[Y - (\theta/100)]$.

10 Prove (7), Section 14.4. (Hint: $-|Z| \leqslant Z \leqslant |Z|$.)

11 Prove the assertion in the last paragraph of Section 14.4.

12 In genetics, one sometimes considers a "truncated" binomial variable X, where

$$P(X = k) = c\binom{n}{k}p^k(1 - p)^{n-k}, \qquad k = 1, \ldots, n$$

The value $k = 0$ is not a possible value of X. For the probabilities to sum to 1, $c = [1 - (1 - p)^n]^{-1}$. Suppose that $X_1, \ldots, X_m$ are independent random variables, each with the preceding truncated binomial distribution with the same n and p.

(a) Show that $X_1 + \cdots + X_m$ is minimal sufficient for p.

(b) Consider the Bayes' procedures for deciding whether $p = p_1$ versus $p = p_2 > p_1$. Use the Neyman–Pearson theorem (Section 13.6) to show that these procedures are of the form

$$\sum_{i=1}^{m} X_i \leqslant c, \qquad \text{decide on } p_1$$

$$\sum_{i=1}^{m} X_i > c, \qquad \text{decide on } p_2$$

NOTES

1 For additional reading about sufficiency, see Section 4.5 of [50] and Sections 8.3 and 8.4 of [51].

2 The concept of sufficiency is due to R. A. Fisher. The fundamental paper [62] should be within the scope of readers of this book. See also Fisher's collected works, [63].

3 David Blackwell (1919–   ) is a contemporary American mathematician.

4 The first rigorous treatment of sufficiency appeared in [64].

Chapter Fifteen

DICHOTOMOUS POPULATIONS

The purpose of this chapter is to apply the preceding material on statistical methods in Chapters 12, 13, and 14 to problems concerning dichotomous populations. A dichotomous population is one having elements of two types. The statistical problems concern estimating or making decisions about the unknown proportion of one of the types. We have attempted to make this chapter as self-contained as possible, although occasional references are made to the general treatment of various concepts in earlier chapters (Note 1).

15.1 DICHOTOMOUS POPULATIONS

In this chapter, we apply some of the statistical ideas from Chapters 12, 13, and 14 to the study of dichotomous populations. It might be helpful to review the probability models for sampling from such populations, as given in Section 3.8. Consider a population consisting of two types of objects. For simplicity, denote these by 1's and 0's, and let p denote the proportion of 1's in the population. Suppose that an investigator wants to know about this p. It may be possible to make a complete survey of the population and thus determine p exactly. However, this process may be too costly or impractical, or even impossible. The best that can be done, then, is to obtain a sample from the population and to make an inference about the unknown p from the sample.

Example 1 A population of voters consists of a proportion p that intend to vote for candidate A. The estimation of the value of p before the actual election is of great interest to candidate A and his political machine, to his opponents, to the press, and to the public at large.

Example 2 A firm produces items that it sells to distributors in lots of 1000. The firm guarantees to its distributors that no lot will contain more than 5% defective items; otherwise, it must pay the distributor a specified penalty. If the firm knew beforehand that a lot contained more than 5% defectives, it would be less costly to weed these out by testing the whole lot than to suffer the penalty. On the other hand, the cost of inspection is sufficiently high that the firm must avoid 100% inspection in general. Associated with each lot is a number p, the proportion of defectives. A quality control plan recommends inspecting a certain number of items. On the basis of this inspection, a decision is to be made as to whether the lot will be sent out to a distributor (Note 2).

Example 3 The circumference of a dial has a certain proportion p marked 1 and the remainder marked 0 (see Figure 15.1). A well-balanced pointer is spun. The experimenter does not know the value of p, but he can buy successive spins of the pointer to try to estimate p. There is no actual population of 1's and 0's at hand here, as in the previous examples. One can conceive of a hypothetical population of all the results of an infinite sequence of spins. According to the frequency theory of probability (Section 1.2), the proportion of 1's among all these spins would be p.

There are several ways of sampling from a dichotomous population. One way is to *sample with replacement*. Suppose that an item is

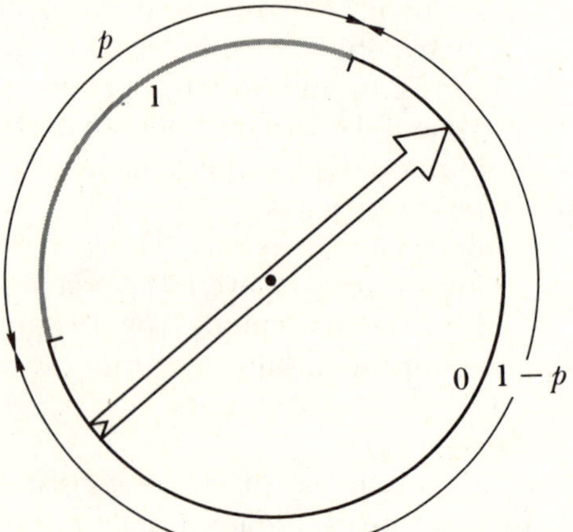

FIGURE 15.1

drawn at random, its type is observed, and the *drawn item is returned to the population.* A second item is drawn at random, and so forth. The mathematical model for this scheme of successive drawings with replacement is the following. Let X_i denote the number obtained on the ith drawing. Then $X_1, X_2, \ldots$ is a sequence of mutually independent indicator random variables, each equaling 1 with probability p and 0 with probability $1-p$. This model has already been described in Section 3.8. Notice that neither the size of the population nor the *number* of 1's enters into the distribution of the indicator random variables, $X_1, X_2, \ldots$. Only p, the *proportion* of 1's in the population, is of importance.

Now consider the same situation as before, except that the successive *items drawn are not replaced.* In the mathematical model for sampling without replacement, the size of the population and the number of 1's becomes relevant. Suppose the population consists of N elements, M of which are 1's and $N-M$ of which are 0's. Items are drawn at random without replacement. Of course, only a maximum of N drawings are possible. A mathematical model for these drawings is provided by $X_1, X_2, \ldots$, a sequence of indicator random variables. The probability distribution of the X_i's is described as follows:

$$P[(X_1 = e_1) \cap \cdots \cap (X_n = e_n)] = \frac{M^{(k)}(N-M)^{(n-k)}}{N^{(n)}}$$

where e_i is 1 or 0 and $k = e_1 + \cdots + e_n$. The X_i's are no longer independent in sampling without replacement, but each of the X_i's individually does have the same distribution. When $n = N$, then the values of $X_1, \ldots, X_n$ represent a *complete census* of the population.

The total number of 1's obtained in n drawings is $Y = X_1 + \cdots + X_n$. In sampling with replacement, Y is *binomially distributed.* In sampling without replacement, Y follows the *hypergeometric distribution.* Explicitly, these distributions are as follows:

With replacement

Binomial

$$P(Y=k) = \binom{n}{k}\left(\frac{M}{N}\right)^k\left(\frac{N-M}{N}\right)^{n-k}$$

$$k = 0, 1, \ldots, n$$

Without replacement

Hypergeometric

$$P(Y=k) = \binom{n}{k}\frac{(M)^{(k)}(N-M)^{(n-k)}}{N^{(n)}}$$

$$= \frac{\binom{M}{k}\binom{N-M}{n-k}}{\binom{N}{n}}$$

$$k = 0, 1, \ldots, n$$

$$n = 1, \ldots, N$$

15.2 ESTIMATING p

Suppose that n drawings are made from a dichotomous population. As previously described, the n random variables $X_1, \ldots, X_n$ serve as a model for the results of the drawings. Suppose that the experimenter wants to estimate $p = M/N$, the proportion of 1's in the population, from the values observed during n drawings. What should estimating p mean? An estimator is a summarizing number obtained from the data, which, "on the average," equal p. Mathematically, one would say that an estimator of p is a function $u(X_1, \ldots, X_n)$ the expected value of which equals p for all p. This definition is not quite good enough—one could choose $u(X_1, \ldots, X_n) \equiv p$, but it would not make sense. The experimenter does not know p and he wants to estimate it from his data, so it is silly to ask him to use p. An estimator should be $u(X_1, \ldots, X_n)$, a function of the observations that does not involve the unknown parameter p. An unbiased estimator was defined in Section 12.2. We repeat the definition within our present setting.

unbiased estimator

$u(X_1, \ldots, X_n)$ is called an *unbiased estimator* of p if
(a) u is completely determined by the values of the X_i's and does not depend on p, and
(b) $E_p[u(X_1, \ldots, X_n)] = p$ identically in p.

Requirement (b) of this definition should be elaborated. The expected value $E_p[u(X_1, \ldots, X_n)]$ is computed from the distribution of the X_i's, which depends on p. For instance, in sampling with replacement,

$$E_p[u(X_1, \ldots, X_n)] = \sum_{e_1, \ldots, e_n} u(e_1, \ldots, e_n) p^{e_1 + \cdots + e_n}(1-p)^{n-(e_1+\cdots+e_n)}$$

where each e_i is 0 or 1. Part (b) requires that whatever the value of p might be, this summation must turn out to equal p.

Example 1 Suppose that two drawings are made (either *with* or *without* replacement). The expected values of

$$X_1, \quad X_2, \quad \frac{X_1 + X_2}{2}, \quad \frac{1}{3}X_1 + \frac{2}{3}X_2$$

all equal p, so each of these is an unbiased estimator of p.

Example 2 Suppose n drawings are made (again either with or without replacement). A common unbiased estimator of p is the "sample proportion," $(X_1 + \cdots + X_n)/n$. It has expected value p, since $E_p(X_i) = p = P(X_i = 1)$. It is certainly reasonable that the proportion of 1's in the population should be estimated by the proportion of 1's in the sample.

Example 3 Consider the same situation as in Example 2. Suppose $a_1, \ldots, a_n$ are constants that sum to 1. Then $a_1X_1 + \cdots + a_nX_n$ is an unbiased estimator of p. The special case of $a_1 = \cdots = a_n = 1/n$ gives the *sample proportion.*

One can extend the idea of estimating p to a broader notion of estimating other functions of p. For instance, an unbiased estimator of p^2 is $h(X_1, \ldots, X_n)$, a function of the X_i's that does not depend on p and the expected value of which is identically equal to p^2.

Example 4 Suppose that the sampling is with replacement. Then $[(X_1 + \cdots + X_n)^2 - (X_1 + \cdots + X_n)]/n(n-1)$ is an unbiased estimator of p^2. (See Section 6.1 and Example 2, Section 14.4.)

15.3 VARIANCE AND OTHER LOSS FUNCTIONS

Suppose that $u(X_1, \ldots, X_n)$ is an unbiased estimator of the proportion p. Since there are many other unbiased estimators of p, how should one measure the effectiveness of this particular one? What does it mean for one estimator to be better than another? A commonly used measure of goodness of an estimator is its variance. If $u(X_1, \ldots, X_n)$ and $u'(X_1, \ldots, X_n)$ are two competing unbiased estimators, and if it happens that for every p

$$\text{Var}[u(X_1, \ldots, X_n)] \leq \text{Var}[u'(X_1, \ldots, X_n)] \qquad (1)$$

with strict inequality for some p's, then, using the variance criterion, one would say that $u(X_1, \ldots, X_n)$ is better than $u'(X_1, \ldots, X_n)$. The variance of an unbiased estimator generally depends on p, since the distribution depends on p. Now, if it happens that for some p's one estimator is better than another, and for some other p's these relations are reversed, then the situation is not so easy. One cannot then say that the one variance is *uniformly better* than the other. However, there exists one unbiased estimator of p, which for fixed n has uniformly smaller variance than all other unbiased estimators. This is true for sampling both with and without replacement (see Section 15.4).

Example 1 Suppose $n = 2$ and the sampling is with replacement. Each of $(X_1 + X_2)/2$ and $(X_1 + 2X_2)/3$ are unbiased estimators of p. The variances of these unbiased estimators are $(1/2)p(1-p)$ and $(5/9)p(1-p)$, respectively. Hence, $(X_1 + X_2)/2$ is a better unbiased estimator than is $(X_1 + 2X_2)/3$ (see Figure 15.2). Then $u(X_1, X_2) = X_1$ is also an unbiased estimator of p. Its variance is $p(1-p)$. This estimator does not make use of the second drawing, and each of the previous two estimators is better than X_1.

Let us compare the variances of $\overline{X}$, the unbiased estimator of p, in sampling with and without replacement. These variances are as follows.

Variance with replacement	*Variance without replacement*
$\dfrac{p(1-p)}{n}$	$\dfrac{N-n}{N-1}\dfrac{p(1-p)}{n}$

(See Examples 7 and 12, Section 10.1.) For $n > 1$, the variance

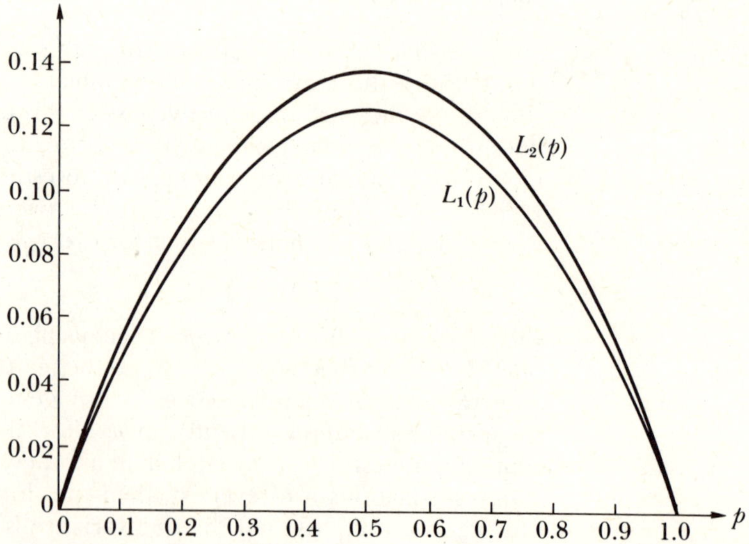

FIGURE 15.2 The variance of $(X_1 + X_2)/2$ is $L_1(p)$. The variance of $(X_1 + 2X_2)/3$ is $L_2(p)$.

without replacement is smaller, uniformly in p. This fact should not be surprising; n observations obtained with replacement are less informative, since the same objects may be sampled several times. The term $(N - n)/(N - 1)$ is called the *finite population factor*. When N is very large compared to n, then there is practically no difference between the two methods.

There is really nothing holy about an estimator being unbiased. If $u(X_1, \ldots, X_n)$ is any estimator of p, whether unbiased or not, one can measure its goodness by the expected loss,

$$E_p\{[u(X_1, \ldots, X_n) - p]^2\} \tag{2}$$

If the estimator happens to be unbiased, then (2) is its variance. In general, (2) equals the variance of the estimator plus a measure of its bias, $[E_p u(X_1, \ldots, X_n) - p]^2$ (see Example 1, Section 12.3).

Example 2 Suppose that the sampling is with replacement. Let $\overline{X} = (X_1 + \cdots + X_n)/n$ and consider the estimator $[n/(n+1)]\overline{X}$. The expected loss as given by (2) is

$$E_p\left[\left(\frac{n}{n+1}\overline{X} - p\right)^2\right] = \left(\frac{n}{n+1}\right)^2 \frac{p(1-p)}{n} + \frac{p^2}{(n+1)^2}$$

The comparison between the expected loss for this estimator and that of the unbiased estimator $\overline{X}$ is shown in Figure 15.3. We leave

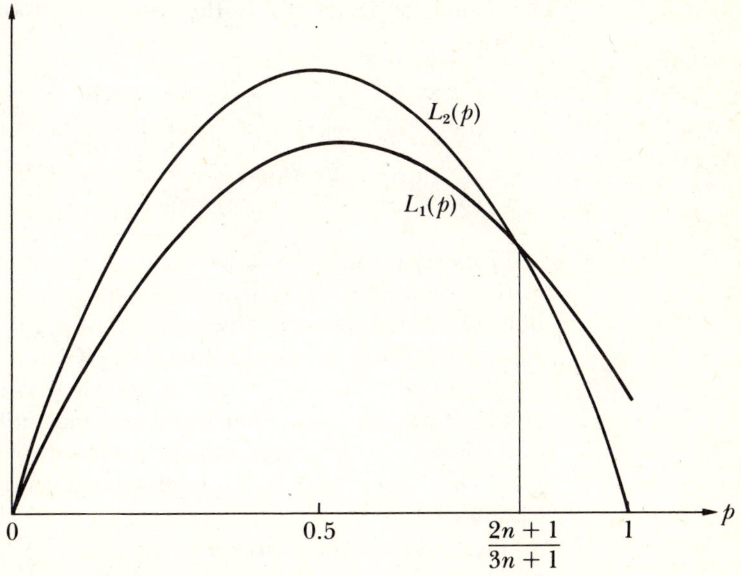

FIGURE 15.3 $L_1(p) = E_p\left[\left(\frac{n}{n+1}\overline{X} - p\right)^2\right]$, $L_2(p) = E_p[(\overline{X} - p)^2]$.

it for the reader to check that for p in the interval $(0, (2n+1)/(3n+1))$, the expected loss using the biased estimator is less than that using the unbiased one. If the experimenter's concern is greater about the p's in this interval than about the remaining p's, he might be better off using the biased estimator.

There are, of course, other measures of expected loss, such as

$$E_p[|u(X_1, \ldots, X_n) - p|], \qquad E_p\{[u(X_1, \ldots, X_n) - p]^4\}$$

The advantage of (2) over other measures is its mathematical tractability. In general, it is difficult for an experimenter to come up with a realistic loss function. However, somehow he must strike a balance between his desire to obtain "good" estimates versus the increasing cost of additional observations. We have been assuming that the number of observations is fixed and preordained. In practice, in addition to deciding how he will estimate p, the experimenter must decide how many observations he can afford to take.

15.4 SUFFICIENT STATISTICS

How should one analyze the "data" $X_1, \ldots, X_n$ in order to estimate p "effectively"? It is sufficient to base the analysis on the single random variable $Y = X_1 + \cdots + X_n$, the individual values of the X_i's being irrelevant; only the sum Y is important. The meaning follows.

the sum is a sufficient statistic

$$P_p[(X_1 = e_1) \cap \cdots \cap (X_n = e_n) | Y = e_1 + \cdots + e_n = k] \qquad (1)$$

does not depend on p.

Property (1) says that once the experimenter knows the value of Y, then knowing the individual values of the X_i's cannot possibly help in estimating p (or any function of p, for that matter). Then, Y is called a *sufficient statistic* for p. (See Chapter 14 for the general treatment of sufficient statistics.) Recall that it was shown in Example 1, Section 14.2, that Y is a minimal sufficient statistic when the sampling is with replacement. For sampling without replacement, the fact that Y is minimal sufficient appears in Exercise 4, Chapter 14.

There is only one unbiased estimator of p based on Y. For sampling with replacement, this fact follows by examining the relationship

$$\sum_{k=0}^{n} a(k)\binom{n}{k}p^k(1-p)^{n-k} = p \quad \text{for all } p \text{ in } (0, 1) \tag{2}$$

where the $a(k)$'s do not involve p. By making the change of variables, $p/(1-p) = u$, the relation (2) becomes

$$\sum_{k=0}^{n} a(k)\binom{n}{k}u^k = u(1+u)^{n-1} \quad \text{for all } u \text{ in } (0, \infty) \tag{3}$$

We know that (2) holds when $a(k) = k/n$, since Y/n is an unbiased estimator. But from (3) we see that these are the only possibilities for $a(k)$, because the coefficients in the polynomial must be uniquely determined. By Blackwell's theorem (Section 14.4), it follows that among all unbiased estimators of p in sampling with replacement, Y/n has variance that is uniformly minimized in p. Exactly the same result holds for sampling without replacement (see Exercise 6).

Using the same approach, we see that if $u(Y)$ is an unbiased estimator of $h(p)$, then it must be the unbiased estimator of uniformly minimum variance. See Example 2, Section 14.4, for the estimator of p^2.

In order for $u(Y)$ to be an unbiased estimator of $h(p)$, we must have

$$\sum_{k=0}^{n} u(k)\binom{n}{k}p^k(1-p)^{n-k} = h(p) \quad \text{for all } p \text{ in } (0, 1) \tag{4}$$

The function u does not depend on p; therefore, the left-hand side of (4) must be a polynomial in p of degree no greater than n. The only functions $h(p)$ for which there exist unbiased estimators based on Y are polynomials of degree no greater than n. (See Exercise 15, Chapter 12.) Similar considerations apply when the sampling is without replacement.

15.5 ANOTHER ESTIMATOR OF p

Another way of estimating p is to use a *random* number of observations, rather than a fixed number. Suppose that the sampling is with replacement. Let X be the number of trials required to obtain m successes. The distribution of X is given as follows (see Section 6.6).

$$P_p(X = k) = \binom{k-1}{m-1}p^m(1-p)^{k-m}, \qquad k = m, m+1, \ldots \tag{1}$$

We claim that $(m-1)/(X-1)$ is an unbiased estimator of p when $m \geqslant 2$. That is,

$$E_p\left(\frac{m-1}{X-1}\right) = p \quad \text{if } m \geqslant 2, \quad \text{for all } p \text{ in } (0, 1) \tag{2}$$

Proof.

$$E_p\left(\frac{1}{X-1}\right) = \sum_{k=m}^{\infty} \left(\frac{1}{k-1}\right)\binom{k-1}{m-1} p^m (1-p)^{k-m}$$

$$= \frac{p}{m-1} \sum_{k=m}^{\infty} \binom{k-2}{m-2} p^{m-1}(1-p)^{k-m} = \frac{p}{m-1}$$

The last sum is the same as the sum of all the probabilities in (1), with m replaced by $m-1$; hence, this sum equals 1. ▮

The proportion of successes among the X trials is m/X. It so happens that this estimator is slightly biased, and that $(m-1)/(X-1)$ is an unbiased one. The number of trials required to make the estimate $(m-1)/(X-1)$ is itself a random variable the distribution of which depends on p. The expected number of trials is

$$E_pX = \frac{m}{p}$$

[By (4), Section 6.4, the expected waiting time for the first head is $1/p$. The expected waiting times for each additional head are the same.] In sampling without replacement, it is still true that $(m-1)/(X-1)$ is an unbiased estimator of p, assuming that there are at least m type 1 objects in the population, but we shall not prove this fact (Note 3).

15.6 CONFIDENCE INTERVAL ESTIMATES

In this section, suppose that the sampling is with replacement. Most of the discussion can be adapted to sampling without replacement as well, but these details will not be presented here.

As we have seen, a good way of estimating p is by means of the "sample proportion," $\hat{p} = (X_1 + \cdots + X_n)/n$. There is a method for asserting "how close" the estimator $\hat{p}$ is to the true $\hat{p}$, which it purports to estimate. Associated with $\hat{p}$ will be two numbers, $p_1 < p_2$, which depend on $\hat{p}$, and a "confidence level," $1-\alpha$, a number "close" to 1, which will have been determined beforehand. Moreover, the probability that the random interval (p_1, p_2) contains the true p is at least $1-\alpha$. Thus, if the experimenter asserts that the true p lies between p_1 and p_2, he may or may not be making a true statement, but the probability of truth is at least $1 - \alpha$. The interval (p_1, p_2) is called a *confidence interval* estimate of p. At this point, you might want to review the previous discussion of confidence intervals in Section 12.7.

Example 1 A professional election predictor has interviewed 1000 voters, 40% of which will vote for candidate A. The published report asserts that the true proportion of A's supporters in the population lies between 0.38 and 0.42 at confidence level 0.95. This statement can be interpreted empirically to mean that over many repetitions of this prediction procedure, the asserted interval would contain the true p at least 95% of the time.

A method for constructing the confidence interval (p_1, p_2) will now be described. Let $1 - \alpha$ be the confidence level—that is, the desired probability of telling the truth. This number $1 - \alpha$ and the sample size n are fixed beforehand. Now construct a chart as in Figure 15.4 in the following way. For every p, let $a(p)$ and $b(p)$ be two numbers such that

$$P_p[a(p) \leqslant \hat{p} \leqslant b(p)] \geqslant 1 - \alpha$$

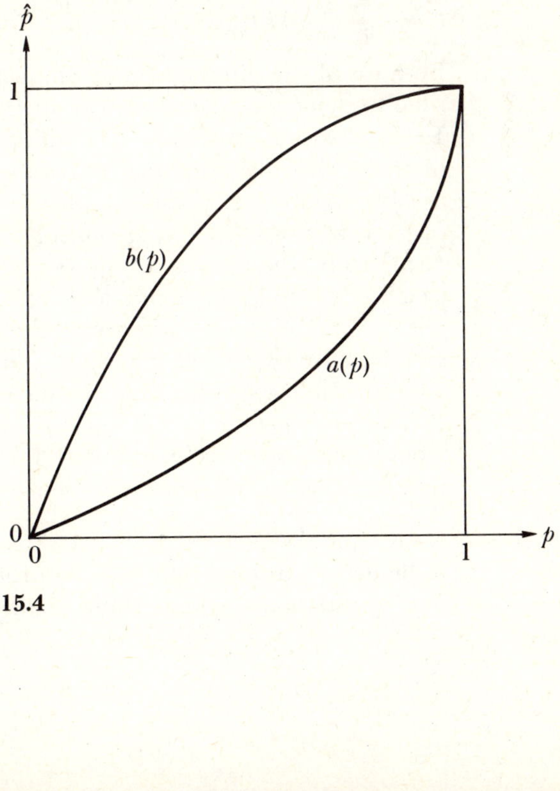

FIGURE 15.4

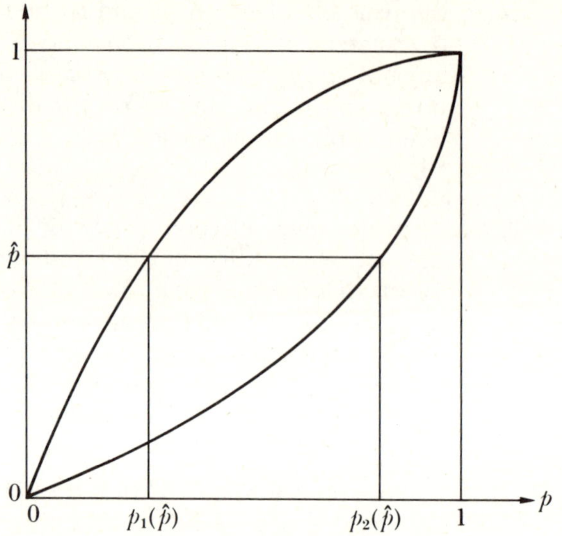

FIGURE 15.5

where the probability is evaluated from the binomial distribution with parameters n and p. (We write P_p to emphasize that the probability depends on p.) Explicitly, the above inequality is,

$$\sum_{a(p)\leq k/n\leq b(p)} \binom{n}{k} p^k(1-p)^{n-k} \geq 1-\alpha \tag{1}$$

Because of the discreteness of the distribution, it may not be possible to determine $a(p)$ and $b(p)$, such that this probability is *exactly* $1-\alpha$. We can determine $a(p)$ and $b(p)$ by inspection from tables of the cumulative binomial distribution. They should be chosen as close together as possible and so that the probability on the left of (1) is as close to $1-\alpha$ as possible. Now define $[p_1, p_2] = [p_1(\hat{p}), p_2(\hat{p})]$ to be the interval of all p's such that $[a(p), b(p)]$ contains $\hat{p}$ (see Figure 15.5). The probability that $[p_1, p_2]$ contains the true p is at least $1-\alpha$ no matter what the true p might be. This fact is true because $[p_1, p_2]$ contains p if, and only if, $\hat{p}$ lies in $[a(p), b(p)]$. The probability of this event happening is at least $1-\alpha$.

The lines $a(p)$ and $b(p)$ enclose what is called a *confidence belt.* These belts have been graphed for various values of n and α (Note 4). Figure 15.6 shows some confidence belts.

Example 2 A biologist wants to estimate the probability of a given genotype in a population. Among 100 specimens, he observes 65 with the

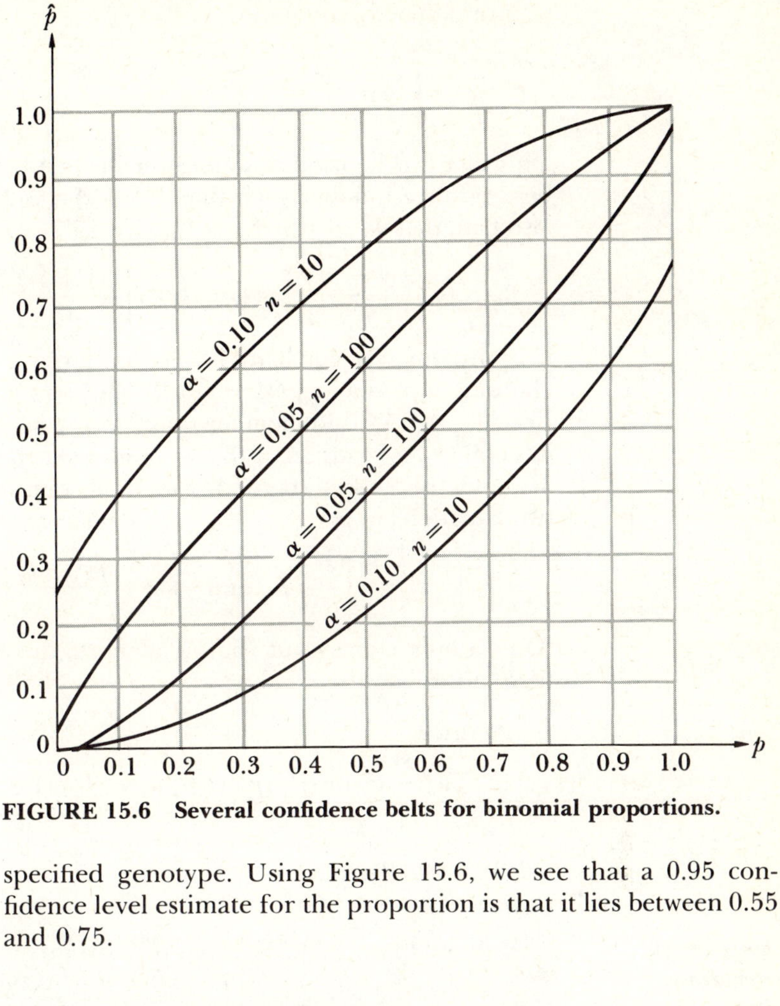

FIGURE 15.6 Several confidence belts for binomial proportions.

specified genotype. Using Figure 15.6, we see that a 0.95 confidence level estimate for the proportion is that it lies between 0.55 and 0.75.

15.7 LARGE SAMPLE CONFIDENCE INTERVALS

Using the central limit theorem, it is possible to give an explicit expression for the confidence interval estimate for p, when the sample size is large. According to the central limit theorem (Section 10.3), the standardized form of $\hat{p}$ is approximately normally distributed, $N(0, 1)$, for large n. Since $E(\hat{p}) = p$ and $\mathrm{Var}(\hat{p}) = p(1-p)/n$, this means that

$$P_p\left(\frac{\hat{p}-p}{[p(1-p)/n]^{1/2}} \leq t\right)$$

is approximately equal to

$$\frac{1}{(2\pi)^{1/2}} \int_{-\infty}^{t} \exp \frac{-x^2}{2}\, dx = F(t) \tag{1}$$

This fact can be used to determine $a(p)$ and $b(p)$ as follows. Let c be a positive number such that $1 - \phi(c) = \alpha/2$. By symmetry, it is also true that $\phi(-c) = \alpha/2$ (see Figure 15.7). Hence, from (1),

$$P_p\left\{p - c\left[\frac{p(1-p)}{n}\right]^{1/2} \leqslant \hat{p} \leqslant p + c\left[\frac{p(1-p)}{n}\right]^{1/2}\right\}$$

is approximately equal to $1 - \alpha$ for large n. This fact suggests choosing $a(p) = p - c[p(1-p)/n]^{1/2}$ and $b(p) = p + c[p(1-p)/n]^{1/2}$ (see Figure 15.8). Remember that for any given p, the confidence interval (p_1, p_2) consists of the set of all p's such that $a(p) \leqslant \hat{p} \leqslant b(p)$. To find the endpoints of this interval, one must then solve the two equations in p,

$$p - c\left[\frac{p(1-p)}{n}\right]^{1/2} = \hat{p} \quad \text{and} \quad p + c\left[\frac{p(1-p)}{n}\right]^{1/2} = \hat{p}$$

After a little clumsy but routine algebra, the endpoints p_1 and p_2 turn out to be

$$(p_1, p_2) = \tag{2}$$

$$\left(\frac{\hat{p} + (c^2/2n) - c\{[\hat{p}(1-\hat{p})/n] + (c^2/4n^2)\}^{1/2}}{1 + (c^2/n)}, \frac{\hat{p} + (c^2/2n) + c\{[\hat{p}(1-\hat{p})/n] + (c^2/4n^2)\}^{1/2}}{1 + (c^2/n)}\right)$$

Let us summarize.

large sample confidence interval for p

> Suppose that $\hat{p}$ is the observed proportion of 1's in a sample of n drawings with replacement from a dichotomous population. Then the interval described by (2) will contain the true p with probability approximately equal to $1 - \alpha$ when n is large.

For n sufficiently large, the interval given by (2) can be replaced by the simpler expression,

$$(p_1, p_2) = \left(\hat{p} - c\left[\frac{\hat{p}(1-\hat{p})}{n}\right]^{1/2},\ \hat{p} + c\left[\frac{\hat{p}(1-\hat{p})}{n}\right]^{1/2}\right) \tag{3}$$

The last approximation to the confidence interval can be justified as follows: $\hat{p}$ is approximately standard normally distributed with expected value p and standard deviation $[p(1-p)/n]^{1/2}$. An esti-

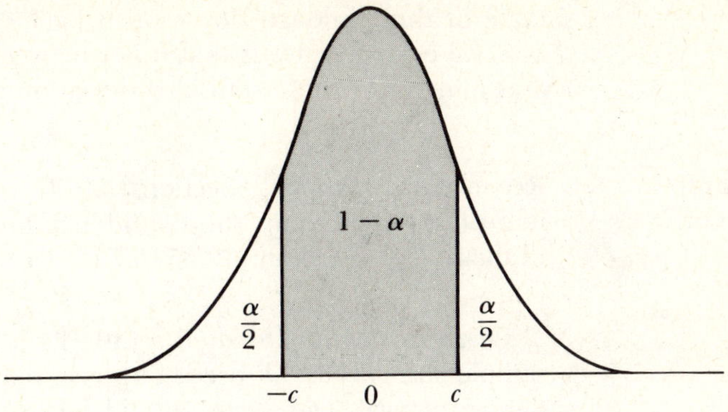

FIGURE 15.7

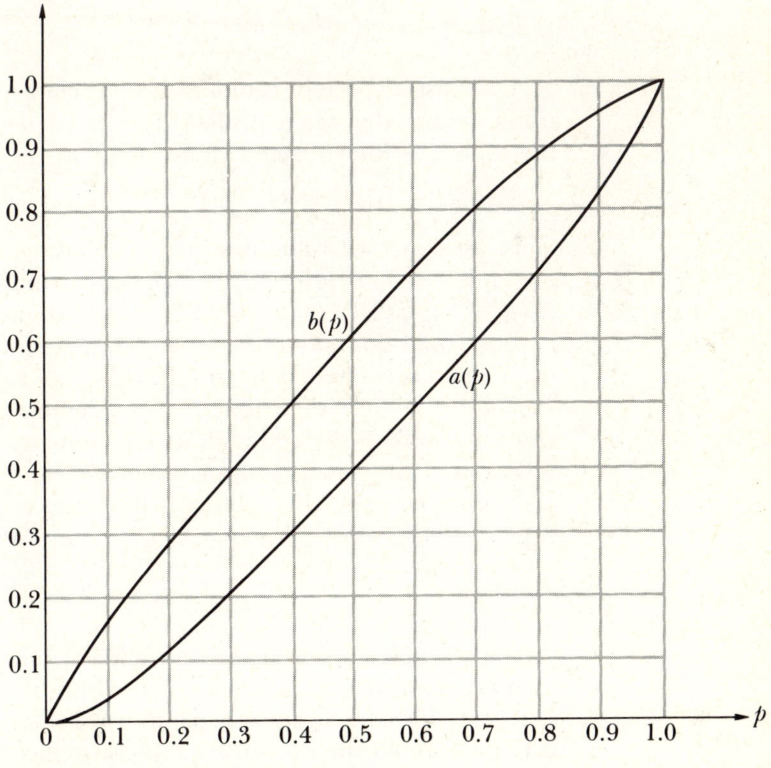

FIGURE 15.8 Curves for $a(p)$ and $b(p)$ for $n = 100$, $\alpha = 0.05$.

mate of this standard deviation is $[\hat{p}(1-\hat{p})/n]^{1/2}$. The confidence interval estimate of p is that it lies between the extremes of p plus and minus c of these standard deviation estimates.

Example 1 Reconsider Example 2, Section 15.6. If we evaluate (2) to two decimal places, we obtain the confidence interval (0.55, 0.74). If we evaluate (3), we obtain, (0.55, 0.75).

One can now try to shed a little light on the question of what sample size should be used. Suppose that the width of the confidence interval, $(p_2 - p_1)$, should be no greater than a specified amount, ϵ, and the probability of telling the truth, $1-\alpha$, is also specified. Now, the width of the confidence interval is a random variable, so one considers the expected value of this random variable—the expected width. Take the approximate confidence interval given by (3) for simplicity. The expected width of the interval is

$$E_p(p_2 - p_1) = \frac{2c}{\sqrt{n}} E_p\{[\hat{p}(1-\hat{p})]^{1/2}\}$$

This last expected value is rather clumsy to compute, but, using the fact that the expected value of a square is never less than the square of the expected value, one has

$$E_p\{[\hat{p}(1-\hat{p})]^{1/2}\} \leqslant \{E_p[\hat{p}(1-\hat{p})]\}^{1/2}$$

This last expected value is easier to compute; its value is $E[\hat{p}(1-\hat{p})] = [(n-1)/n]p(1-p)$, which is less than $p(1-p)$. Hence, $E_p(p_2-p_1) \leqslant (2c/\sqrt{n})[p(1-p)]^{1/2}$.

One could guarantee the desired maximum expected width of the confidence interval by setting $(2c/\sqrt{n})[p(1-p)]^{1/2} = \epsilon$ and solving for n. Of course, this poses a dilemma because p is just the unknown object that one wants to estimate. There are several ways out. One can take the conservative approach of bounding $(2c/\sqrt{n})[p(1-p)]^{1/2}$ by its maximum, $(2c/\sqrt{n})(1/2)$. Setting this equal to ϵ and solving for n gives the criterion that if

$$n = \frac{c^2}{\epsilon^2} \tag{4}$$

then the expected width of the confidence interval is never greater than the desired ϵ.

A less conservative approach is to use prior information that one may have about the range of p's that are likely to prevail. Instead of using 1/2 as the overall maximum of $[p(1-p)]^{1/2}$, one can use instead the possibly smaller maximum of $[p(1-p)]^{1/2}$ within a restricted interval, if there exists the prior knowledge that p lies in this interval. Call this maximum m. Then, solving for n as in (4) gives

$$n = \frac{4c^2m^2}{\epsilon^2} \tag{5}$$

The qualitative import of (4) or (5) is the same: The smaller the width of the confidence interval is to be, the larger the sample size must be. In fact, to halve the width of the confidence interval, one must quadruple the sample size. On the other hand, the greater $1 - \alpha$, the probability of telling the truth, is to be, the bigger the sample size must be. This fact is true because c varies monotonically with $1 - \alpha$. It is all perfectly plausible, of course. One only gets what he pays for. Presumably, observations cost something, and the greater n, the greater the cost. Telling a lie and telling a flabby truth both also cost something. [One could assert that p lies in (0, 1), which is perfectly true but useless!] Smaller ϵ's mean more precise statements. Somehow, this trio of considerations must be balanced against each other and an n, an ϵ, and a c determined. With refined cost accounting, one can try to appraise these various costs systematically, but, in general, the problem is difficult (Note 5).

15.8 DECIDING BETWEEN ALTERNATIVES

Suppose that an experimenter has a choice of two decisions to make. These decisions relate to the unknown p of a dichotomous population; if he knew the true value of p he might know which decision to make—at least for some p's the nature of the decision would be clear.

Example 1 In a quality control situation, p is the proportion of defective items in a lot of goods. If the inspector knew that p was as low as 0.01, he would definitely pass on the lot. If he knew that p was as high as 0.03, he would definitely not send out the lot. For values of p between 0.01 and 0.03, he might have difficulty deciding what to do. For instance, if $p = 0.01$ is acceptable, how does he feel about $p = 0.01001$? In any case, for the ranges beyond the two extreme values of 0.01 and 0.03, his policy is clear.

Example 2 A vaccine has been developed on an experimental basis. Let p be the probability of effectiveness of the vaccine. (That is, if applied to a large number of persons, p is the proportion that would be "sufficiently helped" by the vaccine.) Knowledge of p is crucial to the decision of whether or not the vaccine should be developed commercially. If p is as high as 0.8, the developer would definitely go ahead commercially. What would he do if p is 0.799? 0.700? He might not be sure about these. However, if p is as low as 0.6, he definitely would not produce the vaccine commercially.

The general situation illustrated by these examples follows. Two decisions are available, decision 1 and decision 2. Each relates to the unknown p of a dichotomous population. There are two values of p about which there is strong feeling, p_1 and p_2. For simplicity, say that $p_1 < p_2$. If the true p is less than or equal to p_1, decision 1 is the right decision. If the true p is greater than or equal to p_2, then decision 2 is correct. On the basis of n observations on the population, one proceeds to make a decision. A common procedure is to make the decision using $\hat{p} = (X_1 + \cdots + X_n)/n$, the proportion of 1's in the sample. If $\hat{p}$ is "large" (i.e., is in the direction of p_2), make decision 2; if $\hat{p}$ is "small" (i.e., in the direction of p_1), make decision 1. Specifically, there is a cutoff constant c such that if $\hat{p} \geq c/n$, one makes decision 2, and if $\hat{p} < c/n$, one makes decision 1. Unless the n observations happen to be a complete census of the population, any decision procedure, including this one, may produce mistakes. How should the effectiveness of this decision procedure be judged?

A useful tool in examining the described decision procedure is the *power function.* It is simply the graph of the probability of making one of the decisions (say, decision 1, to be specific) plotted as a function of p. (See Section 13.2 for the general discussion of power functions.)

Example 3 (Quality control.) Three drawings are made without replacement from a lot of $N = 100$ items. If one or more items in the sample are defective, the lot is rejected (decision 2). If all three items pass inspection, the lot is accepted (decision 1). The probability of making decision 1 is

$$\psi(p) = \frac{(N-M)(N-M-1)(N-M-2)}{N(N-1)(N-2)} = \frac{(100-M)^{(3)}}{100^{(3)}}$$

where M is the unknown number of defectives in the lot, and $p = M/N$ (see Figure 15.9). Suppose that $p_1 = 0.1$ and $p_2 = 0.2$. If the true p is actually p_1, the desired decision is 1; but decision 2 will

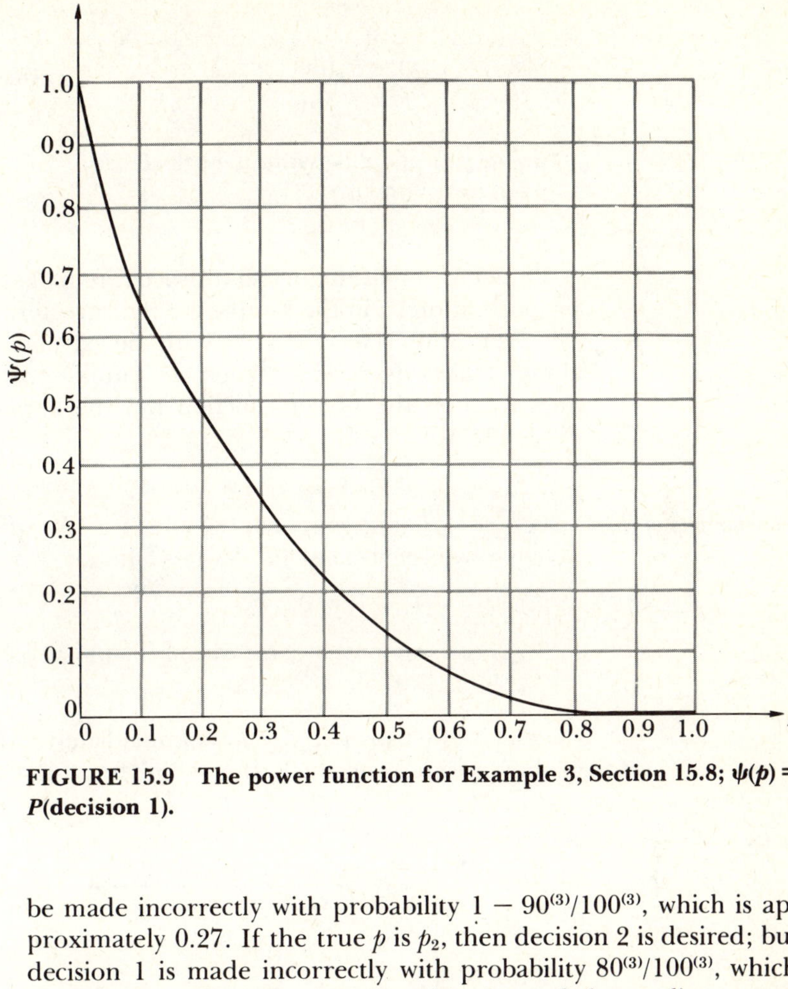

FIGURE 15.9 The power function for Example 3, Section 15.8; $\psi(p) = P$(decision 1).

be made incorrectly with probability $1 - 90^{(3)}/100^{(3)}$, which is approximately 0.27. If the true p is p_2, then decision 2 is desired; but decision 1 is made incorrectly with probability $80^{(3)}/100^{(3)}$, which is approximately 0.51. Are the objectives of the quality control plan consistent with these two kinds of errors? If not, the plan must be changed. Perhaps more items should be inspected.

In general, the power function is given by the following expressions.

power function

$$\psi(p) = P_p(\text{decision } 1) = \sum_{k=0}^{c-1} \binom{n}{k} p^k (1-p)^{n-k} \quad (1)$$

if the sampling is with replacement.

$$\psi\binom{M}{N} = \psi(p) = P_p(\text{decision } 1) = \sum_{k=0}^{c-1} \frac{\binom{M}{k}\binom{N-M}{n-k}}{\binom{N}{n}} \tag{2}$$

if the sampling is without replacement. (Suppose that N is fixed and known.)

Remember that the decision procedure makes decision 2 if the proportion of 1's in the sample is c/n or greater, and decision 1 if this proportion is less than c/n. With the extreme case of $c = 0$, one always makes decision 2. When the cutoff constant c is a *positive* integer, then the power function has the appearance shown in Figure 15.10.

monotonicity of ψ

Let $\psi(p)$ denote the expressions for the power function given by either (1) or (2). Suppose that $0 < c < n$. Then

(a) $\psi(0) = 1$, $\psi(1) = 0$.

(b) ψ is a monotone decreasing function of p.

Proof. The first property follows immediately from the definition of the decision procedure. If $p = 0$, then no 1's can ever be drawn

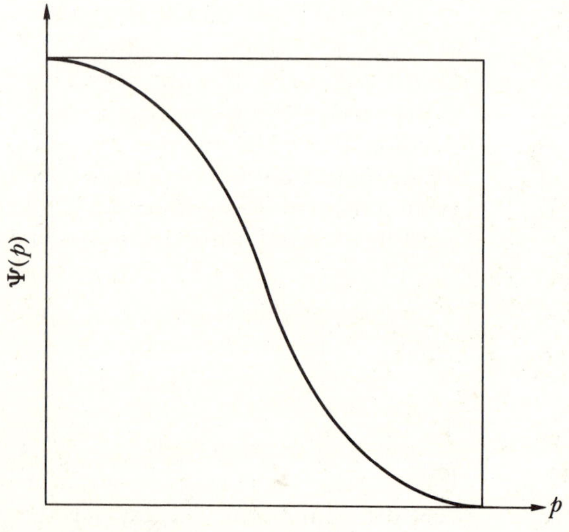

FIGURE 15.10 A typical power function.

in a sample; thus, decision 2 is never made. On the other hand, if $p = 1$, decision 2 is always made. The monotonicity follows from the following facts:

$$\psi(p) = \sum_{k=0}^{c-1} \binom{n}{k} p^k(1-p)^{n-k} = 1 - \int_0^p n\binom{n-1}{c-1} t^{c-1}(1-t)^{n-c}\,dt \quad (3)$$

$$\psi\left(\frac{M}{N}\right) = \psi(p) = \sum_{k=0}^{c-1} \frac{\binom{M}{k}\binom{N-M}{n-k}}{\binom{N}{n}} = \sum_{k>M} \frac{\binom{k-1}{c-1}\binom{N-k}{n-c}}{\binom{N}{n}} \quad (4)$$

The relation (3) is given in (3), Section 6.1. For (4), see Exercise 7. ▮

Suppose, as before, that p_1 and p_2 are the two values of p at which one definitely wants to make decisions 1 and 2, respectively. The probability of making a mistake when $p = p_1$ is $1 - \psi(p_1)$, and $\psi(p_2)$ is the probability of making a mistake when $p = p_2$. The power function sets these error probabilities directly before the experimenter. He must decide whether or not he can tolerate these errors and he must consider what he can do about it. If he changes c, he may make one error smaller at the expense of making the other error larger. If he increases n, he may make both errors smaller but presumably the additional observations cost him something. Somehow the cost of making observations must be balanced with the costs of making errors.

Using the central limit theorem, we can give an approximation to the power curve ψ as follows.

$$\psi(p) \approx 1 - F\left\{\frac{c - np}{[np(1-p)]^{1/2}}\right\} \quad (5)$$

where F is the $N(0, 1)$ cdf. The expression (5) approximates (3) for large n, because $(n\hat{p} - np)/[np(1-p)]^{1/2}$ is approximately $N(0, 1)$ distributed. The approximation also holds for (4) when n is large and N is large with respect to n. In that case, the hypergeometric distribution is approximated by the binomial, which is, in turn, approximated by the normal (Note 6).

SUMMARY

We consider a *dichotomous population* consisting of N elements, M of which are of type 1 and $N - M$ of type 0. The proportion of 1's is $p = M/N$.

If a sample of size n is drawn, the probability distribution of Y, the number of 1's in the sample, depends on whether the sampling is done *with or without replacement.* When the sampling is with replacement, then Y is *binomially distributed* with parameters n and p. When the sampling is without replacement, then Y has a *hypergeometric distribution.*

For either type of sampling, *Y is minimal sufficient* for p, and, as a consequence, $Y/n = \hat{p}$ has *minimum variance* among all *unbiased estimators* of p. If instead of taking a fixed number of observations, one makes repeated observations until a total of m 1's have been obtained, then $(m-1)/(X-1)$ is an unbiased estimator of p, where X is the total number of drawings required and $m \geqslant 2$.

We describe a method for making *confidence interval estimates* for p. For large sample size, it can be based on the central limit theorem. In this case, an explicit estimate is that p lies in the interval $\{\hat{p} - c[\hat{p}(1-\hat{p})/n]^{1/2}, \hat{p} + c[\hat{p}(1-\hat{p})/n]^{1/2}\}$. The probability of telling the truth is approximately the probability that a $N(0, 1)$ random variable lies between $-c$ and c.

We also consider *decision procedures* for deciding whether p is "small" or "large." An intuitively plausible procedure is to base these decisions on the size of the sample proportion $\hat{p}$. The *power function* for such a procedure is monotone. The power function can be used by the experimenter to determine a sample size and a cutoff constant c that give errors consistent with his needs.

EXERCISES

1 A biologist wants to estimate p, the proportion of dumpy wing drosophila in a given drosophila population. He would like the standard deviation of his estimator to be no greater than 0.01.

(a) He knows from past experience that p will never exceed 1/4. How many drosophila should he examine? [Hint: If S_n is the number of specimens with the desired characteristic in a sample of n, then $\text{Var}(S_n/n) = p(1-p)/n$. The maximum of this variance for $0 \leqslant p \leqslant 1/4$ is $3/16n$. Solve for n in $(3/16n)^{1/2} = 0.01$.]

(b) Suppose that the observed proportion, using the above determined n, turn out to be 0.15. Give a 0.95 level confidence interval for p. [Answer: $n = 1875$; confidence interval is (0.134, 0.166), using (3), Section 15.7.]

2 Consider a simple random walk in which a particle makes successive independent steps. Each step is to the right (+1) with probability p and to the left (−1) with probability $1-p$ (see Example 3, Section 11.4). An observer does not know p and he wants to estimate it from his observation of the walk over n steps. Let S_n be the position of the particle after n steps, with the initial position being at 0.

(a) Show that $(1 + S_n/n)/2$ is the minimum variance unbiased estimator for p.

Consider a random walk in the plane. The successive movements of the particle are independent and are given by the following table:

Movement	*Probability*
$+1, +1$	p^2
$+1, -1$	$p(1-p)$
$-1, +1$	$p(1-p)$
$-1, -1$	$(1-p)^2$

Let (S_n, T_n) be the position of the particle after n steps, with the initial position being $(0, 0)$.

(b) Show that $[1 + (S_n + T_n)/2n]/2$ is the minimum variance unbiased estimator of p. [Hint: Let (X_i, Y_i) be the ith movement. Then $X_1, Y_1, \ldots, X_n, Y_n$ are $2n$ independent random variables, each equal to 1 with probability p, and to -1 with probability $1-p$. Then $S_n = X_1 + \cdots + X_n$, $T_n = Y_1 + \cdots + Y_n$, and $S_n + T_n$ is minimal sufficient for p.]

(c) Let Z_n be the number of times the particle makes the movement $(+1, +1)$. Show that Z_n/n is an unbiased estimator of p^2. Can you find an unbiased estimator of p^2 the variance of which is smaller than that of Z_n/n? (Hint : See Example 2, Section 14.4.)

3 A production lot consists of 50 items. Denote the proportion of defectives by p. Consider the following quality control plan. First examine two items. If they are both of acceptable quality, accept the lot (decision 1). If they are both defective, reject the lot (decision 2). Otherwise, examine one more item. If it is defective, reject the lot; otherwise, accept. Determine the power function for this decision procedure.

4 A given population of high school students consists of 10,000 men and 10,000 women. A researcher wants to estimate p, the proportion of students that intend to enter college. His research budget allows him a total of 100 interviews. Suppose that he selects 100 persons at random from the overall population of 20,000 students, and $\hat{p}$ is the observed proportion of students in the sample intending to enter college. (For simplicity, suppose throughout this exercise that sampling is with replacement.) Then $E\hat{p} = p$ and $\text{Var}(\hat{p}) = p(1-p)/100$. On the other hand, suppose that he selects 50 men at random and 50 women at random. Let $\hat{q}$ and $\hat{r}$ be the proportions of men and women in the two samples that intend to enter college. Then $(\hat{q} + \hat{r})/2$ is also an unbiased estimator of p. Show that

$$\text{Var}\left(\frac{\hat{q} + \hat{r}}{2}\right) \leq \text{Var}(\hat{p})$$

with equality if, and only if, $p_1 = p_2$, where p_1 and p_2 are the respective proportions of men and women in the population that intend to enter college. {Hint: After some algebra, conclude that

$$\text{Var}(\hat{p}) - \text{Var}\left(\frac{\hat{q}+\hat{r}}{2}\right) = \frac{(p_1 - p_2)^2}{2} \geq 0$$

This exercise shows an example of efficiencies that can be obtained by *stratified sampling.* By taking advantage of the possibility of sampling from stratified segments of the population, one improves the estimating procedure.

5 A dichotomous population consists of M objects of type 1 and $N-M$ objects of type 2. Suppose items are successively drawn without replacement until m type 1 objects are obtained. (Assume that $M \geq m \geq 2$.) Let X be the number of drawings required. Show that $(m-1)/(X-1)$ is an unbiased estimator of p. [Hint: Follow the style of development of (2), Section 15.5. The distribution of X is given by

$$P(X=k) = \frac{\binom{M}{m-1}\binom{N-M}{k-m}}{\binom{N}{k-1}} \frac{M-m+1}{N-k+1} = \frac{M}{N} \frac{\binom{M-1}{m-1}\binom{N-M}{k-m}}{\binom{N-1}{k-1}}\Bigg]$$

6 Show that in sampling *without replacement* from a dichotomous population, Y/n, the proportion of 1's in the sample, is an unbiased and minimum variance estimator of the population proportion $p = M/n$. Assume that the population size N is known to the experimenter. [Hint: First see that Y (or, equivalently, Y/n) is minimal sufficient (Exercise 4, Chapter 14). Then show that the only unbiased estimator of M/N based on Y is Y/n. Do this by examining the relationship

$$\sum_{k=0}^{n} a(k)\binom{n}{k} \frac{M^{(k)}(N-M)^{(n-k)}}{N^{(n)}} = \frac{M}{N} \quad \text{identically in } M$$

successively for $M = 0, 1, \ldots, N$. For instance, from $M = 0$, conclude that $a(0) = 1$. Then, from $M = 1$, conclude that $a(1) = 1/n$, and so forth. At each step, the next $a(i)$ is determined from the previous ones. The desired conclusion follows from Blackwell's theorem.]

7 Verify the relation (4), Section 15.8. [Hint: Consider fixed positions $1, 2, \ldots, N$. Suppose that n positions are picked at random, and let Y be the number of these that are among $1, \ldots, M$. Also, Y has the hypergeometric distribution. Let $U_1 < \cdots < U_n$ be the ordered values of the positions selected. The event $(Y \leq c-1)$ is exactly the same as the event $(U_c > M)$. Verify that $P(U_c = k) = \binom{k-1}{c-1}\binom{N-k}{n-c}\Big/\binom{N}{n}$, from which (4), Section 15.8, follows.]

NOTES

1 See [65] for statistical procedures relating to making inferences about proportions and other parameters of finite populations.

2 A detailed treatment of quality control procedures appears in [66].

3 The sampling procedure described in Section 15.5 is an example of *sequential sampling.* Sequential statistical procedures require sampling until an "adequate" number of observations have been obtained. A development of this subject, particularly as it relates to decision making, is the work of Abraham Wald. See [67].

4 Charts of confidence belts are given in [68]. See 3.6, [65], for additional material on confidence intervals for estimating proportions in finite populations. Also see [69] for exact confidence intervals when the sampling is without replacement.

5 For additional reading on large-sample confidence intervals, see Section 11.7 of [51].

6 Reference [66] contains many worked-out power curves for various sample sizes and cutoff contants c.

Chapter Sixteen

A REVIEW OF SOME LINEAR ALGEBRA

In this chapter, we review some basic facts about the vector space of n-tuples of real numbers. Then follows a discussion of random vectors, which are n-tuples of random variables. The reader familiar with linear algebra can start with the material on random vectors in Section 16.6. The main purpose of this chapter is to pave the way for the study of linear models and analysis of variance procedures in Chapter 17.

16.1 VECTOR SPACES

(See Note 1.) For our present purposes, a vector will mean an n-tuple of real numbers, $(x_1, \ldots, x_n)$. Later in the chapter, in Section 16.6, we shall also consider n-tuples, the components of which are random variables. We shall sometimes refer to a single real number as a *scalar* to distinguish it from a vector. The basic operations involving vectors will now be reviewed.

vector operations

If $\mathbf{x} = (x_1, \ldots, x_n)$ and $\mathbf{y} = (y_1, \ldots, y_n)$ are two vectors, then their *sum* is defined to be the vector

$$\mathbf{x} + \mathbf{y} = \mathbf{y} + \mathbf{x} = (x_1 + y_1, \ldots, x_n + y_n)$$

If a is a scalar, then the product $a\mathbf{x}$ is defined to be the vector

$$a\mathbf{x} = (ax_1, \ldots, ax_n)$$

By $\mathbf{x} - \mathbf{y}$ is meant the vector

$$\mathbf{x} + (-1)\mathbf{y} = (x_1 - y_1, \ldots, x_n - y_n)$$

The inner product $(\mathbf{x}, \mathbf{y})$ is defined to be the scalar

$(\mathbf{x}, \mathbf{y}) = x_1y_1 + x_2y_2 + \cdots + x_ny_n$

which also equals $(\mathbf{y}, \mathbf{x})$. The length of a vector $\mathbf{x}$ is defined by

$\|\mathbf{x}\| = (\mathbf{x}, \mathbf{x})^{1/2} = (x_1^2 + \cdots + x_n^2)^{1/2}$

If $\|\mathbf{x}\| = 0$, it follows that $\mathbf{x} = (0, \ldots, 0)$. This vector is called the *zero vector.*

It should be emphasized that $\mathbf{x} + \mathbf{y}$ and $a\mathbf{x}$ are vectors, whereas $(\mathbf{x}, \mathbf{y})$ is a scalar.

Example 1 Suppose that $n = 3$ and $\mathbf{x} = (-1, \frac{1}{2}, 3)$, $\mathbf{y} = (1, 7, \pi)$. Then

$$\mathbf{x} + \mathbf{y} = \left(0, \frac{15}{2}, 3 + \pi\right)$$

$$2\mathbf{x} = (-2, 1, 6)$$

$$2\mathbf{x} - \mathbf{y} = (-3, -6, 6 - \pi)$$

$$(\mathbf{x}, \mathbf{y}) = \frac{5}{2} + 3\pi$$

$$\|\mathbf{x}\| = \frac{(41)^{1/2}}{2}$$

Some basic properties follow directly from the preceding definitions.

basic properties

(a) Addition is associative—that is, $\mathbf{x} + (\mathbf{y} + \mathbf{z}) = (\mathbf{x} + \mathbf{y}) + \mathbf{z}$; hence, the sum can simply be denoted as $\mathbf{x} + \mathbf{y} + \mathbf{z}$. If $\mathbf{z} = (z_1, \ldots, z_n)$ and $\mathbf{x}$ and $\mathbf{y}$ are as denoted earlier, then

$\mathbf{x} + \mathbf{y} + \mathbf{z} = (x_1 + y_1 + z_1, \ldots, x_n + y_n + z_n)$

(b) $(a\mathbf{x}, \mathbf{y}) = (\mathbf{x}, a\mathbf{y}) = a(\mathbf{x}, \mathbf{y})$.

(c) $(\mathbf{x} + \mathbf{y}, \mathbf{z}) = (\mathbf{x}, \mathbf{z}) + (\mathbf{y}, \mathbf{z})$.

(d) $\|a\mathbf{x}\| = |a|\,\|\mathbf{x}\|$.

(e) If $\|\mathbf{x}\| \neq 0$ and $a = 1/\|\mathbf{x}\|$, then $\|a\mathbf{x}\| = 1$.

Even if the reader has never met these facts before, he should have no trouble verifying them.

linear combination

Suppose that $\mathbf{x}_1, \ldots, \mathbf{x}_k$ are vectors and $a_1, \ldots, a_k$ are scalars. Then the vector

$$a_1\mathbf{x}_1 + \cdots + a_k\mathbf{x}_k$$

is called a *linear combination* of the $\mathbf{x}_i$'s.

***the vector space* V**

By the vector space V is meant the collection of all n-tuples of real numbers equipped with the vector operations previously described.

vector subspace

Let $\mathbf{x}_1, \ldots, \mathbf{x}_k$ be vectors. By the *vector subspace generated by* $\mathbf{x}_1, \ldots, \mathbf{x}_k$ is meant the set of *all linear combinations of these vectors.* It will be denoted by $V(\mathbf{x}_1, \ldots, \mathbf{x}_k)$.

Notice that a vector subspace is closed under linear combinations. That is, if $\mathbf{x}$ and $\mathbf{y}$ are vectors in the subspace and a and b are scalars, then $a\mathbf{x} + b\mathbf{y}$ are also in the subspace, since it is just another linear combination of $\mathbf{x}_1, \ldots, \mathbf{x}_k$.

Example 2 If $n = 3$ and $\mathbf{x}_1 = (1, 0, 0)$, $\mathbf{x}_2 = (0, 1, 0)$, then $V(\mathbf{x}_1, \mathbf{x}_2)$ consists of all triples of numbers with 0 in the third coordinate. Any vector $(a_1, a_2, 0)$ is expressible as the linear combination

$$(a_1, a_2, 0) = a_1(1, 0, 0) + a_2(0, 1, 0) = a_1\mathbf{x}_1 + a_2\mathbf{x}_2$$

The same vector subspace may be generated by other vectors, however. If $\mathbf{y}_1 = (1, 1, 0)$, $\mathbf{y}_2 = (1, -1, 0)$, then

$$(a_1, a_2, 0) = \left(\frac{a_1 + a_2}{2}\right)\mathbf{y}_1 + \left(\frac{a_1 - a_2}{2}\right)\mathbf{y}_2$$

Hence, the subspace generated by $\mathbf{x}_1, \mathbf{x}_2$ is exactly the same as the subspace generated by $\mathbf{y}_1, \mathbf{y}_2$. That is, $V(\mathbf{x}_1, \mathbf{x}_2) = V(\mathbf{y}_1, \mathbf{y}_2)$.

16.2 LINEAR INDEPENDENCE

Suppose that $\mathbf{x}_1, \ldots, \mathbf{x}_k$ are vectors. If no one of these vectors can be expressed as a linear combination of the others, then $\mathbf{x}_1, \ldots, \mathbf{x}_k$ are said to be *linearly independent.* If $\mathbf{x}_1, \ldots, \mathbf{x}_k$ are not linearly independent, then at least one of the $\mathbf{x}_i$'s can be expressed as a linear combination of the remaining ones. For simplicity, suppose the vectors are so arranged that $\mathbf{x}_k$ is a linear combination of $\mathbf{x}_1, \ldots, \mathbf{x}_{k-1}$. It follows that

$$V(\mathbf{x}_1, \ldots, \mathbf{x}_k) = V(\mathbf{x}_1, \ldots, \mathbf{x}_{k-1})$$

That is, any linear combination of the k vectors can also be expressed as a linear combination of only $k - 1$ of them. This procedure can be continued until one obtains a minimal set of vectors that is now linearly independent and that generates the same vector subspace as did the original set $\mathbf{x}_1, \ldots, \mathbf{x}_k$. This fact is now summarized.

Given vectors $\mathbf{x}_1, \ldots, \mathbf{x}_k$, there exists a set of r linearly independent vectors from this set which generate $V(\mathbf{x}_1, \ldots, \mathbf{x}_k)$. The integer r is called the dimension of the subspace and is, of course, no greater than k.

Example 1 Consider the vectors

$$(1, 0, 0), \quad (0, 1, 0), \quad (0, 0, 1), \quad (1, 1, 0), \quad (1, 1, 1)$$

These vectors are not linearly independent. If the last two vectors are discarded, the remaining three vectors are linearly independent and generate the same vector space as the original five. Other sets of three vectors generate the same space; for instance, (1, 0, 0), (1, 1, 0), (1, 1, 1). (Can you verify this?)

The name *dimension* is justified by the following facts.

dimension, basis

Any set of linearly independent vectors in $V(\mathbf{x}_1, \ldots, \mathbf{x}_k)$ that generates this subspace must consist of the same number of vectors. This number is the dimension r described previously. Any such set of r vectors is called a basis for the subspace. If the original $\mathbf{x}_1, \ldots, \mathbf{x}_k$ are linearly independent, then the dimension is k and these $\mathbf{x}_i$'s are themselves a basis. There are infinitely many bases.

Example 2 Consider Example 2, Section 16.1. That example shows that each of the pairs $\mathbf{x}_1, \mathbf{x}_2$ and $\mathbf{y}_1, \mathbf{y}_2$ generate the same subspace. These pairs are each linearly independent; hence, each is a basis for this subspace, the dimension of which is 2. Among some of the other bases for the same subspace are the following pairs:

$$(3\mathbf{x}_1, -\mathbf{x}_2), \quad (\mathbf{x}_1, \mathbf{y}_1), \quad (\mathbf{y}_1 + 2\mathbf{y}_2, \mathbf{x}_2)$$

(Verify these assertions.)

It should be pointed out that a set of vectors containing the zero vector cannot be linearly independent. For instance, suppose the set is arranged as

$$(0, \ldots, 0), \mathbf{x}_2, \ldots, \mathbf{x}_k$$

Then

$$(0, \ldots, 0) = 0\mathbf{x}_2 + \cdots + 0\mathbf{x}_k$$

and thus the set is not linearly independent (Note 2).

16.3 ORTHOGONALITY

Two vectors $\mathbf{x}$ and $\mathbf{y}$ are said to be orthogonal if $(\mathbf{x}, \mathbf{y}) = 0$. The plausibility of this terminology for $n=2$ is described in Figure 16.1. A set of k vectors $\mathbf{x}_1, \ldots, \mathbf{x}_k$ are called *mutually* orthogonal if $(\mathbf{x}_i, \mathbf{x}_j) = 0$ for every pair of distinct indices, $i \neq j$. One of the im-

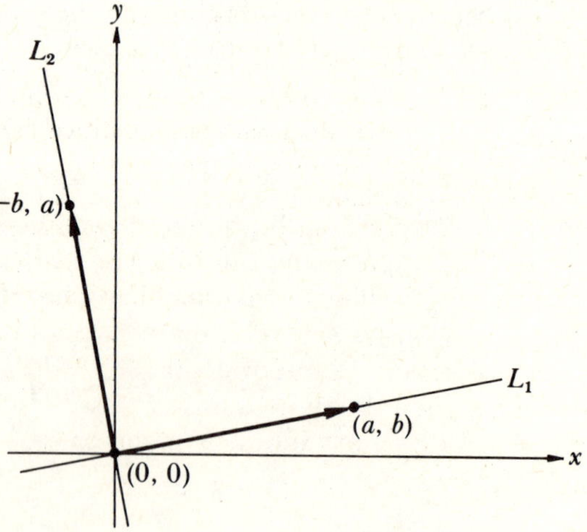

FIGURE 16.1 The slope of L_1 is b/a. The slope of L_2 is $-a/b$. L_1 and L_2 are perpendicular. The inner product of (a, b) and $(-b, a)$ equals zero.

portant uses of orthogonality is in recognizing linear independence.

mutually orthogonal vectors are linearly independent

Suppose that $\mathbf{x}_1, \ldots, \mathbf{x}_k$ are *nonzero* vectors that are mutually orthogonal. Then the $\mathbf{x}_i$'s are linearly independent.

Proof. We argue by contradiction. Suppose that one of the $\mathbf{x}_i$'s can be expressed as a linear combination of the others. Specifically, suppose that

$$\mathbf{x}_k = a_1\mathbf{x}_1 + \cdots + a_{k-1}\mathbf{x}_{k-1}$$

Compute the inner product of both sides with $\mathbf{x}_k$ to obtain

$$(\mathbf{x}_k, \mathbf{x}_k) = \|\mathbf{x}_k\|^2 = a_1(\mathbf{x}_1, \mathbf{x}_k) + \cdots + a_{k-1}(\mathbf{x}_{k-1}, \mathbf{x}_k) = 0$$

Hence, $\mathbf{x}_k$ would have to be the zero vector, which would contradict the hypothesis. By the same argument, no $\mathbf{x}_i$ can be a linear combination of the other $\mathbf{x}_i$'s. ▮

When $\mathbf{x}_i, \ldots, \mathbf{x}_k$ is a set of nonzero, mutually orthogonal vectors, it is usually most convenient to suppose that each $\mathbf{x}_i$ has length 1.

orthonormal set

$\mathbf{x}_1, \ldots, \mathbf{x}_k$ is called an orthonormal set of vectors if

$$(\mathbf{x}_i, \mathbf{x}_i) = 1, \qquad i = 1, \ldots, n$$

$$(\mathbf{x}_i, \mathbf{x}_j) = 0, \qquad i \neq j$$

If $\mathbf{x}_1, \ldots, \mathbf{x}_k$ are nonzero and mutually orthogonal, then $a_1\mathbf{x}_1, \ldots, a_k\mathbf{x}_k$ are also mutually orthogonal and generate $V(\mathbf{x}_1, \ldots, \mathbf{x}_k)$ whenever the a_i's are nonzero scalars. In particular, choosing $a_i = 1/\|\mathbf{x}_i\|$ makes the length of $a_i x_i$ equal to 1 and $a_1\mathbf{x}_1, \ldots, a_k\mathbf{x}_k$ is an orthonormal set of vectors that is a basis for $V(\mathbf{x}_1, \ldots, \mathbf{x}_k)$. Such a basis is called an *orthonormal basis.*

Example 1

(a) (1, 0, 0), (0, 1, 0), (0, 0, 1) is an orthonormal basis for the three-dimensional vector space of all triples of real numbers.

(b) (1, 0, 0), (0, 1, 0) is an orthonormal basis for the two-dimensional subspace of vectors having 0 in the third position.

(c) $(1/\sqrt{2}, 1/\sqrt{2}, 0)$, $(1/\sqrt{2}, -1/\sqrt{2}, 0)$ is an orthonormal basis for the same subspace described in (b).

It is particularly easy to represent a vector that is a linear combination of vectors forming an orthonormal set.

Suppose $\mathbf{x}_1, \ldots, \mathbf{x}_k$ is an orthonormal set of vectors and

$$\mathbf{y} = a_1\mathbf{x}_1 + \cdots + a_k\mathbf{x}_k \tag{1}$$

Then $a_i = (\mathbf{y}, \mathbf{x}_i)$, $i = 1, \ldots, k$. In this setting, the a_i's are called the *Fourier coefficients* of $\mathbf{y}$, with respect to the orthonormal set $\mathbf{x}_1, \ldots, \mathbf{x}_n$.

Proof. Simply compute the inner product of both sides of (1) with $\mathbf{x}_i$. The result is $(\mathbf{y}, \mathbf{x}_i) = a_i(\mathbf{x}_i, \mathbf{x}_i) = a_i$. ∎

Example 2 The set of vectors

$$\mathbf{x}_1 = \left(\frac{1}{\sqrt{2}}, \frac{-1}{\sqrt{2}}, 0\right), \quad \mathbf{x}_2 = \left(\frac{1}{\sqrt{3}}, \frac{1}{\sqrt{3}}, \frac{1}{\sqrt{3}}\right),$$

$$\mathbf{x}_3 = \left(\frac{1}{\sqrt{6}}, \frac{1}{\sqrt{6}}, \frac{-2}{\sqrt{6}}\right)$$

is an orthonormal basis for the vector space of all real triples. (Verify this.) Let us find the representation of the vector $\mathbf{y} = (1, 0, 0)$. Since $(\mathbf{y}, \mathbf{x}_1) = 1/\sqrt{2}$, $(\mathbf{y}, \mathbf{x}_2) = 1/\sqrt{3}$, $(\mathbf{y}, \mathbf{x}_3) = 1/\sqrt{6}$, it follows that

$$\mathbf{y} = \left(\frac{1}{\sqrt{2}}\right)\mathbf{x}_1 + \left(\frac{1}{\sqrt{3}}\right)\mathbf{x}_2 + \left(\frac{1}{\sqrt{6}}\right)\mathbf{x}_3$$

(Verify that this representation is, indeed, correct.)

An earlier assertion says, in effect, that an orthonormal set of vectors must be linearly independent. A converse says that given any linearly independent set of vectors, there exists an orthonormal set that generates the same subspace.

Gram–Schmidt orthogonalization

Suppose that $\mathbf{x}_1, \ldots, \mathbf{x}_k$ are linearly independent vectors. Then there is an orthonormal set $\mathbf{y}_1, \ldots, \mathbf{y}_k$ such that

$$V(\mathbf{x}_1, \ldots, \mathbf{x}_k) = V(\mathbf{y}_1, \ldots, \mathbf{y}_k)$$

One explicit way of constructing the $\mathbf{y}_i$'s is as follows:

$$\mathbf{y}_1 = \frac{1}{\|\mathbf{x}_1\|}\mathbf{x}_1$$

$$\mathbf{y}_2' = \mathbf{x}_2 - (\mathbf{x}_2, \mathbf{y}_1)\mathbf{y}_1, \qquad \mathbf{y}_2 = \frac{1}{\|\mathbf{y}_2'\|}\mathbf{y}_2'$$

$$\vdots$$

$$\mathbf{y}_k' = \mathbf{x}_k - (\mathbf{x}_k, \mathbf{y}_1)\mathbf{y}_1 - \cdots - (\mathbf{x}_k, \mathbf{y}_{k-1})\mathbf{y}_{k-1}, \qquad \mathbf{y}_k = \frac{1}{\|\mathbf{y}_k'\|}\mathbf{y}_k'$$

The procedure described here for constructing an orthonormal basis is called a *Gram–Schmidt orthogonalization procedure.* It is certainly not unique because there are many other ways of constructing an orthonormal basis [see Exercise 3, parts (b) and (c)] (Note 3).

Proof. The first step is to replace $\mathbf{x}_1$ by a multiple of length 1. It is possible because $\mathbf{x}_1$ cannot be the zero vector by the linear independence of the $\mathbf{x}_i$'s. The vector $\mathbf{y}_2'$ is orthogonal to $\mathbf{y}_1$ since $(\mathbf{y}_2', \mathbf{y}_1) = (\mathbf{x}_2, \mathbf{y}_1) - (\mathbf{x}_2, \mathbf{y}_1) = 0$. Also, $\mathbf{y}_2'$ cannot be the zero vector, since $\mathbf{x}_2$ would be a multiple of $\mathbf{x}_1$, which would again contradict the linear independence. Hence, $\mathbf{y}_2'$ can be renormed to $\mathbf{y}_2$, which is a length 1 vector. Thus, $\mathbf{y}_1$, $\mathbf{y}_2$ are the first two vectors of an orthonormal basis, and the verification goes on in the same way for the remaining vectors. It is easy to go backward and express the $\mathbf{x}_i$'s in terms of the $\mathbf{y}_i$'s. Thus, every linear combination of $\mathbf{x}_i$'s is a linear combination of $\mathbf{y}_i$'s, and vice versa. ∎

Example 3 Consider the vectors $\mathbf{x}_1 = (1, 1, 1)$, $\mathbf{x}_2 = (1, 1, 0)$, $\mathbf{x}_3 = (1, 0, 0)$. It is left to the reader to check that these vectors are linearly independent. Applying the preceding orthogonalization procedure, we obtain the following:

$$\mathbf{y}_1 = \left(\frac{1}{\sqrt{3}}, \frac{1}{\sqrt{3}}, \frac{1}{\sqrt{3}}\right)$$

$$\mathbf{y}_2' = \left(\frac{1}{3}, \frac{1}{3}, \frac{-2}{3}\right), \qquad \mathbf{y}_2 = \left(\frac{1}{\sqrt{6}}, \frac{1}{\sqrt{6}}, \frac{-2}{\sqrt{6}}\right)$$

$$\mathbf{y}_3' = \left(\frac{1}{2}, \frac{-1}{2}, 0\right), \qquad \mathbf{y}_2 = \left(\frac{1}{\sqrt{2}}, \frac{-1}{\sqrt{2}}, 0\right)$$

If we label the original set of vectors in reverse order, as $\mathbf{x}_1 = (1, 0, 0)$, $\mathbf{x}_2 = (1, 1, 0)$, $\mathbf{x}_3 = (1, 1, 1)$, we would obtain $(1, 0, 0)$, $(0, 1, 0)$, $(0, 0, 1)$ as the orthonormal set. (Verify this.)

It should be pointed out that the Gram–Schmidt procedure can be profitably employed even if one does not know that the original set of $\mathbf{x}_i$'s are linearly independent. If at the ith stage we obtain a vector $\mathbf{y}_i'$, which is zero, we simply discard the vector $\mathbf{x}_i$ from the original set and continue. The reason is that $\mathbf{y}_i'$ being zero means that $\mathbf{x}_i$ is a linear combination of the preceding $\mathbf{x}_1, \ldots, \mathbf{x}_{i-1}$. The orthonormal set of vectors $\mathbf{y}_1, \ldots, \mathbf{y}_r$ that one finally obtains must generate $V(\mathbf{x}_1, \ldots, \mathbf{x}_k)$. In this way, one learns that the dimension of the subspace is r.

16.4 PROJECTIONS

Suppose that $\mathbf{x}_1, \ldots, \mathbf{x}_k$ is an orthonormal set of vectors. Remember that n is the dimension of the whole vector space. If $n > k$, then it is possible to supplement $\mathbf{x}_1, \ldots, \mathbf{x}_k$ by vectors $\mathbf{x}_{k+1}, \ldots, \mathbf{x}_n$ such that $\mathbf{x}_1, \ldots, \mathbf{x}_k, \mathbf{x}_{k+1}, \ldots, \mathbf{x}_n$ is an orthonormal basis for the whole vector space of real n-tuples. One way of doing so is as follows. Let $\mathbf{u}_1, \ldots, \mathbf{u}_n$ be a basis for the whole space. Now consider the combined set $\mathbf{x}_1, \ldots, \mathbf{x}_k, \mathbf{u}_1, \ldots, \mathbf{u}_n$. Apply the Gram–Schmidt procedure to this set, discarding any vector that is a linear combination of the preceding ones. The result of the first k steps will be the original $\mathbf{x}_1, \ldots, \mathbf{x}_k$, since it is already an orthonormal set. The whole procedure must produce n vectors because n is the dimension of the space. Denote the remaining vectors that follow $\mathbf{x}_1, \ldots, \mathbf{x}_k$ by $\mathbf{x}_{k+1}, \ldots, \mathbf{x}_n$. Let V_1 denote $V(\mathbf{x}_1, \ldots, \mathbf{x}_k)$ and let V_2 denote $V(\mathbf{x}_{k+1}, \ldots, \mathbf{x}_n)$. Any vector in V_1 is orthogonal to any vector in V_2, because

$$(a_1\mathbf{x}_1 + \cdots + a_k\mathbf{x}_k, a_{k+1}\mathbf{x}_{k+1} + \cdots + a_n\mathbf{x}_n) = \sum_{i=1}^{k} \sum_{j=k+1}^{n} a_i a_j (\mathbf{x}_i, \mathbf{x}_j)$$
$$= 0$$

For this reason, V_1 and V_2 are called *orthogonal* subspaces. We summarize in a slightly different form, as follows.

orthogonal subspaces

Let V_1 be a subspace of dimension $k < n$. There exists another subspace V_2 of dimension $n - k$ such that any vector in V_1 is orthogonal to any vector in V_2. V_1 and V_2 are called orthogonal subspaces. If $\mathbf{x}_1, \ldots, \mathbf{x}_k$ is an orthonormal basis for V_1 and $\mathbf{x}_{k+1}, \ldots, \mathbf{x}_n$ is an orthonormal basis for V_2, then the combined set of vectors $\mathbf{x}_1, \ldots, \mathbf{x}_k, \mathbf{x}_{k+1}, \ldots, \mathbf{x}_n$ is an orthonormal basis for the whole space.

If $\mathbf{x}$ is any vector, then its representation in terms of the orthonormal basis $\mathbf{x}_1, \ldots, \mathbf{x}_k, \mathbf{x}_{k+1}, \ldots, \mathbf{x}_n$ is

$$\mathbf{x} = (\mathbf{x}, \mathbf{x}_1)\mathbf{x}_1 + \cdots + (\mathbf{x}, \mathbf{x}_k)\mathbf{x}_k + (\mathbf{x}, \mathbf{x}_{k+1})\mathbf{x}_{k+1} + \cdots + (\mathbf{x}, \mathbf{x}_n)\mathbf{x}_n \tag{2}$$

Denoting $\mathbf{y} = (\mathbf{x}, \mathbf{x}_1)\mathbf{x}_1 + \cdots + (\mathbf{x}, \mathbf{x}_k)\mathbf{x}_k$, and $\mathbf{z} = (\mathbf{x}, \mathbf{x}_{k+1})\mathbf{x}_{k+1} + \cdots + (\mathbf{x}, \mathbf{x}_n)\mathbf{x}_n$, (2) can be written as

$$\mathbf{x} = \mathbf{y} + \mathbf{z} \tag{3}$$

Also, y lies in V_1 and is called the *projection of* $\mathbf{x}$ *on* V_1; $\mathbf{z}$ lies in V_2 and is called the *projection of* $\mathbf{x}$ *on* V_2. The two projections $\mathbf{y}$ and $\mathbf{z}$ are orthogonal, which is illustrated for $n = 2$ in Figure 16.2. If we take the inner product of both sides of (3) with themselves, we have

$$(\mathbf{x}, \mathbf{x}) = \|\mathbf{x}\|^2 = (\mathbf{y} + \mathbf{z}, \mathbf{y} + \mathbf{z}) = (\mathbf{y}, \mathbf{y}) + (\mathbf{z}, \mathbf{z}) = \|\mathbf{y}\|^2 + \|\mathbf{z}\|^2 \tag{4}$$

since $(\mathbf{y}, \mathbf{z}) = 0$. This is really a form of the *Pythagorean theorem.* The representation (3) is unique. That is, if $\mathbf{x} = \mathbf{y}_1 + \mathbf{z}_1$ with $\mathbf{y}_1$ in V_1 and $\mathbf{z}_1$ in V_2, then $\mathbf{y} = \mathbf{y}_1$ and $\mathbf{z} = \mathbf{z}_1$. This relation is true because $(\mathbf{y} - \mathbf{y}_1) + (\mathbf{z} - \mathbf{z}_1)$ equals the zero vector. Because these two parts are orthogonal, we have, as in (4),

$$\|\mathbf{y} - \mathbf{y}_1\|^2 + \|\mathbf{z} - \mathbf{z}_1\|^2 = 0$$

from which it follows that $\mathbf{y} = \mathbf{y}_1$ and $\mathbf{z} = \mathbf{z}_1$. The preceding is summarized as follows.

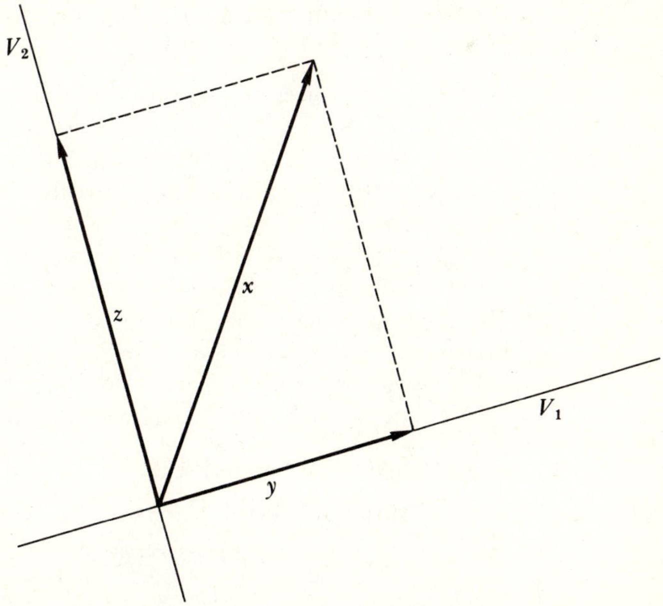

FIGURE 16.2 The projection of x on V_1 is y. The projection of x on V_2 is z.

projections

If V_1 and V_2 are orthogonal subspaces, then any vector $\mathbf{x}$ has a unique representation

$$\mathbf{x} = \mathbf{y} + \mathbf{z}$$

with $\mathbf{y}$ in V_1 and $\mathbf{z}$ in V_2. Projections $\mathbf{y}$ and $\mathbf{z}$ are called the projections on V_1 and V_2, respectively. The fact that

$$\|\mathbf{x}\|^2 = \|\mathbf{y}\|^2 + \|\mathbf{z}\|^2$$

is called the Pythagorean theorem (Note 4).

The preceding ideas extend to more than two subspaces. Suppose that

$$\mathbf{x}_{11}, \ldots, \mathbf{x}_{1k_1}, \ldots, \mathbf{x}_{r1}, \ldots, \mathbf{x}_{rk_r}$$

is an orthonormal basis for the whole n-dimensional space. Let

$$V_1 = V(\mathbf{x}_{11}, \ldots, \mathbf{x}_{1k_1}), \ldots, V_r = V(\mathbf{x}_{r1}, \ldots, \mathbf{x}_{rk_r})$$

These subspaces have dimensions $k_1, \ldots, k_r$, respectively, and $k_1 + \cdots + k_r = n$. Subspaces $V_1, \ldots, V_r$ are called *mutually orthogonal subspaces*, which means that whenever $\mathbf{y}_i$ is in V_i and $\mathbf{y}_j$ is in V_j and $i \neq j$, then $(\mathbf{y}_i, \mathbf{y}_j) = 0$. If $\mathbf{x}$ is any vector, then denote

$$\mathbf{y}_i = (\mathbf{x}, \mathbf{x}_{i1})\mathbf{x}_{i1} + \cdots + (\mathbf{x}, \mathbf{x}_{ik_i})\mathbf{x}_{ik_i}, \qquad i = 1, \ldots, r$$

The representation of $\mathbf{x}$ corresponding to (3) is

$$\mathbf{x} = \mathbf{y}_1 + \cdots + \mathbf{y}_r \tag{5}$$

and $\mathbf{y}_i$ is the projection of $\mathbf{x}$ on V_i. The representation (5) is unique. That is, if $\mathbf{x} = \mathbf{y}_1' + \cdots + \mathbf{y}_r'$, where $\mathbf{y}_i'$ is in V_i, then it must follow that $\mathbf{y}_i = \mathbf{y}_i'$, $i = 1, \ldots, r$. The proof is the same as the earlier proof for two orthogonal subspaces. Finally, it follows from (5) that

$$\|\mathbf{x}\|^2 = \|\mathbf{y}_1\|^2 + \cdots + \|\mathbf{y}_r\|^2 \tag{6}$$

That is, the square of the length of $\mathbf{x}$ is the sum of the squares of the lengths of its projections on the r mutually orthogonal subspaces. As in the two-dimensional case, it is appropriate to call (6) the Pythagorean theorem.

Now we consider the problem of minimizing

$$\|\mathbf{x} - a_1\mathbf{x}_1 - \cdots - a_k\mathbf{x}_k\|^2$$

over all variations of the scalars $a_1, \ldots, a_k$.

$\|\mathbf{x} - a_1\mathbf{x}_1 - \cdots - a_k\mathbf{x}_k\|^2$ is minimized with respect to $a_1, \ldots, a_k$ when $a_1\mathbf{x}_1 + \cdots + a_k\mathbf{x}_k$ is the projection of $\mathbf{x}$ on $V(\mathbf{x}_1, \ldots, \mathbf{x}_k)$.

Proof. Let $\mathbf{y}_1, \ldots, \mathbf{y}_r$ be an orthonormal basis for $V(\mathbf{x}_1, \ldots, \mathbf{x}_k)$ and $\mathbf{y}_1, \ldots, \mathbf{y}_r, \mathbf{y}_{r+1}, \ldots, \mathbf{y}_n$ an orthonormal basis for the whole space. Since any linear combination of $\mathbf{x}_1, \ldots, \mathbf{x}_k$ is an appropriate linear combination of $\mathbf{y}_1, \ldots, \mathbf{y}_r$, we may as well minimize

$$\|\mathbf{x} - b_1\mathbf{y}_1 - \cdots - b_r\mathbf{y}_r\|^2 \tag{7}$$

with respect to the scalars $b_1, \ldots, b_r$. Since $\mathbf{x} = \sum_{i=1}^{n} (\mathbf{x}, \mathbf{y}_i)\mathbf{y}_i$, (7) equals

$$\left\|\sum_{i=1}^{r} [(\mathbf{x}, \mathbf{y}_i) - b_i]\mathbf{y}_i + \sum_{i=r+1}^{n} (\mathbf{x}, \mathbf{y}_i)\mathbf{y}_i\right\|^2 = \sum_{i=1}^{r} [(\mathbf{x}, \mathbf{y}_i) - b_i]^2 + \sum_{i=r+1}^{n} (\mathbf{x}, \mathbf{y}_i)^2 \tag{8}$$

Clearly, (8) is minimized by choosing $b_i = (\mathbf{x}, \mathbf{y}_i)$ and $\sum_{i=1}^{r} (\mathbf{x}, \mathbf{y}_i)\mathbf{y}_i$ is the projection of $\mathbf{x}$ on $V(\mathbf{x}_1, \ldots, \mathbf{x}_k) = V(\mathbf{y}_1, \ldots, \mathbf{y}_r)$. ∎

It should be pointed out that minimizing coefficients $a_i, \ldots, a_k$ in terms of $\mathbf{x}_1, \ldots, \mathbf{x}_k$ are not necessarily unique. However, the minimizing vector $\sum_{i=1}^{r} a_i\mathbf{x}_i$ is unique because it is the projection of $\mathbf{x}$ on $V(\mathbf{x}_1, \ldots, \mathbf{x}_k)$.

Finally, we need to review the Cauchy–Schwarz inequality for inner products. This inequality asserts that for any vectors $\mathbf{x}$ and $\mathbf{y}$

$$|(\mathbf{x}, \mathbf{y})| \leq \|\mathbf{x}\| \, \|\mathbf{y}\| \tag{9}$$

Closely related to (9) is the solution of a certain maximization problem that we need in the next chapter.

Let V_1 and V_2 be orthogonal subspaces. Then, for any vector $\mathbf{y}$,

$$\sup_{\mathbf{x} \text{ in } V_1} \frac{|(\mathbf{x}, \mathbf{y})|}{\|\mathbf{x}\|} = \|\mathbf{u}\| \tag{10}$$

where $\mathbf{u}$ is the projection of $\mathbf{y}$ on V_1.

Proof of (10). By (3), we can write $\mathbf{y} = \mathbf{u} + \mathbf{v}$, where $\mathbf{u}$ is in V_1 and $\mathbf{v}$ is in V_2. Then, since $(\mathbf{x}, \mathbf{v}) = 0$, for any $\mathbf{x}$ in V_1,

$$\frac{(\mathbf{x}, \mathbf{y})}{\|\mathbf{x}\|} = \frac{(\mathbf{x}, \mathbf{u})}{\|\mathbf{x}\|}$$

By the Cauchy–Schwarz inequality, $|(\mathbf{x}, \mathbf{u}| \leq \|\mathbf{x}\| \, \|\mathbf{u}\|$. If we take $\mathbf{x}$ to be proportional to $\mathbf{u}$, as we can because $\mathbf{x}$ varies over all of V_1, we have $|(\mathbf{x}, \mathbf{u})|/\|\mathbf{x}\| = \|\mathbf{u}\|$, which must then be the supremum of the left side of (10). ∎

16.5 ORTHOGONAL MATRICES

Recall the definition of a matrix and the description of the elementary properties of matrices in Section 4.4. We now consider a special type of matrix, called an orthogonal matrix.

orthogonal matrix

An $n \times n$ matrix of real numbers

$$\mathbf{A} = \begin{pmatrix} a_{11} & a_{12} & \cdots & a_{1n} \\ a_{21} & a_{22} & \cdots & a_{2n} \\ \vdots & & & \\ a_{n1} & a_{n2} & \cdots & a_{nn} \end{pmatrix}$$

is called an orthogonal matrix if its rows $(a_{11}, \ldots, a_{1n}), \ldots, (a_{n1}, \ldots, a_{nn})$ form an orthonormal basis for the n-dimensional vector space of real n-tuples.

The preceding definition can be rephrased as follows: A is orthogonal if

$$\begin{pmatrix} a_{11} & a_{12} & \cdots & a_{1n} \\ a_{21} & a_{22} & \cdots & a_{2n} \\ \vdots & & & \\ a_{n1} & a_{n2} & \cdots & a_{nn} \end{pmatrix} \begin{pmatrix} a_{11} & a_{21} & \cdots & a_{n1} \\ a_{12} & a_{22} & \cdots & a_{n2} \\ \vdots & & & \\ a_{1n} & a_{2n} & \cdots & a_{nn} \end{pmatrix} = \begin{pmatrix} 1 & 0 & \cdots & 0 \\ 0 & 1 & \cdots & 0 \\ \vdots & & & \\ 0 & 0 & \cdots & 1 \end{pmatrix} \tag{1}$$

This follows from the inner product calculations

$$((a_{i1}, \ldots, a_{in}), (a_{j1}, \ldots, a_{jn})) = \sum_{k=1}^{n} a_{ik}a_{jk} = \begin{cases} 1 & \text{if } i = j \\ 0 & \text{if } i \neq j \end{cases}$$

If A is a matrix having a (i, j)th element of a_{ij}, then the matrix having a (i, j)th element of a_{ji} is called the transpose of A and is denoted by A'. Equation (1) can be written concisely as

$$\mathbf{AA}' = I \tag{2}$$

where I is the *identity matrix*. (The identity matrix is an $n \times n$ matrix with 1's down the diagonal and 0's elsewhere.)

Example 1 The following are some examples of orthogonal matrices.

$$\begin{pmatrix} 1 & 0 \\ 0 & 1 \end{pmatrix}, \quad \begin{pmatrix} \frac{1}{\sqrt{2}} & \frac{1}{\sqrt{2}} \\ \frac{1}{\sqrt{2}} & \frac{-1}{\sqrt{2}} \end{pmatrix}, \quad \begin{pmatrix} \frac{1}{\sqrt{2}} & \frac{-1}{\sqrt{2}} & 0 \\ \frac{1}{\sqrt{3}} & \frac{1}{\sqrt{3}} & \frac{1}{\sqrt{3}} \\ \frac{1}{\sqrt{6}} & \frac{1}{\sqrt{6}} & \frac{-2}{\sqrt{6}} \end{pmatrix}$$

Orthogonal matrices arise in a natural way in the relations between different orthonormal bases for a vector space. Suppose that $\mathbf{x}_1, \ldots, \mathbf{x}_n$ and $\mathbf{y}_1, \ldots, \mathbf{y}_n$ are two orthonormal bases for the same vector space. Each $\mathbf{y}_i$ is a linear combination of the $\mathbf{x}_i$'s. Suppose that these relationships are explicitly denoted as follows:

$$\begin{aligned} \mathbf{y}_1 &= a_{11}\mathbf{x}_1 + \cdots + a_{1n}\mathbf{x}_n \\ &\vdots \\ \mathbf{y}_n &= a_{n1}\mathbf{x}_1 + \cdots + a_{nn}\mathbf{x}_n \end{aligned} \tag{3}$$

Since

$$(\mathbf{y}_i, \mathbf{y}_j) = \sum_{k=1}^{n} \sum_{r=1}^{n} a_{ik}a_{jr}(\mathbf{x}_k, \mathbf{x}_r) = \sum_{k=1}^{n} a_{ik}a_{jk} = \begin{cases} 1 & \text{if } i = j \\ 0 & \text{if } i \neq j \end{cases} \tag{4}$$

it follows that the matrix of the coefficients a_{ij} in (3) is an orthogonal matrix. Conversely, suppose that $\mathbf{x}_1, \ldots, \mathbf{x}_n$ form an orthonormal basis and the matrix of the a_{ij}'s is orthogonal. If new vectors $\mathbf{y}_1, \ldots, \mathbf{y}_n$ are defined by (3), then the same calculation shows that $\mathbf{y}_1, \ldots, \mathbf{y}_n$ is an orthonormal set.

If $\mathbf{x}_1, \ldots, \mathbf{x}_n$ and $\mathbf{y}_1, \ldots, \mathbf{y}_n$ are two orthonormal bases, then the explicit form of the relationships in (3) must be

$$\begin{aligned} \mathbf{y}_1 &= (\mathbf{y}_1, \mathbf{x}_1)\mathbf{x}_1 + \cdots + (\mathbf{y}_1, \mathbf{x}_n)\mathbf{x}_n \\ &\vdots \\ \mathbf{y}_n &= (\mathbf{y}_n, \mathbf{x}_1)\mathbf{x}_1 + \cdots + (\mathbf{y}_n, \mathbf{x}_n)\mathbf{x}_n \end{aligned} \tag{5}$$

On the other hand, if the $\mathbf{x}_i$'s are expressed in terms of the $\mathbf{y}_i$'s, the relationships must be

$$\begin{aligned} \mathbf{x}_1 &= (\mathbf{x}_1, \mathbf{y}_1)\mathbf{y}_1 + \cdots + (\mathbf{x}_1, \mathbf{y}_n)\mathbf{y}_n \\ &\vdots \\ \mathbf{x}_n &= (\mathbf{x}_n, \mathbf{y}_1)\mathbf{y}_1 + \cdots + (\mathbf{x}_n, \mathbf{y}_n)\mathbf{y}_n \end{aligned} \tag{6}$$

In terms of the a_{ij}'s, (6) can be written

$$\begin{aligned} \mathbf{x}_1 &= a_{11}\mathbf{y}_1 + \cdots + a_{n1}\mathbf{y}_n \\ &\vdots \\ \mathbf{x}_n &= a_{1n}\mathbf{y}_1 + \cdots + a_{nn}\mathbf{y}_n \end{aligned}$$

This relation is so because by (4), $a_{ij} = (\mathbf{y}_i, \mathbf{x}_j)$. Since the roles of the $\mathbf{x}_i$'s and $\mathbf{y}_i$'s can be interchanged in (4), it follows that the transposed matrix

$$\mathbf{A}' = \begin{pmatrix} a_{11} & \cdots & a_{n1} \\ \vdots & & \\ a_{1n} & \cdots & a_{nn} \end{pmatrix}$$

is also an orthogonal matrix. If $\mathbf{A}$ is an orthogonal matrix, not only is it true that $\mathbf{AA}' = I$, as in (2), but also $\mathbf{A}'\mathbf{A} = I$. One important consequence of this fact is that

$$x_1^2 + \cdots + x_n^2 = y_1^2 + \cdots + y_n^2 \tag{7}$$

The reader can verify this by squaring both sides of each of the equations in (3) and summing.

Finally, we ask the reader to review the definition of the *determinant* of a square matrix (Note 5). If A is a square matrix, then $D(A)$ will denote the determinant of A. Recall the alternate notation used in Section 8.1, where it was pointed out that

$$\begin{vmatrix} a & b \\ c & d \end{vmatrix} = D\begin{pmatrix} a & b \\ c & d \end{pmatrix} = ad - bc$$

in the 2×2 case. The determinant of a 3×3 matrix is given by

$$\begin{vmatrix} a & b & c \\ d & e & f \\ g & h & i \end{vmatrix} = D\begin{pmatrix} a & b & c \\ d & e & f \\ g & h & i \end{pmatrix} = aei - ahf - bdi + bgf + cdh - cge$$

A basic property of determinants is that the determinant of a product is the product of the determinants. That is, $D(AB) = D(A)D(B)$. (Of course, A and B must be of the same size for this to make sense.) It is also true that $D(A) = D(A')$ and that $D(I) = 1$. It follows from (2) that if A is an orthogonal matrix, then $D(A)D(A') = [D(A)]^2 = 1$ and

$$|D(A)| = 1 \tag{8}$$

16.6 RANDOM VECTORS

We now want to adapt some of the preceding material concerning vectors to n-tuples, the components of which are random variables.

random vectors

If $X_1, \ldots, X_n$ are random variables, then the n-tuple

$$\mathbf{X} = (X_1, \ldots, X_n)$$

is called a random vector. The vector operations concerning random vectors are exactly the same as those described earlier for n-tuples the components of which are scalars. Thus, if $\mathbf{Y} = (Y_1, \ldots, Y_n)$ is another random vector, then $\mathbf{X} + \mathbf{Y}$ is the random vector defined by

$$\mathbf{X} + \mathbf{Y} = (X_1 + Y_1, \ldots, X_n + Y_n)$$

If a is a scalar, then $a\mathbf{X}$ is the random vector defined by

$$a\mathbf{X} = (aX_1, \ldots, aX_n)$$

The inner product of two random vectors is

$$(\mathbf{X}, \mathbf{Y}) = (X_1Y_1 + \cdots + X_nY_n)$$

Notice that $(\mathbf{X}, \mathbf{Y})$ is a random variable. If $\mathbf{a} = (a_1, \ldots, a_n)$ is a vector of constants, then

$$(\mathbf{a}, \mathbf{X}) = a_1X_1 + \cdots + a_nX_n$$

which also is a random variable.

We now describe how the expected value operation interrelates with vector operations involving random vectors. First, we have the rather natural definition that the expected value of a random vector is the vector of expected values.

expected value of a random vector

If $\mathbf{X} = (X_1, \ldots, X_n)$ is a random vector, then the expected value of $\mathbf{X}$ is defined to be

$$E\mathbf{X} = [E(X_1), \ldots, E(X_n)]$$

As usual, whenever we refer to expected value, we implicitly assume that it exists.

It follows immediately from the definition of expected value and its linearity properties for ordinary random variables, that if $\mathbf{X} = (X_1, \ldots, X_n)$ and $\mathbf{Y} = (Y_1, \ldots, Y_n)$ are random vectors and a and b are scalars,

$$\begin{aligned} E(a\mathbf{X} + b\mathbf{Y}) &= aE\mathbf{X} + bE\mathbf{Y} \\ &= [aE(X_1) + bE(Y_1), \ldots, aE(X_n) + bE(Y_n)] \end{aligned} \tag{1}$$

From now on, the random vectors of interest to us will have the following special property.

uncorrelated components of equal variance

Suppose that $\mathbf{X} = (X_1, \ldots, X_n)$ is a random vector such that

$$\mathrm{Var}(X_i) = \sigma^2, \qquad i = 1, \ldots, n$$

and

$$\mathrm{Cov}(X_i, X_j) = 0, \qquad i \neq j$$

That is, the variance of each component exists and has the same value, and any two different components have covariance zero. Then $\mathbf{X}$ will be said to have *uncorrelated components of equal variance.*

We now consider some basic relations involving inner products.

Suppose that $\mathbf{a} = (a_1, \ldots, a_n)$ and $\mathbf{b} = (b_1, \ldots, b_n)$ are vectors of scalars, and that $\mathbf{X} = (X_1, \ldots, X_n)$ is a random vector with uncorrelated components of equal variance σ^2. Then

(a) $E(\mathbf{a}, \mathbf{X}) = (\mathbf{a}, E\mathbf{X})$

(b) $\mathrm{Cov}[(\mathbf{a}, \mathbf{X}), (\mathbf{b}, \mathbf{X})] = (\mathbf{a}, \mathbf{b})\sigma^2$

(c) $\mathrm{Var}(\mathbf{a}, \mathbf{X}) = \|\mathbf{a}\|^2\sigma^2$

Proof. Since $(\mathbf{a}, \mathbf{X}) = a_1X_1 + \cdots + a_nX_n$, it follows that

$$E(\mathbf{a}, \mathbf{X}) = a_1E(X_1) + \cdots + a_nE(X_n) = (\mathbf{a}, E\mathbf{X})$$

For part (a), we do not need the full hypothesis; only the expected values need exist. The proof of (b) is equally routine.

$$\begin{aligned}\mathrm{Cov}[(\mathbf{a}, \mathbf{X}), (\mathbf{b}, \mathbf{X})] &= \sum_{j=1}^{n}\sum_{i=1}^{n} a_ib_j \,\mathrm{Cov}(X_i, X_j) = \sum_{i=1}^{n} a_ib_i \,\mathrm{Cov}(X_i, X_i)\\ &= (\mathbf{a}, \mathbf{b})\sigma^2\end{aligned}$$

(See Exercise 2, Chapter 10.) Part (c) is a special case of (b) by setting $b = a$. ∎

In the same spirit, we need some facts about projections. First, we point out that if $\mathbf{X}$ is a random vector and V_1 a subspace, then the projection of $\mathbf{X}$ on V_1 is also a random vector. If $\mathbf{u}_1, \ldots, \mathbf{u}_k$ is an orthonormal basis for V_1, then, by Section 16.4, the projection of $\mathbf{X}$ on V_1 is the following vector, $\mathbf{Y}$.

$$\mathbf{Y} = (\mathbf{u}_1, \mathbf{X})\mathbf{u}_1 + \cdots + (\mathbf{u}_k, \mathbf{X})\mathbf{u}_k$$

Each of the inner products $(\mathbf{u}_i, \mathbf{X})$ is a random variable; hence, each of $(\mathbf{u}_i, \mathbf{X})\mathbf{u}_i$ is a random vector. $\mathbf{Y}$ is a sum of random vectors and is a random vector.

Suppose that $\mathbf{X} = (X_1, \ldots, X_n)$ is a random vector with uncorrelated components of equal variance, σ^2. Let V_1 be a subspace of the n-dimensional space of real n-tuples. Let $\mathbf{Y}$ be the random vector, which is the projection of $\mathbf{X}$ on V_1, and let $\mathbf{u}$ be the projection of $E\mathbf{X}$ on V_1.

(a) Then $E\mathbf{Y} = \mathbf{u}$. That is, the expected value of the projection is the projection of the expected value.

(b) Also, $E(\|\mathbf{Y}\|^2) = k\sigma^2 + \|\mathbf{u}\|^2$, where k is the dimension of V_1.

Proof. Let $\mathbf{u}_1, \ldots, \mathbf{u}_k$ be an orthonormal basis for V_1. By Section 16.4, the projection of $\mathbf{X}$ on V_1 is

$$\mathbf{Y} = (\mathbf{u}_1, \mathbf{X})\mathbf{u}_1 + \cdots + (\mathbf{u}_k, \mathbf{X})\mathbf{u}_k \tag{2}$$

Hence, $E\mathbf{Y} = (\mathbf{u}_1, E\mathbf{X})\mathbf{u}_1 + \cdots + (\mathbf{u}_k, E\mathbf{X})\mathbf{u}_k$, which is the projection of the vector $E\mathbf{X}$ on V_1. Thus (a) is proved. Notice that the only assumption required for this part is that $E\mathbf{X}$ exists. From (2), $\|\mathbf{Y}\|^2 = (\mathbf{u}_1, \mathbf{X})^2 + \cdots + (\mathbf{u}_k, \mathbf{X})^2$. Hence, since

$$\begin{aligned} E[(\mathbf{u}_i, \mathbf{X})^2] &= \mathrm{Var}(\mathbf{u}_i, \mathbf{X}) + [E(\mathbf{u}_i, \mathbf{X})]^2 = \|\mathbf{u}_i\|^2\sigma^2 + (\mathbf{u}_i, E\mathbf{X})^2 \\ &= \sigma^2 + (\mathbf{u}_i, E\mathbf{X})^2 \end{aligned}$$

it follows that

$$\begin{aligned} E(\|\mathbf{Y}\|^2) &= k\sigma^2 + (\mathbf{u}_1, E\mathbf{X})^2 + \cdots + (\mathbf{u}_k, E\mathbf{X})^2 \\ &= k\sigma^2 + \|\mathbf{u}\|^2 \quad \blacksquare \end{aligned}$$

Example 1 Suppose that $X_1, \ldots, X_n$ are independent and identically distributed random variables with $E(X_i) = m$ and $\mathrm{Var}(X_i) = \sigma^2$. Let V_2 be the one-dimensional subspace generated by the unit length vector $\mathbf{v} = [(1/\sqrt{n}), (1/\sqrt{n}), \ldots, (1/\sqrt{n})]$, and let V_1 be the $(n-1)$ dimensional subspace of all vectors orthogonal to $\mathbf{v}$. By definition, V_1 and V_2 are orthogonal subspaces. Let $\mathbf{X} = (X_1, \ldots, X_n)$. An orthonormal basis for V_2 is the single vector $\mathbf{v}$; hence, the projection of X on V_2 is $(\mathbf{v}, \mathbf{X})\mathbf{v} = (\overline{X}, \overline{X}, \ldots, \overline{X})$, where $\overline{X} = (X_1 + \cdots + X_n)/n$ is the sample mean. Since it already lies in V_2, $E\mathbf{X} = (m, \ldots, m)$ and its projection on V_2 is just $E\mathbf{X}$; hence, the projection of $E\mathbf{X}$ on V_1 is $(0, 0, \ldots, 0)$. The projection of $\mathbf{X}$ on V_1 is

$$\mathbf{Y} = (X_1 - \overline{X}, X_2 - \overline{X}, \ldots, X_n - \overline{X})$$

because the projections of $\mathbf{X}$ on V_1 and on V_2 must sum to $\mathbf{X}$ itself. Hence, $E\mathbf{Y} = (0, 0, \ldots, 0)$, which agrees with (a) of the last theorem. Also, $\|\mathbf{Y}\|^2 = \Sigma_{i=1}^{n} (X_i - \overline{X})^2$. By (b), we must have that

$$E \sum_{i=1}^{n} (X_i - \overline{X})^2 = (n-1)\sigma^2$$

(See Exercise 2, Chapter 12, for a direct proof of this fact.)

16.7 ORTHOGONAL TRANSFORMATIONS OF NORMAL RANDOM VARIABLES

Suppose that $X_1, X_2, \ldots, X_n$ are independent, $N(0, 1)$ random variables. Let $Y_1, \ldots, Y_n$ be new random variables obtained by the linear transformation

$$\begin{aligned} Y_1 &= a_{11}X_1 + \cdots + a_{1n}X_n \\ &\vdots \\ Y_n &= a_{n1}X_1 + \cdots + a_{nn}X_n \end{aligned} \tag{1}$$

normality is invariant under an orthogonal transformation

If the matrix of the a_{ij}'s in (1) is orthogonal, then $Y_1, \ldots, Y_n$ are also independent $N(0, 1)$.

Proof. The joint density of $X_1, \ldots, X_n$ is

$$\frac{1}{(2\pi)^{n/2}} \exp \frac{-(x_1^2 + \cdots + x_n^2)}{2} = f(x_1, \ldots, x_n)$$

Hence,

$$P[(Y_1, \ldots, Y_n) \text{ in } A] = \int \cdots \int_B f(x_1, \ldots, x_n)\, dx_1, \ldots, dx_n \tag{2}$$

where B is the set of points $(x_1, \ldots, x_n)$ satisfying the conditions that $(y_1, \ldots, y_n)$ lie in the set A, where

$$\begin{aligned} y_1 &= a_{11}x_1 + \cdots + a_{1n}x_n \\ &\vdots \\ y_n &= a_{n1}x_n + \cdots + a_{nn}x_n \end{aligned} \tag{3}$$

Because the matrix of the a_{ij}'s is orthogonal, it follows that $y_1^2 + \cdots + y_n^2 = x_1^2 + \cdots + x_n^2$ [see (7), Section 16.5]. Make a change of variables from $x_1, \ldots, x_n$ to $y_1, \ldots, y_n$, according to (3). The Jacobian is 1 by (8), Section 16.5. Hence, (2) becomes

$$P[(Y_1, \ldots, Y_n) \text{ in } A]$$

$$= \frac{1}{(2\pi)^{n/2}} \int \cdots_A \int \exp \frac{-(y_1^2 + \cdots y_n^2)}{2}\, dy_1 \cdots dy_n$$

We now have, $P[(Y_1, \ldots, Y_n) \text{ in } A] = P[(X_1, \ldots, X_n) \text{ in } A]$, and $Y_1, \ldots, Y_n$ are also independent, $N(0, 1)$. ▮

A special case of the preceding theorem for $n = 2$ appears as Exercise 20, Chapter 9. What makes the proof work is that the value of $f(x_1, \ldots, x_n)$ depends only on the distance of the point $x_1, \ldots, x_n$ from the origin. When the axes are transformed orthogonally, the density function still has this property with respect to the new coordinate system (see Figure 16.3). A useful way of rephrasing this result follows.

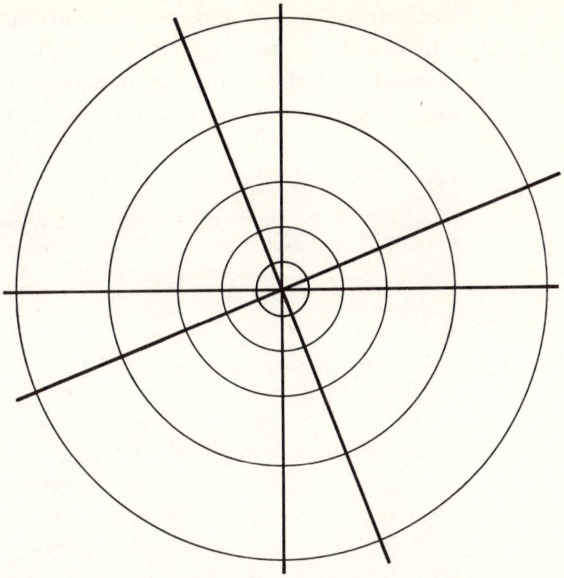

FIGURE 16.3 The integrating density of two independent $N(0, 1)$ random variables is $f(x, y) = (2\pi)^{-1} \exp -[(x^2 + y^2)/2]$. The graph shows the contours on which the surface is constant.

> Let $\mathbf{u}_1, \ldots, \mathbf{u}_n$ be an orthonormal basis for the n-dimensional space of real n-tuples. Let the components of $\mathbf{X} = (X_1, \ldots, X_n)$ be mutually independent, $N(0, 1)$ distributed. Then $\mathbf{Y}_1 = (\mathbf{u}_1, \mathbf{X}), \ldots, Y_n = (\mathbf{u}_n, \mathbf{X})$ are also mutually independent, $N(0, 1)$ distributed.

Proof. This assertion follows by writing

$$\begin{aligned} \mathbf{u}_1 &= (a_{11}, \ldots, a_{1n}) \\ &\vdots \\ \mathbf{u}_n &= (a_{n1}, \ldots, a_{nn}) \end{aligned}$$

Then $Y_1 = (\mathbf{u}_1, \mathbf{X}), \ldots, Y_n = (\mathbf{u}_n, \mathbf{X})$ are exactly the relationships given by (1), and the a_{ij}'s form an orthogonal matrix. ∎

Several computations that are often made in statistical analyses are

$$X = \frac{X_1 + \cdots + X_n}{n}, \qquad S^2 = \frac{(X_1 - \overline{X})^2 + \cdots + (X_n - \overline{X})^2}{n}$$

the sample mean and sample variance. It is an important fact, which follows from the previous theorem, that when the X_i's are normal with the same distribution, then $\bar{X}$ and S^2 are *independent* and nS^2/σ^2 is chi-square distributed with $n-1$ degrees of freedom.

Suppose that $X_1, \ldots, X_n$ are mutually independent, each with the same $N(m, \sigma^2)$ distribution. Then

$$\bar{X} = \frac{X_1 + \cdots + X_n}{n} \quad \text{and} \quad Y = \frac{nS^2}{\sigma^2}$$

$$= \frac{(X_1 - \bar{X})^2 + \cdots + (X_n - \bar{X})^2}{\sigma^2}$$

are independent random variables. Now, $\bar{X}$ has the $N(m, \sigma^2/n)$ distribution and Y is chi-square distributed with $n-1$ degrees of freedom. That is, Y is distributed like a sum of squares of $n-1$ independent $N(0, 1)$ random variables.

Proof. Let us first prove the assertions supposing that $m=0$, $\sigma=1$. Consider the unit length vector $\mathbf{v} = [(1/\sqrt{n}), (1/\sqrt{n}), \ldots, (1/\sqrt{n})]$. Using the same notation as in Example 1, Section 16.6, let V_2 be the one-dimensional subspace generated by $\mathbf{v}$ and let V_1 be the $(n-1)$-dimensional subspace of all vectors that are orthogonal to $\mathbf{v}$. Let $\mathbf{u}_1, \ldots, \mathbf{u}_{n-1}$ be an orthonormal basis for V_1; hence, $\mathbf{v}, \mathbf{u}_1, \ldots, \mathbf{u}_{n-1}$ is an orthonormal basis for the n-dimensional space of real n-tuples. By the previous theorem,

$$(\mathbf{v}, \mathbf{X}), (\mathbf{u}_1, \mathbf{X}), \ldots, (\mathbf{u}_{n-1}, \mathbf{X})$$

are independent, $N(0, 1)$. By (7), Section 16.5,

$$(\mathbf{v}, \mathbf{X})^2 + (\mathbf{u}_1, \mathbf{X})^2 + \cdots + (\mathbf{u}_{n-1}, \mathbf{X})^2 = \sum_{i=1}^{n} X_i^2$$

$$(\mathbf{v}, \mathbf{X})^2 = n(\bar{X})^2$$

$$(\mathbf{u}_1, \mathbf{X})^2 + \cdots + (\mathbf{u}_{n-1}, \mathbf{X})^2 = \sum_{i=1}^{n} X_i^2 - n(\bar{X})^2 = Y$$

The result follows in the special case $m=0$, $\sigma=1$. In general, use the fact that the standardized variables $(X_1 - m)/\sigma, \ldots, (X_n - m)/\sigma$ are independent, $N(0, 1)$ distributed. The sample average for these standardized variables is $(\bar{X} - m)/\sigma$ and the appropriate form of Y is

$$\left[\frac{X_1 - m}{\sigma} - \frac{(\overline{X} - m)}{\sigma}\right]^2 + \cdots + \left[\frac{(X_n - m)}{\sigma} - \frac{(\overline{X} - m)}{\sigma}\right]^2$$

$$= \frac{\sum_{i=1}^{n} (X_i - \overline{X})^2}{\sigma^2}$$

Repeating the previous analysis with the standardized variables shows that $\sqrt{n}(\overline{X} - m)/\sigma$ is $N(0, 1)$ and is independent of $\Sigma_{i=1}^{n} (X_i - \overline{X})^2/\sigma^2$; the latter is distributed like a sum of squares of $(n - 1)$ independent $N(0, 1)$ random variables. ▮

An important application of the preceding theorem in statistics concerns the distribution of a so-called "Student ratio." If U_0, $U_1, \ldots, U_n$ are mutually independent, $N(0, 1)$ random variables, then

$$\frac{U_0}{[(U_1^2 + \cdots + U_n^2)/n]^{1/2}} \tag{4}$$

is said to have the Student-t distribution, with n degrees of freedom. The explicit form of the integrating density of (4) is

$$f(x) = \frac{\Gamma[(n + 1)/2]}{(n\pi)^{1/2}\Gamma(n/2)[1 + (x^2/n)]^{(n+1)/2}} \tag{5}$$

Notice that $U_0^2/(U_1^2 + \cdots + U_n^2)$ is a ratio of independent gamma variables, the general distribution of which is given by (3), Section 9.5 ($\alpha = 1/2$, $\beta = n/2$). From this fact it is now easy to verify that the integrating density of (4) is given by (5); therefore, we leave the details to the reader (Note 6). We now claim that the following is true if $X_1, \ldots, X_n$ are independent, $N(0, 1)$.

$$\frac{n^{1/2}(\overline{X} - m)}{\{[(X_1 - \overline{X})^2 + \cdots + (X_n - \overline{X})^2]/(n - 1)\}^{1/2}} \tag{6}$$

is Student-t distributed with $n - 1$ degrees of freedom.

Proof. By the previous theorem, $n^{1/2}(\overline{X} - m)/\sigma$ and $\{[(X_1 - \overline{X})^2 + \cdots + (X_n - \overline{X})^2]/(n - 1)^{1/2}\sigma\}$ are independent random variables, the first being $N(0, 1)$ and the second $(n - 1)^{1/2}$ times the square root of a chi-square variable with $n - 1$ degrees of freedom. The ratio of these two is just (6), and hence is of the form (4), with $n - 1$ playing the role of n. It should be emphasized that the σ is canceled out in the ratio. Hence, neither (6) nor its distribution depends on σ (Note 7). ▮

There is another consequence of the invariance of normality under an orthogonal transformation that we will need in the next chapter.

Suppose that $X_1, \ldots, X_n$ are mutually independent, normally distributed random variables with parameters $(\theta_1, \sigma), \ldots, (\theta_n, \sigma)$. (The expected values are $\theta_1, \ldots, \theta_n$, but the variance is the same σ^2 for each X_i.) Let $\Delta^2 = \theta_1^2 + \cdots + \theta_n^2$. Then

$$\frac{X_1^2 + \cdots + X_n^2}{\sigma^2}$$

has the same distribution as that of $[U_1 + (\Delta/\sigma)]^2 + U_2^2 + \cdots + U_n^2$, where $U_1, \ldots, U_n$ are mutually independent, $N(0, 1)$ random variables. This distribution is called a noncentral chi-square distribution (Note 8).

Proof. Variables $(X_1 - \theta_1)/\sigma, \ldots, (X_n - \theta_n)/\sigma$ are mutually independent, $N(0, 1)$. Let $\theta = (\theta_1, \ldots, \theta_n)$. We suppose that $\Delta > 0$; otherwise, the result is obvious. Then $\mathbf{a}_1 = (1/\Delta)\boldsymbol{\theta}$ is a vector of length 1. Determine vectors $\mathbf{a}_2, \ldots, \mathbf{a}_n$ such that $\mathbf{a}_1, \mathbf{a}_2, \ldots, \mathbf{a}_n$ is an orthonormal basis for the n-dimensional space of real n-tuples. Then, defining, $\mathbf{X} = (X_1, \ldots, X_n)$,

$$U_1 = \frac{1}{\sigma}(\mathbf{X} - \boldsymbol{\theta}, \mathbf{a}_1),\ U_2 = \frac{1}{\sigma}(\mathbf{X} - \boldsymbol{\theta}, \mathbf{a}_2), \ldots, U_n = \frac{1}{\sigma}(\mathbf{X} - \boldsymbol{\theta}, \mathbf{a}_n)$$

are mutually independent, $N(0, 1)$ random variables. Since $(\boldsymbol{\theta}, \mathbf{a}_i) = 0$, $i = 2, \ldots, n$, it follows that

$$U_2 = \frac{1}{\sigma}(\mathbf{X}, \mathbf{a}_2), \ldots, U_n = \frac{1}{\sigma}(\mathbf{X}, \mathbf{a}_n)$$

Hence,

$$\left(U_1 + \frac{(\theta, a_1)}{\sigma}, U_2, \ldots, U_n\right) = \left(\frac{1}{\sigma}(\mathbf{X}, \mathbf{a}_1), \frac{1}{\sigma}(\mathbf{X}, \mathbf{a}_2), \ldots, \frac{1}{\sigma}(\mathbf{X}, \mathbf{a}_n)\right)$$

Since $(\boldsymbol{\theta}, \mathbf{a}_1) = \Delta$, it follows that

$$\left(U_1 + \frac{\Delta}{\sigma}\right)^2 + U_2^2 + \cdots + U_n^2 = \frac{1}{\sigma^2}[(\mathbf{X}, \mathbf{a}_1)^2 + \cdots + (\mathbf{X}, \mathbf{a}_n)^2] = \frac{\|\mathbf{X}\|^2}{\sigma^2} = \frac{X_1^2 + \cdots + X_n^2}{\sigma^2} \quad \blacksquare$$

SUMMARY

In the first five sections, we review some basic facts about the vector space of n-tuples of real numbers. These relate to *elementary operations* with vectors, *vector subspaces, linear independence, dimension, orthonormal bases,* the *Gram–Schmidt procedure, projections,* and *orthogonal matrices.*

Random vectors are n-tuples of random variables. The basic vector operations for random vectors are exactly the same as those for n-tuples of real numbers. The *expected value* of a random vector is defined to be the vector of expected values. Just as with random variables, we have the linear relation $E(a\mathbf{X} + b\mathbf{Y}) = aE\mathbf{X} + bE\mathbf{Y}$, when $\mathbf{X}$ and $\mathbf{Y}$ are random vectors and a and b are scalars. If the components of $\mathbf{X}$ are uncorrelated and all have the same variance σ^2, then $\text{Cov}[(\mathbf{a}, \mathbf{X}), (\mathbf{b}, \mathbf{X})] = (\mathbf{a}, \mathbf{b})\sigma^2$, where $\mathbf{a}$ and $\mathbf{b}$ are nonrandom vectors. Under the same conditions, the expected value of the projection of a vector on a subspace is the projection of the expected value. If $\mathbf{Y}$ is the projection just referred to, then $E(\|\mathbf{Y}\|^2) = k\sigma^2 + \|E\mathbf{Y}\|^2$, where k is the dimension of the subspace.

If $(X_1, \ldots, X_n)$ consists of independent, $N(0, 1)$ random variables, then any orthogonal transformation of this random vector produces a new vector the components of which are still independent and $N(0, 1)$. An important consequence of this fact is that if the X_i's are independent and each normally distributed with the same parameters, then the sample mean and sample variance are independent. Moreover, the distribution of the latter is related to the chi-square distribution with $n - 1$ degrees of freedom.

EXERCISES

1 Verify the assertions in Examples 1 and 2, Section 16.2.

2 Consider the two orthonormal bases for 3-space, $(1, 0, 0)$, $(0, 1, 0)$, $(0, 0, 1)$, and $(1/\sqrt{3}, 1/\sqrt{3}, 1/\sqrt{3})$, $(1/\sqrt{2}, -1\sqrt{2}, 0)$, $(1/\sqrt{6}, 1/\sqrt{6}, -2/\sqrt{6})$. Denoting the first set by $\mathbf{x}_1, \mathbf{x}_2, \mathbf{x}_3$ and the second set by $\mathbf{y}_1, \mathbf{y}_2, \mathbf{y}_3$, determine the matrix of elements a_{ij} where $\mathbf{x}_i = a_{i1}\mathbf{y}_1 + a_{i2}\mathbf{y}_2 + a_{i3}\mathbf{y}_3$, $i = 1, 2, 3$, and check directly that it is an orthogonal matrix.

3 Consider the vectors $\mathbf{x}_1 = (1, 1, 1, 1, 0)$, $\mathbf{x}_2 = (1, 1, 1, 0, 0)$, $\mathbf{x}_3 = (1, 1, 0, 0, 0)$, $\mathbf{x}_4 = (1, 0, 0, 0, 0)$.

(a) Show that these vectors form a basis for the vector subspace of 5-tuples having 0 in the fifth coordinate.

(b) Follow through the Gram–Schmidt orthogonalization procedure to obtain an orthonormal basis.

(c) Write the vectors in reverse order, $\mathbf{x}_4$, $\mathbf{x}_3$, $\mathbf{x}_2$, $\mathbf{x}_1$, and apply the Gram–Schmidt orthogonalization procedure.
(d) Find the projection of (1, 2, 3, 4, 5) on $V(\mathbf{x}_1, \mathbf{x}_2, \mathbf{x}_3, \mathbf{x}_4)$.

4 Suppose that W, X, Y, and Z are independent random variables and D is the determinant

$$\begin{vmatrix} W & X \\ Y & Z \end{vmatrix}$$

Show that ED is the determinant of the expected values.

5 Suppose that $X_1, \ldots, X_n, Y_1, \ldots, Y_n$ are mutually independent $N(0, 1)$ random variables. Let $\mathbf{X} = (X_1, \ldots, X_n)$, $\mathbf{Y} = (Y_1, \ldots, Y_n)$. Show that $(\mathbf{X}, \mathbf{Y})$ is distributed like $U - V$, where U and V are independent random variables, each chi-square distributed with n degrees of freedom. (Hint: Compare Exercise 21a, Chapter 9.)

6 Suppose that X, Y, and Z are independent random variables, each distributed $N(m, \sigma^2)$ (m and σ are the same for each). Show that

$$X + Y + Z, \quad X - Y, \quad \text{and } X + Y - 2Z$$

are independent random variables. What are their distributions?

7 Suppose that $X_1, \ldots, X_n$ are independent random variables, each $N(0, 1)$. Let $\overline{X} = (X_1 + \cdots + X_n)/n$ and let Y be a function of the $\binom{n}{2}$ differences, $X_i - X_j$, $1 \leq i < j \leq n$. Show that $\overline{X}$ and Y are independent random variables. {Hint: Let $\mathbf{u}_1 = [(1/\sqrt{n}), \ldots, (1/\sqrt{n})]$ and let $\mathbf{u}_1, \mathbf{u}_2, \ldots, \mathbf{u}_n$ be an orthonormal basis for n-space. Let $U_i = (\mathbf{u}_i, \mathbf{X})$, $i = 1, \ldots, n$. If we express the X_i's in terms of U_i's, we have

$$\begin{pmatrix} X_1 \\ \vdots \\ X_n \end{pmatrix} = \begin{pmatrix} \dfrac{1}{\sqrt{n}} & a_{21} & \cdots & a_{n1} & U_1 \\ \vdots & & & \vdots & \vdots \\ \dfrac{1}{\sqrt{n}} & a_{2n} & \cdots & a_{nn} & U_n \end{pmatrix}$$

where the $n \times n$ matrix is orthogonal.
Premultiply both sides of this matrix relationship by

$$\begin{pmatrix} \dfrac{1}{\sqrt{n}} & \dfrac{1}{\sqrt{n}} & \dfrac{1}{\sqrt{n}} & \cdots & & \dfrac{1}{\sqrt{n}} \\ 1 & -1 & 0 & & & 0 \\ 0 & 1 & -1 & & \cdot & 0 \\ 0 & 0 & 1 & & \cdot & 0 \\ \vdots & \vdots & \vdots & & \vdots & \vdots \\ \cdot & \cdot & \cdot & & -1 & 0 \\ \cdot & \cdot & \cdot & & 1 & -1 \end{pmatrix}$$

to obtain

$$\begin{pmatrix} \sqrt{n}\bar{X} \\ X_1 - X_2 \\ \vdots \\ X_{n-1} - X_n \end{pmatrix} = \begin{pmatrix} 1 & 0 & \cdots & 0 \\ 0 & b_{22} & \cdots & b_{2n} \\ \vdots & & & \\ 0 & b_{n2} & \cdots & b_{nn} \end{pmatrix} \begin{pmatrix} U_1 \\ \vdots \\ U_n \end{pmatrix}$$

The specific values of the b_{ij}'s are not relevant.}

8 Show that the result of Exercise 7 holds if the X_i's are each $N(m, \sigma^2)$.

9 Suppose that $X_1, \ldots, X_n$ are independent random variables, each with the same $N(m, \sigma^2)$ distribution. Let

$$R = \max_{1 \le i \le n} X_i - \min_{1 \le i \le n} X_i$$

(R is called the *range* of the observations.) Show that $\bar{X}$ and R are independent.

10 Suppose that $\mathbf{X} = (X_1, X_2, X_3, X_4)$ is a vector of independent random variables, each with the same $N(m, \sigma^2)$ distribution. Define $\mathbf{x}_1 = (1, 1, 1, 1)$, $\mathbf{x}_2 = (1, -1, 1, -1)$. Suppose that $c\mathbf{x}_1, d\mathbf{x}_2, \mathbf{x}_3, \mathbf{x}_4$ is an orthonormal basis for 4-space. Let $\mathbf{T}$ be the projection of $\mathbf{X}$ on $V(\mathbf{x}_1, \mathbf{x}_2)$ and let $\mathbf{U}$ be its projection on $V(\mathbf{x}_3, \mathbf{x}_4)$.
(a) Evaluate $E\mathbf{T}$, $E\mathbf{U}$, $E(\|\mathbf{T}\|^2)$, and $E(\|\mathbf{U}\|^2)$.
(b) Describe completely the joint distribution of $(\mathbf{x}_1, \mathbf{X})$ and $(\mathbf{x}_2, \mathbf{X})$.
(c) Show that $\|\mathbf{U}\|^2/\sigma^2$ is chi-square distributed with two degrees of freedom.

11 Suppose that $\mathbf{X} = (X_1, X_2, X_3, X_4)$ is a vector of independent random variables, each with the same variance σ^2. Suppose that $E(X_1) = 1$, $E(X_2) = 2$, $E(X_3) = 3$, $E(X_4) = 4$. Let $\mathbf{x}_1 = (1, 2, 3, 4)$ and suppose that $c\mathbf{x}_1, \mathbf{x}_2, \mathbf{x}_3, \mathbf{x}_4$ is an orthonormal basis for 4-space. Let $\mathbf{T}$ be the projection of $\mathbf{X}$ on $V(\mathbf{x}_2, \mathbf{x}_3, \mathbf{x}_4)$.
(a) Show that $E\mathbf{T}$ is the 0 vector.
(b) Show that $E(\|\mathbf{T}\|^2) = 3\sigma^2$.
(c) If the X_i's are normally distributed, show that $\|\mathbf{T}\|^2/\sigma^2$ is chi-square distributed with three degrees of freedom.

NOTES

1 For a review of elementary linear algebra and vector spaces, see Chapter 13 of [2] and Chapter 15 of [3].

2 Some additional discussion of linear independence and dependence is given in Section 13.4 of [2] and Sections 15.7 and 15.8 of [3].

3 The Gram–Schmidt procedure is covered on p. 450 of [2] and in Section 15.13 of [3].

4 For additional material on orthogonal subspaces and projections, see Section 15.14 of [3].

5 Determinants are covered in Appendix 1, [2].

6 The Student-*t* distribution is widely tabulated; most applied statistics books contain some of these tables. See [19].

7 "Student" was the pseudonym of an English chemist named W. S. Gosset. While employed for the Guiness breweries, he discovered the distribution of the ratio given by (6), Section 16.7. The ratio is called "Student's ratio" in Gosset's honor, and its use has been very significant in applied statistics. Gosset's proof was nonrigorous and partly based on empirical methods. (See the original 1908 article in [70].) The first rigorous derivation of the distribution of Student's ratio is due to R. A. Fisher.

8 For references to tables of the noncentral chi-square distributions, see [19].

Chapter Seventeen

LINEAR MODELS

Many important experimental models in statistics are of the type called *linear models.* Statistical procedures for linear models are usually called the *analysis of variance.* Such procedures probably account for most of the applications of statistics to the experimental sciences. In this chapter, we use the methods of linear algebra described in Chapter 16 to devise methods of estimating and making decisions in problems concerning linear models (Note 1).

17.1 THE LINEAR MODEL

Many problems of statistics are concerned with special cases of what we shall call the *linear model.*

linear model

Suppose that $\mathbf{X} = (X_1, \ldots, X_n)$ is a random vector. Let

$$\mathbf{a}_1 = (a_{11}, \ldots, a_{1n})$$
$$\vdots$$
$$\mathbf{a}_k = (a_{k1}, \ldots, a_{kn})$$

be k vectors. Each vector is an n-tuple of scalars. The distribution of $\mathbf{X}$ is said to satisfy a *linear model* if

$$E\mathbf{X} = \beta_1\mathbf{a}_1 + \cdots + \beta_k\mathbf{a}_k \tag{1}$$

where $\beta_1, \ldots, \beta_k$ are scalar parameters unknown to the experimenter.

We should emphasize that in a practical context, the constants a_{ij} are known to the experimenter. They are called *structural con-*

stants, which are determined by the nature of the experiment. What (1) asserts is that $E\mathbf{X}$ is some linear combination of the known structural vectors $\mathbf{a}_1, \ldots, \mathbf{a}_k$. However, the particular linear combination that gives $E\mathbf{X}$ is not known, since the parameters $\beta_1, \ldots, \beta_k$, which give the correct linear combination, are *not* known to the experimenter. The purpose of this chapter is to describe some statistical methodology for obtaining information about these parameters. A useful way of restating the linear model follows.

The distribution of $\mathbf{X}$ is said to satisfy a *linear model* if

$E\mathbf{X}$ lies in $V(\mathbf{a}_1, \ldots, \mathbf{a}_k)$

the subspace generated by $\mathbf{a}_1, \ldots, \mathbf{a}_k$, where these vectors are known to the experimenter.

It may be useful to write (1) as a matrix multiplication:

$$\begin{pmatrix} EX_1 \\ \vdots \\ EX_n \end{pmatrix} = \begin{pmatrix} a_{11} & \cdots & a_{k1} \\ \vdots & & \\ a_{1n} & & a_{kn} \end{pmatrix} \begin{pmatrix} \beta_1 \\ \vdots \\ \beta_k \end{pmatrix} \tag{2}$$

or, equivalently,

$$[E(X_1), \ldots, E(X_n)] = (\beta_1, \ldots, \beta_k) \begin{pmatrix} a_{11} & \cdots & a_{1n} \\ \vdots & & \\ a_{k1} & \cdots & a_{kn} \end{pmatrix} \tag{3}$$

We leave it to the reader to check that (2) or (3) is equivalent to (1).

Example 1 Suppose n independent observations are made on the same population. These observations are described mathematically by $\mathbf{X} = (X_1, \ldots, X_n)$, a vector of independent and identically distributed random variables.

$$EX_1 = \cdots = EX_n = \beta$$

but the common expected value β is unknown to the experimenter. The linear model form of this situation is that

$$E\mathbf{X} = \beta(1, \ldots, 1)$$

There is a single vector of structural constants, $\mathbf{a} = (1, \ldots, 1)$.

Example 2 Suppose m observations are made on one population and r observations are made on a second population. Mathematically, these are described by

$$\mathbf{X} = (X_1, \ldots, X_m, X_{m+1}, \ldots, X_{m+r})$$

These are $m + r$ independent random variables, the first m having one distribution and the last r another distribution. The expected values will be denoted by

$$E(X_i) = \begin{cases} \beta_1 & \text{if } i = 1, \ldots, m \\ \beta_2 & \text{if } i = m + 1, \ldots, m + r \end{cases}$$

and these are unknown to the experimenter. The linear model is that

$$E\mathbf{X} = \beta_1(1, \ldots, 1, 0, \ldots, 0) + \beta_2(0, \ldots, 0, 1, \ldots, 1)$$

The structural vectors are

$$\mathbf{a}_1 = (1, \ldots, 1, 0, \ldots, 0) \qquad (m \text{ 1's, } r \text{ 0's})$$

$$\mathbf{a}_2 = (0, \ldots, 0, 1, \ldots, 0) \qquad (m \text{ 0's, } r \text{ 1's})$$

Example 3 (Linear regression.) Consider a physical law relating two variables, x and t, of the form

$$x = \beta_1 + \beta_2 t$$

Suppose that what the experimenter observes for a specified t is not exactly equal to $\beta_1 + \beta_2 t$ but is a random variable that, on the average, equals this value. The experimenter chooses to make observations $\mathbf{X} = (X_1, \ldots, X_n)$ at predetermined t values, $t_1, \ldots, t_n$. This is described by the following mathematical model.

$$E\mathbf{X} = \beta_1(1, \ldots, 1) + \beta_2(t_1, \ldots, t_n)$$

The structural vectors are

$$\mathbf{a}_1 = (1, \ldots, 1) \quad \mathbf{a}_2 = (t_1, \ldots, t_n)$$

These vectors are known to the experimenter. The unknown parameters are β_1 (intercept) and β_2 (slope) (see Figure 17.1).

Example 4 Example 4 is a more specific version of Example 3. An experimenter observes the velocity of a particle moving linearly. His observations are made at times of his own choosing, $t_1, \ldots, t_n$. If the movement of the particle is determined by an initial velocity β_1 and an acceleration β_2 and no new forces are applied during the ensuing motion, then the velocity at time t is given by

$$x = \beta_1 + \beta_2 t$$

The experimenter does not know β_1 and β_2, and he must obtain

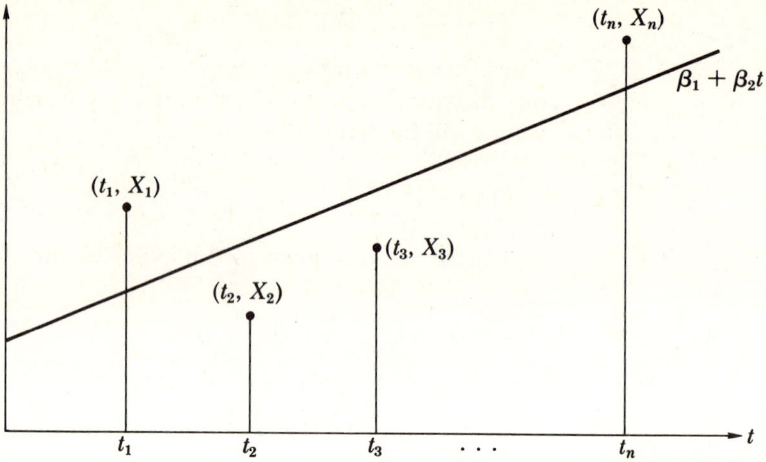

FIGURE 17.1 Points (t_1, EX_1), (t_2, EX_2), . . . , (t_n, EX_n) all lie on the straight line $\beta_1 + \beta_2 t$. The actual observed values (t_1, X_1), (t_2, X_2), . . . , (t_n, X_n) fluctuate about this line.

information about them from observations $X_1, \ldots, X_n$, made at times $t_1, \ldots, t_n$.

Example 5 (Quadratic regression.) This example is similar to Example 3, except now the relation between x and t is

$$x = \beta_1 + \beta_2 t + \beta_3 t^2$$

If the experimenter makes his observations at predetermined t values $t_1, \ldots, t_n$, then the linear model specifies that

$$(EX_1, \ldots, EX_n) = \beta_1(1, \ldots, 1) + \beta_2(t_1, \ldots, t_n) + \beta_3(t_1^2, \ldots, t_n^2)$$

The structural vectors are

$$\mathbf{a}_1 = (1, \ldots, 1), \quad \mathbf{a}_2 = (t_1, \ldots, t_n), \quad \mathbf{a}_3 = (t_1^2, \ldots, t_n^2)$$

These are known to the experimenter, whereas the parameters $\beta_1, \beta_2, \beta_3$ are unknown.

Example 6 Consider the following special case of the last example, following the lines of Example 4. What is now observed is the location of the particle on the straight line along which it moves. As before, the times of observation, $t_1, \ldots, t_n$ are chosen by the experimenter. The unknown parameters are

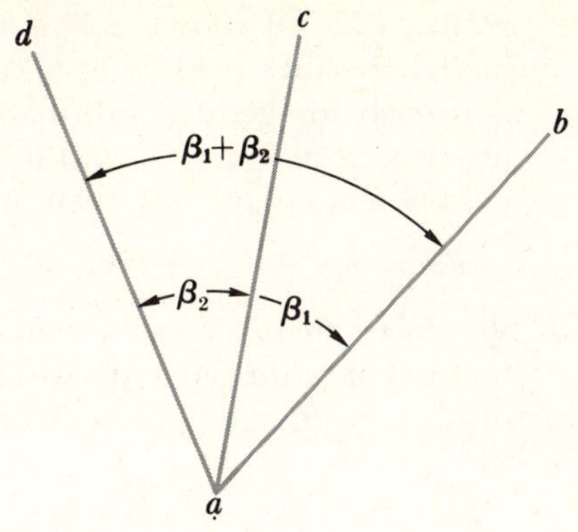

FIGURE 17.2

$$\beta_1 = \text{initial position}, \quad \beta_2 = \text{initial velocity}, \quad \beta_3 = \text{acceleration}$$

The physical law, according to Newtonian mechanics, is

$$x = \beta_1 + \beta_2 t + \frac{\beta_3}{2} t^2$$

where x is the location of the particle. The observations $X_1, \ldots, X_n$, measured at times $t_1, \ldots, t_n$, are governed by the model

$$(EX_1, \ldots, EX_n) = \beta_1(1, \ldots, 1) + \beta_2(t_1, \ldots, t_n) + \beta_2\left(\frac{t_1^2}{2}, \ldots, \frac{t_n^2}{2}\right)$$

Example 7 An observer stationed at a point a measures the angles between the lines $\overline{ab}$, $\overline{ac}$, and $\overline{ad}$, having the true values β_1, β_2, and $\beta_1 + \beta_2$, respectively (see Figure 17.2). Suppose that the measurements are subject to random errors but, on the average, are equal to the true values. Denoting the measurements by $\mathbf{X} = (X_1, X_2, X_3)$, the linear model is

$$E\mathbf{X} = \beta_1(1, 0, 1) + \beta_2(0, 1, 1)$$

(Note 2).

17.2 ESTIMATING LINEAR FUNCTIONS WITH LINEAR FUNCTIONS

Statistical problems with linear models are concerned with getting information about the unknown expected value of the random vector, $E\mathbf{X}$. Of course, $E\mathbf{X}$ is not completely unknown since the model specifies it to lie in a known subspace $V(\mathbf{a}_1, \ldots, \mathbf{a}_k)$. One approach for getting information about $E\mathbf{X}$ is to try to estimate just where in $V(\mathbf{a}_1, \ldots, \mathbf{a}_k)$ it lies. If one knew the values of the parameters $\beta_1, \ldots, \beta_k$, then one would know $E\mathbf{X}$, since

$$E\mathbf{X} = \beta_1\mathbf{a}_1 + \cdots + \beta_k\mathbf{a}_k$$

We could then try to estimate the individual values $\beta_1, \ldots, \beta_k$. Somewhat more generally, we could try to estimate linear functions of $\beta_1, \ldots, \beta_k$,

$$c_1\beta_1 + \cdots + c_k\beta_k \tag{1}$$

where $c_1, \ldots, c_k$ are scalars specified by the experimenter. The theory for estimating such linear functions is most developed when the estimators are themselves *linear* functions of $X_1, \ldots, X_n$. By a linear function of $X_1, \ldots, X_n$ is meant

$$u_1X_1 + \cdots + u_nX_n \tag{2}$$

where $u_1, \ldots, u_n$ are constants known to the experimenter. It is convenient to represent (1) and (2) in vector notation.

Let

$$\mathbf{c} = (c_1, \ldots, c_k), \quad \boldsymbol{\beta} = (\beta_1, \ldots, \beta_k), \quad \mathbf{u} = (u_1, \ldots, u_n)$$

Then the linear function to be estimated is the inner product $(\mathbf{c}, \boldsymbol{\beta})$, and the linear function to be the estimator is the inner product $(\mathbf{u}, \mathbf{X})$. We should warn the reader that these are vectors from two different vector spaces, of possibly different dimensions. The vectors $\mathbf{c}$ and $\boldsymbol{\beta}$ lie in the vector space of real k-tuples, whereas $\mathbf{u}$ and $\mathbf{X}$ lie in the vector space of real n-tuples.

unbiased linear estimator

Let $(\mathbf{c}, \boldsymbol{\beta})$ be a linear function of the unknown parameters and $(\mathbf{u}, \mathbf{X})$ a linear function of $X_1, \ldots, X_n$. Function $(\mathbf{u}, \mathbf{X})$ is said to be an *unbiased linear estimator* of $(\mathbf{c}, \boldsymbol{\beta})$ if

$$E(\mathbf{u}, \mathbf{X}) = (\mathbf{c}, \boldsymbol{\beta}) \tag{3}$$

identically in $\beta_1, \ldots, \beta_k$

This definition of unbiased estimator is exactly the same as that given earlier in (3), Section 12.2. Following the ideas of Chapter 12, we can denote the distribution of $\mathbf{X}$ by an identifying label θ. The collection of all distributions is Ω. For the time being, we suppose that the only knowledge about Ω concerns $E\mathbf{X}$; namely, $E\mathbf{X}$ satisfies the linear model described in Section 17.1. In keeping with the notation used in Chapter 12, we can write (3) as

$$E_\theta(\mathbf{u}, \mathbf{X}) = (\mathbf{c}, \boldsymbol{\beta}) \tag{4}$$

Equation (4) means that whatever distribution θ is used, within the confines of the linear model, the expected value of $(\mathbf{u}, \mathbf{X})$ equals $(\mathbf{c}, \boldsymbol{\beta})$. Also, β determines θ, and unbiasedness means that (4) is an identity in θ. We shall restrict ourselves to the simpler notation in (3). We emphasize again that the vectors $\mathbf{c}$ and $\mathbf{u}$ are assumed known by the experimenter.

We shall soon see that not every linear function $(\mathbf{c}, \boldsymbol{\beta})$ can be estimated. By Section 16.6,

$$E(\mathbf{u}, \mathbf{X}) = (\mathbf{u}, E\mathbf{X})$$

Hence, the requirement that $(\mathbf{u}, \mathbf{X})$ be unbiased is that

$$(\mathbf{u}, E\mathbf{X}) = (\mathbf{u}, \beta_1\mathbf{a}_1 + \cdots + \beta_k\mathbf{a}_k) = (\mathbf{c}, \boldsymbol{\beta}) \tag{5}$$

identically in β. Equation (5) requires that

$$\beta_1(\mathbf{u}, \mathbf{a}_1) + \cdots + \beta_k(\mathbf{u}, \mathbf{a}_k) = c_1\beta_1 + \cdots + c_k\beta_k \tag{6}$$

identically in $\beta_1, \ldots, \beta_k$. From (6), there follows the criterion for estimability.

criterion for estimability

Let

$$\mathbf{A} = \begin{pmatrix} a_{11} & \cdots & a_{k1} \\ \vdots & & \\ a_{1n} & \cdots & a_{kn} \end{pmatrix}$$

The columns of A are the structural vectors previously denoted as $\mathbf{a}_1, \ldots, \mathbf{a}_k$. Consider now the rows of A,

$$\begin{aligned} \boldsymbol{\alpha}_1 &= (a_{11}, \ldots, a_{k1}) \\ &\vdots \\ \boldsymbol{\alpha}_n &= (a_{1n}, \ldots, a_{kn}) \end{aligned}$$

A necessary and sufficient condition that there exist an n-tuple $\mathbf{u}$ such that $(\mathbf{u}, \mathbf{X})$ be an unbiased estimator of $(\mathbf{c}, \boldsymbol{\beta})$ is that $\mathbf{c}$ lie in the subspace generated by $\boldsymbol{\alpha}_1, \ldots, \boldsymbol{\alpha}_n$.

Proof. First suppose that $\mathbf{c}$ is some linear combination of $\boldsymbol{\alpha}_1, \ldots, \boldsymbol{\alpha}_n$. That is, there exist $u_1, \ldots, u_n$ such that

$$\mathbf{c} = u_1\boldsymbol{\alpha}_1 + \cdots + u_n\boldsymbol{\alpha}_n \tag{7}$$

If $\mathbf{u} = (u_1, \ldots, u_n)$, then we claim that $(\mathbf{u}, \mathbf{X})$ is an unbiased estimator of $(\mathbf{c}, \boldsymbol{\beta})$. We realize this fact by writing (5) in matrix form, using (2), Section 17.1.

$$E(u, X) = (u, EX)$$

$$= (u_1, \ldots, u_n)\begin{pmatrix} a_{11} & \cdots & a_{k1} \\ \vdots & & \\ a_{1n} & \cdots & a_{kn} \end{pmatrix}\begin{pmatrix} \beta_1 \\ \vdots \\ \beta_k \end{pmatrix} = (\mathbf{c}, \boldsymbol{\beta}) \tag{8}$$

The reason is that (7) asserts

$$\mathbf{c} = (u_1, \ldots, u_n)\begin{pmatrix} a_{11} & \cdots & a_{k1} \\ \vdots & & \\ a_{1n} & \cdots & a_{kn} \end{pmatrix}$$

Conversely, suppose there is a $\mathbf{u} = (u_1, \ldots, u_n)$ such that (8) holds identically for all $\beta_1, \ldots, \beta_k$. Hence, (8) must hold for each of the β vectors

$$(1, 0, \ldots, 0), \qquad (0, 1, \ldots, 0), \ldots, (0, 0, \ldots, 1)$$

By successively substituting each of these vectors for $\boldsymbol{\beta}$ in (8), we obtain,

$$(u_1, \ldots, u_n)\begin{pmatrix} a_{11} & \cdots & a_{k1} \\ \vdots & & \\ a_{1n} & \cdots & a_{kn} \end{pmatrix} = (c_1, \ldots, c_k) \tag{9}$$

which says that

$$u_1\boldsymbol{\alpha}_1 + \cdots + u_n\boldsymbol{\alpha}_n = \mathbf{c} \qquad \blacksquare$$

For later reference, we point out that (9) is equivalent to

$$(\mathbf{u}, \mathbf{a}_i) = c_i, \qquad i = 1, \ldots, k \tag{10}$$

Example 1 Consider the following linear model.

$$(EX_1, EX_2, EX_3, EX_4) = \beta_1(1, 0, 1, 0) + \beta_2(0, 1, 0, 1) + \beta_3(1, 1, 0, 0) + \beta_4(0, 0, 1, 1)$$

The linear function $1\beta_1 + 0\beta_2 + 0\beta_3 + 0\beta_4 = \beta_1$ cannot be estimated. To determine this fact, we examine

$$\mathbf{A} = \begin{pmatrix} 1 & 0 & 1 & 0 \\ 0 & 1 & 1 & 0 \\ 1 & 0 & 0 & 1 \\ 0 & 1 & 0 & 1 \end{pmatrix}$$

and check whether $(1, 0, 0, 0)$ is a linear combination of

$$\boldsymbol{\alpha}_1 = (1, 0, 1, 0), \quad \boldsymbol{\alpha}_2 = (0, 1, 1, 0), \quad \boldsymbol{\alpha}_3 = (1, 0, 0, 1),$$
$$\boldsymbol{\alpha}_4 = (0, 1, 0, 1)$$

We leave it to the reader to verify that such a linear combination is *not* possible. Hence, β_1 cannot be estimated. Similarly, neither can the individual values of β_2, β_3, or β_4 be estimated. However, $\beta_1 - \beta_2$ can be determined, because

$$(1, -1, 0, 0) = \tfrac{1}{2}\alpha_1 - \tfrac{1}{2}\alpha_2 + \tfrac{1}{2}\alpha_3 - \tfrac{1}{2}\alpha_4$$

(See Exercise 4.)

Example 2 Consider the following linear model.

$$(EX_1, \ldots, EX_n) = \beta_1(1, \ldots, 1) + \beta_2(1, \ldots, 1)$$

The structural vectors $\boldsymbol{\alpha}_1$ and $\boldsymbol{\alpha}_2$ both are n-tuples of 1's. The n $\boldsymbol{\alpha}_i$ vectors are all of the form (1, 1). It follows that the only linear functions of β_1 and β_2 that can be estimated are $\beta_1 + \beta_2$ or multiples of this sum. This fact is certainly plausible intuitively. The model specifies that each expected value has the same value β as in Example 1, Section 17.1. Breaking β up into a sum $\beta_1 + \beta_2$ is completely artificial and has nothing to do with the underlying population.

Example 3 In Example 2, Section 17.1, every linear function of β_1 and β_2 can be estimated.

Example 4 Consider Example 3, Section 17.1. Suppose that there are at least two distinct t_i's, and, for simplicity, say these are t_1 and t_2. Then, among the $\boldsymbol{\alpha}_i$ vectors are $(1, t_1)$ and $(1, t_2)$. These vectors are linearly independent since $t_1 \neq t_2$; hence, the $\boldsymbol{\alpha}_i$'s generate the two-dimensional vector space of real pairs. Consequently, all linear functions of β_1 and β_2 can be estimated. On the other hand, suppose that $t_1 = \cdots = t_n$. In other words, all observations are made at a common value t. Then each $\boldsymbol{\alpha}_i$ equals $(1, t)$ and the only linear functions of the parameters that can be estimated are multiples of $\beta_1 + t\beta_2$, which is the height of the straight line at t. The individual slope β_2 or intercept β_1 cannot be estimated. Of course, this fact should not be surprising. How can one obtain the slope or intercept of a straight line by only knowing one point on the line?

Example 5 Consider Example 5, Section 17.1. The situation here is similar to that in the previous example. If there are at least three distinct t_i's, then every linear function of β_1, β_2, and β_3 can be estimated.

Otherwise, only some linear functions can be estimated. We leave the details to the reader (see Exercise 5). Again, the intuitive reason should be clear: To get full information about a quadratic function, we need to know at least three distinct points on the curve.

17.3 UNBIASED LINEAR ESTIMATORS OF MINIMUM VARIANCE

In Section 17.2, we gave a criterion for the existence of an unbiased linear estimator of $(\mathbf{c}, \boldsymbol{\beta})$. In general, there are many unbiased estimators of $(\mathbf{c}, \boldsymbol{\beta})$. How should one choose the "best one?" *Suppose that we use the variance of an unbiased estimator as our criterion of goodness* (see Example 1, Section 12.3). A linear estimator is, by definition, of the form $(\mathbf{u}, \mathbf{X})$, where $\mathbf{u}$ is an n-tuple of scalars chosen by the experimenter. To deal with the variance of $(\mathbf{u}, \mathbf{X})$, we shall henceforth have to impose some additional assumptions on the linear model.

assumption that components are uncorrelated with equal variance

We shall hereafter assume that the components of $\mathbf{X} = (X_1, \ldots, X_n)$ are uncorrelated and have equal variance σ^2. That is,

$$\operatorname{Cov}(X_i, X_j) = \begin{cases} 0 & \text{if } i \neq j \\ \operatorname{Var}(X_i) = \sigma^2 & \text{if } i = j \end{cases}$$

The preceding assumption will always be satisfied when $X_1, \ldots, X_n$ are mutually independent, with equal variances.

It follows from Section 16.6 that the variance of $(\mathbf{u}, \mathbf{X})$ is $\|\mathbf{u}\|^2\sigma^2$. What we are looking for, then, is a vector $\mathbf{u}$ such that $E(\mathbf{u}, \mathbf{X}) = (\mathbf{c}, \boldsymbol{\beta})$ identically in $\boldsymbol{\beta}$, and, subject to this requirement, $\|\mathbf{u}\|^2$ is as small as possible. Such an estimator will be called a *linear unbiased estimator of minimum variance.* For short, we shall designate it as a *LUV estimator.*

There is a nice way of characterizing LUV estimators in terms of the structural vectors $\mathbf{a}_1, \ldots, \mathbf{a}_k$, as follows.

LUV estimators

Let $V_1 = V(\mathbf{a}_1, \ldots, \mathbf{a}_k)$ be the vector subspace generated by the structural vectors $\mathbf{a}_1, \ldots, \mathbf{a}_k$. If there exists a linear unbiased estimator of $(\mathbf{c}, \boldsymbol{\beta})$, then there exists one and only one vector $\mathbf{u}$ in V_1 such that $(\mathbf{u}, \mathbf{X})$ is an unbiased estimator. Moreover, $(\mathbf{u}, \mathbf{X})$ is the LUV estimator.

Proof. First we prove uniqueness. There can be only one unbiased estimator $(\mathbf{u}, \mathbf{X})$ with $\mathbf{u}$ in V_1. Suppose there are two vectors $\mathbf{u}$ and $\mathbf{u}'$ such that $(\mathbf{u}, \mathbf{X})$ and $(\mathbf{u}', \mathbf{X})$ are both unbiased. By the structure of a linear model,

$$E(\mathbf{u}, \mathbf{X}) = (\mathbf{u}, E\mathbf{X}) = \beta_1(\mathbf{u}, \mathbf{a}_1) + \cdots + \beta_k(\mathbf{u}, \mathbf{a}_k)$$

$$E(\mathbf{u}', \mathbf{X}) = (\mathbf{u}', E\mathbf{X}) = \beta_1(\mathbf{u}', \mathbf{a}_1) + \cdots + \beta_k(\mathbf{u}', \mathbf{a}_k)$$

identically in $\beta_1, \ldots, \beta_k$. By successively setting one β_i equal to 1, and the remaining ones equal to 0, we have that

$$(\mathbf{u} - \mathbf{u}', \mathbf{a}_i) = 0, \qquad i = 1, \ldots, k$$

But $\mathbf{u} - \mathbf{u}'$ lies in V_1. Since it is orthogonal to each of $\mathbf{a}_1, \ldots, \mathbf{a}_k$, it must be orthogonal to any vector in V_1; hence it is orthogonal to itself. Therefore, $\|\mathbf{u} - \mathbf{u}'\| = 0$ and $\mathbf{u} = \mathbf{u}'$. Now suppose that $(\mathbf{v}, \mathbf{X})$ is any linear unbiased estimator. (By hypothesis, there exists at least one.) Write

$$\mathbf{v} = \mathbf{u} + \mathbf{w}$$

where $\mathbf{u}$ is the projection of $\mathbf{v}$ on V_1 and $\mathbf{w}$ is orthogonal to every vector in V_1 [see (3), Section 16.4.] We have

$$E(\mathbf{v}, \mathbf{X}) = E(\mathbf{u}, \mathbf{X}) + E(\mathbf{w}, \mathbf{X})$$

and

$$E(\mathbf{w}, \mathbf{X}) = (\mathbf{w}, E\mathbf{X}) = \beta_1(\mathbf{w}, \mathbf{a}_1) + \cdots + \beta_k(\mathbf{w}, \mathbf{a}_k) = 0$$

since $\mathbf{w}$ is orthogonal to every vector in V_1. Hence,

$$E(\mathbf{u}, \mathbf{X}) = E(\mathbf{v}, \mathbf{X}) = (\mathbf{c}, \boldsymbol{\beta})$$

and $(\mathbf{u}, \mathbf{X})$ is also an unbiased estimator.

Now, $\mathrm{Var}(\mathbf{v}, \mathbf{X}) = \mathrm{Var}(\mathbf{u} + \mathbf{w}, \mathbf{X}) = \|\mathbf{u} + \mathbf{w}\|^2\sigma^2 = \|\mathbf{u}\|^2\sigma^2 + \|\mathbf{w}\|^2\sigma^2$, since $(\mathbf{u}, \mathbf{w}) = 0$. Hence,

$$\mathrm{Var}(\mathbf{v}, \mathbf{X}) \geqslant \mathrm{Var}(\mathbf{u}, \mathbf{X}) \tag{1}$$

Actually, there is strict inequality in (1) unless $\mathbf{w}$ is the zero vector. By the first part of the proof, the vector $\mathbf{u}$ is uniquely determined—no matter what $\mathbf{v}$ we start with the same $\mathbf{u}$ results, since there is only one $\mathbf{u}$ in V_1 such that $(\mathbf{u}, \mathbf{X})$ is an unbiased estimator. We have shown that any linear unbiased estimator different from $(\mathbf{u}, \mathbf{X})$ is improved on by $(\mathbf{u}, \mathbf{X})$. ∎

Example 1 Consider Example 1, Section 17.1. The subspace V_1 consists of all n-tuples of the form $(a, a, \ldots, a)$, since there is only one structural vector $\mathbf{e} = (1, \ldots, 1)$. Let $\bar{X} = (X_1 + \cdots + X_n)/n$ denote the sample average. Since $E(\bar{X}) = \beta$ and $\bar{X} = ((1/n)\mathbf{e}, \mathbf{X})$, it follows that the sample average is the LUV estimator of β.

Example 2 Consider Example 2, Section 17.1. Let $\bar{X}_1 = (X_1 + \cdots + X_m)/m$, $\bar{X}_2 = (X_{m+1} + \cdots + X_{m+r})/r$ be the sample averages from each of the populations. We leave it to the reader to verify that $c_1\bar{X}_1 + c_2\bar{X}_2$ is the LUV estimator of $c_1\beta_1 + c_2\beta_2$.

Example 3 Consider Example 3, Section 17.1. Suppose that $t_1, \ldots, t_n$ are not all the same. Otherwise, as shown in Example 4, Section 17.2, not all functions of β_1 and β_2 are estimable. The structural vectors are

$$\mathbf{a}_1 = (1, \ldots, 1), \quad \mathbf{a}_2 = (t_1, \ldots, t_n)$$

Let $\bar{t} = (t_1 + \cdots + t_n)/n$ be the average of the t_i values at which observations are taken. Then $\mathbf{a}_3 = [(t_1 - \bar{t}), \ldots, (t_n - \bar{t})] = \mathbf{a}_2 - \bar{t}\mathbf{a}_1$ lies in V_1 and $\mathbf{a}_1$ and $\mathbf{a}_3$ also generate V_1. Since $t_1 - \bar{t} + \cdots + \mathbf{t}_n - \bar{t} = 0$, it follows that $\mathbf{a}_1$ and $\mathbf{a}_3$ are orthogonal. Since $E\mathbf{X} = \beta_1\mathbf{a}_1 + \beta_2\mathbf{a}_2$, it follows that

$$E(\mathbf{a}_1, \mathbf{X}) = (\mathbf{a}_1, \mathbf{a}_1)\beta_1 + (\mathbf{a}_1, \mathbf{a}_2)\beta_2, \qquad E(\mathbf{a}_3, \mathbf{X}) = (\mathbf{a}_2, \mathbf{a}_3)\beta_2 \tag{2}$$

Notice that $(\mathbf{a}_3, \mathbf{a}_3) = (\mathbf{a}_3, \mathbf{a}_2)$. Define

$$\mathbf{u}_2 = \frac{1}{(\mathbf{a}_3, \mathbf{a}_3)}\mathbf{a}_3, \qquad \mathbf{u}_1 = \frac{1}{n}\mathbf{a}_1 - \bar{t}\mathbf{u}_2$$

It follows from (2) that

$$E(\mathbf{u}_2, \mathbf{X}) = \beta_2, \qquad E(\mathbf{u}_1, \mathbf{X}) = \beta_1$$

Since $\mathbf{u}_1$ and $\mathbf{u}_2$ lie in V_1, it follows that $(\mathbf{u}_1, \mathbf{X})$ and $(\mathbf{u}_2, \mathbf{X})$ are the LUV estimators of β_1 and β_2, respectively. These estimators are sometimes denoted as $\hat{\beta}_1$ and $\hat{\beta}_2$. Explicitly,

$$\hat{\beta}_1 = \frac{\sum_{i=1}^{n} X_i}{n} - \bar{t}\hat{\beta}_2$$

$$\hat{\beta}_2 = \frac{\sum_{i=1}^{n} X_i(t_i - \bar{t})}{\sum_{i=1}^{n} (t_i - \bar{t})^2}$$

Example 4 Consider Example 7, Section 17.1. Subspace V_1 is generated by the vectors

$$\mathbf{a}_1 = (1, 0, 1), \qquad \mathbf{a}_2 = (0, 1, 1)$$

Since $E\mathbf{X} = \beta_1\mathbf{a}_1 + \beta_2\mathbf{a}_2$, it follows that

$$\begin{aligned} E(\mathbf{a}_1, \mathbf{X}) &= \beta_1(\mathbf{a}_1, \mathbf{a}_1) + \beta_2(\mathbf{a}_1, \mathbf{a}_2) = 2\beta_1 + \beta_2 \\ E(\mathbf{a}_2, \mathbf{X}) &= \beta_1(\mathbf{a}_1, \mathbf{a}_2) + \beta_2(\mathbf{a}_2, \mathbf{a}_2) = \beta_1 + 2\beta_2 \end{aligned} \tag{3}$$

Define

$$\mathbf{u}_1 = \tfrac{2}{3}\mathbf{a}_1 - \tfrac{1}{3}\mathbf{a}_2\,, \qquad \mathbf{u}_2 = \tfrac{2}{3}\mathbf{a}_2 - \tfrac{1}{3}\mathbf{a}_1$$

It follows from (3) that $E(\mathbf{u}_1, \mathbf{X}) = \beta_1$ and $E(\mathbf{u}_2, \mathbf{X}) = \beta_2$. Since $\mathbf{u}_1$ and $\mathbf{u}_2$ lie in V_1, $(\mathbf{u}_1, \mathbf{X})$ and $(\mathbf{u}_2, \mathbf{X})$ are the LUV estimators of β_1 and β_2.

Example 5 (The two-way layout.) (See Note 3.) The following design is of considerable interest in practical experimentation. Observed variables are associated with two kinds of "effects." For simplicity, we call these *column* effects and *row* effects. The model specifies that the column and row effects combine additively to give the expected value of the variable. We now spell this out in greater detail. The components of $\mathbf{X}$ are km random variables X_{ij}, $i = 1, \ldots, k$, $j = 1, \ldots, m$. Denote $n = km$ and suppose that the X_{ij}'s are arranged in some specified order; the particular order need not concern us, but use the same arrangement consistently with all n-tuples. The linear model specifies that

$$E(X_{ij}) = \gamma_i + \delta_j\,, \qquad i = 1, \ldots, k, \qquad j = 1, \ldots, m \tag{4}$$

The γ_i's and the δ_j's are the column and row effects and play the roles of the parameters β_i. That is,

$$\boldsymbol{\beta} = (\gamma_1, \ldots, \gamma_k, \delta_1, \ldots, \delta_m)$$

We now describe the structural vectors. Let $\mathbf{g}_i$ be a km-tuple with 1's in those positions in which the entries $X_{i1}, \ldots, X_{im}$ appear, and 0's elsewhere. Let $\mathbf{d}_j$ be a km-tuple with 1's in those positions in which the entries $X_{1j}, \ldots, X_{kj}$ appear, and 0's elsewhere. In terms of the structural vectors, (4) becomes

$$E\mathbf{X} = \gamma_1\mathbf{g}_1 + \cdots + \gamma_k\mathbf{g}_k + \delta_1\mathbf{d}_1 + \cdots + \delta_m\mathbf{d}_m \tag{5}$$

We denote

$$\bar{X}_{i\cdot} = \frac{X_{i1} + \cdots + X_{im}}{m} = \frac{1}{m}(\mathbf{g}_i, \mathbf{X})$$

$$\bar{X}_{\cdot j} = \frac{X_{1j} + \cdots + X_{kj}}{k} = \frac{1}{k}(\mathbf{d}_j, \mathbf{X})$$

The value $\bar{X}_{i\cdot}$ is the sample average associated with the ith column effect, and $\overline{X}_{\cdot j}$ is the sample average associated with the jth row effect. The expected values of these sample averages are

$$E(\overline{X}_{i\cdot}) = \frac{1}{m}(\mathbf{g}_i, E\mathbf{X}) = \gamma_i + \frac{\delta_1 + \cdots + \delta_m}{m} \tag{6}$$

$$E(\overline{X}_{\cdot j}) = \frac{1}{k}(\mathbf{d}_j, E\mathbf{X}) = \frac{\gamma_1 + \cdots + \gamma_k}{k} + \delta_j \tag{7}$$

These expected values follow from (5) and from the following inner products:

$$(\mathbf{g}_i, \mathbf{g}_j) = \begin{cases} m & \text{if } i = j \\ 0 & \text{if } i \neq j \end{cases}, \qquad (\mathbf{d}_i, \mathbf{d}_j) = \begin{cases} k & \text{if } i = j \\ 0 & \text{if } i \neq j \end{cases}, \qquad (\mathbf{g}_i, \mathbf{d}_j) = 1$$

It follows from (6) that

$$E(\overline{X}_{i\cdot} - \overline{X}_{j\cdot}) = \gamma_i - \gamma_j = \frac{1}{m} E(\mathbf{g}_i - \mathbf{g}_j, \mathbf{X})$$

Since $(1/m)(\mathbf{g}_i - \mathbf{g}_j)$ lies in V_1, it follows that $X_{i\cdot} - X_{j\cdot}$ is the LUV estimator of $\gamma_i - \gamma_j$. Similarly, the LUV estimator of $\delta_i - \delta_j$ is $X_{\cdot i} - X_{\cdot j}$. Hence, differences between column or row effects are estimated by the corresponding differences between the column or row sample averages. Not every linear function of the unknown parameters can be estimated. For instance, the values of the individual γ_i's or δ_i's are not estimable. (See Exercise 6 for the special case $k = m = 2$.)

There is another way of characterizing LUV estimators—the *least squares criterion.* This method was already suggested by Gauss.

least squares criterion

Minimize $\|\mathbf{X} - \Sigma_{i=1}^k (\beta_i \mathbf{a}_i\|^2$ with respect to $\beta_1, \ldots, \beta_k$. Let $\hat\beta_1, \ldots, \hat\beta_k$ be a minimizing set of values. Estimate $\Sigma_{i=1}^k c_i\beta_i$ by $\Sigma_{i=1}^k c_i\hat\beta_i$. $\Sigma_{i=1}^k c_i\hat\beta_i$ is called a *least squares estimator.*

If there exists an unbiased linear estimator of $\Sigma_{i=1}^k c_i\beta_i$, then the least squares estimator of $\Sigma_{i=1}^k c_i\beta_i$ is the same as the LUV estimator (Note 4).

Proof. By Section 16.4, $\Sigma_{i=1}^k \hat\beta_i \mathbf{a}_i$, where $\hat\beta_1, \ldots, \hat\beta_k$ minimize $\|\mathbf{X} - \Sigma_{i=1}^k \beta_i \mathbf{a}_i\|^2$, is the projection of $\mathbf{X}$ on $V(\mathbf{a}_1, \ldots, \mathbf{a}_k)$. Hence,

$$\mathbf{X} = \sum_{i=1}^{n} \hat{\beta}_i \mathbf{a}_i + \mathbf{Y} \tag{8}$$

where $\mathbf{Y}$ is orthogonal to each $\mathbf{a}_i$. Let $(\mathbf{u}, \mathbf{X})$ be the LUV estimator of $(\mathbf{c}, \boldsymbol{\beta})$. By (10), Section 17.2, $(\mathbf{u}, \mathbf{a}_i) = c_i$, and by the preceding theorem, $\mathbf{u}$ lies in $V(\mathbf{a}_1, \ldots, \mathbf{a}_k)$. Hence, from (8),

$$(\mathbf{u}, \mathbf{X}) = \left(\mathbf{u}, \sum_{i=1}^{k} \hat{\beta}_i \mathbf{a}_i + \mathbf{Y}\right) = \sum_{i=1}^{k} c_i \hat{\beta}_i \quad ▮$$

Example 6 Consider the linear regression model, Example 3, Section 17.1. Suppose that $t_1, \ldots, t_n$ are not all the same, which guarantees, by Example 4, Section 17.2, that β_1 and β_2 can be estimated. Let us find the values of β_1 and β_2 that minimize

$$f(\beta_1, \beta_2) = \sum_{i=1}^{n} (X_i - \beta_1 - \beta_2 t_i)^2 \tag{9}$$

Using standard calculus techniques, we compute

$$\frac{\partial f}{\partial \beta_1} = -2 \sum_{i=1}^{n} (X_i - \beta_1 - \beta_2 t_i)$$

$$\frac{\partial f}{\partial \beta_2} = -2 \sum_{i=1}^{n} t_i(X_i - \beta_1 - \beta_2 t_i)$$

A little algebra shows that

$$\frac{\partial f}{\partial \beta_1} = \frac{\partial f}{\partial \beta_2} = 0$$

implies that the solutions in β_1 and β_2 are

$$\hat{\beta}_1 = \frac{\sum_{i=1}^{n} X_i}{n} - \bar{t}\hat{\beta}_2 \qquad \hat{\beta}_2 = \frac{\sum_{i=1}^{n} X_i(t_i - \bar{t})}{\sum_{i=1}^{n} (t_i - \bar{t})^2}$$

We leave it to the reader to check that these values of $\hat{\beta}_1$ and $\hat{\beta}_2$ do, indeed, give a minimum. The minimizing values agree with the LUV estimator previously found in Example 3, Section 17.3.

17.4 ESTIMATING σ^2

There is a standard procedure for estimating σ^2 in a linear model. Suppose that $\mathbf{y}_1, \ldots, \mathbf{y}_r$ is an orthonormal basis for $V(\mathbf{a}_1, \ldots, \mathbf{a}_k) =$

V_1, and that r, the dimension of V_1, is less than n. Let $\mathbf{y}_1, \ldots, \mathbf{y}_r, \mathbf{y}_{r+1}, \ldots, \mathbf{y}_n$ be an orthonormal basis of the whole n-dimensional space of real n-tuples. Defining $V_2 = V(\mathbf{y}_{r+1}, \ldots, \mathbf{y}_n)$, we now have that V_1 and V_2 are orthogonal subspaces.

sum of squares due to error

Suppose we decompose $\mathbf{X}$ as $\mathbf{X} = \mathbf{Y} + \mathbf{Z}$ with $\mathbf{Y}$ in V_1 and $\mathbf{Z}$ in V_2. Then

$$E(\|\mathbf{Z}\|^2) = (n - r)\sigma^2 \tag{1}$$

regardless of what the values of $\beta_1, \ldots, \beta_k$ may be. In other words, $\|\mathbf{Z}\|^2/(n - r)$ is an unbiased estimator of σ^2. The value $\|\mathbf{Z}\|^2$ is often called *the sum of squares due to error.*

Proof. By Section 16.6, $E(\|\mathbf{Z}\|^2) = (n - r)\sigma^2 + \|E\mathbf{Z}\|^2$, and $E\mathbf{Z}$ is the projection of $E\mathbf{X}$ on V_2. The linear model specifies that $E\mathbf{X}$ lies in V_1; hence, $E\mathbf{Z}$ is the zero vector and (1) holds. ∎

A useful way of computing $\|\mathbf{Z}\|^2$ often is to use the relationship

$$\|\mathbf{X}\|^2 = \|\mathbf{Y}\|^2 + \|\mathbf{Z}\|^2 \tag{2}$$

Example 1 Consider Example 1, Section 17.1. The dimensions of V_1 and V_2 are 1 and $n - 1$, respectively.

$$\mathbf{e} = \left(\frac{1}{\sqrt{n}}, \ldots, \frac{1}{\sqrt{n}}\right)$$

is a length-one vector that generates V_1. Hence,

$$\mathbf{Y} = (\mathbf{X}, \mathbf{e})\mathbf{e}, \qquad \|\mathbf{Y}\|^2 = \frac{\left(\sum_{i=1}^{n} X_i\right)^2}{n} = n(\bar{X})^2$$

$$\|\mathbf{Z}\|^2 = \|\mathbf{X}\|^2 - \|\mathbf{Y}\|^2 = \sum_{i=1}^{n} X_i^2 - n(\bar{X})^2 = \sum_{i=1}^{n} (X_i - \bar{X})^2$$

We have already seen in Example 1, Section 16.6, and Exercise 2, Chapter 12, that $\Sigma_{i=1}^{n} (X_i - \bar{X})^2/(n - 1)$ is an unbiased estimator of σ^2.

Example 2 (The k population problem.) The experimenter makes m observations on each of k populations. The observations on the ith population are denoted $X_{i1}, \ldots, X_{im}$. The mathematical model is that the components of the vector $\mathbf{X} = (X_{11}, \ldots, X_{1m}, \ldots, X_{k1}, \ldots, X_{km})$ are $km = n$ mutually independent random variables with common variance σ^2 and $E(X_{i1}) = \cdots = E(X_{im}) = \beta_i$. In vector notation, the linear model specifies that

$$E\mathbf{X} = \beta_1\mathbf{a}_1 + \cdots + \beta_k\mathbf{a}_k$$

(The special case of $k = 2$ is Example 2, Section 17.1.) The structural vector $\mathbf{a}_1$ has 1's in the first m positions and 0's elsewhere; $\mathbf{a}_2$ has 1's in the next m positions and 0's elsewhere; and so forth. The vectors $\mathbf{a}_1, \ldots, \mathbf{a}_k$ are already mutually orthogonal and $(1/\sqrt{m})\mathbf{a}_1, \ldots, (1/\sqrt{m})\mathbf{a}_k$ form an orthonormal basis for V_1. Hence,

$$\|\mathbf{Y}\|^2 = \sum_{i=1}^{k} \left(\frac{1}{\sqrt{m}}\mathbf{a}_i, \mathbf{X}\right)^2 = m \sum_{i=1}^{k} (\overline{X}_i)^2$$

where $\overline{X}_i = (X_{i1} + \cdots + X_{im})/m$ is the sample average of the observations on the ith population. The sum of squares due to error is

$$\|\mathbf{Z}\|^2 = \|\mathbf{X}\|^2 - \|\mathbf{Y}\|^2 = \sum_{i=1}^{k}\sum_{j=1}^{m} X_{ij}^2 - m \sum_{i=1}^{m} (\overline{X}_i)^2 = \sum_{i=1}^{k}\sum_{j=1}^{m} (X_{ij} - \overline{X}_i)^2$$

The dimension of V_1 is k, and the dimension of V_2 is $km - k$.

Example 3 Consider the linear regression model, Example 3, Section 17.1. We refer to the vectors $\mathbf{a}_1$ and $\mathbf{a}_3$ described in Example 3, Section 17.3, and, as before, we suppose that $t_1, \ldots, t_n$ are not all the same. The square of the length of the projection of X on V_1 is

$$\|\mathbf{Y}\|^2 = \frac{1}{(\mathbf{a}_1, \mathbf{a}_1)}(\mathbf{X}, \mathbf{a}_1)^2 + \frac{1}{(\mathbf{a}_3, \mathbf{a}_3)}(\mathbf{X}, \mathbf{a}_3)^2 = n(\overline{X})^2 + \sum_{i=1}^{n} (t_i - \overline{t})^2(\hat{\beta}_2)^2$$

The sum of squares due to error is

$$\|\mathbf{Z}\|^2 = \|\mathbf{X}\|^2 - \|\mathbf{Y}\|^2 = \sum_{i=1}^{n} X_i^2 - n(X)^2 - \sum_{i=1}^{n} (t_i - t)^2(\beta_2)^2$$

We leave it to the reader to check that we also have

$$\|\mathbf{Z}\|^2 = \sum_{i=1}^{n} (X_i - \hat{\beta}_1 - \hat{\beta}_2 t_i)^2 \qquad (3)$$

(see Figure 17.3). The dimension of V_1 is 2, and the dimension of V_2 is $n - 2$; β_1, β_2 are the LUV estimates of β_1, β_2. The expression (3) is especially meaningful experimentally. It describes $\|\mathbf{Z}\|^2$ as the sum of the squared deviations of the observations from the least squares line, which predicts the true line.

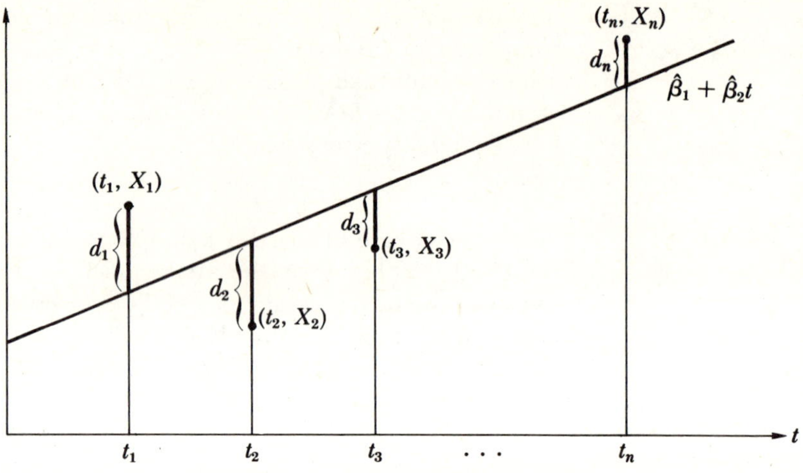

FIGURE 17.3 The least squares line is $\hat{\beta}_1 + \hat{\beta}_2 t$. The sum of squares of vertical deviations from this line is $\|Z\|^2$, where $\|Z\|^2 = d_1^2 + \cdots + d_n^2$.

17.5 A TWO-DECISION PROBLEM FOR LINEAR MODELS

Of considerable interest in the study of linear models is a certain two-decision problem (see Chapter 13). The linear model specifies that $E\mathbf{X}$ lies in a known subspace, V_1, of n-dimensional space. This subspace is broken up into two parts, and the experimenter wants to decide in which of them the expectation vector $E\mathbf{X}$ actually lies. It is surprising that this rather abstract formulation covers most of the decision problems of interest in practical applications of linear models.

two-decision problem

Let W_1 be a vector subspace of V_1. [In other words, for certain vectors $\mathbf{b}_1, \ldots, \mathbf{b}_m$ in V_1, $W_1 = V(\mathbf{b}_1, \ldots, \mathbf{b}_m)$.] In the linear model, the experimenter knows that EX lies in V_1. Suppose that he wants to decide further whether or not EX lies in W_1. This is called the two-decision problem in the linear model.

Example 1 The components of $\mathbf{X} = (X_1, \ldots, X_n)$ are n independent random variables with common expected value β and common variance

σ^2. The subspace of all multiples of $(1, \ldots, 1)$ is V_1. Suppose that the experimenter wants to decide whether or not $\beta = 0$. It is convenient to think of the single vector $(0, \ldots, 0)$ as constituting a 0-dimensional subspace. Denoting this subspace as W_1, the abstract formulation of this decision problem is to decide whether or not $E\mathbf{X}$ lies in W_1.

Example 2 Consider the k-population problem described in Example 2, Section 17.4. The vector subspace V_1 is the m-dimensional space generated by the m orthogonal vectors $\mathbf{a}_1, \ldots, \mathbf{a}_k$. A hypothesis of considerable experimental interest is that the k population means are the same; that is, $\beta_1 = \cdots = \beta_k$. The hypothesis specifies that the components of $E\mathbf{X}$ are all the same. Hence, W_1 is the one-dimensional subspace of all multiples of $(1, \ldots, 1)$. Deciding whether the hypothesis is correct is the same as deciding whether or not $E\mathbf{X}$ lies in W_1.

Example 3 Consider the two-way layout linear model described in Example 5, Section 17.3. A hypothesis of interest is that the column effects (or the row effects) are not relevant. That is, the hypothesis is that $\gamma_1 = \gamma_2 = \cdots = \gamma_k$. Since $\mathbf{g}_1 + \cdots + \mathbf{g}_k = \mathbf{d}_1 + \cdots + \mathbf{d}_m$ (a vector all of whose entries are 1's), it follows from (5), Section 17.3, that the hypothesis specifies that $E\mathbf{X}$ lies in the subspace $W_1 = V(\mathbf{d}_1, \ldots, \mathbf{d}_m)$. It may be helpful to consider a specific version of this situation. A tire company wants to study the longevity in miles of two different types of tires, under the two different conditions of city driving and highway driving. Suppose that the two-way linear model is appropriate and that γ_1 and γ_2 are the effects of the types of tire and δ_1 and δ_2 are the effects of the types of conditions. The hypothesis $\gamma_1 = \gamma_2$ is that the type of tire does not affect wear, although there may or may not be differences due to driving conditions. The hypothesis $\delta_1 = \delta_2$ is that differences between city and highway driving are not relevant, although there may or may not be differences due to tire type. Can you think of some other hypotheses of interest?

17.6 THE ANALYSIS OF VARIANCE TEST

Let us now consider the two-decision problem for linear models described in the previous section. The hypothesis to be tested is that the expectation vector $E\mathbf{X}$, which is known to lie in V_1, actually lies in the smaller subspace W_1. There is a standard procedure by which the experimenter tests this hypothesis called the *analysis of variance test.* This test seems to have been developed separately for the various special examples of linear models. Before looking at some of these examples, we shall first describe the test in abstract form.

Let $\mathbf{y}_1, \ldots, \mathbf{y}_q$ be an orthonormal basis for W_1, where q is the dimension of W_1. We suppose that $q < r =$ dimension of V_1, and that $\mathbf{y}_1, \ldots, \mathbf{y}_q, \mathbf{y}_{q+1}, \ldots, \mathbf{y}_r$ is an orthonormal basis for W_1. We also suppose that $r < n$, and that

$$\mathbf{y}_1, \ldots, \mathbf{y}_q, \mathbf{y}_{q+1}, \ldots, \mathbf{y}_r, \mathbf{y}_{r+1}, \ldots, \mathbf{y}_n \tag{1}$$

is an orthonormal basis for the whole n-dimensional space of real n-tuples. The subspace $V_2 = V(\mathbf{y}_{r+1}, \ldots, \mathbf{y}_n)$ was introduced earlier in Section 17.4 in connection with estimating σ^2. We also define $W_2 = V(\mathbf{y}_{q+1}, \ldots, \mathbf{y}_r)$. We then summarize these definitions.

The vectors given by (1) are an orthonormal basis for the whole space of real n-tuples. $W_1 = V(\mathbf{y}_1, \ldots, \mathbf{y}_q)$, $W_2 = V(\mathbf{y}_{q+1}, \ldots, \mathbf{y}_r)$, $V_2 = V(\mathbf{y}_{r+1}, \ldots, \mathbf{y}_n)$ are mutually orthogonal subspaces. Also remember that $V_1 = V(\mathbf{y}_1, \ldots, \mathbf{y}_r)$. That is, W_1 and W_2 together generate all of V_1.

Let us now review the two-decision problem in terms of this notation. The linear model asserts that $E\mathbf{X}$ lies in V_1. The hypothesis to be tested is that $E\mathbf{X}$ actually lies in W_1, a smaller part of V_1. Equivalently, *the hypothesis is that the projection of* $E\mathbf{X}$ *on* W_2 *is the zero vector.* Let us break $\mathbf{X}$ up as

$$\mathbf{X} = \mathbf{S} + \mathbf{T} + \mathbf{Z} \tag{2}$$

where $\mathbf{S}$ is the projection of $\mathbf{X}$ on W_1, $\mathbf{T}$ is the projection of $\mathbf{X}$ on W_2, and $\mathbf{Z}$ is the projection of $\mathbf{X}$ on V_2. Remember that $\mathbf{Z}$ was introduced earlier in Section 17.4. There we denoted the projection of $\mathbf{X}$ on V_1 by $\mathbf{Y}$, so we have $\mathbf{Y} = \mathbf{S} + \mathbf{T}$. It was shown in (1), Section 17.4, that

$$E(\|\mathbf{Z}\|^2) = (n - r)\sigma^2 \tag{3}$$

It follows from Section 16.6 that

$$E(\|\mathbf{T}\|^2) = (r - q)\sigma^2 + \|E\mathbf{T}\|^2 \tag{4}$$

where $E\mathbf{T}$ is the projection of $E\mathbf{X}$ on W_2. As we have said, the hypothesis is that $E\mathbf{T}$ is the zero vector.

A rather roundabout form of the hypothesis that follows immediately from (3) and (4) is that

$$\frac{E(\|\mathbf{T}\|^2)/(r-q)}{E(\|\mathbf{Z}\|^2)/(n-r)} = 1 \tag{5}$$

The analysis of variance test examines the ratio

$$\frac{\|\mathbf{T}\|^2/(r-q)}{\|\mathbf{Z}\|^2/(n-r)} \tag{6}$$

and uses it as a clue toward deciding whether (5) holds. Of course, it is not true that the expected value of a ratio is the ratio of expected values; nevertheless, large values of (6) would suggest that the hypothesis should be rejected.

analysis of variance test

The analysis of variance test for the hypothesis that $E\mathbf{X}$ lies in W_1 follows; a cutoff constant c is determined. If the ratio (6) exceeds c, then the hypothesis is rejected; otherwise, it is accepted.

One cannot say much about the choice of the constant c or about the power function of the decision procedure without specifying the distribution of $\mathbf{X}$ in greater detail. When the components of $\mathbf{X} = (X_1, \ldots, X_n)$ are assumed to be normally distributed, then the power function can be described completely in a computable form; this is the usual approach and the one that we shall follow. First, let us look at several examples (Note 5).

Example 1 Consider the k-population problem introduced in Example 2, Section 17.4. The hypothesis of interest is that the population means are the same—that is, $\beta_1 = \beta_2 = \cdots = \beta_k$. As pointed out in Example 2, Section 17.5, W_1 is the one-dimensional subspace of multiples of $\mathbf{e} = (1, \ldots, 1)$. We shall explicitly compute the test ratio (6). In Example (2), Section 17.4, we showed that $\|\mathbf{Y}\|^2 = m \sum_{i=1}^{k} (\bar{X}_i)^2$. We also have for $\mathbf{S}$, the projection of $\mathbf{X}$ on W_1,

$$\mathbf{S} = \left(\mathbf{X}, \frac{1}{\sqrt{n}}\mathbf{e}\right)\frac{1}{\sqrt{n}}\mathbf{e}, \qquad \|\mathbf{S}\|^2 = \frac{1}{n}(\mathbf{X}, \mathbf{e})^2 = n\bar{X}^2$$

where $\bar{X}$ is the overall sample average,

$$\bar{X} = \frac{\sum_{i=1}^{k}\sum_{j=1}^{m} X_{ij}}{n} = \frac{\bar{X}_1 + \cdots + \bar{X}_k}{k}$$

and $n = km$. Using the fact that $\|\mathbf{Y}\|^2 = \|\mathbf{S}\|^2 + \|\mathbf{T}\|^2$, we have

$$\|\mathbf{T}\|^2 = m \sum_{i=1}^{k} (\bar{X}_i)^2 - km(\bar{X})^2 = m \sum_{i=1}^{k} (\bar{X}_i - \bar{X})^2$$

It was also previously shown that

$$\|\mathbf{Z}\|^2 = \sum_{i=1}^{k} \sum_{j=1}^{m} (X_{ij} - \bar{X}_i)^2$$

We have the following dimensions values:

dimension of $V_1 = k$, dimension of $W_1 = 1$

Hence,

dimension of $W_2 = k - 1$, dimension of $V_2 = km - k$

The test ratio (6) is

$$\frac{m \sum_{i=1}^{k} (\bar{X}_i - \bar{X})^2/(k-1)}{\sum_{i=1}^{k} \sum_{j=1}^{m} (X_{ij} - \bar{X}_i)^2/k(m-1)}$$

Example 2 In Example 3, Section 17.5, we discussed the hypothesis $\gamma_1 = \cdots = \gamma_k$, that there is no difference among column effects in the two-way layout model. We now proceed to compute the test ratio for this hypothesis. Let $\mathbf{e}$ be the n-tuple all having components of 1, $(n = km)$. We have that $\mathbf{e} = \mathbf{g}_1 + \cdots + \mathbf{g}_k = \mathbf{d}_1 + \cdots + \mathbf{d}_m$. Let $\mathbf{e}, \mathbf{y}_1, \ldots, \mathbf{y}_{k-1}$ be an orthonormal basis for $V(\mathbf{g}_1, \ldots, \mathbf{g}_k)$, and let $\mathbf{e}, \mathbf{z}_1, \ldots, \mathbf{z}_{m-1}$ be an orthonormal basis for $V(\mathbf{d}_1, \ldots, \mathbf{d}_m)$. Each $\mathbf{y}_i$ is a linear combination of $\mathbf{g}_i$'s. If $\mathbf{y}_i = a_1\mathbf{g}_1 + \cdots + a_k\mathbf{g}_k$, then, since $(\mathbf{e}, \mathbf{y}_i) = 0$ and $(\mathbf{e}, \mathbf{g}_1) = \cdots = (\mathbf{e}, \mathbf{g}_k) = m$, it follows that $a_1 + \cdots + a_k = 0$. By a similar argument, if $\mathbf{z}_j = b_1\mathbf{d}_1 + \cdots + b_m\mathbf{d}_m$, it must be true that $b_1 + \cdots + b_m = 0$. Hence,

$$(\mathbf{y}_i, \mathbf{z}_j) = (a_1\mathbf{g}_1 + \cdots + a_k\mathbf{g}_k, b_1\mathbf{d}_1 + \cdots + b_m\mathbf{d}_m)$$

$$= \sum_{i,j} a_i b_j (\mathbf{g}_i, \mathbf{d}_i) = (a_1 + \cdots + a_k)(b_1 + \cdots + b_m) = 0$$

since $(\mathbf{g}_i, \mathbf{d}_j) = 1$. We are now prepared to compute the projections needed for the test ratio. We use the following facts:

$$W_2 = V(\mathbf{z}_1, \ldots, \mathbf{z}_{m-1}), \; V_1 = V(\mathbf{g}_1, \ldots, \mathbf{g}_k, \mathbf{d}_1, \ldots, \mathbf{d}_m)$$

$$= V(\mathbf{e}, \mathbf{y}_1, \ldots, \mathbf{y}_{k-1}, \mathbf{z}_1, \ldots, \mathbf{y}_{m-1})$$

The projection of $\mathbf{X}$ on the one-dimensional space generated by $\mathbf{e}$ is $[\mathbf{X}, (1/\sqrt{n})\mathbf{e}](1/\sqrt{n})\mathbf{e}$. The square of the length of this projection is $n(\bar{X}^2)$, where $\bar{X} = \Sigma_{i,j} X_{ij}/n$ is the overall sample average. If

$\mathbf{U}$ is the projection of $\mathbf{X}$ on $V(\mathbf{d}_1, \ldots, \mathbf{d}_m)$, then

$$\|\mathbf{U}\|^2 = \frac{1}{k}(\mathbf{X}, \mathbf{d}_1)^2 + \cdots + \frac{1}{k}(\mathbf{X}, \mathbf{d}_m)^2 = k(\overline{X}^2_{\cdot 1}) + \cdots + k(X^2_{\cdot m})$$

since the $\mathbf{d}_i$'s are mutually orthogonal and $(\mathbf{d}_i, \mathbf{d}_i) = k$. Hence, the square of the length of the projection of $\mathbf{X}$ on W_2 is

$$\|\mathbf{T}\|^2 = k \sum_{i=1}^{m} (\overline{X}_{\cdot i})^2 - km(\overline{X}^2) = k \sum_{i=1}^{m} (\overline{X}_{\cdot i} - \overline{X})^2$$

Similarly, the square of the length of the projection of $\mathbf{X}$ on $V(\mathbf{z}_1, \ldots, \mathbf{z}_{m-1})$ is $\|\mathbf{R}\|^2 = m \sum_{j=1}^{k} (\overline{X}_{j\cdot} - \overline{X})^2$. It follows that the sum of squares due to error is

$$\|\mathbf{Z}\|^2 = \sum_{i,j} X_{ij} - n(\overline{X}^2) - \|\mathbf{T}\|^2 - \|\mathbf{R}\|^2 = \sum_{i,j} (X_{ij} - \overline{X})^2 - \|\mathbf{T}\|^2 - \|\mathbf{R}\|^2$$

Example 3 In Example 7, Section 17.1, suppose that we test the hypothesis that the angles β_1 and β_2 are the same. The linear model is

$$E\mathbf{X} = \beta_1(1, 0, 1) + \beta_2(0, 1, 1)$$

and under the hypothesis $\beta_1 = \beta_2$, $E\mathbf{X}$ lies in the one-dimensional subspace W_1 generated by $(1, 0, 1) + (0, 1, 1) = (1, 1, 2)$. The vector $(1, 0, 1) - (0, 1, 1) = (1, -1, 0)$ lies in V_1 and is orthogonal to W_1. Hence, W_2 is generated by $(1, -1, 0)$. The value of $\|\mathbf{T}\|^2$ is

$$\|\mathbf{T}\|^2 = \frac{1}{2}[(X_1, X_2, X_3), (1, -1, 0)]^2 = \frac{(X_1 - X_2)^2}{2}$$

The vector $(1, 1, -1)$ is orthogonal to each of the structural vectors $(1, 0, 1)$ and $(0, 1, 1)$. Hence,

$$\|\mathbf{Z}\|^2 = \frac{1}{3}((X_1, X_2, X_3), (1, 1, -1))^2 = \frac{(X_1 + X_2 - X_3)^2}{3}$$

The dimensions of W_2 and V_2 are each 1, and the test ratio is

$$\frac{\|\mathbf{T}\|^2}{\|\mathbf{Z}\|^2} = \frac{(X_1 - X_2)^2/2}{(X_1 + X_2 - X_3)^2/3}$$

17.7 THE ANALYSIS OF VARIANCE TEST WITH NORMAL DISTRIBUTIONS

The usual assumption is that the components of $\mathbf{X} = (X_1, \ldots, X_n)$ are mutually independent, normally distributed random variables, with common variances σ^2. The expected values are, of course, specified by the linear model. We suspect that the main reason for the wide use of this assumption is that it makes possible explicit probability computations that would not be so easy with other distributions. There seems to be some evidence that the analysis of variance test is quite useful, even when the distributions are not normal. The following are the basic facts relating to the distribution of the analysis of variance test ratio.

Suppose that the components of $\mathbf{X} = (X_1, \ldots, X_n)$ are mutually independent, normally distributed random variables, each with the same variance σ^2. Using the notation of Section 17.6, let $\mathbf{T}$ be the projection of $\mathbf{X}$ on W_2 and let $\mathbf{Z}$ be the projection of $\mathbf{X}$ on V_2. Then:

(a) $\|\mathbf{T}\|^2$ and $\|\mathbf{Z}\|^2$ are independent random variables.
(b) $\|\mathbf{Z}\|^2/\sigma^2$ is chi-square distributed with $n - r$ degrees of freedom.
(c) $\|\mathbf{T}\|^2/\sigma^2$ is distributed like $[U_1 + (\Delta/\sigma)]^2 + U_2^2 + \cdots + U_{r-q}^2$, where $U_1, \ldots, U_{r-q}$ are mutually independent, $N(0, 1)$ random variables, and

$$\Delta = \|E\mathbf{T}\|$$

in which $E\mathbf{T}$ is the projection of $\mathbf{X}$ on W_2. As pointed out in Section 16.7, this is a noncentral chi-square distribution. In particular, when $\Delta = 0$, $\|\mathbf{T}\|^2/\sigma^2$ is chi-square distributed with $r - q$ degrees of freedom.
(d) When the hypothesis that $E\mathbf{X}$ is in W_1 is true (or, equivalently, that $\Delta = 0$), then the distribution of the test ratio

$$\frac{\|\mathbf{T}\|^2/(r-q)}{\|\mathbf{Z}\|^2/(n-r)}$$

does not depend on the parameter σ.

Proof. Let $\mathbf{y}_1, \ldots, \mathbf{y}_n$ be the orthonormal basis described in (1), Section 17.6. Then, by Section 16.7, $(\mathbf{X}, \mathbf{y}_1), \ldots, (\mathbf{X}, \mathbf{y}_n)$ are mutually independent, normally distributed random variables, with common variances σ^2 and with expected values $(E\mathbf{X}, \mathbf{y}_1), \ldots, (E\mathbf{X}, \mathbf{y}_n)$. Since the linear model specifies that $E\mathbf{X}$ is orthogonal to each of $\mathbf{y}_{r+1}, \ldots, \mathbf{y}_n$, it follows that $(\mathbf{X}, \mathbf{y}_{r+1})/\sigma, \ldots, (\mathbf{X}, \mathbf{y}_n)/\sigma$ are each $N(0, 1)$, and

$$\frac{\|\mathbf{Z}\|^2}{\sigma^2} = \frac{(\mathbf{X}, \mathbf{y}_{r+1})^2}{\sigma^2} + \cdots + \frac{(\mathbf{X}, \mathbf{y}_n)^2}{\sigma^2}$$

is chi-square distributed with $n - r$ degrees of freedom. By the last theorem of Section 16.7,

$$\frac{\|\mathbf{T}\|^2}{\sigma^2} = \frac{(\mathbf{X}, \mathbf{y}_{q+1})^2 + \cdots + (\mathbf{X}, \mathbf{y}_r)^2}{\sigma^2}$$

is distributed like $[U_1 + (\Delta/\sigma)]^2 + U_2^2 + \cdots + U_{r-q}^2$, where $\mathbf{U}_1, \ldots, \mathbf{U}_{r-q}$ are mutually independent, $N(0, 1)$ and

$$\Delta^2 = (E\mathbf{X}, \mathbf{y}_{q+1})^2 + \cdots + (E\mathbf{X}, \mathbf{y}_r)^2 = \|E\mathbf{T}\|^2$$

which proves part (c). The independence of $\|\mathbf{T}\|^2$ and $\|\mathbf{Z}\|^2$ asserted in (a) follows from the mutual independence of

$$(\mathbf{X}, \mathbf{y}_{q+1}), \ldots, (\mathbf{X}, \mathbf{y}_r), (\mathbf{X}, \mathbf{y}_{r+1}), \ldots, (\mathbf{X}, \mathbf{y}_n)$$

Part (d) follows from (a), (b), and (c), and from the fact that

$$\frac{\|\mathbf{T}\|^2}{\|\mathbf{Z}\|^2} = \frac{\|\mathbf{T}\|^2/\sigma^2}{\|\mathbf{Z}\|^2/\sigma^2} \qquad \blacksquare$$

We emphasize again that the hypothesis that $E\mathbf{X}$ lies in W_1 is equivalent to $\Delta = 0$. When $\Delta = 0$, then $\|T\|^2/\|Z\|^2$ is distributed like a ratio of independent gamma variables [see (3), Section 9.5]. The distribution of the slightly different form

$$\frac{\|\mathbf{T}\|^2/(r-q)}{\|\mathbf{Z}\|^2/(n-r)}$$

is called the F distribution, with $r - q$, $n - r$ degrees of freedom (Note 6). Table 17.1 shows a section of the F distribution, in which Y stands for an F-distributed random variable with the indicated numbers of freedom. This distribution is used to determine the value of the cutoff constant c described in Section 17.6. When $\Delta = 0$,

$$P\left(\frac{\|\mathbf{T}\|^2/(r-q)}{\|\mathbf{Z}\|^2/(n-r)} < c\right)$$

is the probability of accepting the hypothesis when it should be ac-

TABLE 17.1

$r-q$	$n-r$	c	$P(Y < c)$	$r-q$	$n-r$	c	$P(Y < c)$
5	5	5.05	0.95	5	5	10.97	0.99
5	10	3.33	0.95	5	10	5.64	0.99
10	5	4.74	0.95	10	5	10.05	0.99
10	10	2.98	0.95	10	10	4.85	0.99
20	5	4.56	0.95	20	5	9.55	0.99
20	10	2.77	0.95	20	10	4.41	0.99
20	20	2.12	0.95	20	20	2.94	0.99

cepted. It is called the *significance level* of the test (Section 13.7). A desired significance level determines the value of c. It is an important fact from part (d) of the last theorem that the determination of c does not depend on σ. This is significant, since σ is usually unknown to the experimenter.

Let us denote by $\psi(\Delta/\sigma)$ the probability of rejecting the hypothesis when Δ is the value of $\|E\mathbf{T}\|$. The last theorem shows that this probability depends on $E\mathbf{X}$ only through the single value of Δ/σ, so the notation $\psi(\Delta/\sigma)$ makes sense. We can interpret Δ/σ as a *measure of how untrue the hypothesis is.* The *power function* of the analysis of variance test is $1-\psi$ (Note 7). Figure 17.4 shows some graphs of these power functions.

Example 1 We look again at the k-population problem, which was last discussed in Example 1, Section 17.6. It is instructive to determine the value of Δ explicitly. Recall that $\Delta = \|E\mathbf{T}\|$ where $E\mathbf{T}$ is the projection of $E\mathbf{X}$ on W_2. We can use some previous computations as follows. We know that the projection of X on W_2 is $m\,\Sigma_{i=1}^{k}\,(\bar{X}_i-\bar{X})^2$. We also know that $E\mathbf{X}=(\beta_1, \ldots, \beta_1, \beta_2, \ldots, \beta_2, \ldots, \beta_k, \ldots, \beta_k)$. (This is a string of m β_1's, m β_2's, etc.) The projection of $E\mathbf{X}$ on W_2

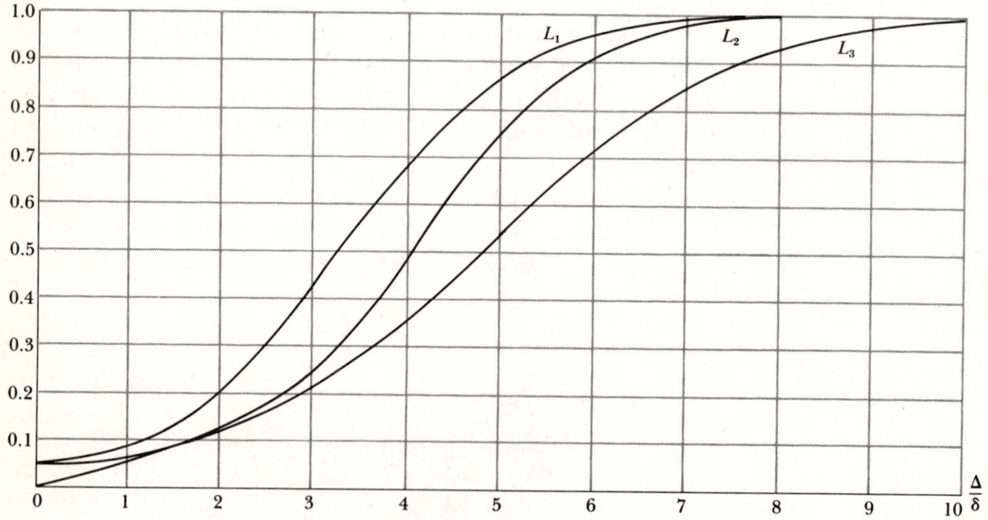

FIGURE 17.4 Some power functions of analysis of variance tests.
L_1: significance level = 0.95, $r-q=3$, $n-r=6$
L_2: significance level = 0.99, $r-q=3$, $n-r=10$
L_3: significance level = 0.95, $r-q=8$, $n-r=6$

can be obtained from the projection of X on W_2 by replacing $\mathbf{X}_i$ by $E\mathbf{X}_i$ in every coordinate. Thus,

$$\frac{\Delta^2}{\sigma^2} = \frac{m \sum_{i=1}^{k} (\beta_1 - \bar{\beta})^2}{\sigma^2}$$

where $\bar{\beta}$ is the average of the β_i's. A reasonable measure of how untrue is the hypothesis, is Δ/σ. It is in a sense the average distance among the population means β_i, measured in terms of the basic unit σ. In particular, if $k = 2$, then

$$\frac{\Delta}{\sigma} = \left(\frac{m}{2}\right)^{1/2} \frac{|\beta_1 - \beta_2|}{\sigma}$$

is the measure of how far apart β_1 and β_2 are.

17.8 CONFIDENCE INTERVAL ESTIMATES IN A LINEAR MODEL

The analysis required for making confidence interval estimates for functions of the unknown parameters in a linear model is similar to that used in the analysis of variance test. We still assume that the components of $\mathbf{X} = (X_1, \ldots, X_n)$ are mutually independent, normally distributed, random variables with common variance σ^2. Before describing a general approach, we first look at several examples.

Example 1 Consider Example 1, Section 17.1, in which n independent observations are made on the same population having a mean of β. The random variable $(\bar{X} - \beta)\sqrt{n}/\sigma$ is $N(0, 1)$ distributed. We know from Section 16.7 that this random variable is independent of $(nS^2/\sigma^2) = \Sigma_{i=1}^{n} (X_i - \bar{X})^2/\sigma^2$, which is chi-square distributed with $n - 1$ degrees of freedom. The distribution of

$$\frac{(\bar{X} - \beta)\sqrt{n}/\sigma}{\sqrt{n}S/(n-1)^{1/2}\sigma} = \frac{(\bar{X} - \beta)}{S/(n-1)^{1/2}} \tag{1}$$

is called the Student-t distribution with $n - 1$ degrees of freedom. Table 17.2 shows a section of a tabulation of the Student-t distribution (Note 6), in which Y stands for a Student-t distributed random variable with the indicated number of degrees of freedom. It is important that neither the value of the random variable (1) nor the form of its distribution depends on the unknown parameter σ. (Why not?) Let $(-a, a)$ be an interval containing a prescribed

TABLE 17.2

degrees of freedom	c	$P(Y < c)$
5	1.476	0.90
5	2.015	0.95
5	3.365	0.99
10	1.372	0.90
10	1.812	0.95
10	2.764	0.99
20	1.325	0.90
20	1.725	0.95
20	2.528	0.99

amount of probability (say, $1 - \alpha$) of the Student-t distribution with $n - 1$ degrees of freedom. Then the random variable (1) lies between $-a$ and a with probability $1 - \alpha$, no matter what the unknown parameters β and σ may be. This implies that

$$\bar{X} - \frac{aS}{(n-1)^{1/2}} \leqslant \beta \leqslant \bar{X} + \frac{aS}{(n-1)^{1/2}}$$

holds true with probability $1 - \alpha$. If the experimenter asserts that β lies in the random interval

$$\left(\bar{X} - \frac{aS}{(n-1)^{1/2}}, \bar{X} + \frac{aS}{(n-1)^{1/2}}\right)$$

then α is the probability with which he will fail to tell the truth. The constant a is typically chosen so that α is "small", with values such as 0.01 or 0.05. Compare this example with Examples 1 and 2, Section 12.7, in which σ was supposed to be known and equal to 1.

Example 2 Consider the linear regression model introduced as Example 3, Section 17.1, and most recently discussed in Example 3, Section 17.4. For t any real number, the LUV estimator of $\beta_1 + \beta_2 t$ is $\hat{\beta}_1 + \hat{\beta}_2 t$. The explicit form of this estimator is $(\mathbf{u}, \mathbf{X})$ where $\mathbf{u} = (1/n)\mathbf{a}_1 + [(t - \bar{t})/(\mathbf{a}_3, \mathbf{a}_3)]\mathbf{a}_3$, using the notation introduced in Example 3, Section 17.3. Since $\mathbf{u}$ lies in V_1, it follows that $(\mathbf{u}, \mathbf{X})$ and $\|\mathbf{Z}\|^2$, the sum of squares due to error, are independent random variables, by the theorem of Section 17.7. We also have that

$$\operatorname{Var}(\mathbf{u}, \mathbf{X}) = \|\mathbf{u}\|^2\sigma^2 = \left[\frac{1}{n} + \frac{(t - \bar{t})^2}{\sum_i (t_i - \bar{t})^2}\right]\sigma^2$$

and

$$\frac{\hat{\beta}_1 + \hat{\beta}_2 t - \beta_1 - \beta_2 t}{\|\mathbf{u}\| \, \|\mathbf{Z}\|/(n-2)^{1/2}} \tag{2}$$

is Student-t distributed with $n - 2$ degrees of freedom. This is so because (2) is the ratio of the independent random variables $(\hat{\beta}_1 + \hat{\beta}_2 t - \beta_1 - \beta_2 t)/\|\mathbf{u}\|\sigma$, which is $N(0, 1)$, and $\|\mathbf{Z}\|/\sigma(n-2)^{1/2}$, where $\|\mathbf{Z}\|^2/\sigma^2$ is chi-square with $n - 2$ degrees of freedom. If $(-a, a)$ includes $1 - \alpha$ probability of the Student-t distribution, then (2) lies between $-a$ and a with probability $1 - \alpha$ no matter what the values of β_1, β_2, and σ may be. This implies that

$$\hat{\beta}_1 + \hat{\beta}_2 t - \frac{a\|\mathbf{u}\| \, \|\mathbf{Z}\|}{(n-2)^{1/2}} \leqslant \beta_1 + \beta_2 t \leqslant \hat{\beta}_1 + \hat{\beta}_2 t + \frac{a\|\mathbf{u}\| \, \|\mathbf{Z}\|}{(n-2)^{1/2}} \tag{3}$$

with probability $1 - \alpha$. The extreme sides of (3) define a confidence interval estimate for $\beta_1 + \beta_2 t$. The probability that this interval fails to cover $\beta_1 + \beta_2 t$ is α. It should be emphasized that this analysis assumes that we deal with *one fixed t*.

This procedure can be extended to provide a confidence estimate appropriate for *all t simultaneously*. Remember that the numerator of (2) is $(\mathbf{u}, \mathbf{X})$, and let us consider

$$\sup_t \frac{|(\mathbf{u}, \mathbf{X})|}{\|\mathbf{u}\|} \tag{3}$$

We claim that the value of (3) is $\|\mathbf{Y}\|$, where $\mathbf{Y}$ is the projection of $\mathbf{X}$ on V_1. This follows from (9), Section 16.4, and from the fact that as t varies, then $\mathbf{u}$ varies over vectors that generate all of V_1. Hence,

$$\sup_t \frac{|\hat{\beta}_1 + \hat{\beta}_2 t - \beta_1 - \beta_2 t|/\sqrt{2}}{\|\mathbf{u}\| \, \|\mathbf{Z}\|/(n-2)^{1/2}} = \frac{\|\mathbf{Y}\|/\sqrt{2}}{\|\mathbf{Z}\|/(n-2)^{1/2}} \tag{4}$$

By a slight variation on the theorem of Section 17.7, $(\|\mathbf{Y}\|^2/2)/[\|\mathbf{Z}\|^2/(n-2)]$ is F distributed with $2, n-2$ degrees of freedom. This uses the fact that V_1 has dimension 2 and $\|\mathbf{Y}\|^2/\sigma^2$ and $\|\mathbf{Z}\|^2/\sigma^2$ are independent chi-square random variables the degrees of freedom of which are 2 and $n - 2$. Let $(0, a)$ include $1 - \alpha$ probability of the F distribution with $2, n - 2$ degrees of freedom. Then the left side of (4) lies in $(0, \sqrt{a})$ with probability $1 - \alpha$, no matter what the values of β_1, β_2, and σ might be. It is equivalent to the fact that for all t, $\beta_1 + \beta_2 t$ lies in the interval

$$\left(\hat{\beta}_1 + \hat{\beta}_2 t - \frac{\sqrt{a}\|\mathbf{u}\| \, \|\mathbf{Z}\|}{\sqrt{2}(n-2)^{1/2}}, \hat{\beta}_1 + \hat{\beta}_2 t + \frac{\sqrt{a}\|\mathbf{u}\| \, \|\mathbf{Z}\|}{\sqrt{2}(n-2)^{1/2}}\right) \tag{5}$$

If the experimenter asserts that $\beta_1 + \beta_2 t$ lies in the interval (5) *for every t*, then the probability of his failing to tell the truth is α. It should be emphasized that this probability relates to the assertions for *all values of t simultaneously*. For this reason, (5) is said to describe a set of *simultaneous confidence intervals*. Define the function

$$f(t) = \frac{\sqrt{a}\|\mathbf{u}\|\,\|\mathbf{Z}\|}{\sqrt{2}(n-2)^{1/2}} = \frac{\sqrt{a}\|\mathbf{Z}\|}{\sqrt{2}(n-2)^{1/2}}\left(\frac{1}{n} + \frac{(t-\bar{t})^2}{\sum (t_i - \bar{t})^2}\right)^{1/2} \tag{6}$$

Then the assertion is that the whole line $\beta_1 + \beta_2 t$ lies in the belt between $\hat{\beta}_1 + \hat{\beta}_2 t - f(t)$ and $\hat{\beta}_1 + \hat{\beta}_2 t + f(t)$. The probability that this is true for the whole line is $1 - \alpha$. See Figure 17.5.

Example 3 This example concerns the k-population problem, which was last discussed in Example 1, Section 17.6. We shall describe a method for simultaneously estimating all linear functions of the unknown parameters, $\Sigma_{i=1}^{k} b_i\beta_i$. The analysis is based on the fact that

$$\sup_{b_1,\ldots,b_k} \frac{\left|\sum_{i=1}^{k} b_i(\bar{X}_i - \beta_i)\right|}{\left(\sum_{i=1}^{k} b_i^2\right)^{1/2}} = \left[\sum_{i=1}^{k} (\bar{X}_i - \beta_i)^2\right]^{1/2} \tag{7}$$

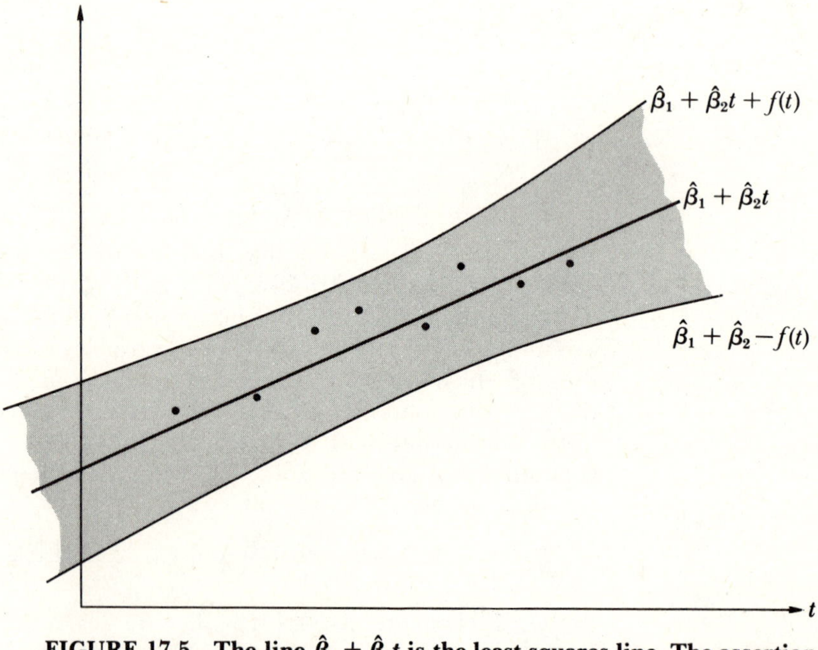

FIGURE 17.5 The line $\hat{\beta}_1 + \hat{\beta}_2 t$ is the least squares line. The assertion that the true line $\beta_1 + \beta_2 t$ lies wholly within the shaded region will be true with probability $1 - \alpha$.

which follows from the Cauchy-Schwarz inequality, (9), Section 16.4. Hence, from (7),

$$\sup_{b_1,\ldots,b_k} \frac{\left|\sum_{i=1}^{k} b_i(\bar{X}_i - \beta_i)\right|(m/k)^{1/2}}{\left(\sum_{i=1}^{k} b_i^2\right)^{1/2} \|\mathbf{Z}\|/[k(m-1)]^{1/2}} = \left[\frac{\sum_{i=1}^{k} (\bar{X}_i - \beta_i)^2(m/k)}{\|\mathbf{Z}\|^2/k(m-1)}\right]^{1/2} \tag{8}$$

The square of (8) is F distributed with $k, k(m-1)$ degrees of freedom, because $(\bar{X}_i - \beta_i)/(\sigma/\sqrt{m})$ is $N(0, 1)$. Hence, $\Sigma_{i=1}^{k} (\bar{X}_i - \beta_i)^2 m/\sigma^2$ is chi-square distributed with k degrees of freedom. This random variable is independent of $\|\mathbf{Z}\|^2/\sigma^2$, which is chi-square distributed with $k(m-1)$ degrees of freedom. Let $(0, a)$ contain $1-\alpha$ probability of the F distribution with $k, k(m-1)$ degrees of freedom. Then the left side of (8) lies in $(0, \sqrt{a})$ with probability $1-\alpha$ regardless of the values of $\beta_1, \ldots, \beta_k$, and σ. It follows that $\Sigma_{i=1}^{k} b_i\beta_i$ lies in the interval

$$\left(\sum_{i=1}^{k} b_i\bar{X}_i - \left[\frac{a}{(m-1)m}\sum_{i=1}^{k} b_i^2\right]^{1/2}\|\mathbf{Z}\|, \sum_{i=1}^{k} b_i\bar{X}_i + \left[\frac{a}{(m-1)m}\sum_{i=1}^{k} b_i^2\right]^{1/2}\|\mathbf{Z}\|\right) \tag{9}$$

with probability $1-\alpha$. This probability holds for all $b_1, \ldots, b_k$ simultaneously. If the experimenter asserts that $\Sigma_{i=1}^{k} b_i\beta_i$ lies in the interval (9) for all possible $b_1, \ldots, b_k$, then he fails to tell the truth with probability α.

The preceding special examples of confidence interval estimation can be placed in a general setting as follows. Let W be a subspace of V_1. In some examples, W may be V_1 itself or W_1 or W_2. Consider the following fact:

$$\sup_{\mathbf{b} \text{ in } W} \frac{|(\mathbf{b}, \mathbf{X}) - (\mathbf{b}, E\mathbf{X})|}{\|\mathbf{b}\|} = \sup_{\mathbf{b} \text{ in } W} \frac{|(\mathbf{b}, \mathbf{X} - E\mathbf{X})|}{\|\mathbf{b}\|} = \|\mathbf{R}\| \tag{10}$$

where $\mathbf{R}$ is the projection of $\mathbf{X} - E\mathbf{X}$ on W. It follows from (10), Section 16.4. Let d be the dimension of W. Then $\|\mathbf{R}\|^2/\sigma^2$ is chi-square distributed with d degrees of freedom, $\|\mathbf{R}\|$ and $\|\mathbf{Z}\|$ are independent random variables, and

$$\frac{\|\mathbf{R}\|^2/d}{\|\mathbf{Z}\|^2/(n-r)} \tag{11}$$

is F distributed with $d, n-r$ degrees of freedom. These facts follow from the theorem in Section 17.7. Let $(0, a)$ contain $1-\alpha$ probability of the F distribution with $d, n-r$ degrees of freedom. By dividing both sides of (10) by $[d\|\mathbf{Z}\|^2/(n-r)]^{1/2}$, we conclude that

$$\sup_{\mathbf{b} \text{ in } W} \frac{|(\mathbf{b}, \mathbf{X}) - (\mathbf{b}, E\mathbf{X})|/\sqrt{d}}{\|\mathbf{b}\| \, \|\mathbf{Z}\|/(n-r)^{1/2}} \tag{12}$$

equals the square root of (11), and hence lies between 0 and $\sqrt{a}$ with probability $1 - \alpha$. This fact implies that

$$(\mathbf{b}, \mathbf{X}) - \left(\frac{a\|\mathbf{b}\|^2\|\mathbf{Z}\|^2 d}{n-r}\right)^{1/2} \leqslant (\mathbf{b}, \mathbf{EX}) \leqslant (\mathbf{b}, \mathbf{X}) + \left(\frac{a\|\mathbf{b}\|^2\|\mathbf{Z}\|^2 d}{n-r}\right)^{1/2} \tag{13}$$

for *all* vectors $\mathbf{b}$ in W with probability $1 - \alpha$. The intervals (13) constitute a set of *simultaneous confidence interval estimates for all linear functions* $(\mathbf{b}, E\mathbf{X})$, $\mathbf{b}$ in W. In making these assertions, the probability of failing to tell the truth is α.

The reader should review Examples 1, 2, and 3 in the light of this general setting. In each of these, the role of W is played by V_1. We now give a version of Example 3 in which W is smaller than V_1.

Example 4 We consider Example 3 concerning the k-population problem again. Instead of estimating arbitrary linear forms $\Sigma_{i=1}^k b_i\beta_i$, we shall consider only those linear forms where the b_i's are subject to the restriction that $b_1 + \cdots + b_k = 0$. Such linear forms are called *contrasts* among the β_i's. For instance, $\beta_1 - \beta_2$ is a contrast that compares the first population mean with the second. Also, $\beta_1 - (\beta_2 + \beta_3)/2$ is a contrast comparing the first population mean with the average of the second and third population means. If we repeat the analysis in (7) and (8), we will need the modification that

$$\sup_{\substack{b_1,\ldots,b_k \\ b_1+\cdots+b_k=0}} \frac{\left|\sum_{i=1}^k b_i(\bar{X}_i - \beta_i)\right| [m/(k-1)]^{1/2}}{\left(\sum_{i=1}^k b_i^2\right)^{1/2} \|\mathbf{Z}\|/[k(m-1)]^{1/2}} = \left\{\frac{\sum_{i=1}^k (\bar{X}_i - \beta_i - \bar{X} + \beta)^2 [m/(k-1)]}{\|\mathbf{Z}\|^2/k(m-1)}\right\}^{1/2} \tag{14}$$

Now the square of the right side of (14) is F distributed with $k-1$, $k(m-1)$ degrees of freedom, instead of k, $k(m-1)$ degrees of freedom, as it was earlier. (Why?) The counterpart of the confidence intervals (9) are now that $\Sigma_{i=1}^k b_i\beta_i$ lie in the interval

$$\left(\sum_{i=1}^k b_i\bar{X} - \left[\frac{a(m-1)(k-1)}{km(m-1)} \sum_{i=1}^k b_i^2\right]^{1/2} \|\mathbf{Z}\|,\right.$$
$$\left.\sum_{i=1}^k b_i\bar{X}_i + \left[\frac{a(m-1)(k-1)}{km(m-1)} \sum_{i=1}^k b_i^2\right]^{1/2} \|\mathbf{Z}\|\right) \tag{15}$$

simultaneously for all $b_1, \ldots, b_k$ that sum to 0. The constant a is now chosen from the F distribution with $k-1$, $k(m-1)$ degrees of freedom. It is a fact we shall not prove here, that the widths of the intervals (15) are narrower than the intervals (9), whenever the α is the same for both. It is certainly plausible, since fewer linear functions are being estimated by (15) than by (9). The confidence intervals (15) follow from the general formulation by taking for W the subspace of all vectors that are orthogonal to $(1, \ldots, 1)$. This $(k-1)$-dimensional subspace was denoted as W_2 in Example 1, Section 17.6. We leave the details to the reader.

SUMMARY

A *linear model* specifies that $E\mathbf{X} = (EX_1, \ldots, EX_n)$ lies in a given vector subspace of n space. Explicitly, $\mathbf{a}_1, \ldots, \mathbf{a}_k$ are given *structural vectors* and $E\mathbf{X}$ is known to be a linear combination of these vectors,

$$E\mathbf{X} = \beta_1\mathbf{a}_1 + \cdots + \beta_k\mathbf{a}_k$$

The parameters $\beta_1, \ldots, \beta_k$ are unknown to the experimenter.

The problems of estimation concern estimating linear functions of the parameters, of the form $c_1\beta_1 + \cdots + c_k\beta_k = (\mathbf{c}, \boldsymbol{\beta})$. The estimators are restricted to be linear functions of the X_i's, of the form $u_1X_1 + \cdots + u_nX_n = (\mathbf{u}, \mathbf{X})$. A criterion is given that is necessary and sufficient for the existence of unbiased linear estimators.

In what follows, we assume that the components of $\mathbf{X} = (X_1, \ldots, X_n)$ are uncorrelated and have equal variances. If there exists a $(\mathbf{u}, \mathbf{X})$ that is an unbiased estimator of $(\mathbf{c}, \boldsymbol{\beta})$, then there is a unique such estimator the *variance of which is minimum*, the *LUV estimator.* The LUV estimator is completely determined by the requirements that it be unbiased and that $\mathbf{u}$ lie in $V(\mathbf{a}_1, \ldots, \mathbf{a}_k)$. The LUV estimator can also be described by a *least squares* criterion.

The decision problems concern a specified subspace W_1 of $V(\mathbf{a}_1, \ldots, \mathbf{a}_k)$. The experimenter wants to decide whether or not $E\mathbf{X}$ lies in W_1. Suppose that $\mathbf{y}_1, \ldots, \mathbf{y}_q, \mathbf{y}_{q+1}, \ldots, \mathbf{y}_r, \mathbf{y}_{r+1}, \ldots, \mathbf{y}_n$ is an orthonormal basis for all of n space where

$$\mathbf{y}_1, \ldots, \mathbf{y}_q \quad \text{is a basis for } W_1$$

and

$$\mathbf{y}_1, \ldots, \mathbf{y}_q, \mathbf{y}_{q+1}, \ldots, \mathbf{y}_r \quad \text{is a basis for } V(\mathbf{a}_1, \ldots, \mathbf{a}_k)$$

The analysis of variance requires the computation of

$$\|\mathbf{T}\|^2 = (\mathbf{X}, \mathbf{y}_{q+1})^2 + \cdots + (\mathbf{X}, \mathbf{y}_r)^2,$$

$$\|\mathbf{Z}\|^2 = (\mathbf{X}, \mathbf{y}_{r+1})^2 + \cdots + (\mathbf{X}, \mathbf{y}_n)^2$$

These are the squared lengths of the projections of $\mathbf{X}$ on $V(\mathbf{y}_{q+1}, \ldots, \mathbf{y}_r)$ and $V(\mathbf{y}_{r+1}, \ldots, \mathbf{y}_n)$, respectively. The value $\|\mathbf{T}\|^2$ is often called the "sum of squares due to hypothesis" and $\|\mathbf{Z}\|^2$ is often referred to as the "sum of squares due to error." If

$$\frac{\|\mathbf{T}\|^2/(r-q)}{\|\mathbf{Z}\|^2/(n-r)}$$

is "small," one decides in favor of the hypothesis that $E\mathbf{X}$ lies in $W_1 = V(\mathbf{y}_1, \ldots, \mathbf{y}_q)$; otherwise, one decides against this hypothesis. This test is the *analysis of variance test.* One usually assumes that the X_i's are normally distributed, which allows an explicit evaluation of the power function of this decision procedure.

There is also given a confidence procedure allowing one to make *confidence interval estimates, simultaneously,* for broad classes of linear functions $(\mathbf{c}, \boldsymbol{\beta})$.

EXERCISES

1 The three objects *a, b,* and *c* have true weights of β_1, β_2, and β_3, respectively. The scale used for weighing these objects is subject to error and has the property that no matter what β (true weight) might be, the reading X satisfies $EX = \beta$ and $\text{Var}(X) = \sigma^2$ where σ^2 does not depend on β. We assume that readings from repeated weighings are uncorrelated. The use of the scale is costly and only four weighings are allowed to determine the value of β_1, β_2, and β_3.

(a) Suppose each object is weighed separately. Also, all three objects are weighed together. Denote the results of these weighings by X_1, X_2, X_3, X_4 (X_4 is the combined weight for all three objects). Find the LUV estimators of the true weights. [Partial answer: $\hat{\beta}_1 = (3X_1 - X_2 - X_3 + X_4)/4$.]

(b) Consider an alternate scheme for weighing. The objects are weighed together two at a time to give three weighings, and all three are weighed together to give a fourth weighing. Denote the readings as follows: *a* and *b*, X_1; *a* and *c*, X_2; *b* and *c*, X_3; *a, b,* and *c*, X_4. Determine LUV estimators for this procedure. [Partial answer: $\hat{\beta}_2 = (3X_1 - 4X_2 + 3X_3 + X_4)/7$.]

(c) From the standpoint of variance, which of the two weighing schemes is better?

2 A surveyor stationed at a point *a* observes positions *b, c,* and *d.* He measures various angles between lines as follows:

Angle	*Measurements*
between $\overline{ab}$ and $\overline{ac}$	X_1, X_2
between $\overline{ac}$ and $\overline{ad}$	X_3, X_4
between $\overline{ab}$ and $\overline{ad}$	X_5

Suppose that $E(X_1, X_2, X_3, X_4, X_5) = (\beta_1, \beta_1, \beta_2, \beta_2, \beta_1 + \beta_2)$, and that the X_i's are uncorrelated and of equal variance. (Compare Example 7, Section 17.1, and Figure 17.2.)

(a) Find the LUV estimators of β_1 and β_2. [Answers: $\hat{\beta}_1 = (3X_1 + 3X_2 - X_3 - X_4 + 2X_5)/8$, $\hat{\beta}_2 = (-X_1 - X_2 + 3X_3 + 3X_4 + 2X_5)/8$.]

(b) Suppose we know that $\beta_1 = \beta_2 = \beta$. Find the LUV estimator of β. [Answer: $(\hat{\beta}_1 + \hat{\beta}_2)/2$.]

3 A paper manufacturer produces two thicknesses of paper. Denote the "true" weights (in ounces) per sheet of each of these types by β_1 and β_2. To estimate the values of β_1 and β_2, one must weigh stacks of paper consisting of varying quantities of *both* kinds of paper. The number of sheets of each kind in any stack are known, but the stacks cannot be broken up or in any way tampered with. Suppose one examines four stacks of paper with compositions and weights as follows:

Number of sheets of first type	*Number of sheets of second type*	*Weight in ounces*
t_1	u_1	X_1
t_1	u_2	X_2
t_2	u_1	X_3
t_2	u_2	X_4

We suppose that the weighing device is such that the X_i's are uncorrelated and have equal variances.

(a) Formulate an appropriate linear model.

(b) Show that one can find linear, unbiased estimators of β_1 and β_2 if, and only if, $t_1 \neq t_2$ *or* $u_1 \neq u_2$. Otherwise, which linear functions of β_1 and β_2 are estimable?

(c) Suppose that $t_1 = 100$, $t_2 = 200$, $u_1 = 200$, $u_2 = 100$. Determine the LUV estimators of β_1 and β_2. [Partial answer: $\hat{\beta}_1 = (-8X_1 + X_2 + 2X_3 + 11X_4)/1900$.]

4 (a) Verify the assertions in Example 1, Section 17.2, that the individual β_i's are not estimable. [Hint: $(1, 0, 0, 0) = c_1\alpha_1 + c_2\alpha_2 + c_3\alpha_3 + c_4\alpha_4$ implies that

$$c_1 + c_3 = 1, \quad c_2 + c_4 = 0, \quad c_1 + c_2 = 0, \quad c_3 + c_4 = 0$$

Sum the last three equations to conclude that $c_1 + c_3 = 0$, which is a contradiction, proving that β_1 is not estimable. The other cases are similar.]

(b) It was shown in Example 1, Section 17.2, that $\beta_1 - \beta_2$ is estimable. Show that $\beta_3 - \beta_4$ is also estimable.

5 Suppose that t_1, t_2, t_3 are distinct numbers. Consider the determinant

$$p(t) = \begin{vmatrix} 1 & t & t^2 \\ 1 & t_2 & t_2^2 \\ 1 & t_3 & t_3^2 \end{vmatrix} = a_0 + a_1 t + a_2 t^2$$

(a) Show that $p(t_3) = p(t_2) = 0$, and, hence,

$$p(t_1) = \begin{vmatrix} 1 & t_1 & t_1^2 \\ 1 & t_2 & t_2^2 \\ 1 & t_3 & t_3^2 \end{vmatrix} \neq 0$$

[Hint: If two rows of a determinant are equal, its value is 0. Also, $p(t)$ can have at most two distinct roots.]

(b) We ask the reader to accept the fact that three vectors (a_1, a_2, a_3), (b_1, b_2, b_3), (c_1, c_2, c_3) are linearly independent if

$$\begin{vmatrix} a_1 & a_2 & a_3 \\ b_1 & b_2 & b_3 \\ c_1 & c_2 & c_3 \end{vmatrix} \neq 0$$

Use this and (a) to prove the assertion in Example 5, Section 17.2. (See also Example 5, Section 17.1.) Namely, in the model

$$E(X_i) = \beta_1 + \beta_2 t_i + \beta_3 t_i^2, \qquad i = 1, 2, \ldots, n$$

every linear function of β_1, β_2, and β_3 can be estimated if there are at least three distinct values among $t_1, \ldots, t_n$.

6 In Example 5, Section 17.3, consider the special case that $k = m = 2$. Make the change of notation,

$$X_1 = X_{11}, \quad X_2 = X_{21}, \quad X_3 = X_{12}, \quad X_4 = X_{22}$$

$$\beta_1 = \delta_1, \quad \beta_2 = \delta_2, \quad \beta_3 = \gamma_1, \quad \beta_4 = \gamma_2$$

Observe that the linear model is equivalent to the one in Example 1, Section 17.2; hence, according to Exercise 4, the individual δ_i's or γ_i's are not estimable.

7 Show that $\Sigma_{i=1}^{n} (X_i - c)^2 = f(c)$ is minimized when $c = \overline{X} = (X_1 + \cdots + X_n)/n$. Hence, for the linear model in Example 1, Section 17.1, and Example 1, Section 17.3, the least squares estimator is the LUV estimator. This fact illustrates the general theorem about least squares estimators in Section 17.3.

8 Consider the linear model,

$$(EX_1, EX_2, EX_3) = \beta_1(1, 1, 1) + \beta_2(-1, 2, -1)$$

(a) Determine the analysis of variance test of the hypothesis that $\beta_1 = \beta_2$. [Answer: The test statistic is $(X_1 + X_3)^2/(X_1 - X_3)^2$.]

(b) Show that the following is a $1 - \alpha$ confidence interval for $\beta_1 - \beta_2$:

$$\frac{X_1 + X_3}{2} - \frac{c}{2}|X_1 - X_3| \leq \beta_1 - \beta_2 \leq \frac{X_1 + X_3}{2} + \frac{c}{2}|X_1 - X_3|$$

where $(-c, c)$ includes $1 - \alpha$ probability of the t distribution with 1 degree of freedom.

9 Consider Exercises 1, parts (a) and (b). Determine the analysis of variance tests for the hypothesis that $\beta_1 = \beta_2 = \beta_3$. The answer for (a) is based on the following quantities:

Sum of squares	*Degrees of freedom*
Due to hypothesis:	
$\|T\|^2 = X_1^2 + X_2^2 + X_3^2 + X_4^2 - \dfrac{(X_1 + X_2 + X_3 - X_4)^2}{4} - \dfrac{(X_1 + X_2 + X_3 + 3X_4)^2}{12} = \dfrac{(X_1 - X_2)^2}{2} + \dfrac{(X_1 + X_2 - 2X_3)^2}{6}$	2
due to error:	
$\|Z\|^2 = \dfrac{(X_1 + X_2 + X_3 - X_4)^2}{4}$	1

The answer for (b) is based on the following quantities:

Sum of squares	*Degrees of freedom*
due to hypothesis:	
$\|T\|^2 = X_1^2 + X_2^2 + X_3^2 + X_4^2 - \dfrac{(X_1 + X_2 + X_3 - 2X_4)^2}{7} - \dfrac{(2X_1 + 2X_2 + 2X_3 + 3X_4)^2}{21} = \dfrac{(X_1 - X_2)^2}{2} + \dfrac{(X_1 + X_2 - 2X_3)^2}{6}$	2
due to error:	
$\|Z\|^2 = \dfrac{(X_1 + X_2 + X_3 - 2X_4)^2}{7}$	1

10 Consider the two-way layout linear model described in Example 5, Section 17.3, and Example 3, Section 17.5. Use the theory in Section 17.8 to develop simultaneous confidence intervals for all "contrasts"

$$b_1\gamma_1 + \cdots b_k\gamma_k, \quad \text{subject to } b_1 + \cdots + b_k = 0$$

(Use Example 4, Section 17.8, as a guide.)

NOTES

1 For additional reading on linear models and their analyses, see Chapter 13 of [51].

2 Example 7, Section 17.1, is suggestive of some of the earliest versions of linear models. The method of least squares, described in Section 17.3, was already used by Gauss to reconcile discordant measurements in astronomy and surveying.

3 The two-way layout model, described in Example 5, Section 17.3, probably first arose in agricultural experimentation. With various coworkers, R. A. Fisher gave great stimulus to the use and development of such models in this field. See [71] and [72].

4 The fact that the least squares estimator coincides with the LUV estimator is known as the Gauss–Markov theorem. See Theorems 13.2 and 13.3 of [51].

5 Reference [73] is a collection of a large number of linear models with explicit calculations of the analysis of variance test ratios for various hypotheses of interest.

6 Tables of the F distribution and the Student-t distribution can be found in most experimental statistics books. For instance, see [74]. For a guide to such tables, see [19].

7 A large collection of graphs of the power function of the analysis of variance test appears in Appendix 13, [74]. In using these charts, you will have to compute ϕ^2, which in our notation is given by $\phi^2 = \Delta^2/\sigma^2(r - q + 1)$. For additional charts and useful references, see [75]. See also [19].

Chapter Eighteen

NONPARAMETRIC STATISTICS

Nonparametric methods allow for estimation or hypothesis testing procedures that are valid under fairly broad assumptions. These are appropriate when the experimenter is unable to assume that the populations he observes are described by closely specified families of distributions, such as normal, exponential, and so forth. This chapter develops some of the main examples of such nonparametric procedures.

18.1 NONPARAMETRIC PROCEDURES

The statistical procedures in Chapters 12 through 17 depend to a large extent on being able to specify the form of the distributions of the underlying random variables observed by the experimenter. For instance, the analysis of variance procedures in Chapter 17 assumes that the random variables are normally distributed, if one is to be able to compute significance levels or power functions. Of course, one could try to develop methods to allow the computation of significance levels or power functions when the random variables have some other class of distributions, such as exponential, for instance. The fact is, however, that often the experimenter is not in a position to know very much about the functional form of the distribution of his random variables. There has developed over the last several decades a class of statistical procedures that depend to a lesser extent on a close specification of the functional form of the distributions of the observed random variables. These procedures are called *nonparametric*. Typically, for testing hypotheses with nonparametric decision procedures, one has to know relatively little about the form of the distributions in order to determine significance levels. However, to evaluate a power function, one usually has to specify the form of the distributions quite concretely. In general, both this specification as well as the power computation may be quite difficult; therefore, the term "nonparametric" may be somewhat of a misnomer in this respect.

Nonparametric procedures are widely used, are usually simple computationally, and tend to be based on intuitive criteria rather than on rigorous principles. In what follows, we do not mean to give an exhaustive review of the subject. Rather, we look at a few of the dominant examples and we hope that the interested reader will pursue the subject with other reading (Note 1).

18.2 INTERVAL OF MEDIANS

In common parlance, a median of a population of numbers is a number such that half the population is above it and half is below it. For instance, the median of the set of numbers 1, 3, 4, 5, 7 is 4. However, if the set of numbers is 1, 3, 4, 5, 7, 8, the situation is less clear. The number 4.1 has the property that half the numbers of the set are smaller than it and half are greater. But so also do 4.11, 4.5, and 4.99; any number greater than 4 and less than 5 has this property. Accordingly, we shall refer to the interval (4, 5) as the interval of medians. Our interest will be restricted to populations described by integrating density functions.

Suppose that a random variable X has an integrating density f. Then the cdf F given by

$$F(t) = \int_{-\infty}^{t} f(x)\,dx = P(X < t)$$

rises continuously from height 0 to height 1. There is then at least one value t where $F(t) = 1/2$. There may be more than one however, but there is a largest such t and a smallest one.

intervals of medians

If a and b are the smallest and largest values of t such that $F(t) = 1/2$, then $[a, b]$ is called the interval of medians for X. If there is a unique $t = a$ such that $F(t) = 1/2$, then a is called *the* median of X.

Example 1 (a) Suppose that X is uniformly distributed over $[0, 1]$. Then the median of X is $1/2$.

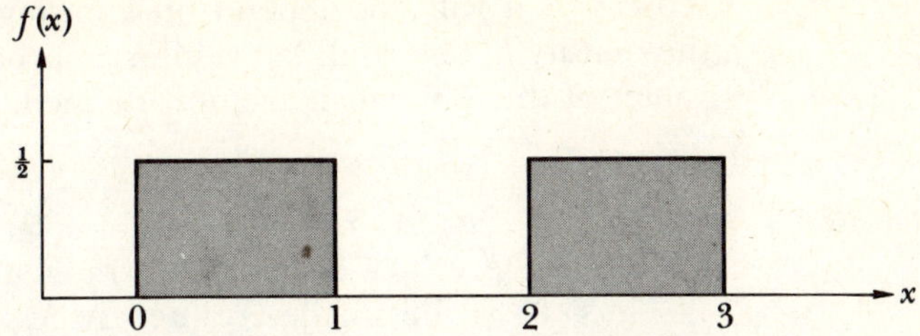

FIGURE 18.1 Any number in [1, 2] is a median.

(b) Suppose that

$$f(x) = \begin{cases} 2x & \text{if } 0 \leq x \leq 1 \\ 0 & \text{otherwise} \end{cases}$$

Then

$$F(t) = \begin{cases} 0 & \text{if } t \leq 0 \\ t^2 & \text{if } 0 \leq t \leq 1 \\ 1 & \text{if } t \geq 1 \end{cases}$$

The median is uniquely determined and equals $1/\sqrt{2}$.

(c) Suppose that

$$f(x) = \begin{cases} \frac{1}{2} & \text{if } 0 \leq x \leq 1 \\ \frac{1}{2} & \text{if } 2 \leq x \leq 3 \\ 0 & \text{otherwise} \end{cases}$$

Then the interval of medians is [1, 2] (see Figure 18.1).

One could argue that in most "practical" situations there is a unique median, and the extension to an interval of medians is an unnecessary generalization. The statistical procedures described in the next several sections usually assume that there is a unique median. However, these procedures go through without change for an interval of medians. Thus, we deal with the somewhat more general situation.

18.3 CONFIDENCE INTERVAL ESTIMATES FOR THE INTERVAL OF MEDIANS

Suppose that $X_1, \ldots, X_n$ are n mutually independent random variables, each having the same integrating density $f(x)$. We shall describe a method for making confidence interval estimates for the interval of medians. It will be a nonparametric method, in the sense that it will not depend on knowing the particular form of the density f. The confidence interval procedure uses the ordered values of the X_i's, which are now defined.

order statistics

If $X_1, \ldots, X_n$ are random variables, let $U_1 \leq U_2 \leq \cdots \leq U_n$ denote the values of the X_i's arranged in increasing order. Thus, U_1 is the smallest value of the X_i's, U_2 is the second smallest value, . . . , U_n is the largest value. The values $U_1, \ldots, U_n$ are called order statistics.

[For example, if $(X_1, X_2, X_3, X_4) = (2.1, -1, 3.5, 2.1)$, then $(U_1, U_2, U_3, U_4) = (-1, 2.1, 2.1, 3.5)$.]

It is an interesting fact that for $i < j$, the random interval $[U_i, U_j]$ will contain the interval of medians $[a, b]$ with a probability that depends only on n, i, and j and not on f. This depends on the assumption that the X_i's have an integrating density function and are not discrete. Let us illustrate with the extreme interval $[U_1, U_n]$. The calculation is as follows: $P[(U_1 \leq a) \cap (U_n \geq b)]$ is the probability that $[a, b]$ is contained in the random interval. The complement of the event $(U_1 \leq a) \cap (U_n \geq b)$ is $(U_1 > a) \cup (U_n < b)$, and $P[(U_1 > a) \cap (U_n < b)] = 0$, since $P(X_i \text{ in } [a, b]) = 1/2 - 1/2 = 0$. Hence,

$$P[(U_1 \leq a) \cap (b \leq U_n)] = 1 - P(U_1 > a) - P(U_n < b)$$

$$= 1 - (\tfrac{1}{2})^n - (\tfrac{1}{2})^n = 1 - (\tfrac{1}{2})^{n-1}$$

The last step follows from the facts that

$$P(U_1 > a) = P[(X_1 > a) \cap \cdots \cap (X_n > a)]$$

$$= P(X_1 > a) \cdots P(X_n > a)$$

$$P(U_n < b) = P[(X_1 < b) \cap \cdots \cap (X_n < b)]$$

$$= P(X_1 < b) \cdots P(X_n < b)$$

and $P(X_i > a) = 1/2 = P(X_i < b)$. Thus, the experimenter can assert that the random interval $[U_1, U_n]$ covers all medians, and he will be telling the truth with probability $1 - (1/2)^{n-1}$.

We now describe a more general situation in which a certain type of symmetric intervals is used. It is an interval of the form $[U_1, U_n]$, $[U_2, U_{n-1}]$, and so forth.

confidence intervals for medians

The probability that the random interval $[U_k, U_{n-k+1}]$ covers the interval of medians $[a, b]$ is

$$1 - 2 \sum_{i=n-k+1}^{n} \binom{n}{i} (\tfrac{1}{2})^n \qquad (1)$$

(We assume that $k < n - k + 1$ or, equivalently, $k < (n+1)/2$.)

Proof. As in the previous proof for $[U_1, U_n]$, we have that

$$P([U_k, U_{n-k+1}] \text{ covers } [a, b]) = 1 - P(U_k > a) - P(U_{n-k+1} < b)$$

Now use the facts that

$$\begin{aligned}(U_k > a) &= (\text{fewer than } k\ X_i\text{'s are less than } a)\\ &= (\text{more than } n - k\ X_i\text{'s are less than } a)\end{aligned}$$

$$\begin{aligned}(U_{n-k+1} < b) &= (\text{more than } k\ X_i\text{'s are greater than } b)\\ &= (\text{fewer than } n - k\ X_i\text{'s are greater than } b)\end{aligned}$$

The desired result follows from the fact that the number of X_i's less than a is binomially distributed with parameters n, 1/2, as is the number of X_i's greater than b. ∎

Example 1 The probability that $[U_1, U_n]$ covers the medians is $1 - (1/2)^{n-1}$. The probability that $[U_2, U_{n-1}]$ covers the medians is $1 - (n+1)/2^{n-1}$. The probability that $[U_3, U_{n-2}]$ covers the medians is $1 - (n^2 + n + 2)/2^n$. The larger is k, the narrower is the interval $[U_k, U_{n-k+1}]$. On the other hand, the larger is k, the smaller is the probability of covering the medians. The experimenter must strike a balance between his desire to have a narrow interval and his desire to be correct with high probability. It should be pointed out that although the probability that $[U_k, U_{n-k+1}]$ covers the medians does not depend on f, the expected length of the interval—namely, $E(U_k) - E(U_{n-k+1})$—definitely does depend on f (see Exercise 4).

Example 2 Ten persons are selected at random from the population of white, urban high school seniors and given a mathematics ability test. Suppose that the ten test scores arranged in order are 27.36, 31.12, 33.14, 37.91, 39.32, 41.62, 43.76, 45.90, 46.83, and 47.13. Assume that the population of all possible scores is reasonably described by an integrating density function. The confidence interval [33.14, 45.90] = $[U_3, U_8]$ will contain the medians of the population with probability $1 - 112/2^9 = 0.781$. Empirically, according to the frequency interpretation of probability, this fact means that over a large number of repetitions of testing 10 persons from the same population, the interval $[U_3, U_8]$ will actually contain the medians about 78.1% of the time.

18.4 THE SIGN TEST

Suppose that an experimenter wants to decide whether a number m is a median of a given population. We suppose as before that the mathematical model describes the population in terms of an integrating density f, and that the observations are $X_1, \ldots, X_n$, n mutally independent random variables, each with density f. Let

Y = number of X_i's that exceed m

The sign test prescribes that the decision that m is a median be based on the value of Y alone. When m is a median, then Y is binomially distributed with parameters n and $p = 1/2$. If the observed value of Y "does not seem consistent" with this distribution, then the experimenter should decide that m is not a median. To appraise the reasonableness of a decision procedure, we shall consider its power function (see Section 13.2). The computations turn out to be extremely easy because Y is binomially distributed whether m is a true median or not; only the parameter p is affected.

Y is binomially distributed

Let $\theta = P(X_i > m)$. Then Y is binomially distributed with parameters n and $p = \theta$. When m is a median, then $\theta = 1/2$.

The power of a decision procedure based on Y depends on the true state of nature only through the value of θ. We shall describe two procedures for deciding about θ. Let Ω denote the set of all possible distributions of $X_1, \ldots, X_n$. (Remember that the X_i's are independent and identically distributed with the same integrating density f.) Let Ω_1 be all those distributions for which $\theta = \int_m^\infty f(x)\,dx \leq 1/2$, and let Ω_2 be the complementary class in which $\theta > 1/2$. Thus, in Ω_2 the median is m or greater, and in Ω_1, the median is smaller than m. Suppose that the experimenter wishes to decide whether the true distribution is in Ω_1 (decision 1) or in Ω_2 (decision 2). The following decision procedure is intuitively plausible.

one-sided sign test

If $Y < c$, make decision 1. If $Y \geq c$, make decision 2.

The cutoff constant c may be chosen so that the power function has a prescribed value when the true median is m. By Eq. (1), Section 13.2, the power function is given by

$$\psi(\theta) = P_\theta\,(\text{making decision 1}) = P_\theta(Y < c) \tag{1}$$

where P_θ denotes the probability evaluation when θ is the true value of $P(X_i > m)$. Since the distribution of Y is completely determined by θ, the notation P_θ in (1) makes sense. The explicit expression for the power function is

$$\psi(\theta) = \sum_{i=0}^{c-1} \binom{n}{i} \theta^i (1-\theta)^{n-i} = 1 - \sum_{i=c}^{n} \binom{n}{i} \theta^i (1-\theta)^{n-i} \tag{2}$$

A power function of this form has already arisen previously in Section 15.8. We refer to that section for its properties. Figure 18.2, which relates to Example 1, shows the typical appearance of the power function for $c \geqslant 1$. Also, $\psi(1/2) = \Sigma_{i=0}^{c-1} \binom{n}{i} (1/2)^n$ is the probability of deciding that the median is no greater than m when it really is equal to m, and is called the *significance level* of the procedure (see Section 13.7). The one-sided test is suitable when the experimenter does not want to make decision 1 when $\theta > 1/2$ (i.e., when the true median is less than m) and is anxious to make decision 1 when $\theta < 1/2$ (i.e., when the true median is greater than or equal to m).

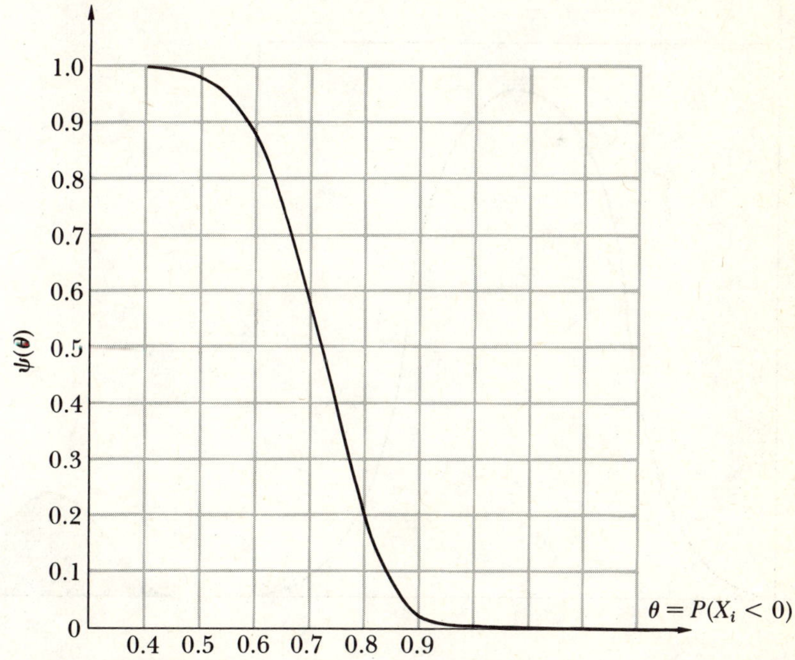

FIGURE 18.2 Probability of deciding drug is not useful is $\psi(\theta)$. Probability of a negative blood pressure reading is θ.

Example 1 A medical researcher gives a drug that is supposed to reduce high blood pressure to 20 subjects who suffer from the disease. The change in reading after use of the drug for the ith subject is X_i. ($X_i < 0$ means the pressure went down; $X_i \geqslant 0$ means it did not.) The experimenter wants to decide whether the median of the distribution of X_i's is nonnegative (decision 2) or negative (decision 1). Decision 1 suggests that the drug is useful, and decision 2 suggests that the drug is not useful. Using the sign test, he would simply count Y, the number of nonnegative X_i's—that is, the number of subjects who show no improvement. If he chooses the cutoff constant c to be 15, then the significance level of the test is, by (2), equal to $\left[\binom{20}{0} + \cdots + \binom{20}{14}\right] \Big/ 2^{20} = 0.98$. Thus, when the true median is nonnegative, the probability of making decision 1 is 0.98 or greater. (See Figure 18.2 for the power function.) By using the power function, the experimenter must appraise whether $1 - \psi(\theta)$, the probability of making decision 2, is sufficiently close to 1 for values of θ greater than 1/2. If not, he may need more than 20 observations.

It may be that a more suitable way of breaking up Ω is as follows.

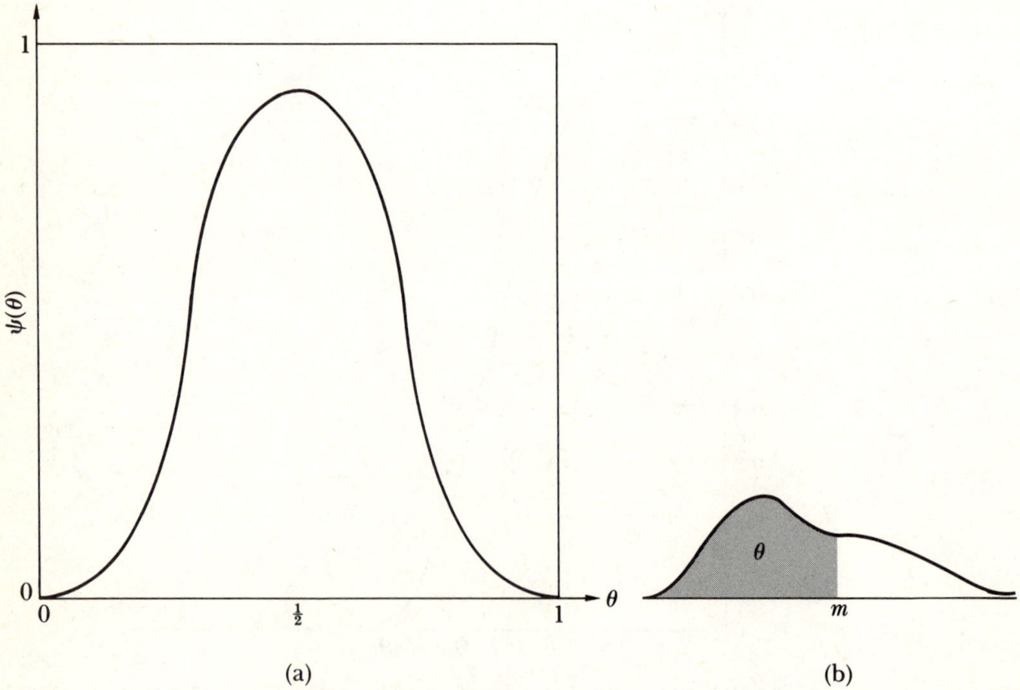

FIGURE 18.3 Power function of the two-sided sign test.

First, Ω_1 equals those distributions for which $\theta = P(X_i > m) = 1/2$ (decision 1), and, second, Ω_2 equals those distributions for which $\theta \neq 1/2$ (decision 2). An intuitively plausible decision procedure for this situation follows.

two-sided sign test

> If $|Y - n/2| \leq c$, make decision 1; otherwise, make decision 2.

If the true median is m, then Y is binomially distributed with parameters n and $p = 1/2$ and is symmetrically distributed about the value $n/2$. The two-sided test leads to decision 1 if Y is "close" to the expected value $EY = n/2$, which prevails when the distribution is in Ω_1. A typical power function for this test is shown in Figure 18.3. It is left to the reader to verify that the power function has the indicated appearance.

18.5 COMPARING TWO POPULATIONS

Suppose that an experimenter is confronting two populations and he wants to determine whether the form of both populations is the same. We assume that the mathematical model appropriate to this situation is the following. An observation on the first population is described by a random variable X with an integrating density function f, and an observation on the second population is described by a random variable Y with an integrating density function g. Suppose that $X_1, \ldots, X_m, Y_1, \ldots, Y_n$ are mutually independent random variables representing m observations on the first population and n observations on the second.

Suppose first, for simplicity, that $m = n$. (The case $m \neq n$ will be covered in the next section.) One way of deciding whether both populations have the same form is to use the sign test discussed in Section 18.4. Consider the n differences $X_1 - Y_1, \ldots, X_n - Y_n$. When the density functions f and g are the same, then $X_i - Y_i$ has a median equal to 0. In fact, 0 is a unique median (see Exercise 3). Let Z be the number of positive differences among the n independent random variables $X_1 - Y_1, \ldots, X_n - Y_n$. When $f = g$, then Z is binomially distributed with parameters n and $p = 1/2$. An appropriate test follows.

sign test for differences

> If $|Z - n/2| \leq c$, decide that both populations are the same (decision 1); otherwise, decide that they are not the same (decision 2).

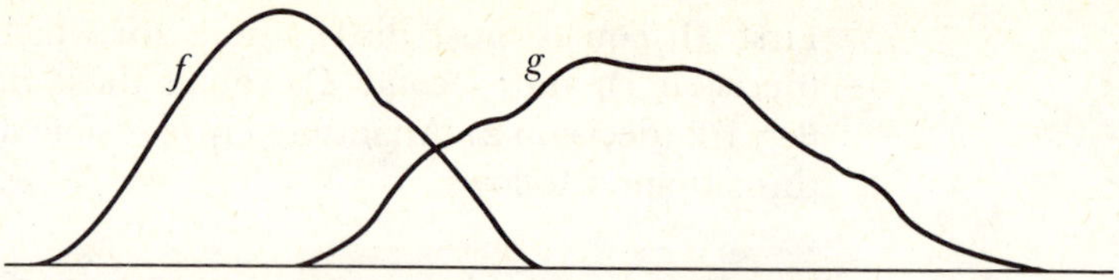

FIGURE 18.4 That $P(X < Y) > 1/2$ suggests that g is shifted to the right of f. The density of X is f and the density of Y is g.

The cutoff constant c is chosen so that the probability of making decision 1 has a specified value when, in fact, both populations are the same. (Compare the discussion of the two-sided test in Section 18.4.) The power of this decision procedure depends on the parameter $\theta = P(X > Y)$. Explicitly,

$$P_\theta \text{ (making decision 1)} = \psi(\theta) = \sum_{(n/2)-c \leq i \leq (n/2)+c} \binom{n}{i} \theta^i (1-\theta)^{n-i}$$

The shape of this power function appears in Figure 18.3. It should be pointed out that this sign test actually decides whether $P(X > Y) = 1/2$. It is true that when $f = g$, $P(X > Y) = P(X < Y) = 1/2$. However it is not true that $P(X > Y) = 1/2$ implies necessarily that X and Y have the same density functions (see Exercise 3b). Strictly speaking, this decision procedure should be understood as one that discriminates between the state of nature that $P(X > Y) = 1/2$ (decision 1 should be made in this case) and the state of nature that $P(X > Y) \neq 1/2$ (decision 2). Like the sign test discussed in Section 18.4, this test has the great advantage of simplicity. The procedure does not depend on knowing the explicit forms of the densities f and g. It should also be pointed out that the appropriateness of the procedure does not depend on a full requirement of independence of the random variables $X_1, \ldots, X_n, Y_1, \ldots, Y_n$. All that is necessary is that the n differences $X_1 - Y_1, \ldots, X_n - Y_n$ be independent, which is a simpler assumption and more relevant to practical situations. Example 1 is relevant to this point.

In addition to a two-sided test, it may be useful to consider a one-sided test to decide whether $P(X < Y) \leq 1/2$ or $P(X < Y) > 1/2$. The situation that $P(X < Y) > 1/2$ suggests that the second population is "shifted to the right" of the first population (see Figure 18.4). The appropriate test in this case is the following.

one-sided test

If $Z < c$, decide that $P(X < Y) > \frac{1}{2}$ (decision 2).

If $Z \geq c$, decide that $P(X < Y) \leq \frac{1}{2}$ (decision 1).

Example 1 Reconsider Example 1 in Section 18.4. Suppose that Y_i is the blood pressure reading for the ith subject before the drug and X_i is the reading after the drug. Then $X_i - Y_i$ is the change in blood pressure (previously described as X_i). It is not likely that X_i, Y_i are independent, since they relate to the same subject. However, it is reasonable to assume that the n differences $X_1 - Y_1, \ldots, X_n - Y_n$ are independent. If the experimenter's concern is whether the drug makes for a reduction in blood pressure, he should use the one-sided test. In effect, this test is the same as that described previously in Example 1, Section 18.4.

18.6 THE WILCOXON TEST

The procedure described in Section 18.5 has several shortcomings. One is the assumption that $m = n$—namely, that the number of observations from each population is the same. Another is that the observations must be paired off as $(X_1, Y_1), (X_2, Y_2), \ldots, (X_n, Y_n)$. It may be that there is no one natural or preferred way of pairing off the data. If $i_1, \ldots, i_n$ is any permutation of $1, 2, \ldots, n$, one could just as well use the pairs $(X_1, Y_{i_1}), (X_2, Y_{i_2}), \ldots, (X_n, Y_{i_n})$ and there would be no change in the described procedure. Moreover, one gets the feeling that by looking at only some of the pairs, he is throwing away some useful information; why not compare all possible pairs? The sign test is then not appropriate, but the procedure we shall describe now is. It looks at all mn pairs (X_i, Y_j) and counts the number of times that X_i exceeds Y_j. We shall call this count W (Note 2).

the statistic W

Let

$$c(x, y) = \begin{cases} 1 & \text{if } x \geqslant y \\ 0 & \text{if } x < y \end{cases}$$

Define

$$W = \sum_{i=1}^{m} \sum_{j=1}^{n} c(X_i, Y_j)$$

Statistic W is the counterpart of the random variable Z discussed in Section 18.5; W is based on all mn possible pairs (X_i, Y_j), whereas Z is based only on some pairs. There is another way of looking at the count W in terms of *ranks*. If X_i is the smallest of all the $m + n$ observations, we say that $R(X_i)$, the rank of X_i, equals 1. [Of course, $R(Y_i) = 1$ means that Y_i is the smallest of all $m + n$ observations.]

That $R(X_i) = 2$ means X_i is the second smallest among all $m + n$ observations, and so forth.

Example 1 Suppose that $m = 4$, $n = 6$, and the values of the observations are the following:

$$(X_1, X_2, X_3, X_4, Y_1, Y_2, Y_3, Y_4, Y_5, Y_6)$$
$$= (-1.2, 2.4, 1.7, 3.4, -3.2, 1.4, 2.5, -1.0, 2.3, 4.1)$$

Then the ranks are the following:

$$R(Y_1) = 1, \quad R(X_1) = 2, \quad R(Y_4) = 3, \quad R(Y_2) = 4, \quad R(X_3) = 5,$$
$$R(Y_5) = 6, \quad R(X_2) = 7, \quad R(Y_3) = 8, \quad R(X_4) = 9, \quad R(Y_6) = 10$$

Now, instead of considering W, we can just as well consider $R(X_1) + \cdots + R(X_m)$, the sum of the ranks of the X_i's, since the two quantities are linearly related as follows:

$$R(X_1) + \cdots + R(X_m) = W + \frac{m(m+1)}{2} \tag{1}$$

Proof. Using the previously defined notation $c(x, y)$, we have that

$$R(X_i) = \sum_{j=1}^{n} c(X_i, Y_j) + \sum_{j \neq i} c(X_i, X_j) + 1 \tag{2}$$

since this counts the number of elements of the $m + n$ observations that are not greater than X_i. Summing (2) over $i = 1, \ldots, m$ gives a value of $W + m(m + 1)/2$. The reason is that the sum over i of the first term on the right of (2) equals W, and $c(X_i, X_j) + c(X_j, X_i) = 1$ for every i and j. ∎

The relation (1) produces some other relationships as well, because of the fact that

$$R(X_1) + \cdots + R(X_m) + R(Y_1) + \cdots + R(Y_n)$$
$$= 1 + 2 + \cdots + (m + n) = \frac{(m+n)(m+n-1)}{2}$$

Among these is the following:

$$\left[\frac{R(X_1) + \cdots + R(X_m)}{m}\right] - \left[\frac{R(Y_1) + \cdots + R(Y_n)}{n}\right] = aW + b \tag{3}$$

where a and b are appropriate constants depending on m and n (see Exercise 5). Notice that (3) is like a difference between the average of the X_i's and the average of the Y_i's, except that the actual values of the observations have been replaced by their ranks.

Using W [or any linear function of W, such as (3)], we can now prescribe tests similar to the ones given in Section 18.5.

one-sided Wilcoxon test (Note 3)

> If $W < c$, decide that $P(X_i < Y_j) > \frac{1}{2}$ (decision 2).
>
> If $W \geq c$, decide that $P(X_i < Y_j) \leq \frac{1}{2}$ (decision 1).

As usual, the cutoff constant c is to be determined so that the power function has a prescribed value when the densities of X_i and Y_j are the same.

It should be emphasized that when the densities of the X_i's and the Y_i's are the same, then the distribution of W does not depend on the form of the density. The reason is that each of the orderings of $X_1, \ldots, X_m, Y_1, \ldots, Y_n$ has the same probability. These orderings completely determine the value of W. For instance, if $m = n = 2$, the six patterns of orderings of X's and Y's are the following:

$$X < X < Y < Y \quad X < Y < X < Y \quad X < Y < Y < X$$

$$Y < X < X < Y \quad Y < X < Y < X \quad Y < Y < X < X$$

The values of W for these orderings are the following:

$$0, \quad 1, \quad 2, \quad 2, \quad 3, \quad 4$$

Equalities between variables can be ignored because such patterns occur with probability 0. Hence, when the X_i's and Y_j's have the same distributions, then

$$P(W = 0) = \tfrac{1}{6}, \quad P(W = 1) = \tfrac{1}{6}, \quad P(W = 2) = \tfrac{2}{6}, \quad P(W = 3) = \tfrac{1}{6},$$

$$P(W = 4) = \tfrac{1}{6}$$

In addition to the one-sided test, there exists a two-sided test.

two-sided Wilcoxon test (Note 3)

> If $|W - mn/2| < c$, decide that $P(X_i < Y_j) = 1/2$ (decision 1). Otherwise, decide that $P(X_i < Y_j) \neq 1/2$ (decision 2).

It is a fact that when the distributions of the X_i's and the Y_i's are the same, then $EW = mn/2$. In fact, the distribution of W is symmetric about $mn/2$. We leave the verification for the reader. The two-sided test is appropriate if the experimenter is concerned with both possibilities that $P(X_i < Y_j) < 1/2$ and $P(X_i < Y_j) > 1/2$, as alternatives to the hypothesis that $P(X_i < Y_j) = 1/2$.

The computation of the power function of the analogous tests based on Z in Section 18.6 was very easy because of the fact that Z is binomially distributed. The situation is more complicated now because $W = \Sigma_{j=1}^{n} \Sigma_{i=1}^{m} c(X_i, Y_j)$ is a sum of random variables that are not mutually independent. For instance, $c(X_1, Y_1)$ and $c(X_1, Y_2)$ are not independent, and there are many such dependent terms in the sum for W. Fortunately, however, there is a central limit theorem for W that asserts the following:

For large m and n, $(W - EW)/[\mathrm{Var}(W)]^{1/2}$ is approximately $N(0, 1)$ distributed.

We shall not prove this assertion here (Note 4). To use the approximate normality of W in computing the power function of the decision procedure based on it, we must know the values of EW and $\mathrm{Var}(W)$. To evaluate these, we introduce the following notation:

$$P(X_i > Y_j) = \theta$$

$$P(X_i > Y_j > Y_k) = \rho \qquad (j \neq k)$$

$$P(X_i > X_j > Y_k) = \lambda \qquad (i \neq j)$$

EW *and Var*(W)

$$EW = mn\theta$$

$$\mathrm{Var}(W) = mn\theta(1-\theta) + mn(n-1)(2\rho - \theta^2) + m(m-1)n(2\lambda - \theta^2) \tag{4}$$

Proof. Since $W = \Sigma_{j=1}^{n} \Sigma_{i=1}^{m} c(X_i, Y_j)$, it follows that $EW = mnE[c(X_i, Y_j)] = mn\theta$. Similarly,

$$\mathrm{Var}(W) = \sum_{i,j,k,l} \mathrm{Cov}[c(X_i, Y_j), c(X_k, Y_l)]$$

If the indices i, j, k, and l are all distinct, then $c(X_i, Y_j)$ and $c(X_k, Y_l)$ are independent and have covariance 0. There are mn terms of the form

$$\mathrm{Cov}[c(X_i, Y_j), c(X_i, Y_j)] = \mathrm{Var}[c(X_i, Y_j)] = \theta(1-\theta)$$

There are $mn(n-1)$ terms of the form $\mathrm{Cov}[c(X_i, Y_j), c(X_i, Y_k)]$, $j \neq k$. Using the computational formula for covariance [(9), Section 10.1],

$$\begin{aligned}\text{Cov}[c(X_i, Y_j), c(X_i, Y_k)] &= E[c(X_i, Y_j)c(X_i, Y_k)] - \theta^2 \\ &= 2\rho - \theta^2\end{aligned}$$

the reason being

$$c(X_i, Y_j)c(X_i, Y_k) = \begin{cases} 1 & \text{if } X_i > Y_j > Y_k \\ & \text{or } X_i > Y_k > Y_j \\ 0 & \text{otherwise} \end{cases}$$

A similar computation shows that the $m(m-1)n$ terms of the form $\text{Cov}[c(X_i, Y_k), c(X_j, Y_k)]$, $i = j$, each equals $2\lambda - \theta^2$. ∎

When X_i and Y_j have the same integrating density functions,

$$\theta = \frac{1}{2}, \qquad \rho = \frac{1}{6}, \qquad \lambda = \frac{1}{6}$$

These values of ρ and λ follow from the fact that each of the six ordered arrangements $(X_i > Y_j > Y_k)$, $(Y_j > X_i > Y_k)$, . . . , has the same probability when the densities are the same. The union of these disjoint events must have probability 1, since events like $(Y_j = X_i > Y_k)$ have probability 0. Hence, when X and Y have the same densities, after a little algebra we have the following.

$$EW = \frac{mn}{2}, \qquad \text{Var}(W) = \frac{mn(m+n+1)}{12} \tag{5}$$

Notice that, in general, the asymptotic distribution of W depends not only on θ but on ρ and λ as well. Hence, the power function is, asymptotically, a function of the triple of parameters (θ, ρ, λ).

Example 2 Consider the data given in Example 1. The value of W is 13. When X_i and Y_j have the same density, then, using (5), $(W - 12)/(22)^{1/2}$ is approximately $N(0, 1)$ distributed. Suppose we want the power function to have a specified value—say, 0.98, to be specific—when the densities are the same. Since, by Table 9.2, Section 9.7,

$$\frac{1}{(2\pi)^{1/2}} \int_{-\infty}^{2.1} \left(\exp \frac{-x^2}{2}\right) dx \approx 0.98$$

the value of c is determined as follows:

$$P(W < c) = P\left(\frac{W-12}{(22)^{1/2}} < \frac{c-12}{(22)^{1/2}}\right) = P\left(\frac{W-12}{(22)^{1/2}} < 2.1\right)$$

Hence, $(c - 12)/(22)^{1/2} = 2.1$ implies that $c = 2.1(22)^{1/2} + 12 = 21.8$.

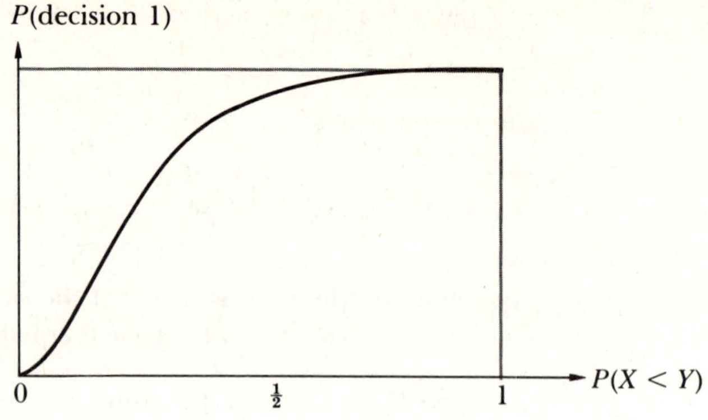

FIGURE 18.5 Power function of one-sided Wilcoxon test.

Since $W = 13$, we make decision 2; namely, we decide that $P(X_i < Y_j) > 1/2$ [or, equivalently, that $P(X_i > Y_j) < 1/2$]. Qualitatively, this fact means that we decide that the X distribution is "shifted to the left" of the Y distribution, as in Figure 18.4. The general appearance of the power function is shown in Figure 18.5. The three parameters (θ, ρ, λ) jointly are a measure of how unlike the distributions of X_i and Y_j are; however, in general, it may be quite difficult for the experimenter to appraise the meaning of this measure.

18.7 THE EMPIRICAL cdf

Suppose that $X_1, X_2, \ldots, X_n$ are independent and identically distributed random variables with a common integrating density function f. Let F denote the cdf of the X_i's. That is,

$$F(x) = \int_{-\infty}^{x} f(u)\, du, \qquad -\infty < x < \infty$$

(see Section 7.3). We want to define a function $F_n(x)$, the empirical cdf, which is determined by $X_1, \ldots, X_n$ that mimics the behavior of the cdf $F(x)$.

***empirical* cdf**

For any real x, let

$F_n(x) =$ proportion of the values of $X_1, \ldots, X_n$ that are $\leq x$

Then $F_n(x)$ is called the *empirical cdf* determined by $X_1, \ldots, X_n$.

Example 1 Suppose that the values of X_1, X_2, X_3, X_4 are 1.3, −7.1, −1.1, and −3.0, respectively. Then the empirical cdf is the following:

$$F_n(x) = \begin{cases} 0 & \text{if } x < -7.1 \\ \frac{1}{4} & \text{if } -7.1 \leq x < -3.0 \\ \frac{2}{4} & \text{if } -3.0 \leq x < -1.1 \\ \frac{3}{4} & \text{if } -1.1 \leq x < 1.3 \\ 1 & \text{if } 1.3 \leq x \end{cases}$$

(See Figure 18.6.)

The empirical cdf F_n is a *random step function.* Suppose, as in Section 18.3, the ordered values of the X_i's are denoted by $U_1 < U_2 < \cdots < U_n$. Then

$$F_n(x) = \begin{cases} 0 & \text{if } x < U_1 \\ \dfrac{1}{n} & \text{if } U_1 \leq x < U_2 \\ \vdots & \\ \dfrac{n-1}{n} & \text{if } U_{n-1} \leq x < U_n \\ 1 & \text{if } U_n \leq x \end{cases}$$

In other words, the empirical cdf jumps in steps of $1/n$ from height 0 to height 1. The successive jumps occur at the sequence of ordered values of the n random variables. If there happened to be several X_i's assuming the same value, then there would be a jump greater than $1/n$. However, since the X_i's have an integrating density, the probability of such ties is 0 and the situation can be ignored theoretically.

The following basic properties relate the empirical cdf F_n, to the "theoretical" cdf F.

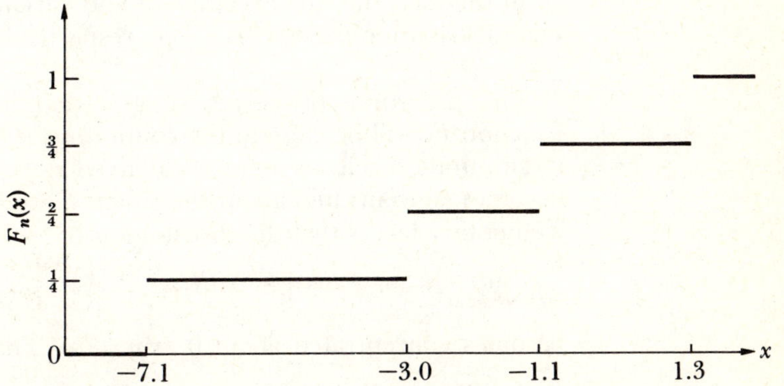

FIGURE 18.6 The empirical cdf for Example 1, Section 18.7.

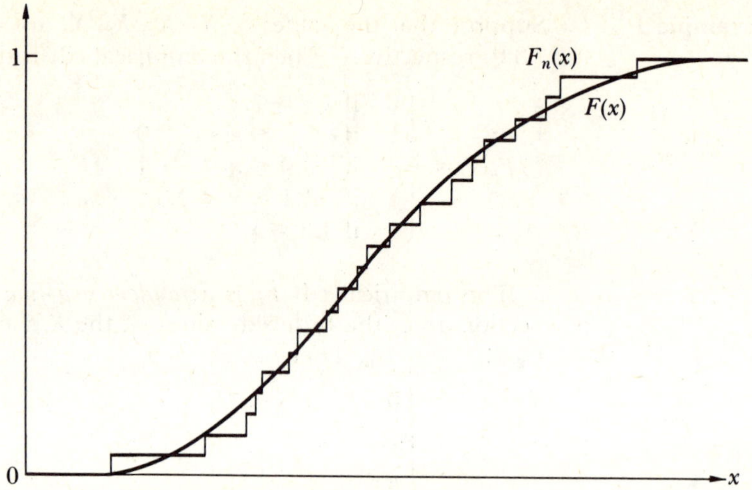

FIGURE 18.7 For large n, F_n is close to F.

***properties of* $\mathbf{F}_n$**

(a) For any fixed x, $nF_n(x)$ is binomially distributed with parameters n and $p = F(x)$.

(b) $E[F_n(x)] = F(x)$

(c) $\text{Var}[F_n(x)] = E[F_n(x) - F(x)]^2 = \dfrac{F(x)[1 - F(x)]}{n}$

Proof. Each X_i is either less than or equal to x (success) or not (failure). The probability of a success on any single trial is $P(X_i \leq x) = F(x)$. Hence, the total number of successes is binomially distributed with parameters n and $p = F(x)$. Parts (b) and (c) follow from the facts that the exceptation and variance for the n, p binomial distribution are np, $np(1 - p)$, respectively. ∎

The preceding properties suggest that for large n, the step function F_n will be close to the continuous function F. The law of large numbers tells us that for any fixed x, the distribution of $F_n(x)$ becomes concentrated about the single value $F(x)$ as $n \to \infty$. A less elementary fact is that the distribution of

$$\sup_{-\infty<x<\infty} |F_n(x) - F(x)| \equiv \Delta_n(X_1, \ldots, X_n) = \Delta_n$$

becomes concentrated about 0 as $n \to \infty$. That is, for any $\epsilon > 0$,

$$\lim_{n\to\infty} P(\Delta_n > \epsilon) = 0$$

but we shall not prove this fact here. The point is, however, that for large n, F_n tends to be very close to F (see Figure 18.7 and Note 5).

18.8 COMPARING EMPIRICAL cdf's

Suppose that $X_1, \ldots, X_m, Y_1, \ldots, Y_n$ are independent and identically distributed random variables, each with the same integrating density function, f. Let F_m be the empirical cdf of $X_1, \ldots, X_m$ and let G_n be the empirical cdf of $Y_1, \ldots, Y_n$. We shall be interested in studying some "measures of closeness," which are used to compare F_m and G_n.

D^+ and D (Note 6)

Define

$$D^+ = \sup_{-\infty<x<\infty} [F_m(x) - G_n(x)]$$

$$D = \sup_{-\infty<x<\infty} |F_m(x) - G_n(x)|$$

Example 1 Suppose the values of X_1, X_2, X_3, X_4 are $-0.7, -7.1, -1.1,$ and -3.0, respectively, and the values of Y_1, Y_2, Y_3, Y_4 are -1.0, 0.0, 1.4, and 2.0, respectively. The two empirical cdf's are shown in Figure 18.8. In this example, $D^+ = 3/4$ and $D = 3/4$.

It is an interesting fact that the distributions of the random variables D^+ and D do not depend on the form of the density f. In fact, the values of D^+ and D do not depend on the actual values of

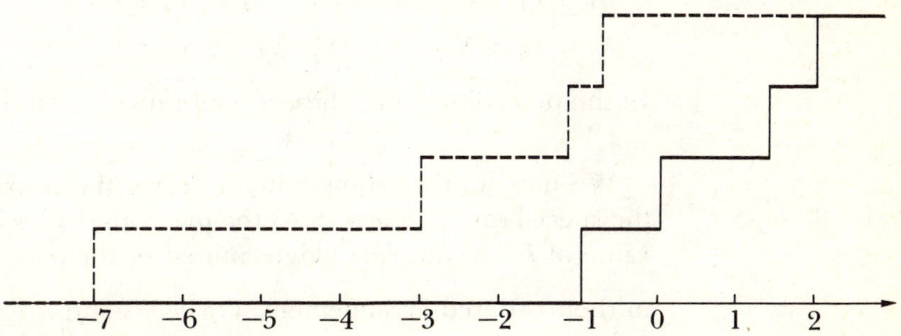

FIGURE 18.8 The dashed line denotes $F_4(x)$. The solid line denotes $G_4(x)$, $D = D^+ = 3/4$.

the X_i's and Y_i's, only on their relative orderings. Thus, for $m = n = 2$, if $X_1 < Y_2 < Y_1 < X_2$, we have that

$$F_m(x) - G_n(x) = \begin{cases} 0 & \text{if } x < X_1 \\ \frac{1}{2} - 0 = \frac{1}{2} & \text{if } X_1 \leq x < Y_2 \\ \frac{1}{2} - \frac{1}{2} = 0 & \text{if } Y_2 \leq x < Y_1 \\ \frac{1}{2} - 1 = -\frac{1}{2} & \text{if } Y_1 \leq x < X_2 \\ 1 - 1 = 0 & \text{if } X_2 \leq x \end{cases}$$

(Draw a picture!) Hence, for this ordering of X_i's and Y_i's, the values of D^+ and D are $D^+ = D = 1/2$. Notice that the actual values of the X_i's and Y_i's do not enter into the computation. The values of D^+ and D for the special case $m = n = 2$ are given in the following example.

Example 2 There are $\binom{4}{2} = 6$ patterns for arranging two X's and two Y's. The following is a list of these patterns and the associated values of D^+ and D.

Pattern	D^+	D
$X < X < Y < Y$	1	1
$X < Y < X < Y$	1/2	1/2
$X < Y < Y < X$	1/2	1/2
$Y < X < X < Y$	1/2	1/2
$Y < X < Y < X$	0	1/2
$Y < Y < X < X$	0	1

Notice that the values of D^+ and D are determined by the pattern of X's and Y's and not by their numerical values or by the particular indices of the X_i's and Y_i's within the pattern. Thus, the values of D^+ and D are the same for each of the following four arrangements:

$$X_1 < Y_1 < X_2 < Y_2, \quad X_2 < Y_1 < X_1 < Y_2,$$

$$X_1 < Y_2 < X_2 < Y_1, \quad X_2 < Y_2 < X_1 < Y_1$$

In the preceding table, these are all represented by $X < Y < X < Y$.

We now want to show how to derive the distribution of D^+ in the special case that $m = n$. As the previous discussion indicates, the value of D^+ is completely determined by the pattern of X's and Y's in their ordered arrangement. For n X's and n Y's, there are $\binom{2n}{n}$ distinct patterns. For instance, in Example 1, we displayed $\binom{4}{2} =$

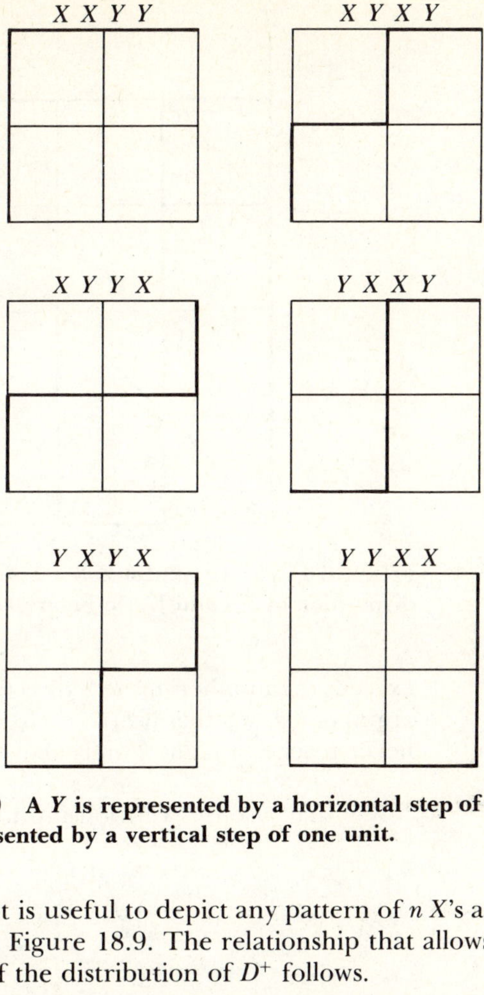

FIGURE 18.9 A Y is represented by a horizontal step of one unit, and an X is represented by a vertical step of one unit.

6 patterns. It is useful to depict any pattern of n X's and n Y's by a graph, as in Figure 18.9. The relationship that allows for an easy derivation of the distribution of D^+ follows.

> An ordered arrangement of n X's and n Y's has a value of D^+ such that $D^+ \geq k/n$ if, and only if, the graph ever reaches the line L, which is k units above the diagonal (k is a positive integer). (See Figure 18.10.)

Proof. Suppose that for some x, $F_n(x) - G_n(x) \geq k/n$. This fact means that the number of X's less than x exceeds the number of Y's less than x by at least k. That part of the graph representing these X's and Y's that are less than x must reach a point k units above the diagonal, since to this point the number of steps in the X direction

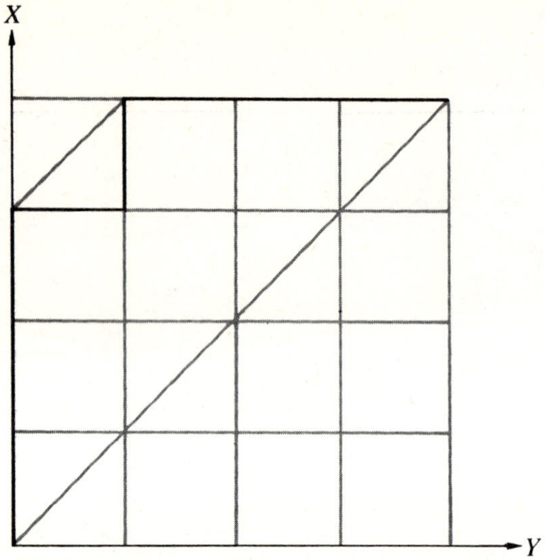

FIGURE 18.10 Graph for *X X X Y X Y Y Y*. This describes the relative disposition of *X*'s and *Y*'s in Figure 18.8 ($D^+ = 3/4$).

exceeds the number in the Y direction by at least k units. The same argument shows that if $F_n(x) - G_n(x) < k/n$ for all x, then the graph never reaches a point k units above the diagonal. ▮

We shall use this relationship to derive the distribution of D^+.

distribution of D^+

$$P\left(D^+ \geqslant \frac{k}{n}\right) = \binom{2n}{n+k} \Big/ \binom{2n}{n}, \qquad k = 0, 1, \ldots, n \tag{1}$$

Proof. There are $\binom{2n}{n}$ graphs altogether representing the distinct orderings of the n X's and n Y's. We must count the number of graphs that ever reach the line L, which is k units above the diagonal. Since the assertion is obviously true for $k = 0$, suppose that k is a positive integer. We shall count the number of graphs by a "reflection" device. A graph that reaches L must reach it at some first step. Interchange the X's and Y's in the graph up to that point to obtain a new graph that proceeds from $0'$ to P (see Figure 18.11). Conversely, consider any graph from $0'$ to P. It must intersect L at some point. Interchange the X's and Y's of the graph up to that point to obtain a graph from 0 to P. Thus, there is a one-to-one correspondence between graphs from 0 to P that ever reach L and

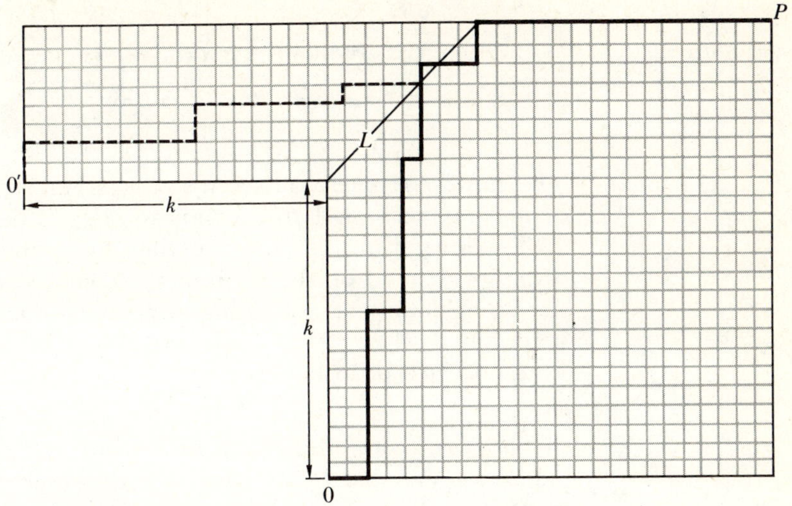

FIGURE 18.11

all graphs from 0′ to P. There are $\binom{2n}{n+k}$ possible graphs from 0′ to P, since any such graph has $n+k$ X steps and $n-k$ Y steps. Since any of the $\binom{2n}{n}$ graphs from 0 to P has the same probability $1\Big/\binom{2n}{n}$, the proof is complete. ▮

It is also possible to derive the distribution of D along similar but somewhat more complicated lines, which we shall not pursue here. When $m \neq n$, the distribution of D^+ is more complicated, but it is possible to determine general expressions, which we also shall not do here (Note 6).

18.9 COMPARING TWO POPULATIONS, USING D or D^+

Consider the same situation as in Section 18.5. That is, we suppose that there are n observations on each of two populations, and the experimenter wants to decide whether or not the populations are the same. The mathematical model is that the observations are described by $X_1, \ldots, X_n, Y_1, \ldots, Y_n$, which are mutually independent random variables, the X_i's having an integrating density f and the Y_i's an integrating density g. A procedure for deciding whether $f = g$ can be based on the random variable D, as follows.

two-sided test

If $D < c$, decide that $f = g$ (decision 1).

If $D \geqslant c$, decide that $f \neq g$ (decision 2).

The cutoff constant c is chosen so that when $f = g$, the probability of making decision 1 (the significance level) has a desired value. This follows from the fact that when $f = g$, the distribution of D does not depend on the form of the common density. Values of the probabilities corresponding to various values of c and n can be determined when $f = g$. Table 18.1 (Note 7) shows some sample values of such probabilities.

The preceding described procedure is certainly intuitively plausible. Variable D measures the difference between the two empirical cdf's of the data from each of the two populations. These empirical cdf's reflect the behaviors of the theoretical cdf's,

$$F(x) = \int_{-\infty}^{x} f(u)\, du, \qquad G(x) = \int_{-\infty}^{x} g(u)\, du$$

A small value of D means that decision 1 is appropriate.

Similarly, a one-sided procedure may be in order if the experimenter wants to decide whether one population tends to be

TABLE 18.1

$n = m$	c	$P(D < c)$
4	2/4	0.23
4	3/4	0.77
4	4/4	0.97
5	2/5	0.13
5	3/5	0.64
5	4/5	0.92
5	5/5	0.99
10	5/10	0.83
10	6/10	0.95
10	7/10	0.99
20	6/20	0.83
20	7/20	0.92
20	8/20	0.97
20	9/20	0.99

"shifted to one side of the other population" or whether the relation is reversed. Mathematically, this statement means that if $F(x) = G(x)$ for all x, the experimenter wants to make one decision, and, if this is not true, then he knows that $F(x) \geqslant G(x)$ for all x and he wants to make another decision. This situation is described in Example 1, Section 18.4, and Example 1, Section 18.5. In such a case, a decision procedure based on D^+ is appropriate.

one-sided test

If $D^+ < c$, make decision 1.

If $D^+ \geqslant c$, make decision 2.

The cutoff constant c is chosen so that when $f = g$, then the probability of making decision 1 has a desired level. As in the case of the two-sided test based on D, this probability does not depend on the common form of these densities. In fact, as we saw in Section 18.8, Eq. (1), this probability has the following simple expression:

$$P(D^+ < c) = 1 - \binom{2n}{n+k} \Big/ \binom{2n}{n} \qquad (1)$$

when $f = g$ and $c = k/n$. If we denote the power function of this procedure by $\psi(f, g)$, then (1) gives the value of $\psi(f, f)$. The test is "nonparametric" in the sense that $\psi(f, f)$ does not depend on f. This fact allows the experimenter to determine c to give a desired value of the significance level $\psi(f, f)$ and to make a decision based on his observed data. In general, however, it is not easy to give simple and usable expressions for the power function $\psi(f, g)$ for arbitrary densities f and g. These same comments apply to the two-sided test.

18.10 COMPARING EMPIRICAL AND THEORETICAL cdf's

Suppose that $X_1, \ldots, X_n$ are independent random variables, each with the same integrating density function f. As before, denote the cdf by $F(x) = \int_{-\infty}^{x} f(u)\, du$. Let F_n denote the empirical cdf of the X_i's. We consider the following measures of difference between F and F_n.

$$\Delta^+ = \sup_{-\infty < x < \infty} [F_n(x) - F(x)]$$

$$\Delta = \sup_{-\infty < x < \infty} |F_n(x) - F(x)|$$

In other words, Δ^+ measures the amount by which the empirical cdf exceeds the theoretical cdf and Δ measures the absolute value of the difference between them (Note 6). It is an interesting fact that the distributions of Δ^+ and Δ do not depend on the form of f.

Theorem

> Whatever the form of the integrating density f, the distributions of Δ^+ and Δ are the same as if f is the uniform density over (0, 1).

Proof. Let us outline the proof for Δ^+. The proof for Δ is the same and is left to the reader. The proof rests on the fact that if $X_1, \ldots, X_n$ are mutually independent, each with the same cdf F, then $F(X_1), \ldots, F(X_n)$ are also mutually independent and each is uniformly distributed over (0, 1). (See the last theorem of Section 9.2.) Suppose that for a given sample point w, the values of $X_1(w), \ldots, X_n(w)$ arranged in increasing order are $x_1 < x_2 < \cdots < x_n$. For the same sample point, the values of $F[X_1(w)], \ldots, F[X_n(w)]$, arranged in order, are $F(x_1) < F(x_2) < \cdots < F(x_n)$. The reason is F is a monotone function and does not disturb the ordering. The value of Δ^+ for the n numbers $x_1 < \cdots < x_n$ is

$$\max\left[\frac{1}{n} - F(x_1), \frac{2}{n} - F(x_2), \ldots, \frac{n}{n} - F(x_n)\right] \tag{1}$$

(see Figure 18.12). On the other hand, the value of Δ^+ for the n numbers $F(x_1) < \cdots < F(x_n)$ is

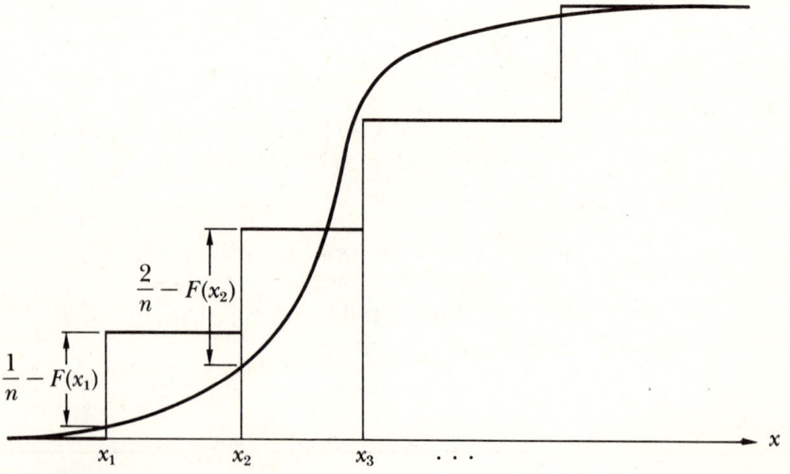

FIGURE 18.12

TABLE 18.2

n	c	$P(\Delta^+ < c)$	$P(\Delta < c)$
5	0.45	0.90	0.80
5	0.51	0.95	0.90
10	0.32	0.90	0.80
10	0.37	0.95	0.90
20	0.23	0.90	0.80
20	0.26	0.95	0.90
50	0.15	0.90	0.80

$$\max\left\{\frac{1}{n} - G[F(x_1)], \frac{2}{n} - G[F(x_2)], \ldots, \frac{n}{n} - G[F(x_n)]\right\} \tag{2}$$

where G is the cdf of the uniform distribution. But since $G(u) = u$ for u in (0, 1), the values of (1) and (2) are the same. Hence, the distribution of Δ^+ is the same in general as it is in the special case that f is the uniform density on (0, 1). ▮

The point of the theorem is the following. To find the distributions of Δ^+ and Δ, one may as well do this for the special case of uniformly distributed random variables. We shall not derive these distributions here. Table 18.2 (Note 7) shows some selected values in the distributions of these random variables. Notice that Δ^+ cannot be negative since (1) includes the nonnegative term $1 - F(x_n)$.

18.11 SOME STATISTICAL PROCEDURES BASED ON Δ^+ AND Δ

Suppose that the experimenter wants to decide whether his observations are being made on a specified population. Mathematically, these observations are represented by the independent random variables $X_1, \ldots, X_n$, and the problem is to decide whether the common density function is one particularly specified density f. If the true density is this particular f, then the experimenter wants to make decision 1; otherwise, he will make decision 2. A reasonable decision procedure based on Δ is the following. (As before, F is the cdf associated with the density f.)

test based on Δ

If $\Delta < c$, make decision 1.

If $\Delta \geq c$, make decision 2.

This procedure compares the empirical cdf with the theoretical cdf. If Δ, the measure of difference between these two cdf's is less than the cutoff constant c, then one decides that the observations are, indeed, from the population described by the cdf F; otherwise, one makes the contrary decision. When F is the correct cdf, *the probability of making decision 1 does not depend on the form of* F. This fact follows from the theorem of Section 18.10. Suppose that $\psi_f(g)$ denotes the power function of the described procedure. The variable argument runs over integrating densities g, and f is fixed. The probability of making decision 1 is $\psi_f(g)$ (that is, deciding that f is the correct integrating density), when actually g is the correct density. The procedure is nonparametric in the sense that $\psi_f(f)$ does not depend on the particular form of f. However, in general, useful explicit values of $\psi_f(g)$ are hard to compute. There do exist useful bounds, which we shall not discuss here.

In a similar way, we can consider a one-sided procedure based on Δ^+. It may be that the experimenter knows that the true cdf G either has the property that $G(x) \leq F(x)$ for all x, in which case he wants to make decision 1, or that $G(x) \geq F(x)$ for all x, in which case he wants to make decision 2. A reasonable test is then based on Δ^+.

one-sided test

If $\Delta^+ < c$, make decision 1.

If $\Delta^+ \geq c$, make decision 2.

Again it follows from the theorem in Section 18.10 that when f is the true density, the probability of making decision 1 does not depend on the form of f. As before, we denote the power function by $\psi_f(g)$. It is the probability of making decision 1 when g is the correct density. The cutoff constant c is shown so that $\psi_f(f)$ has a specified value, the significance level of the test. In general, except when $g = f$, explicit calculations of the power function are difficult.

The preceding decision procedures are usually not too practical since they require that the experimenter has some highly specialized information—namely, whether the population before him is specifically of the form f.

On the other hand, it is possible to describe a confidence procedure for the cdf that may be quite useful. Let

$$1 - \alpha = P\left(\sup_{-\infty < x < \infty} |F_n(x) - F(x)| < c\right),$$

where c is a fixed constant. We know that this probability does not depend on the particular form of F.

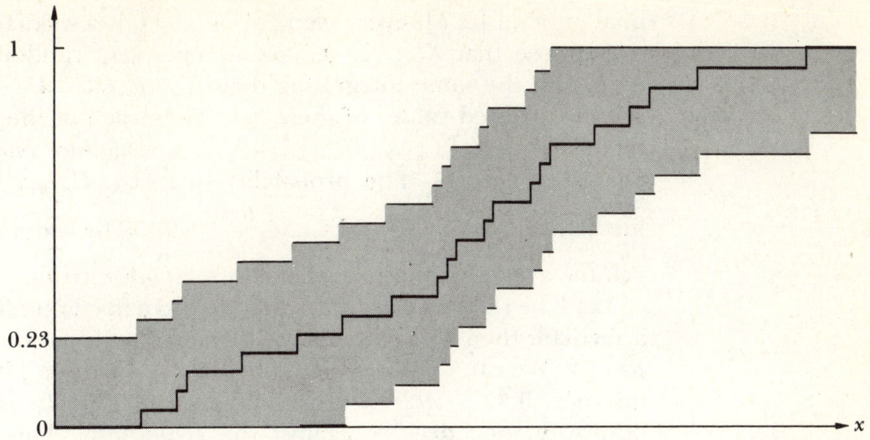

FIGURE 18.13 The band contains the true cdf $F(x)$ with probability 0.80.

confidence band for F

Suppose that the experimenter asserts that the true cdf F lies within the band

$$F_n(x) - c, F_n(x) + c$$

for all c. The probability that this random band actually contains the entire true cdf is $1 - \alpha$.

Proof. This assertion follows immediately from the theorem in Section 18.10. Whatever the true cdf may be, it will lie between $F_n(x) - c$ and $F_n(x) + c$ for all x only if $\Delta^+ < c$. This event is of probability $1 - \alpha$. ▮

Example 1 Suppose an experimenter has obtained 20 observations and he wants to estimate the cdf of the population from which he is sampling. Assume that the population is adequately described by an integrating density function. Figure 18.13 shows a confidence band guaranteed to contain the true cdf with probability 0.80. The width of the belt is $2(0.23) = 0.46$ (see Table 18.2).

SUMMARY

A *median* of a random variable X is a number m such that $P(X < t) = 1/2$. If X has an integrating density function, then there is a maximal interval $[a, b]$, every element of which is a median.

Suppose that $X_1, \ldots, X_n$ are independent random variables, each with the same integrating density, and $U_1 < U_2 < \cdots < U_n$ are the ordered values of these X_i's. Then each of the random intervals $[U_1, U_n]$, $[U_2, U_{n-1}], \ldots,$ is a *confidence interval for the interval of medians.* The probability that $[U_k, U_{n-k+1}]$ covers the median is given by $1 - 2\sum_{i=n-k+1}^{n} \binom{n}{i} (1/2)^n$. The wider is the interval, the greater is the probability of telling the truth.

Let Y be the number of X_i's that exceed a fixed number m. If m is a median, then Y is binomially distributed with parameters n and $p = 1/2$. We can use Y to test the hypothesis that m is a median. For instance, if $|Y - n/2| \leqslant c$, one decides that m really is a median; otherwise, one decides against this hypothesis. This test is two sided. Similarly, there are one-sided tests if the experimenter is only concerned with the alternative that the median is to one side of m. The tests based on Y are called *sign tests.*

Suppose that $X_1, \ldots, X_m, Y_1, \ldots, Y_n$ are independent random variables, the X_i's with one integrating density f and the Y_i's with a possibly different density g. Compare the mn pairs (X_i, Y_j) and let W count the number of such pairs in which $X_i > Y_j$. When $f = g$, then $EW = mn/2$, and the distribution of W does not depend on f and g. We can use W to test the hypothesis that $f = g$, as follows. If $|W = mn/2| < c$, one decides in favor of the hypothesis; otherwise, the hypothesis is rejected. Probability calculations for the significance level and for the power function are aided by the fact that for large m and n, W is approximately normally distributed. There are also one-sided tests if one wants to test the hypothesis that f "tends to be shifted" to one side of g. The tests based on W are called Wilcoxon tests.

Suppose that $X_1, \ldots, X_n$ are random variables with a common integrating density function. Denote the ordered values of the X_i's by $U_1 < U_2 < \cdots < U_n$ and define the *empirical cdf* to be the following random step function:

$$F_n(x) = \begin{cases} 0 & \text{if } x \leqslant U_1 \\ \dfrac{1}{n} & \text{if } U_1 \leqslant x < U_2 \\ \vdots & \\ \dfrac{n}{n} & \text{if } U_n \leqslant x \end{cases}$$

For any x, $nF_n(x)$ is binomially distributed with parameters n and $p = E[F_n(x)] = F(x)$, where F is the cdf of each X_i.

Suppose that $X_1, \ldots, X_m, Y_1, \ldots, Y_n$ are independent random variables, the X_i's with one density integrating f and the Y_i's with a possibly different density g. Another test for deciding whether $f=g$ is to compare the empirical cdf's. Let F_m be the empirical cdf of the X_i's and G_n the empirical cdf of the Y_i's. Define a measure of closeness between F_m and G_n to be

$$D = \sup_{-\infty<x<\infty} |F_m(x) - G_n(x)|$$

When $D < c$, one decides in favor of the hypothesis that $f=g$, otherwise, the hypothesis is rejected. When $f = g$, then the probability distribution does not depend on the form of f and g. There exist analogous one-sided tests based on the one-sided measure

$$D^+ = \sup_{-\infty<x<\infty} [F_m(x) - G_m(x)]$$

Again, when $f = g$, the distribution of D^+ does not depend on f and g.

Similarly, one can measure the distance between an empirical cdf F_n and a "theoretical" cdf, F. If $X_1, \ldots, X_n$ are independent random variables with a common integrating density, let

$$\Delta = \sup_{-\infty<x<\infty} |F_n(x) - F(x)|$$

where F is the cdf of each X_i. The distribution of Δ does not depend on the distribution of the X_i's. If $P(\Delta < c) = 1 - \alpha$, then one can make the *confidence band* assertion that

$$F_n(x) - c \leq F(x) \leq F_n(x) + c$$

The probability that this assertion is true for all x simultaneously is $1 - \alpha$. There are tests based on Δ of the hypothesis that the true cdf is a given F. There are also tests based on the one-sided measure

$$\Delta^+ = \sup_{-\infty<x<\infty} [F_n(x) - F(x)]$$

EXERCISES

1 Suppose that 10 independent observations are made on a population described by an integrating density function. The values of the observations are -0.71, 1.11, 1.31, 1.21, -0.50, -0.32, 1.07, 1.71, 0.04, -0.16.

(a) Use Section 18.3 to determine a confidence interval for the medians, with probability of telling the truth being $1 - (1/2^9) - (10/2^9) - (45/2^9) \approx 1 - 0.11 = 0.89$. (Answer: $[-0.32, 1.21]$.)

(b) Consider the hypothesis that 0 is a median. Would this be accepted or rejected, using the two-sided sign test with significance level 0.89?

2 (a) Consider the confidence interval estimate for the medians $[U_k, U_{n-k+1}]$, described in Section 18.3. Show that each of the random intervals $[U_k, U_{n-k+1}]$, $(U_k, U_{n-k+1}]$, $[U_k, U_{n-k+1})$, (U_k, U_{n-k+1}) covers the true medians with exactly the same probability as given by (1), Section 18.3. In other words, including or excluding endpoints gives essentially equivalent procedures. [Hint: Examine the proof of (1), Section 18.3, and observe that $P(U_k > a) = P(U_k \geqslant a)$.]

(b) Consider the confidence interval for the medians $[U_k, U_{n-k+1})$, and the following related test of the hypothesis that m is a median. If $[U_k, U_{n-k+1})$ contains m, accept the hypothesis (decision 1). Otherwise, reject the hypothesis (decision 2). Show that this decision procedure is exactly the same as the two-sided sign test of Section 18.4, with $c = (n/2) - k$. [Hint: As in Section 18.4, let Y be the number of X_i's that exceed m. Show that $[U_k, U_{n-k+1})$ contains m if, and only if, $k \leqslant Y \leqslant n - k$. (Draw a picture!) This event is the same as $|Y - (n/2)| \leqslant (n/2) - k$.]

3 (a) Suppose that X and Y are independent random variables, each with the same integrating density function f. Show that $X - Y$ has a unique median equal to 0. [Hint: $X - Y$ and $Y - X$ have the same distributions. Hence, $X - Y$ is symmetrically distributed about 0 and 0 is a median for $X - Y$. For every $\epsilon > 0$, there is some interval $(a, a + \epsilon)$ such that $\int_a^{a+\epsilon} f(x)\, dx > 0$. (Why?) Since X in $(a, a + \epsilon)$ and Y in $(a, a + \epsilon)$ implies that $X - Y$ is in $(-\epsilon, \epsilon)$, it follows that

$$P(-\epsilon < X - Y < \epsilon) \geqslant P(a < X < a + \epsilon)P(a < Y < a + \epsilon) > 0$$

Since $X - Y$ is symmetrically distributed about 0, it is true that $P(0 < X - Y < \epsilon) = P(-\epsilon < X - Y < 0)$; hence, these are positive. It follows that

$$P(X - Y < -\epsilon) < P(X - Y < 0) = \tfrac{1}{2} < P(X - Y < \epsilon)$$

Therefore, the interval of medians is contained in $(-\epsilon, \epsilon)$. Since ϵ can be an arbitrarily small positive number, the result follows.]

(b) It is possible for X and Y to be independent random variables with *different* integrating density functions and for $X - Y$ to have a unique median equal to 0. For instance, show that this fact is true when X is uniformly distributed over $[0, 1]$ and Y is uniformly distributed over $[1/4, 3/4]$.

4 Show that the expected value of the kth order statistic is given by

$$E(U_k) = n\binom{n-1}{k-1} \int_{-\infty}^{\infty} x[F(x)]^{k-1}[1 - F(x)]^{n-k} f(x)\, dx$$

(Hint: Use Example 3, Section 9.6, as a guide.)

5 Verify Eq. (3), Section 18.6, and determine the constants a and b.

6 Consider the one-sided Wilcoxon test described in Section 18.6. Suppose that $m = n = 3$, and the significance level is to be 0.95. Show that decision 2 is made if, and only if, all three observed Y_i's exceed all three observed X_i's.

7 Suppose that $m = n = 3$. List all $\binom{6}{3} = 20$ patterns of orderings of the six symbols X, X, X, Y, Y, Y, and determine the distributions of the random variables D^+ and D defined in Section 18.8. Check the distribution of D^+ against formula (1), Section 18.8. The X_i's and Y_i's are assumed to have the same integrating density functions.

8 Let D^+ be as defined in Section 18.8 and suppose that $m = n$ and that the X_i's and Y_i's have the same integrating density functions.

(a) Show that

$$E(D^+) = \left[\frac{2^{2m-1}}{\binom{2m}{m}} - \frac{1}{2}\right] \bigg/ m$$

{Hint: Use (3), Section 2.10, and (1), Section 18.8, to give

$$E(mD^+) = \left[\binom{2m}{m+1} + \cdots + \binom{2m}{2m}\right] \bigg/ \binom{2m}{m}$$

Now use the fact that $\Sigma_{i=0}^{2m} \binom{2m}{i} = 2^{2m}$.}

(b) Check the preceding expected value against a direct computation from the distribution of D^+ for $m = 2$ and $m = 3$ from Example 2, Section 18.8, and Exercise 7.

9 (a) Suppose that $m = n = 2$. Let the X_i's have the integrating density

$$f(x) = \begin{cases} 1 & \text{if } 0 \leqslant x \leqslant 1 \\ 0 & \text{otherwise} \end{cases}$$

and let the Y_i's have the integrating density

$$g(y) = \begin{cases} (1 + \alpha)y^\alpha & \text{if } 0 \leqslant y \leqslant 1 \\ 0 & \text{otherwise} \end{cases}$$

Determine the probabilities of each of the six orderings of X, X, Y, Y listed in Example 2, Section 18.8. [Partial answer:

$$P(X, X, Y, Y) = \frac{(2!)(2!)(1 + \alpha)^2}{(1)(2)(3 + \alpha)(4 + 2\alpha)}$$

$$P(X, Y, X, Y) = \frac{(2!)(2!)(1 + \alpha)^2}{(1)(2 + \alpha)(3 + \alpha)(4 + 2\alpha)}$$

(b) Consider the same situation as in (a), but now $m = n = 3$. Show that the probability of the pattern X, X, X, Y, Y, Y is

$$r(\alpha) = \frac{(3!)(3!)(1 + \alpha)^3}{(1)(2)(3)(4 + \alpha)(5 + 2\alpha)(6 + 3\alpha)}$$

(c) Notice that the power function of the test described in Exercise 6 is $1 - r(\alpha)$, $\alpha \geq 0$, when the distributions of X_i and Y_j are those given in (a).

10 Let $F_n(x)$ be the empirical cdf of n independent random variables, each of which is uniformly distributed over $[0, 1]$. It is a remarkable fact that

$$P(F_n(x) < \gamma x, 0 < x < 1) = 1 - \frac{1}{\gamma}$$

for any $n = 1, 2, 3, \ldots$, where $\gamma > 1$. [The preceding probability is that F_n stays below the line γx throughout the whole interval $(0, 1)$.] Verify this fact for $n = 1$ and $n = 2$. [Hint: For $n = 2$, the required probability is

$$P\left(\frac{1}{2\gamma} < X_1 < X_2, \frac{1}{\gamma} < X_2\right) + P\left(\frac{1}{2\gamma} < X_2 < X_1, \frac{1}{\gamma} < X_1\right)$$

(Draw a picture!)]

11 Consider Δ^+ as defined in Section 18.10.
(a) Show that for $n = 1$, Δ^+ is uniformly distributed over $[0, 1]$.
(b) Show that for $n = 2$, Δ^+ has the integrating density

$$h(x) = \begin{cases} 1 + 2x & \text{if } 0 \leq x \leq \frac{1}{2} \\ 2(1 - x) & \text{if } \frac{1}{2} < x \leq 1 \\ 0 & \text{otherwise} \end{cases}$$

12 Consider the confidence belt procedure described in Section 18.11. Determine a 0.95 probability confidence level belt for the cdf F, using the data given in Exercise 1. [It is a fact that for $n = 10$, $P(\Delta < 0.41)$ is approximately equal to 0.95 (Note 7).]

NOTES

1 For additional reading in nonparametric methods, see Chapter 16 of [51]. Several books on the subject are [76], [77], [78]. There also exists an extensive bibliography of papers and books in [79].

2 Statistic W is named for a contemporary American chemist, Frank Wilcoxon, (1892–1965).

3 The tests we call Wilcoxon tests were suggested by Wilcoxon with W expressed as a sum of ranks as in (1), Section 18.6. The same test, with W as defined in the beginning of Section 18.6, was suggested by Mann and Whitney [80], who first proved the asymptotic normality of W when the X_i's and Y_i's have the same distributions.

4 Statistic W is an example of a rank order statistic. An extensive treatment is given in [81]. This book contains proofs of the asymptotic normality of W and of more general rank order statistics.

5 The fact that the maximum deviation between an empirical cdf and a theoretical cdf "converges to zero" is related to the Cantelli–Glivenko lemma. See [82] for references.

6 The various tests based on D, D^+, Δ, and Δ^+ are due to the Russian mathematicians Kolmogorov and N. V. Smirnov (1887–   ). These random variables are commonly called Kolmogorov–Smirnov statistics. Reference [82] is a review article on these statistics.

7 Use [19] as a guide to various tables of the distributions of the Kolmogorov–Smirnov statistics. See also Tables M and N in [77].

REFERENCES

1 Ore, O., *Cardano, The Gambling Scholar* (with a translation from the Latin of Cardano's *Games of Chance,* by S. H. Gould), Princeton University Press, Princeton, New Jersey, 1953.

2 Thomas, G. B., Jr., *Calculus and Analytic Geometry,* Addison-Wesley, Reading, Massachusetts, 1968, 4th ed.

3 Apostol, T. M., *Calculus,* Vol. I, Blaisdell, Waltham, Massachusetts, 1967, 2nd ed.

4 Apostol, T. M., *Calculus,* Vol. II, Blaisdell, Waltham, Massachusetts, 1962, 2nd ed.

5 Weaver, W., "Probability," *Scientific American,* October 1950.

6 Kac, M., "Probability," *Scientific American,* September 1964.

7 Ayer, A. J., "Chance," *Scientific American,* October 1965.

8 von Mises, R., *Mathematical Theory of Probability and Statistics,* Academic, New York, 1964.

9 Feller, W., *An Introduction to Probability Theory and Its Applications,* Vol. I, John Wiley, New York, 1968, 3rd ed.

10 Dwass, M., *First Steps in Probability,* McGraw-Hill, New York, 1967.

11 Whitworth, W. A., *Choice and Chance,* Hafner, New York, 1948, 5th ed.

12 Whitworth, W. A., *DCC Exercises in Choice and Chance,* Hafner, New York, 1959.

13 Todhunter, I., *A History of the Theory of Probability from the Time of Pascal to that of Laplace,* Chelsea, New York, 1949.

14 Parzen, E., *Modern Probability Theory and Its Applications,* John Wiley, New York, 1960.

15 Kemeny, J. G., Snell, J. L., and Knapp, A. W., *Denumerable Markov Chains,* Van Nostrand, Princeton, New Jersey, 1966.

16 *Tables of the Binomial Probability Distribution,* Applied Mathematics Series 6, National Applied Mathematics Laboratories of the National Bureau of Standards, U.S. Government Printing Office, Washington, D.C., 1950.

17 Romig, H. G., *50–100 Binomial Tables,* John Wiley, New York, 1953.

18 "Tables of the Cumulative Binomial Probability Distribution," *Annals of the Computation Laboratory of Harvard University,* Vol. XXXV, Harvard University Press, Cambridge, Massachusetts, 1955.

19 Greenwood, J. A., and Hartley, H. O., *Guide to Tables in Mathematical Statistics,* Princeton University Press, Princeton, New Jersey, 1962.

20 Pearson, K., *Tables of the Incomplete Beta-Function,* University Press, Cambridge, 1934.

21 Lieberman, G. J., and Owen, D. B., *Tables of the Hypergeometric Distribution,* Stanford University Press, Stanford, California, 1961.

22 Williamson, E., and Bretherton, M. H., *Tables of the Negative Binomial Distribution,* John Wiley, New York, 1963.

23 Syski, R., *Introduction to Congestion Theory in Telephone Systems,* Oliver and Boyd, Edinburgh, 1960.

24 Fry, T. C., *Probability and Its Engineering Uses,* Van Nostrand, Princeton, New Jersey, 1968.

25 Molina, E. C., *Poisson's Exponential Binomial Limit,* Van Nostrand, Princeton, New Jersey, 1942.

26 Kolmogorov, A. N., *Foundations of the Theory of Probability,* Chelsea, New York, 1956, 2nd ed.

27 Halmos, P. R., *Measure Theory,* Van Nostrand, Princeton, New Jersey, 1950.

28 Krickeberg, K., *Probability Theory,* Addison-Wesley, Reading, Massachusetts, 1965.

29 Loéve, M., *Probability Theory,* Van Nostrand, Princeton, New Jersey, 1963, 3rd ed.

30 Feller, W., *An Introduction to Probability Theory and Its Applications,* Vol. II, John Wiley, New York, 1966.

31 Cramer, H., *Mathematical Methods of Statistics,* Princeton University Press, Princeton, New Jersey, 1946.

32 Widder, D. V., *Advanced Calculus,* Prentice-Hall Englewood Cliffs, New Jersey, 1961, 2nd ed.

33 Widder, D. V., *The Laplace Transform,* Princeton University Press, Princeton, New Jersey, 1946.

34 Rand Corporation, *A Million Random Digits with 100,000 Normal Deviates,* The Free Press, Glencoe, Illinois, 1955.

35 Federal Works Agency, WPA for City of New York, *Tables of Probability Functions,* Vols. I and II, 1941.

36 Gnedenko, B. V., and Kolmogorov, A. N., *Limit Distributions for Sums of Independent Random Variables* (translated from the Russian by K. L. Chung), Addison Wesley, Reading, Massachusetts, 1954.

37 Trotter, H. F., "An elementary proof of the central limit theorem," *Archiv der Mathematik,* **9,** 226–234, 1959.

38 Parzen, E., *Stochastic Processes,* Holden-Day, San Francisco, California, 1960.

39 Feynman, R. P., Leighton, R. B., and Sands, M., *The Feynman Lectures on Physics,* Vol. I, Addison-Wesley, Reading, Massachusetts, 1963.

40 Bush, R. R., and Mosteller, F., *Stochastic Models for Learning,* John Wiley, New York, 1955.

41 Bush, R. R., and Estes, W. K., eds., *Studies in Mathematical Learning Theory,* Stanford University Press, Stanford, California, 1959.

42 Riordan, J., *Stochastic Service Systems,* John Wiley, New York, 1962.

43 Arley, N., *On the Theory of Stochastic Processes and Their Application to the Theory of Cosmic Radiation,* Gads Forlag, Copenhagen, 1943.

44 Bharucha-Reid, A. T., *Elements of the Theory of Markov Processes*

and Their Applications, McGraw-Hill, New York, 1960.

45 Haight, F. A., *Mathematical Theories of Traffic Flow,* Academic, New York, 1963.

46 Bailey, N. T. J., *The Mathematical Theory of Epidemics,* Griffin, London, 1957.

47 Bailey, N. T. J., *The Elements of Stochastic Processes with Applications to the Natural Sciences,* John Wiley, New York, 1964.

48 Bailey, N. T. J., *The Mathematical Approach to Biology and Medicine,* John Wiley, New York, 1967.

49 Bartlett, M. S., *Stochastic Population Models in Ecology and Epidemiology,* Methuen, London, 1960.

50 Lindgren, B. W., *Statistical Theory,* Macmillan, New York, 1968, 2nd ed.

51 Mood, A. M. and Graybill, F. A., *Introduction to the Theory of Statistics,* McGraw-Hill, New York, 1963, 2nd ed.

52 Fisher, R. A., "On the mathematical foundations of theoretical statistics," *Phil. Trans. Roy. Soc. London,* Series A, **222,** 309–368, 1922.

53 van Neuman, J. and Morgenstern, O., *Theory of Games and Economic Behavior,* Princeton University Press, Princeton, New Jersey, 1953, 2nd ed.

54 McKinsey, J. C. C., *Introduction to the Theory of Games,* McGraw-Hill, New York, 1952.

55 Wald, A., *Statistical Decision Functions,* John Wiley, New York, 1950.

56 Blackwell, D. and Girshick, M. A., *Theory of Games and Statistical Decisions,* John Wiley, New York, 1954.

57 Ferguson, T. S., *Mathematical Statistics, a decision theoretic approach,* Academic Press, New York, 1967.

58 Savage, L. J., *The Foundations of Statistics,* John Wiley, New York, 1954.

59 Raiffa, H. and Schlaifer, R., *Applied Statistical Decision Theory,* Graduate School of Business Administration, Harvard University, Cambridge, Massachusetts, 1961.

60 Neyman, J., "Two recent breakthroughs in statistics: The

theory of statistical decision making," *Rev. Intern. Stat. Inst.* **30,** 11–27, 1962.

61 Neyman, J. and Pearson, E. S., *Joint Statistical Papers by J. Neyman and E. S. Pearson,* University of California Press, Berkeley, 1967.

62 Fisher, R. A., "Theory of statistical estimation," *Proc. Cambridge Phil. Soc.,* Vol. XXIII, pt. 5, pp. 700–725, 1925.

63 Fisher, R. A., *Contributions to Mathematical Statistics,* John Wiley, New York, 1950.

64 Halmos, P. and Savage, L. J., "Application of the Radon-Nikodym theorem to the theory of sufficient statistics," *Ann. Math. Stat.,* **20,** 225–241, 1948.

65 Cochran, W. G., *Sampling Techniques,* John Wiley, New York, 1953.

66 Columbia University, *Statistical Research Group, Sampling Inspection: Principles, Procedures and Tables,* McGraw-Hill, New York, 1948.

67 Wald, A., *Sequential Analysis,* John Wiley, New York, 1947.

68 Clopper, C. J., and Pearson, E. S., "The use of confidence or fiducial limits illustrated in the base of the binomial," *Biometrika,* **26,** 404–413, 1934.

69 Chung, J. and De Lury, D. B., *Confidence Limits for the Hypergeometric Distribution,* University of Toronto Press, Toronto, Canada, 1950.

70 Student, "The Probable Error of a Mean," *Biometrika,* **6,** 1–25, 1908.

71 Fisher, R. A., *The Design of Experiments,* Hafner, New York, 1949, 5th ed.

72 Fisher, R. A., *Statistical Methods for Research Workers,* Oliver and Boyd, Edinburgh, 1948, 10th ed.

73 Cochran, W. G., and Cox, G. M., *Experimental Designs,* John Wiley, New York, 1950.

74 Dixon, W. J., and Massey, F. J., Jr., *Introduction to Statistical Analysis,* McGraw-Hill, New York, 1969, 3rd ed.

75 Fox, M., "Charts of the power of the F-test," *Ann. Math. Stat.,* **27,** 484–497, 1956.

76 Fraser, D. A. S., *Nonparametric Methods in Statistics,* John Wiley, New York, 1957.

77 Kraft, C. H. and van Eeden, C., *A Nonparametric Introduction to Statistics,* Macmillan, New York, 1968.

78 Noether, G. E., *Elements of Nonparametric Statistics,* John Wiley, New York, 1967.

79 Savage, I. R., Bibliography of *Nonparametric Statistics,* Harvard University Press, Cambridge, Massachusetts, 1962.

80 Mann, H. B. and Whitney, D. R., "On a test of whether one of two random variables is stochastically larger than the other," *Ann. Math. Stat.,* **18,** 50–60, 1947.

81 Hajek, J., *Theory of Rank Tests,* Academic Press, New York, 1967.

82 Darling, D. A., "The Kolmogorov-Smirnov, Cramer-von Mises tests," *Ann. Math. Stat.,* **28,** 823–838, 1957.

INDEX

SHENANDOAH COLLEGE
LIBRARY
WINCHESTER, VA.

SHENANDOAH COLLEGE
LIBRARY
WINCHESTER. VA.